This collection of expert reviews emphasizes the importance of scientific method in the understanding and clinical management of neurological disease. Covering a range of rapidly advancing areas in neurology, it concentrates particularly on pathophysiology as the basis of disordered function.

The selection of topics has been influenced by the contributions to clinical neuroscience of Professor James Lance, to whom the book is dedicated. They range from molecular approaches in neurology, through sensory, motor and autonomic function, to clinical problems such as headache and epilepsy. A number of new diagnostic tests are considered, including PET scanning and motor-evoked potentials. The volume concludes with an authoritative overview by Lord Walton on the changing face of neuroscience, concentrating particularly on the contribution of molecular genetics.

This is an important reference work for established neurologists and those in training and will appeal also to a wider audience of practitioners including psychologists and those involved in rehabilitation.

SCIENCE AND PRACTICE IN CLINICAL
NEUROLOGY

SCIENCE AND PRACTICE IN CLINICAL NEUROLOGY

Edited by
S.C. GANDEVIA, D. BURKE AND M. ANTHONY
Institute of Neurological Sciences, The Prince Henry and Prince of Wales Hospitals, Sydney, NSW, Australia

Published by the Press Syndicate of the University of Cambridge
The Pitt Building, Trumpington Street, Cambridge CB2 1RP
40 West 20th Street, New York, NY 10011-4211, USA
10 Stamford Road, Oakleigh, Melbourne 3166, Australia

© Cambridge University Press 1993

First published in 1993

Printed in Great Britain at the University Press, Cambridge

A catalogue record for this book is available from the British Library

Library of Congress cataloguing in publication data
Science and practice in clinical neurology / edited by S. C. Gandevia,
D. Burke, and M. Anthony.
   p.   cm.
ISBN 0-521-43119-0 (hardback)
1. Neurology–Practices–Congresses.   2. Neurosciences–Congresses.
I. Gandevia, Simon C.   II. Burke, David C.   III. Anthony, M. (Michael)
[DNLM: 1. Nervous System Diseases–physiopathology–congresses.
2. Nervous System–physiology–congresses. WL 102 S4155 1994]
RC327.S35   1994
616.8–dc20
DNLM/DLC
for Library of Congress   93-19637 CIP

ISBN 0 521 43119 0 hardback

Contents

List of contributors	page	ix
Preface		xiii
Foreword: A personal perspective of 40 years of science and practice in clinical neurology *J. W. Lance*		xv

Part I: Normal and abnormal sensation

1. Aspects of proprioception *D. I. McCloskey and S. C. Gandevia* — 3
2. The pathophysiological basis of paresthesiae *D. Burke* — 20
3. Spinal pain: backache and neck pain *N. Bogduk* — 39

Part II: Normal motor control and its disorders

4. Corticospinal control of movement *R. Porter* — 61
5. Assessment of corticofugal output: strength testing and transcranial stimulation of the motor cortex *S. C. Gandevia* — 75
6. Muscle spindles, muscle tone and the fusimotor system *D. Burke and S. C. Gandevia* — 89
7. The neurophysiology of human spinal spasticity *P. Ashby* — 106
8. Future strategies for the treatment of Parkinson's disease *S. Fahn* — 130
9. Understanding the genetics of classical idiopathic torsion dystonia *S. Fahn* — 139
10. Functional imaging in motor disorders *J. G. Colebatch* — 154
11. Tonic stretch reflex in normal subjects and in cerebral palsy *P. D. Neilson* — 169
12. Measurement of physiological and essential tremor *J. D. Gillies* — 191

Part III: Autonomic function and vascular disorders

13. Disorders of the autonomic nervous system *J. G. McLeod* — 205
14. Autonomic innervation of the face *P. D. Drummond* — 223
15. Stroke and antiphospholipid antibodies *K. M. A. Welch and S. R. Levine* — 243

Part IV: Migraine and other headaches

16. Experimental headaches. A useful tool in the investigation of migraine mechanisms   *J. Oleson and H. Iversen*   253
17. Unilateral headache   *M. Anthony*   266
18. Pathways for headache   *G. A. Lambert*   284
19. Pathophysiology of migraine   *P. J. Goadsby*   303
20. Serotonin, sumatriptan and migraine   *P. P. A. Humphrey, W. Feniuk, M. J. Perren and A.W. Oxford*   323
21. The treatment of primary headache   *F. Clifford Rose and M. Anthony*   334

Part V: Viral and immunological disorders

22. Neurovirology: the evolution of new challenges   *R. T. Johnson*   363
23. Multiple sclerosis: current concepts and Australian studies   *J. G. McLeod*   374

Part VI: Epilepsy

24. Partial epilepsies   *R. A. Mackenzie*   397
25. The changing face of antiepileptic drug therapy   *M. J. Eadie*   416

Part VII: Clinical myology and conclusion

26. The changing face of neuroscience   *Lord Walton of Detchant*   431

    Index   448

List of contributors

M. Anthony
The Institute of Neurological Sciences, The Prince Henry and Prince of Wales Hospitals and School of Medicine, University of New South Wales, Sydney, Australia.

P. Ashby
Playfair Neuroscience Unit, The Toronto Hospital (Western Division), University of Toronto, Canada M5T 2S8

N. Bogduk
Faculty of Medicine, University of Newcastle, Callaghan NSW, Australia

D. Burke
Department of Clinical Neurophysiology, The Prince Henry and Prince of Wales Hospitals and Prince of Wales Medical Research Institute, University of New South Wales, Sydney NSW, Australia

F. Clifford Rose
Academic Unit of Neuroscience, Charing Cross Hospital, London, England

J. G. Colebatch
Department of Neurology, Institute of Neurological Sciences, The Prince Henry and Prince of Wales Hospitals, Sydney NSW, Australia

P. D. Drummond
Department of Psychology, Murdoch University, Perth WA, Australia

M. J. Eadie,
Department of Medicine, The University of Queensland, St Lucia, Queensland, Australia

S. Fahn
Dystonia Clinical Research Center, Department of Neurology, Columbia University College of Physicians and Surgeons and The Neurological Institute of New York, Columbia-Presbyterian Medical Center, New York, NY, USA

W. Feniuk
Glaxo Group Research Ltd, Park Road, Ware, Herts SG12 0DP, UK

S. C. Gandevia
Prince of Wales Medical Research Institute, University of New South Wales, Sydney NSW, Australia

J. D. Gillies
Department of Neurology, Institute of Neurological Sciences, The Prince Henry and Prince of Wales Hospitals, Sydney, NSW, Australia

P. J. Goadsby
Department of Neurology, The Prince Henry and Prince of Wales Hospitals, Little Bay, Sydney NSW, Australia

P. A. Humphrey
Glaxo Group Research Ltd., Park Road, Ware, Herts. SG12 0DP, UK

H. Iversen
Department of Neurology, University of Copenhagen Gentofte Hospital, DK-2900 Hellerup, Denmark

R. T. Johnson
The Johns Hopkins University School of Medicine, Baltimore, Maryland, USA

G. A. Lambert
Institute of Neurological Sciences, The Prince Henry and Prince of Wales Hospitals and School of Medicine, University of New South Wales, Sydney, Australia 2036

R. Levine
Center for Stroke Research, Department of Neurology, Henry Ford Hospital and Health Sciences Center, Detroit, Michigan, USA

D. I. McCloskey
Prince of Wales Medical Research Institute, University of New South Wales, Sydney NSW, Australia

R. A. Mackenzie
Comprehensive Epilepsy Programme, The Prince Henry and Prince of Wales Hospitals, Little Bay, Sydney NSW, Australia

J. G. McLeod
Department of Medicine, University of Sydney, Sydney NSW, Australia

P. D. Neilson
Cerebral Palsy Research Unit, Institute of Neurological Sciences, The Prince Henry and Prince of Wales Hospitals and School of Electrical Engineering, University of New South Wales, Sydney NSW, Australia

List of contributors

J. Oleson
Department of Neurology, University of Copenhagen, Gentofte Hospital, DK-2900 Hellerup, Denmark

A.W. Oxford
Glaxo Group Ltd, Park Road, Ware, Herts SG12 0DP, UK

M. J. Perren
Glaxo Group Research Ltd, Park Road, Ware, Herts SG12 0DP UK

R. Porter
Faculty of Medicine, Monash University, Clayton, Melbourne VIC, Australia

Lord Walton of Detchant
13 Norham Gardens, Oxford OX2 6PS, UK

K. M. A. Welch
Center for Stroke Research, Department of Neurology, Henry Ford Hospital and Health Sciences Center, Detroit, Michigan, USA

Preface

This book developed from a symposium held in honour of James W. Lance, AO, CBE, MD, DSc, FRCP, FRACP, FAA, the Foundation Chairman of the Department of Neurology of The Prince Henry and Prince of Wales Hospitals. Professor Lance was the first Professor of Neurology in Australia, and is one of Australia's most respected clinicians and distinguished medical scientists. His interests and influence have been truly eclectic, and this is reflected in the breadth and depth of this book.

In choosing the title 'Science and Practice in Clinical Neurology', the editors wish to emphasize the virtues of scientific method over an empirical approach in clinical practice, because this is the way in which Jim Lance approaches a clinical problem.

We are grateful to Glaxo for a contribution towards the production costs of this volume. In addition we thank Mrs Mary Sweet for invaluable secretarial assistance throughout all stages of preparation of the volume.

S. C. Gandevia
David Burke
Michael Anthony

January 19, 1992

Foreword

A personal perspective of 40 years of science and practice in clinical neurology

James W. Lance, AO, CBE

Emeritus Professor of Neurology, University of New South Wales, Sydney, NSW, Australia

Galen's views influenced medicine for 1500 years and are still being quoted. If the oral tradition during those stagnant years had been documented, he would have the highest citation index of all time. Current views will fortunately prove less durable because medicine is now scientifically based, and treatments are constantly undergoing critical assessment. Nevertheless, our shared experience is steadily building a firmer base for neurological practice. We are fortunate to be living in a logarithmic phase of expansion in scientific knowledge: it has been suggested that 90% of the scientists in world history are still living, a fact that must seem abundantly clear to any clinician trying to keep abreast of the scientific literature. A professional lifetime of 40 years is short by comparison with thousands of years of recorded human endeavour, but the past 40 years encompass some of the greatest scientific advances.

Neurological practice has undergone radical changes during my professional life. Many of these changes are addressed in the chapters of this book but the selection is slightly biased, reflecting the major interests of my Department. I apologise to any whose work may have been underrepresented or those whose special interests have been omitted.

A scientific approach to any clinical disorder implies an attempt to understand the pathogenesis of the disease, and how that disturbance can alter normal physiological mechanisms to produce the patient's symptoms. My first research interest was the corticospinal system, specifically the pyramidal tract, the executive pathway for volitional movement. It is therefore pleasing that recent developments on the primate corticospinal system are reviewed by Porter, that its functional assessment in human subjects is considered by Gandevia, that Burke and Gandevia discuss the fusimotor system and spinal reflex mechanisms, and that the pathophysiology of motor disturbances such as spasticity, cerebral palsy and tremor are discussed, respectively, by Ashby, Neilson and Gillies. The basis of proprioception is covered by McCloskey and Gandevia, and the abnormal sensations of paraesthesiae by Burke. Pain is the most important symptom to be considered in any form of clinical practice, and chronic or recurrent pain still presents a therapeutic challenge. Bogduk addresses pain of spinal origin, and a number of authors tackle craniofacial pain: experimental headache (Olesen & Iversen), persistent unilateral headache (Anthony), humoral factors in pathogenesis (Humphrey, Feniuk, Perren & Oxford), neural mechanisms (Lambert; Goadsby) and headache treatment (Clifford Rose & Anthony).

The advances of the last four decades do not all revolve around pathophysiology and neither does this book, though its emphasis is on disordered function. Our understanding of vascular disorders has undergone major changes with the recognition of new stroke syndromes (Welch & Levine) and of the mechanisms and assessment of autonomic dysfunction (Drummond; McLeod). Epidemiology and detailed case analysis have made major contributions to our concepts of, for example, multiple sclerosis (McLeod), dystonia (Fahn) and disease transmission (Johnson). Our ability to image brain structure and function has undergone a revolution which has irrevocably altered neurological practice, not always for the better.

In the 1950s, the structure of the brain had to be inferred from cerebral angiography and pneumoencephalography. The former was done by direct puncture of the carotid and vertebral arteries and the latter by causing a headache of intense severity. If patients did not experience a headache during this procedure, one could almost be sure that they had cerebral atrophy before the X-ray films were processed! Before the days of image intensification, it was obligatory to don red glasses for half an hour before performing myelography to prepare one's eyes for the dim underwater world on the screen where a black column of oily contrast medium slid slowly through a deep green sea. In the 1950s, electroencephalography was the only available test of brain function. Radionuclide scanning had a brief burst of glory, as a form of functional imaging but then subsided until technological advances led to its re-emergence in the form of single photon emission computerized tomography (SPECT). The ability to assess regional cerebral blood flow and metabolic activity by SPECT and positron emission tomography (PET) has led to new understanding of cortical localization and the sequence of events in disorders like epilepsy and migraine. The labelling of neurotransmitters such as dopamine has provided physiological evidence of the site of origin of dystonia, and insight into the pathophysiology of movement disorders (Colebatch, this volume).

The easy access to computerized tomography and magnetic resonance imaging has had one adverse effect on neurological practice. Any variation from normal, however irrelevant, on CT or MR scanning leads to direct referral from general practitioner to neurosurgeon. The availability of Doppler studies and digital subtraction angiography promoted referral of patients with transient ischemic attacks to vascular surgeons, thus bypassing the neurologist whose function should be to think out problems from the beginning and advise patients on the best medical and surgical options. The chapter by Welch and Levine, emphasises the importance of the physicians' role in the assessment of strokes. Intracranial surgery, a risky adventure 40 years ago, is now responsible for a happy outcome in most instances, aided by operating microscopes, ventriculoscopes, laser beams and high standards of anaesthesia.

In the late 1940s, George Dawson at the National Hospital, Queen Square, was pioneering the use of averaging techniques to record sensory evoked potentials. Since then, visual, auditory and somatosensory evoked responses have become routine procedures. The increased sophistication of clinical neurophysiology now confirms or corrects the supposition of the neurologist concerning central as well as peripheral pathology and gives practical aid to the surgeon who operates near the optic nerves or aims to straighten a bent spine without damaging its precious contents. It is possible to record the action potentials of single axons in human peripheral nerves, and this has given a wealth of information about how nerves function under natural circumstances, including recordings of muscle spindle discharge (Burke & Gandevia) and the ectopic sensory impulses that give rise to paraesthesiae (Burke).

Genetics has advanced considerably since Mendel observed generations of peas in his monastic garden. Once the DNA molecule was unravelled, the ability to invade the chromosome by the technique of divide and conquer has pinpointed the faulty gene in a variety of neurological disorders, has improved genetic counselling and offers the hope of manipulating genes in the future to eliminate some hereditary diseases (Fahn; Walton). Forty years ago, immunology recognized antigens, antibodies, complement and allergy but the microcosm of killer T cells, helper cells and the rest of the gruesome band of bodily defences was unknown.

In the 1950s, streptomycin had just joined penicillin and the sulphonamides in the antibiotic family. Sir Francis Walshe (1947) in the fifth edition of his textbook *Diseases of the Nervous System* noted tersely: 'tuberculous meningitis is invariably fatal'. The proliferation of antibiotics was about to start. Now one has to struggle to remember the name of the latest cephalosporin so as not to appear outdated. In 1961, when I joined the staff of The Prince Henry Hospital, the last poliomyelitis epidemic in this State was in progress. Because the hospital was a centre for infectious disease, each ward round included up to 30 patients with suspected polio, many of whom had other motor disorders, real or imagined. The 'iron lung', which proved to be an iron coffin for many patients with bulbar as well as respiratory paralysis, was being phased out in favour of positive pressure respiration. Poliomyelitis has now disappeared in those countries with efficient immunization programmes, but AIDS has arrived with even more sinister implications, and the future will bring other challenges (Johnson).

In the 1950s, the pharmacological treatment of Parkinson's disease was limited to anticholinergics, and the only remedies for epilepsy were phenytoin and the barbiturates. Who can forget the excitement of seeing patients who had been wheelchair-bound with Parkinson's disease walking again when treated with L-dopa? Who can forget the dismay when triumph turned to tragedy as the large doses of L-dopa administered before decarboxylase inhibitors were available finally took effect and produced dyskinesiae so wild that patients fell heavily, breaking hips and legs? Advances over the past 40 years have certainly been satisfying, but there are still patients who are incapacitated by pain of spinal origin (Bogduk), Parkinson's disease (Fahn), epilepsy (Mackenzie; Eadie) and migraine (Clifford Rose & Anthony). There remain conditions like motor neurone disease, dystonia (Fahn) and multiple sclerosis (McLeod), the natural history of which remain largely unscathed by our therapeutic arrows.

Without research, medical practice would still be guided by Galen. It is of concern that economic pressures and the duration and rigidity of training in specialist medical disciplines are reducing the number of young men and women prepared to deviate from the ordained clinical path to undertake research. They must be reassured that they are not leaving the mainstream of medicine to do research; they are entering the mainstream.

It is paradoxical that, in this scientific age, there should be so much enthusiasm for fringe medicine of various kinds. Naturopaths, herbalists, iridologists and the like have never had to confront an ethics committee, let alone carry out a double-blind trial. The best means of combating this drift into folklore is to improve our own performance in handling not only life-threatening disease but the many disorders that impair or destroy the enjoyment of life. This book documents some of our attempts to grapple with these issues by bringing a scientific approach to neurological practice.

Part I

Normal and abnormal sensation

1
Aspects of proprioception

D. I. McCLOSKEY AND S. C. GANDEVIA

Prince of Wales Medical Research Institute,
University of New South Wales,
Sydney NSW, Australia

Introduction

The sense of the positions and actions of the limbs was called *the sixth sense* by Sir Charles Bell (1833). Much the same thing was referred to by Bastian (1888) when he described as *kinaesthetic* 'the body of sensations which result from or are directly occasioned by movements'. The Greek root and Bastian's definition stress the dynamic aspects of the sense. Sherrington (1906) introduced the term *proprioception* (from the Latin *proprius*, meaning one's own) to refer to the sensory processes responsible for the conscious appreciation of posture and movement, and also the many sensory inputs involved in unconscious, reflex adjustments of balance. In the present chapter we are mostly concerned with the former aspects of proprioception, especially the perception of the location and action of the body's parts.

It is of interest that Sherrington himself referred to the sense of limb movement and position as 'the muscular sense' (Sherrington, 1900), and he clearly believed that the discharges from muscle receptors reached consciousness. Physiologists and neurologists who came after him, however, did not share this view. They believed that proprioceptive sensibility could be attributed entirely to the discharges of sensory nerve endings in and around the capsules and ligaments of joints. This belief was based, in part, on the apparent suitability of joint receptors for this role as revealed by electrophysiological recordings said to be of their behaviour (although it later became apparent that some recordings were made, mistakenly, from intramuscular receptors). The belief was reinforced by the apparent unsuitability of intramuscular receptors for conscious proprioception. While stretch-sensitive receptors, such as muscle spindles, in muscles operating around a joint might have seemed appropriate to provide a signal from which the central nervous system could compute joint position or velocity of movement, these receptors could also be made to discharge by fusimotor (γ) activation without any change of joint position. It seemed that such discharges could not signal joint position unambiguously. Furthermore, intramuscular receptors were thought not to project to the cerebral cortex. For these and other reasons intramuscular receptors were believed to be 'private' to the muscles and to deal only with reflex adjustments at a subconscious level.

Sources of proprioceptive signals

The classes of afferent fibre that are candidates for subserving position and movement senses are those from the muscles and tendons, the joint capsules and ligaments, and the skin.

Muscle receptors

So entrenched was the view of the 'insentience' of muscles that, when a cortical projection from muscles was first discovered (Amassian & Berlin, 1958; Oscarsson and Rosén, 1963; Phillips, Powell & Wiesendanger , 1971), it was taken to indicate that access of sensory information to the cortex need not involve conscious awareness of its message. This notion opened the way to the proposal that some motor reflexes may be transcortical (see Phillips, 1969). Subsequently, however, it was shown in a variety of experiments on normal subjects that proprioceptive signals based on intramuscular receptors can be perceived. These experiments have been reviewed elsewhere (e.g. Goodwin, McCloskey & Matthews, 1972; Goodwin, 1976; Matthews, 1977, 1988; McCloskey, 1978) but briefly include the following.

1. Anaesthesia of the joints and skin of a finger or thumb, or of the whole hand, is a procedure which does not include the muscles which flex and extend the digits and which lie in the forearm away from the anesthetized area. However, this procedure does not abolish a subject's ability to detect flexion or extension movements imposed on a digit, an ability which must therefore be ascribed to discharges of sensory receptors in the unaffected muscles.
2. A simpler experiment is to pull upon an exposed tendon in a conscious subject so as to stretch its muscle; this can be done while immobilizing the joint which the muscle normally moves. This simple experiment has sometimes failed to give this result and the subject is said to get no sensation from tendon pulling (Gelfan & Carter, 1967; Moberg, 1972, 1983).

However, such experiments were carried out on patients during surgery and it is possible that they were not sufficiently at ease and attentive to perceive rather subtle sensations. When the experiment is performed under laboratory conditions, and in other circumstances on surgical patients, however, the subject reports that

Aspects of proprioception

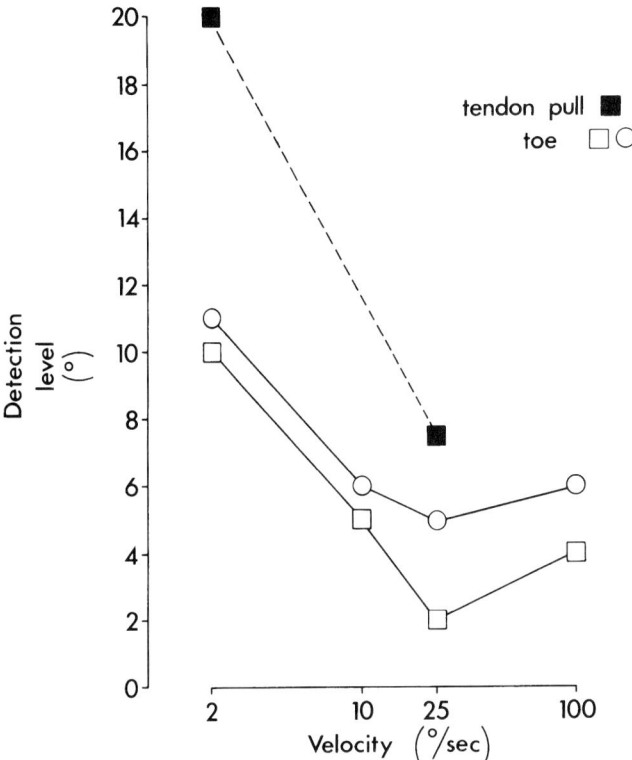

Fig. 1.1. This shows the relation between 70% detection levels of joint rotation and the angular velocity of rotation, for the interphalangeal joint of the big toe. The open symbols show the relation for the intact toe: circles show data for imposed ramp displacements into dorsiflexion and squares for plantar flexion. The filled squares show data obtained subsequently in the experiment on the subject's exposed, transected extensor hallucis longus tendon. Each point represents the smallest angle (or equivalent angle) of imposed displacement, at the velocity shown, for which the subject correctly nominated the direction of joint rotation in seven or more of ten trials. Velocity is plotted on a log scale. In the experiment on the exposed tendon, ramp displacments at 2.5 mm/s (equivalent to 25°/s) were tested in two directions: the 'pulls', detected as plantar flexions, are plotted; 'pushes' at the same velocity, even when 2 mm (20°) in amplitude, were not detected. McCloskey et al. (1983a) reproduced with permission.

the joint which is usually operated by the lengthened muscle seems to move, and to move in the direction which would normally stretch the muscle (McCloskey *et al.*, 1983a; Moberg, 1983). In one of these studies (Moberg, 1983), the proprioceptive sensation was reported only when the muscle was pulled near to the end of its physiological range. In the other, finely graded proprioceptive sensations were evoked, and the detection thresholds of these corresponded quite well to the thresholds for detection of displacements imposed on the same joint prior to the tendon-pulling experiment (McCloskey *et al.*, 1983a; Fig. 1.1).

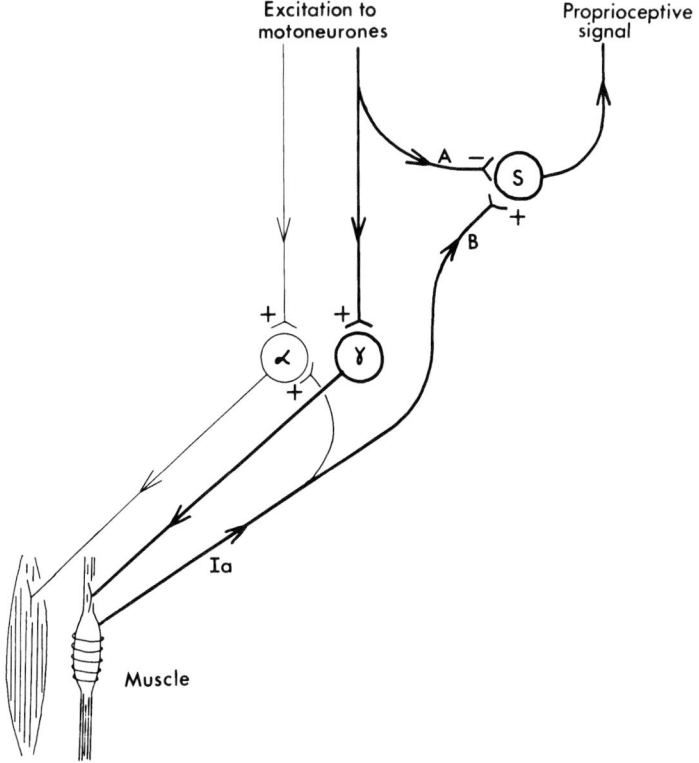

Fig. 1.2. Model showing how a kinaesthetic signal might be extracted from the discharge of muscle spindle afferents. The Figure shows descending pathways activating spinal α- and γ-MNs. Signals, shown here in pathway A, and related to the input to the γ-MN, provide an inhibitory input to cell S. The afferent pathway from the muscle spindles (Ia), as well as providing excitation to α-MNs, is shown to provide an excitatory input to cell S along pathway B. Subtractive interaction between A and B at cell S could extract a signal of kinesthetic significance from the spindle input. Pathway A carries an efference copy which cancels from the total afference in pathway B, the reafference due to fusimotor activation: A kinesthetic signal (exafference) results. McCloskey et al. (1983b) reproduced with permission.

The intramuscular mechanoreceptors most likely to subserve position and movement sense are muscle spindles. Both the spindle primary and secondary receptors are exquisitely sensitive to small changes in length, and the spindle primaries have a high dynamic sensitivity well suited to signalling velocity. Importantly, high frequency vibration applied to human muscle tendons, a stimulus known to excite muscle spindle afferents (Burke *et al.*, 1976; Roll & Vedel, 1982), evokes powerful illusions of joint movement and altered joint position (Goodwin *et al.*, 1972; McCloskey *et al.*, 1983*a*). Furthermore, when thresholds for detection of movements of relaxed human joints are expressed in terms of the lengths of the muscle fascicles that operate those joints, the thresholds are close to the known limits of resolution of the muscle spindles, and below those of conceivable alternative detectors (Hall & McCloskey, 1983).

A kinesthetic role for muscle spindles would require a solution to the problem of the potential ambiguity contained in their discharges because of their sensitivity to both stretch and to fusimotor activity. The answer seems to be that the central nervous system informs itself internally of the level of fusimotor drive in any given circumstance and that only the spindle discharge which exceeds what would be appropriate to this has kinesthetic significance. Such a simple computation could be easily achieved by the central nervous system. Essentially, it may amount to nothing more than an appropriately scaled subtraction, as indicated in Fig. 1.2, and some evidence has been advanced to suggest that it is, in fact, done this way (McCloskey, 1973). Alternatively, a process that does not require subtraction, or negation, of the fusimotor-induced part of the signal might be used. The essential requirement would be that the central nervous system is able to distinguish a proprioceptively significant element of spindle discharge from a fusimotor-induced one.

Electrophysiological assessment of a muscle afferent projection

Electrophysiological evidence exists in experimental animals for a cortical projection for specialized muscle afferents, both for muscle spindle endings and Golgi tendon organs. Even in subcortical relay nuclei, input from stretch and tension receptors is segregated (McIntyre, Proske & Rawson, 1984, 1985). Translation of this information into an effective electrophysiological test for the integrity of human kinesthetic pathways has proceeded slowly. An obvious approach, namely to record the cortical potentials produced by limb or joint movement, is flawed, because movement activates proprioceptors, not only in muscles but also receptors within the skin and joint, cortical potentials evoked by movement are impossible to ascribe to a specific afferent class. An alternative approach is to stimulate the muscle afferents at or near the motor point using insulated electrodes. This will activate spindle and tendon organ afferents from the muscle under study together with only a small number of nonmuscle afferents coursing in the 'muscle' nerve. Using this technique, it is possible to map the cerebral potentials evoked by stimulation of muscle afferents from a variety of distal, proximal, and truncal muscles (e.g. Gandevia & Burke, 1988; Macefield, Burke & Gandevia, 1989*a*). Latencies obtained in this way have been used to estimate the cortical processing time in the transcortical muscle stretch reflex (Deuschl *et al.*, 1989). Although the amplitude of the initial cortical component for most muscles is relatively small (~1 μV), the potentials are recordable from naive patients under normal laboratory conditions (e.g. Gandevia & Burke, 1988). Intraneural stimulation with microneurography is not practicable for recording cerebral potentials in patients because it can be both painful and time consuming, but it was used initially to define unequivocally a cortical projection from human muscle afferents in the lower and upper limbs (Starr *et al.*, 1981; Burke, Skuse & Lethlean, 1981; Gandevia, Burke & McKeon, 1984).

Assessment of the projection from muscle afferents is particularly important because, for the lower limb, the relevant spinal pathway (via Clarke's column, and the dorsal spinocerebellar tract) differs from that travelled by specialized cutaneous afferents (via dorsal columns) (for review, see York, 1985). Thus a test of the integrity of the muscle afferent pathway from the lower limb would expose deficits in a different long spinal tract from that conventionally believed to mediate the earliest component of the cortical potential evoked from a cutaneous nerve (namely the dorsal column pathway, see Halliday & Wakefield, 1963).

Stimulation of mixed nerves in the lower limb offers an indirect approach to assess the muscle afferent projection because the cortical potential produced by the posterior tibial nerve at the ankle is dominated by the muscle rather than the cutaneous contribution (e.g. Burke *et al.*, 1981; Macefield *et al.*, 1989*a*; *cf.* Jones & Power, 1984). Thus deficits in this cortical potential correlate better with deficits in joint movement than with deficits of cutaneous sensation. Originally, this domination of the posterior tibial cortical potential was partly ascribed to a difference in peripheral conduction times. However, as observed originally by Dawson in the upper limb (1956), and now documented formally for both upper and lower limbs, the conduction velocity of the bulk of rapidly conducting muscle afferents does not exceed that of cutaneous afferents at least for the input from the hand and foot (Macefield, Gandevia & Burke, 1989*b*). It may eventually be possible to study the central projection of Golgi tendon organs in patients using a technique which minimizes the electrical activation of muscle spindle afferents (e.g. Rossi, Mazzachio & Parlanti, 1991).

Joint

Joint afferents were once considered so critical for detection of limb movement that the term *joint* position sense seemed clinically appropriate. However, with the appreciation that muscle spindle afferents subserve this sensation (see above) the potential kinesthetic role for joint afferents was dismissed. A combination of psychophysical and electrophysiological studies in human subjects has reaffirmed a possible kinesthetic role for some specialized joint receptors.

The hand can be positioned so that the distal interphalangeal joint of the middle finger is effectively disengaged from its muscular attachments (Gandevia & McCloskey, 1976; see later). When this is done, applied movements within the normal movement range cannot be detected at all if the digital nerves of the finger are anesthetized (Goodwin et al., 1972). This implies that joint and cutaneous receptors subserve the residual kinesthetic sensation, but it does not distinguish between these two afferent species. This distinction was made when subjects, with the hand postured to eliminate a contribution from muscle afferents, detected movements applied by a motor far less effectively when the joint was anesthetized by injection of a small volume of lignocaine than when the joint was unanesthetized but distended by a volume of a plasma expander. These data are discussed further, below in the context of 'multi-component' proprioceptive performance.

For other joints in the body, a similar role for joint receptors has not yet been found. Indeed, anesthesia of the knee joint does not impair the detection of extremely slow movements (~1°/min; Clark et al., 1979); nor does replacement of diseased joints impair movement detection (see McCloskey, 1978). In the former studies, muscle afferents were still able to provide kinesthetic signals, and in the latter, disease may have already eliminated useful kinesthetic input from joint receptors. Anesthesia of the proximal interphalangeal joint in human subjects systematically distorts its perceived position particularly at the extremes of an angular range (Ferrell & Smith, 1988; 1989). In the cat, acute anesthesia of the knee joint produces an obvious impairment in gait (Ferrell et al., 1985).

Microneurography has corroborated a possible kinesthetic role for some digital joint receptors. Although the majority of slowly adapting receptors discharged at the end of an angular range and often responded to more than one axis of movement (e.g. flexion/extension, adduction/abduction), some discharged with movement through the mid range (Burke, Gandevia & Macefield, 1988). Microstimulation through the recording electrode of putative single joint afferents produced illusory movements: either small movements within the usual range of motion or sensations that the joint was held at one angular extreme (Macefield, Gandevia & Burke, 1990). Thus, it appears that minimal spatial summation is required for joint afferent stimulation to evoke a percept. Muscle spindle endings behaved quite differently: stimulation of single afferents in isolation went undetected.

Cutaneous receptors

Despite the obvious cutaneous displacements and distortions produced by movement, the role of the skin in detection of movements is less well established than that of the other afferent groups. Perhaps we can look back as far as Sherrington to see how a proprioceptive role for cutaneous receptors has been neglected in favour of their obvious exteroceptive role. A number of clues point to a kinesthetic role for specialised cutaneous receptors. First, both slowly and rapidly adapting receptors in the skin of the human hand discharge during passive (e.g. Knibestöl, 1975; Burke et al., 1988; Edin & Abbs, 1991) and active movements of the fingers (Hulliger et al., 1979). The discharge of the slowly adapting receptors increases towards the extremes of flexion and extension, much as noted for joint receptors (Hulliger et al., 1979; Burke et al., 1988). Recordings from the radial nerve of afferents innervating the skin on the dorsal aspect of the hand have emphasized the potential for cutaneous receptors, some centimetres proximal to the joint, to encode passive and active movements. Unlike receptors within the territory of digital nerves, these receptors may discharge to movement of more than one digit and more than one joint within a digit. Commonly, discharge increases with flexion of the fingers and stretch of the skin on the dorsum of the hand. By analogy with spindle endings in muscles which cross several joints, the decoding of cutaneous signals for specification of individual joint movement would require reference to other kinesthetic inputs.

Microstimulation of some cutaneous afferents innervating the skin over the distal interphalangeal joint provides a perception of joint movement consistent with the passive responses of the afferent (Macefield et al., 1990). Stimulation of digital nerves also produces illusions of joint movement and abnormal joint positions, presumably attributable to joint and cutaneous inputs (Gandevia, 1985).

Cutaneous (and other) afferents with a tonic discharge may affect kinesthesia more indirectly by 'facilitation' of the detection. Removal of the afferent inputs from the *index and ring* finger by local anesthesia impairs detection of movements of the *middle* finger (Gandevia & McCloskey, 1976). Anesthesia of the fingertip impairs detection of movements about the proximal interphalangeal joint (Clark, Grigg & Chapin, 1989; cf. Ferrell & Smith, 1988), an effect due either to the loss of a general facilitatory input or to a specific but 'remote' input from cutaneous receptors located distally.

Against the positive evidence above, cutaneous anesthesia around large joints fails to impair detection of slow movements when all other afferents are intact (Clark et al., 1979). Anesthesia of a small area of skin on the dorsum of a distal joint of the finger does not impair movement detection (Ferrell, Gandevia & McCloskey, 1987). Cutaneous anesthesia can have a seemingly paradoxical effect on movement detection: it may improve detection, perhaps by removing some unnatural cutaneous input generated by the attachment of the motor to the limb (Clark *et al.*, 1979). This phenomenon itself is evidence of movement sensation attributable to cutaneous inputs. We have also observed a comparable phenomenon at the distal interphalangeal joint of the finger.

In summary, cutaneous afferents have a complex kinesthetic role involving some capacity to signal the occurrence of movements near, and even remote from, their receptors and an ability to facilitate kinesthetic judgements involving intramuscular receptors.

Kinesthetic performance of proximal versus distal muscles

Performance of the distal muscles is often believed to be more precise or accurate than that of proximal and truncal muscles. This has been felt to reflect both the need for precise movement of the finger tips and the distribution of corticomotoneuronal control exerted via pyramidal and other pathways. However, the two examples given below will highlight the problems with this simple notion: if kinesthetic performance is measured in ways which take into account the different physical properties of muscle (e.g. length and cross-sectional area), apparent differences between proximal and distal musculature are diminished.

The first example is derived from the suggestions of Goldscheider (1889) who found that detection of movement occurs for smaller angular rotations of proximal than distal joints. In contrast, if one compares how far these angular rotations move the tip of the finger, the opposite trend occurs: movements at distal joints are detected when they produce a smaller displacement of the finger tip. If signals from muscle spindle afferents provide the major indicator that a joint moved, then it would be reasonable to express the thresholds for detection of applied movements not in angular terms but in terms of the relative change in muscle length. When this is done, the difference in movement detection *between* joints in the upper limb is much less (Hall & McCloskey, 1983). While the calculations above involve assumptions about the changes in muscle length, tendon compliance and the behaviour of muscle spindle afferents, they reveal that comparative kinesthetic performance depends on how it is expressed. As indicated below, comparative performance in a passive test may carry over to an active tracking task. It is notable that this form of analysis has not been extended to include the intrinsic muscles of the hand which have relatively short tendons and high densities of muscle spindles (Buchthal & Schmalbruch, 1980).

The second example involves the judgement of force or perceived heaviness by different muscle groups. When subjects match a reference weight lifted by a muscle (or muscle group) on one side with a variable one lifted similarly on the other side, judgements are biased by the 'effort' or centrally generated motor command rather than signals of absolute force (e.g. McCloskey, Ebeling & Goodwin, 1974; Gandevia & McCloskey, 1977a,b; for review see McCloskey, 1981; Gandevia, 1987; Cafarelli, 1988). Repetitions of the task allow calculation of the coefficient of variation, an index of accuracy or reproducibility. If this is performed for several muscle groups in the upper limb (elbow flexors, extrinsic and intrinsic muscles for the thumb and other digits) with the reference weights chosen to be specific fractions of the maximal voluntary strength of the test muscle, then accuracy is not highest for the intrinsic muscles of the hand (Gandevia & Kilbreath, 1990; Kilbreath & Gandevia, 1992). Yet, these muscles receive an especially dense corticomotoneuronal innervation in sub-human primates (Phillips & Porter, 1977; see also Rothwell et al., 1987) and move the densely innervated digital skin (e.g. Johansson & Vallbo, 1979).

Proprioceptive performance

Multi component performance

One method through which insight has been gained into how the components of proprioception perform individually, and combine together, has involved an anatomical peculiarity to 'disconnect' muscles from their attachment at a joint. If all the joints of the index, ring and little fingers are extended, and then held extended while the middle finger is flexed maximally at its first interphalangeal joint, the terminal joint of the middle finger becomes impossible to move by voluntary muscular effort. This phenomenon has been known to anatomists for very many years. The flexor and extensor of the joint attach also to adjacent fingers and to more proximal joints of the middle finger and, in the posture described, are held by these attachments at lengths which are inappropriate for their action on the terminal joint of the middle finger. Nevertheless, it remains easy to impose movements on the joint. Because the muscles cannot operate the joint, imposed movements do not pull upon the muscles, and proprioceptive acuity can be tested when only joint and cutaneous receptors can contribute to performance. By extending the middle finger, muscular action is restored at the terminal joint, and proprioceptive acuity can be tested again when all receptor species (intramuscular, joint and cutaneous) can be stimulated.

After injection of local anaesthetic around the digital nerves of the middle finger, afferents from joint and cutaneous receptors are blocked, leaving only the receptors in the long flexor and extensor muscles available for proprioception (Gandevia & McCloskey, 1976; Gandevia et al., 1983). Injection of local

Aspects of proprioception 13

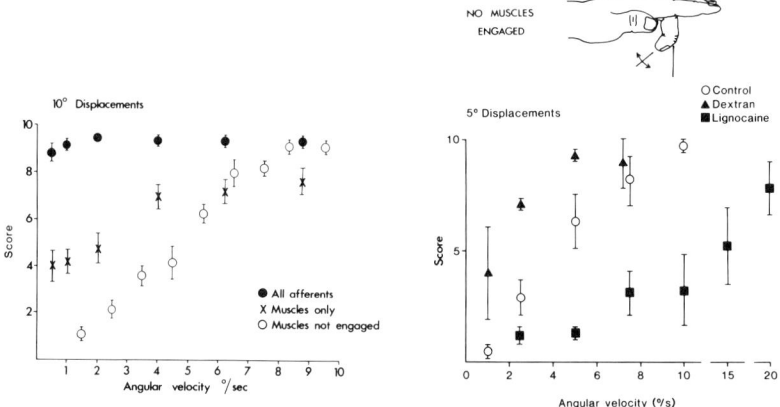

Fig. 1.3. *Left*: Detections made by eight normal subjects of 10° displacements imposed on the terminal joint of the middle finger at various angular velocities. Each subject was presented with sets of ten flexions and ten extensions, randomly mixed, and the "score" of detections out of ten for each was recorded: detection required correct nomination of the direction of imposed movement. This Figure shows the mean (±SEM) detection scores for subjects with all proprioceptive machinery intact (filled circles); with only muscle afferents available to contribute to kinesthesia: with both flexor and extensor engaged during digital anesthesia (crosses); and, with muscles disengaged, leaving only joint and cutaneous afferents available to contribute to kinaesthesia (open circles). Gandevia *et al*. (1983) reproduced with permission.

*Righ*t: Detection scores for subjects in whom sets of ten flexion and ten extension, randomly mixed, were imposed on the terminal joint of middle finger while the hand was postured so as to disengage effective muscular action on joint (see inset). Detection required correct nomination of direction. Two scores were obtained for each of three subjects, and means ±SEM are plotted. Circles: performance when joint and/or cutaneous afferents were available; triangles: enhancement of performance resulting from expanding joint capsule by injection of a dextran solution; squares: reduced performance resulting from anesthesia of joint capsular receptors by intraarticular injection of lignocaine. In this condition only cutaneous afferents were available to provide proprioceptive information. Based on data in Ferrell *et al*. (1987).

anesthetic into the joint capsule will impair joint receptors but spare cutaneous (and muscle) receptors (Ferrell *et al*., 1987). Figure 1.3 illustrates some of the findings using these experimental procedures.

Proprioceptive acuity with all receptors available is superior to that when intramuscular receptors cannot contribute or when only they can contribute (Gandevia & McCloskey, 1976; Gandevia *et al*., 1983). By itself, anesthesia of joint receptors does little to blunt acuity provided that intramuscular receptors can be stimulated (i.e. muscles 'connected'), but causes significant deficits when they cannot (i.e. muscles 'disconnected') (Ferrell *et al*., 1987). This implies some redundancy of proprioceptive function between these two receptor classes. When both cutaneous and joint receptors are anesthetized, proprio-

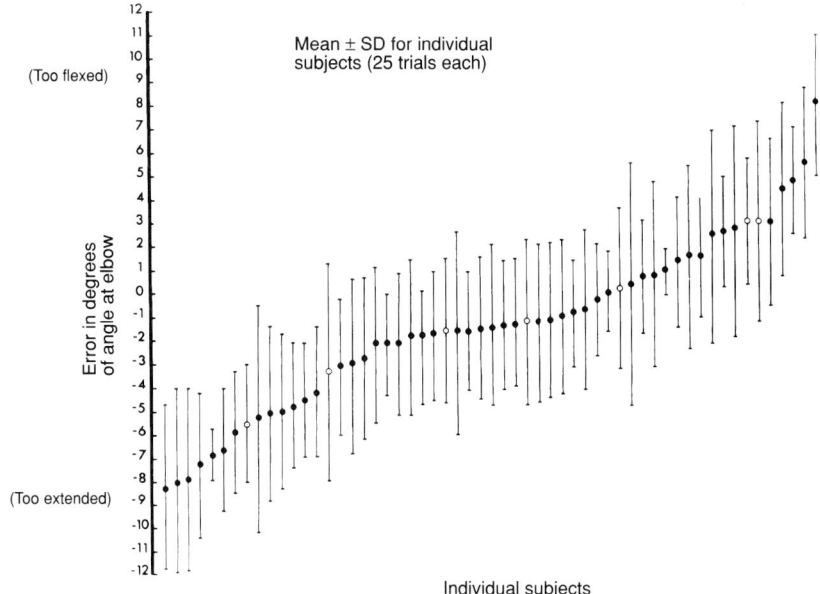

Fig. 1.4. Results of tests on position sense of the elbow joint measured in 50 normal sighted subjects (filled circles) and 7 normal congenitally blind subjects (open circles). Each subject made five attempts at each of 10°, 20°, 50°, 70° and 80° from full extension to touch the index fingers together. A perspex board prevented the fingers from touching. The results give the error of alignment in degrees of angle of the elbow, showing mean ±SD for 25 trials in each of the subjects. McCloskey (1980) reproduced with permission.

ceptive performance deteriorates (Gandevia & McCloskey, 1976; Gandevia *et al.*, 1983). As cutaneous receptors alone probably have a limited *specific* role in proprioceptive performance, this probably indicates an important nonspecific background *facilitatory* role of cutaneous receptors (i.e. facilitating inputs from specific proprioceptive inputs), at least for the fingers.

Proprioception in everyday tasks

There seems little doubt that all movements depend critically upon proprioceptive inputs. For slower movements, these may well be crucial *during* the task, while for faster ones proprioceptive input probably define start and end points, and provide *post-hoc* knowledge of results.

Simple daily tasks that seem to have a clear proprioceptive component are executed with variable accuracy. It has been shown, for example, that human subjects have a remarkable capacity to discriminate the thicknesses of objects held between finger and thumb (John, Goodwin & Darian-Smith, 1989) in circumstances in which possible contributions from non proprioceptive inputs have been carefully excluded as the basis of discrimination. From this, it might be concluded that proprioception is a sense of high acuity. However, this contrasts with the variable performance within and between subjects in simply approximating the tips of the index fingers of the two hands without looking (Fig. 1.4). From this, it might be concluded that proprioception is a sense of relatively low acuity when compared, for example, to vision or hearing.

It needs to be borne in mind that tasks with a strong dependence on proprioception may not depend solely on proprioception, and may use proprioceptive cues in various ways. Touching the index fingers together without vision, for example, requires not only accuracy of signalling of the position of each finger tip independently (an accuracy that in turn depends upon accuracy of signals about positions of several joints), but a 'matching' of the calibrations of these two signals centrally, together with an accuracy of motor control on each side to achieve the movements necessary to bring the two composite position signals into register. The potential for small errors to summate in such a multi-component system is large.

Conclusion

When testing proprioceptive acuity in the clinical situation at distal joints, muscle, joint and perhaps even cutaneous afferents can contribute to the detection of the movement. Certainly each of these afferent classes will discharge in response to applied movement. Based on their documented responsiveness to very small changes in length and the proprioceptive illusions produced by their relatively selective activation, muscle spindle afferents play a major role in kinesthesia. Given that subjects often inadvertently contract their muscles during such testing, cues from muscle receptors are likely to be further improved. Optimally, proprioception should be assessed when the patient is as relaxed as possible. When doubt about a deficit exists, or the patient contracts to "follow" the movements applied by the tester, or a more complete examination is relevant, the hand can be positioned by the examiner as in Fig. 1.3 to eliminate the influence of muscle contraction and the contribution from muscle afferents.

References

Amassian, V. E. & Berlin, L. (1958). Early cortical projection of Group I afferents in the forelimb muscle nerves of cat. *Journal of Physiology* (*London*), **143**, 61P.
Bastian, H. C. (1888). The 'muscular sense'; its nature and cortical localisation. *Brain*, **10**, 1–137.
Bell, C. (1833). *The Hand – Its Mechanism and Vital Endowments as Evincing Design*. London: Wiliam Pickering.

Buchthal, F. & Schmalbruch, H. (1980). Motor unit of mammalian muscle. *Physiological Reviews*, **60**, 90–142.

Burke, D., Gandevia, S. C. & Macefield, G. (1988). Responses to passive movement of receptors in joint, skin and muscle of the human hand. *Journal of Physiology (London)*, **402**, 347–61.

Burke, D., Hagbarth, K.-E., Löfstedt, L. & Wallin, B. G. (1976). The responses of human muscle spindle endings to vibration of non-contracting muscles. *Journal of Physiology (London)*, **261**, 673–94.

Burke, D., Skuse, N. F. & Lethlean, A. K. (1981). cutaneous and muscle afferent components of the cerebral potential evoked by electrical stimulaiton of human peripheral nerves. *Electroencephalography and Clinical Neurophysiology*, **51**, 579–88.

Cafarelli, D. (1988). Force sensation in fresh and fatigued human skeletal muscle. *Exercise and Sports Science Review*, **16**, 139–68.

Clark, F. J., Grigg, P. & Chapin, J. W. (1989). The contribution of articular receptors to proprioception with the fingers in humans. *Journal of Neurophysiology*, **61**, 186–93.

Clark, F. J., Horch, K. W., Bach, S. M. & Larson, G. F. (1979). Contributions of cutaneous and joint receptors to static knee-position sense in man. *Journal of Neurophysiology*, **42**, 877–88.

Dawson, G. D. (1956). The relative excitability and conduction velocity of sensory and motor nerve fibres in man. *Journal of Physiology (London)*, **131**, 436–51.

Deuschl, G., Ludolph, A., Schenck, E. Lücking, C. H. (1989). The relations between long-latency reflexes in hand muscles, somatosensory evoked potentials and transcranial stimulation of motor tracts. *Electroencephalography and Clinical Neurophysiology*, **74**, 425–30.

Edin, B. B. & Abbs, J. H. (1991). Finger movement responses of cutaneous mechanoreceptors in the dorsal skin of the human hand. *Journal of Neurophysiology*, **65**, 657–70.

Ferrell, W. R. (1987). The effect of acute joint distension on mechanoreceptor discharge in the knee of the cat. *Quarterly Journal of Experimental Physiology*, **72**, 493–9.

Ferrell, W. R., Baxendale, R. H., Carnachan, C. & Hart, I. K. (1985). The influence of joint afferent discharge on locomotion, proprioception and activity in conscious cats. *Brain Research*, **347**, 41–8.

Ferrell, W. R., Gandevia, S. C. & McCloskey, D. I. (1987). The role of joint receptors in human kinaesthesia when intramuscular receptors cannot contribute. *Journal of Physiology (London)*, **386**, 63–71.

Ferrell, W. R. & Smith, A. (1988). Position sense at the proximal interphalangeal joint of the human index finger. *Journal of Physiology (London)*, **399**, 49–61.

Ferrell, W. R. & Smith, A. (1989). The effect of loading on position sense at the proximal interphalangeal joint of the human index finger. *Journal of Physiology (London)*, **418**, 145–61.

Gandevia, S. C. (1985). Illusory movements produced by electrical stimulation of low-threshold muscle afferents from the hand. *Brain*, **108**, 965–81.

Gandevia, S. C. (1987). Roles for perceived voluntary motor commands in motor control. *Trends in Neurosciences*, **10**, 81–5.

Gandevia, S. C. & Burke, D. (1988). Projection to the cerebral cortex from proximal and distal muscles in the human upper limb. *Brain*, **111**, 389–403.

Gandevia, S. C., Burke, D. & McKeon, B. (1984). The projection of muscle afferents from the hand to cerebral cortex in man. *Brain*, **107**, 1–13.

Gandevia, S. C., Hall, L. A., McCloskey, D. I. & Potter, E. K. (1983). Proprioceptive sensation at the terminal joint of the middle finger. *Journal of Physiology (London)*, **355**, 507–17.

Gandevia, S. C. & Kilbreath, S. (1990). Accuracy of weight estimation for weights lifted by proximal and distal muscles of the human upper limb. *Journal of Physiology (London)*, **423**, 299–310.

Gandevia, S. C., Macefield, G., Burke, D. & McKenzie, D. K. (1990). Voluntary activation of human motor axons in the absence of muscle afferent feedback: the control of the deafferented hand. *Brain*, **113**, 1563–81.

Gandevia, S. C. & McCloskey, D. I. (1976). Joint sense, muscle sense, and their combination as position sense measured at the distal interphalangeal joint of the middle finger. *Journal of Physiology (London)*, **260**, 387–407.

Gandevia, S. C. & McCloskey, D. I. (1977*a*). Sensations of heaviness. *Brain*, **100**, 345–54.

Gandevia, S. C. & McCloskey, D. I. (1977*b*). Effects of related sensory inputs on motor performances in man studied through changes in perceived heaviness. *Journal of Physiology (London)*, **272**, 653–72.

Gelfan, S. & Carter, S. (1967). Muscle sense in man. *Experimental Neurology*, **18**, 469–73.

Goldscheider, A. (1889). See Sherrington (1900).

Goodwin, G. M. (1976). The sense of limb position and movement. *Exercise and Sport Science Review*, **4**, 87–124.

Goodwin, G. M., McCloskey, D. I. & Matthews, P. B. C. (1972). The contribution of muscle afferent to kinaesthesia shown by vibration induced illusions of movement and by the effects of paralysing joint afferents. *Brain*, **95**, 705–48.

Hall, L. A. & McCloskey, D. I. (1983). Detections of movements imposed on finger, elbow and shoulder joints. *Journal of Physiology (London)*, **335**, 519–33.

Halliday, A. M. & Wakefield, G. S. (1963). Cerebral evoked potentials in patients with dissociated sensory loss. *Journal of Neurology, Neurosurgery, and Psychiatry*, **26**, 211–19.

Hulliger, M., Nordh, E., Thelin, A.-E. & Vallbo, Å. B. (1979). The responses of afferent fibres from the glabrous skin of the hand during voluntary finger movements in man. *Journal of Physiology (London)*, **291**, 233–49.

Johansson, R. S. & Vallbo, Å. B. (1979). Tactile sensibility in the human hand: relative and absolute densities of four types of mechanoreceptive units in glabrous skin. *Journal of Physiology (London)*, **286**, 283–300.

John, K. T., Goodwin, A. W. & Darian-Smith, I. (1989). Tactile discrimination of thickness. *Experimental Brain Research*, **78**, 62–8.

Jones, S. J. & Power, C. N. (1984). Scalp topography of human somatosensory evoked potentials: the effect of interfering tactile stimulation applied to the hand. *Electroencephalography and Clinical Neurophysiology*, **58**, 25–36.

Knibeström, M. (1975). Stimulus response functions of slowly adapting mechanoreceptors in the human glabrous skin area. *Journal of Physiology (London)*, **243**, 63–80.

Kilbreath, S. & Gandevia, S. C. (1992). Is voluntary control of human thumb muscles special? *Proceedings of the Australian Neuroscience Society*, **3**, 76.

McCloskey, D. I. (1973). Differences between the senses of movement and position shown by the effects of loading and vibration of muscles in man. *Brain Research*, **61**, 119–31.

McCloskey, D. I. (1978). Kinesthetic sensibility. *Physiological Reviews*, **58**, 763–820.

McCloskey, D. I. (1980). Kinaesthetic sensations and motor commands in man. In *Spinal and Supraspinal Mechanisms of Voluntary Motor Control and Locomotion. Progress in Clinical Neurophysiology,* vol. 8, ed J. E. Desmedt, pp. 203–14. Basel: Karger.

McCloskey, D. I. (1981). Corollary discharges: motor commands and perception. In *Handbook of Physiology. The Nervous System. Vol. III. Motor Control*, ed. V. B. Brooks, pp. 1415–47. Bethesda, MD: American Physiological Society.

McCloskey, D. I., Cross, M. J., Honner, R. & Potter, E. K. (1983*a*). Sensory effects of pulling or vibrating exposed tendons in man. *Brain*, **106**, 21–37.

McCloskey, D. I., Ebeling, P. & Goodwin, G. M. (1974). Estimation of weights and tensions and apparent involvement of a 'sense of effort'. *Experimental Neurology*, **42**, 220–32.

McCloskey, D. I., Gandevia, S., Potter, E. K. & Colebatch, J. G. (1983*b*). Muscle sense and effort: motor commands and judgements about muscular contractions. In *Motor Control Mechanisms in Health and Disease. Advances in Neurology*, vol. 39, ed. J. E. Desmedt, pp. 151–67. New York: Raven.

McIntyre, A. K., Proske, U. & Rawson, J. A. (1984). Cortical projection of afferent information from tendon organs in the cat. *Journal of Physiology (London)*, **354**, 395–406.

McIntyre, A. K., Proske, U. & Rawson, J. A. (1985). Pathway to the cerebral cortex for impulses from tendon organs in the cat's hindlimb. *Journal of Physiology (London)*, **369**, 115–26.

Macefield, G., Burke, D. & Gandevia, S. C. (1989*a*). The cortical distribution of muscle and cutaneous afferent projections from the human foot. *Electroencephalography and Clinical Neurophysiology*, **72**, 518–28.

Macefield, G., Gandevia, S. C. & Burke, D. (1989*b*). Conduction velocities of muscle and cutaneous afferents in the upper and lower limbs of human subjects. *Brain*, **112**, 1519–32.

Macefield, G., Gandevia, S. C. & Burke, D. (1990). Perceptual responses to microstimulation of single afferents innervating joints, muscles and skin of the human hand. *Journal of Physiology (London)*, **429**, 113–29.

Matthews, P. B. C. (1977). Muscle afferents and kinaesthesia. *British Medical Bulletin*, **33**, 137–42.

Matthews, P. B. C. (1988). Proprioceptors and their contribution to somatosensory mapping: complex messages require complex processing. *Canadian Journal of Physiology and Pharmacology*, **66**, 403–38.

Moberg, E. (1972). Fingers were made before forks. *Hand*, **4**, 201–6.

Moberg, E. (1983). The role of cutaneous afferents in position sense, kinaesthesia and motor function of the hand. *Brain*, **106**, 1–19.

Oscarsson, O. & Rosén, I. (1963). Projection to cerebral cortex of large muscle-spindle afferents in forelimb nerves of the cat. *Journal of Physiology (London)*, **169**, 924–45.

Phillips, C. G. (1969). Motor apparatus of the baboon's hand. *Proceedings of the Royal Society of London, Series B*, **173**, 141–74.

Phillips, C. G. & Porter, R. (1977). *Corticospinal Neurones*. London: Academic Press.

Phillips, C. G., Powell, T. P. S. & Wiesendanger, M. (1971). Projection from low-threshold muscle afferents of hand and forearm to area 3a of baboon's cortex. *Journal of Physiology (London)*, **217**, 419–46.

Roll, J. P. & Vedel, J. P. (1982). Kinaesthetic role of muscle afferents in man, studied by tendon vibration and microneurography. *Experimental Brain Research*, **47**, 177–90.

Rossi, A., Mazzacchio, R. & Parlanti, S. (1991). Cortical projection of putative group Ib afferent fibres from the human forearm. *Brain Research*, **547**, 62–8.

Rothwell, J. C., Thompson, P. D., Day, B. L., Dick, J. P. R., Kachi, T., Cowan, J. M. A. & Marsden, C. D. (1987). Motor cortex stimulation in intact man. *Brain*, **110**, 1173–90.

Sherrington, C. S. (1900). The muscular sense. In *Text-Book of Physiology*, vol. 2, ed. E. A. Schäfer, pp. 1002–25. Edinburgh: Young J. Pentland.
Sherrington, C. S. (1906). On the proprio-ceptive system, especially in its reflex aspects. *Brain*, **29**, 467–82.
Starr, A., McKeon, B., Skuse, N. & Burke, D. (1981). Cerebral potentials evoked by muscle stretch in man. *Brain*, **104**, 149–66.
York, D. H. (1985). Somatosensory evoked potentials in man: differentiation of spinal pathways responsible for conduction from the forelimb vs hindlimb. *Progress in Neurobiology*, **25**, 1–25.

2

The pathophysiological basis of paresthesiae

D. BURKE

Department of Clinical Neurophysiology,
The Prince Henry and Prince of Wales Hospitals
and
Prince of Wales Medical Research Institute,
University of New South Wales,
Sydney NSW, Australia

In 1979 Lance described a number of families afflicted by neuromyotonia, and in them it could be shown that ectopic activity in motor axons was the cause of the cramp-like after-contraction, myokymia and fasciculation (Lance, Burke & Pollard, 1979a,b). In one of the families, inflation of a sphygmomanometer cuff around the arm produced carpo-pedal spasm and intense paresthesiae and, in parallel with this, spontaneous activity developed in cutaneous sensory axons (Lance et al., 1979a). This was one of the first published demonstrations in human subjects that spontaneous activity in individual sensory axons caused paresthesiae. Since then whenever an adequate study of paresthesiae has been possible, they have been found to be due to ectopic impulse activity in disturbed cutaneous afferents (Hagbarth, Torebjörk & Wallin, 1984). Ectopic impulse activity has been recorded in patients experiencing paresthesiae due to nerve root compression (Fig. 2.1; Nordin et al., 1984), a neuroma (Nyström & Hagbarth, 1981), and Lhermitte's symptom (Nordin et al., 1984), and in normal volunteers following nerve compression/ischemia (Ochoa & Torebjörk, 1980), prolonged tetanic stimulation of cutaneous afferents (Ochoa & Torebjörk, 1983; Burke & Applegate, 1989) and hyperventilation (Macefield & Burke, 1991a).

Figure 2.1 shows a recording of the ectopic sensory discharge that occurred in a male patient aged 44 suffering from S1 root compression when a straight-leg raising test was performed (Nordin et al., 1984). The focus of ectopic activity was presumably the site of compression of the S1 root, and the ectopic discharge propagated in both directions along the affected axons away from the site of initiation. The antidromic impulses were recorded from the peripheral nerve; the orthodromic impulses entered the spinal cord and evoked the sensation of paresthesiae. Whenever the manoeuvre produced the abnormal neural discharge the patient experienced paresthesiae (Fig. 2.1b); when it did not, he did not (Fig. 2.1c).

Pathophysiological basis of paresthesiae

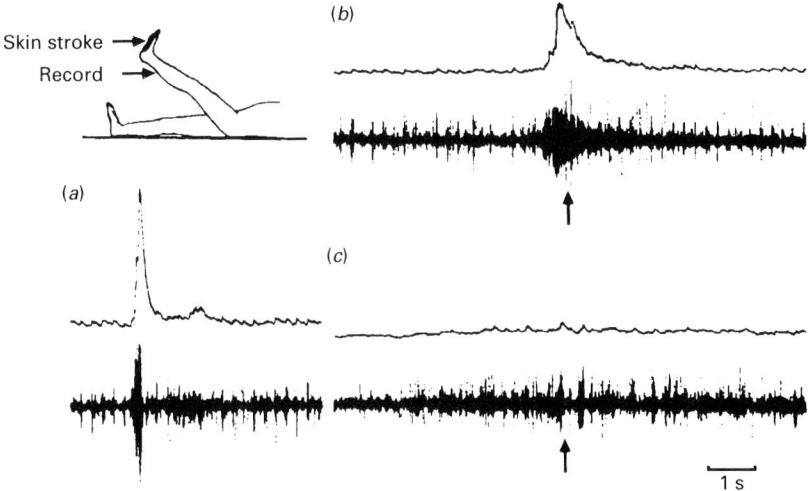

Fig. 2.1. Development of ectopic impulse activity in a patient with S1 nerve root compression during a straight leg raising test. Multiunit recording from the sural nerve with receptive field indicated. (*a*) shows the afferent response to stroking the skin of the lateral side of the foot with a probe (marked by a bar). The original neurogram is in the lower trace and the mean voltage neurogram in upper trace. In (*b*), a burst of neural activity accompanies the straight-leg raising test. In (*c*), a straight-leg raising test was performed at the moment indicated by an arrow. This time, no paresthesiae were evoked and there is no distinct burst in the neurogram. Nordin et al. (1984) reproduced with permission.

This chapter addresses the pathophysiological basis of paresthesiae, with specific attention to the following questions. Why do axons in a peripheral nerve develop foci of ectopic activity? What are the rate-limiting factors in the ectopic discharge? What are the factors that limit the duration of the ectopic train? Before these issues can be considered, it is important to consider some of the mechanisms responsible for the initiation and propagation of nerve impulses in normal axons.

Impulse initiation and conduction in normal cutaneous afferents

In sensory receptors, an increasingly strong stimulus results in increasing depolarization such that, within limits, the receptor potential is proportional to stimulus strength (Fig. 2.2). This graded change in membrane potential is converted into the all-or-none propagated action potential at a pacemaker region, and for myelinated fibres this is probably the first node of Ranvier. Thereafter, the impulse is conducted in a saltatory manner with depolarization occurring only at nodes of Ranvier.

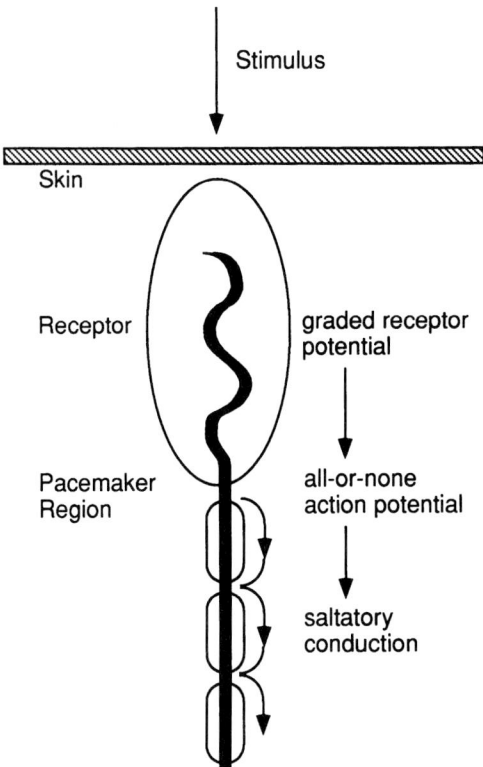

Fig. 2.2. Impulse initiation in cutaneous mechanoreceptors.

Depolarization of the nodal membrane is due to the opening of voltage-dependent sodium channels, which occur in high density at the node (Fig. 2.3). Such channels can be blocked by tetrodotoxin: in patients poisoned by consumption of puffer-fish, there is impaired action-potential electrogenesis (Oda et al., 1989). In mammalian axons, repolarization does not depend on potassium currents. Indeed, there are very few potassium channels with the fast kinetics required for them to play a role in repolarization at the nodes of Ranvier of mammalian axons (Chiu et al., 1979; Brismar, 1980; Bostock, Sears & Sherratt, 1981; Röper & Schwartz, 1989). Instead, repolarization depends almost exclusively on leakage currents and inactivation of the sodium conductance. The latter mechanism is also largely responsible for the subsequent refractoriness of the axon to being reexcited. In the internodal region of large myelinated axons, there are very few sodium channels under the myelin sheath (see Black, Kocsis & Waxman, 1990), insufficient to sustain continuous conduction in the majority of demyelinated axons. When large axons are

Pathophysiological basis of paresthesiae

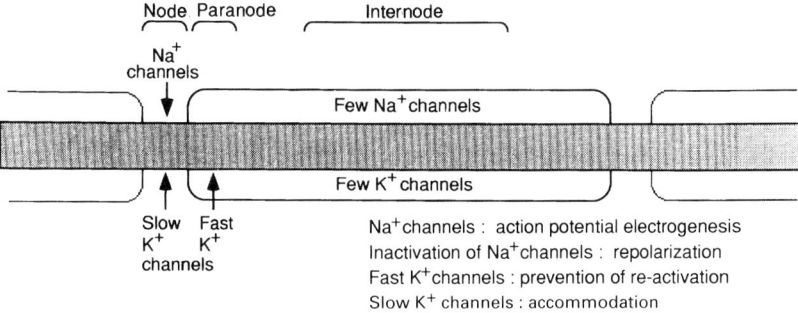

Fig. 2.3. Ion channel distribution in mammalian myelinated axons.

chronically demyelinated, the internodal membrane may develop isolated clumps of sodium channels, but the number and distribution are usually inadequate to support continuous conduction, at least in large axons (Meiri, Baum & Rosenthal, 1989; Black *et al.*, 1990, 1991). If conduction can be sustained in these axons despite the leakage of current across the internodal region, it will occur with a reduced safety margin but will remain saltatory.

Axonal excitability following single impulses

When a single impulse passes along an axon, the membrane potential undergoes a number of fluctuations, such that its susceptibility to being reexcited changes. Initially the axon will be refractory, first absolutely refractory such that it cannot be reexcited no matter how strong the stimulus, and subsequently relatively refractory, such that a stronger stimulus than normal must be used to excite the axon. The durations of these refractory phases vary directly with temperature and inversely with fibre size, with absolute refractoriness lasting 0.5–1.5 ms and relative refractoriness up to 3–4 ms. Following recovery from refractoriness, the axon becomes more easily excited as it passes into the supernormal period, which commonly lasts some 20 ms. The supernormal period is then followed by a late phase of subnormal excitability which can last for 100 ms before the axon regains its control excitability.

These properties can be studied in human cutaneous afferents, using relatively simple techniques (e.g. Gilliatt & Willison, 1963; Stöhr, 1981*a*; Ng, Burke & Al-Shehab, 1987; Stys & Ashby, 1990). Figure 2.4 illustrates data obtained with one of these techniques, that of measuring the changes in amplitude of the compound sensory action potential produced by a test stimulus of constant intensity derived from a constant-current source, when preceded by a supramaximal conditioning stimulus to the same axons (Ng *et al.*, 1987).

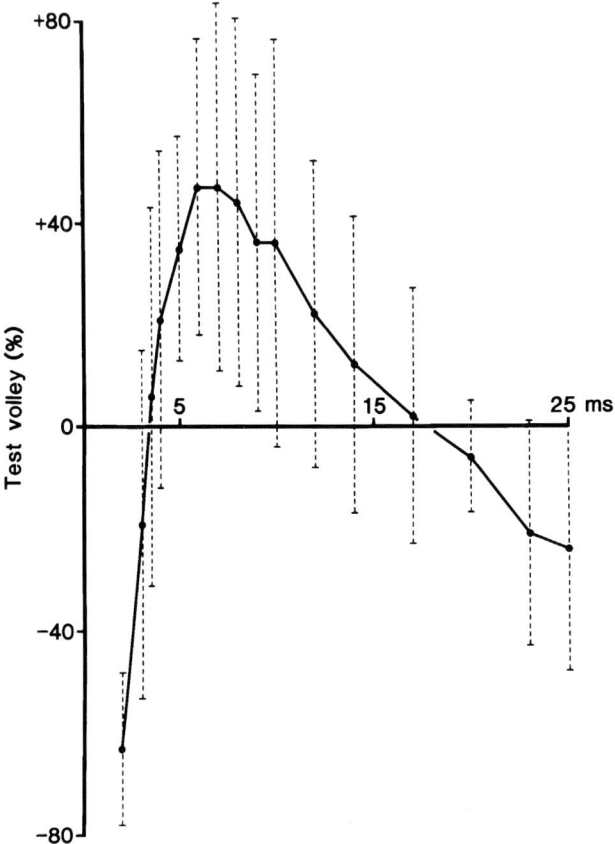

Fig. 2.4. The recovery cycle of human cutaneous afferents after a single discharge. A constant test stimulus (producing a test volley of 30–40% of maximum when delivered in isolation) was delivered to cutaneous afferents innervating the index finger after a maximal conditioning stimulus at conditioning test intervals from 2 ms to 25 ms. The change in amplitude of the test volley is expressed as a percentage of the unconditioned test volley on the vertical axis, with the conditioning test interval on the horizontal axis. Mean (±SD) of data for 20 normal subjects. Refactoriness lasts approximately 3–4 ms, and is followed by supernormality which peaks at 6–7 ms, decaying into a late phase of subnormality at approximately 20 ms. Ng et al. (1987) reproduced with permission.

The supernormal period is due to the passive discharge of capacitative current stored in the internodal membrane (Barrett & Barrett, 1982; Blight & Someya, 1985). This passive depolarization could be sufficient to reexcite the nodal membrane and generate another impulse but for the presence of fast potassium channels located in the paranodal region (Black et al., 1990). Indeed, when fast potassium channels are blocked by 4-aminopyridine and 3,4-diaminopyridine, drugs which can be useful clinically in treating the

Eaton–Lambert myasthenic syndrome, patients commonly develop paresthesiae. In immature or demyelinated axons, these agents can gain access to paranodal K^+ channels when added to the bathing medium. They then produce a prolonged after-depolarization in sensory axons with repetitive spike discharges (Eng *et al.*, 1988), and this has been suggested to be the mechanism underlying paresthesiae in patients receiving these drugs (Kocsis, Bowe & Waxman, 1986).

Axonal excitability following trains of impulses

When an axon discharges a train of impulses, compensatory mechanisms are activated to limit the spike train, and this occurs whether the spike train is set up naturally by stimulation of the receptor, electrically by stimulation of the axon or abnormally from an ectopic focus. The subnormal period increases in depth, the nodal membrane hyperpolarizes, and the axon becomes more difficult to excite. There are two mechanisms for this hyperpolarization (Gasser, 1935, 1958; Gasser & Grundfest, 1936; Graham & Lorente de Nó, 1938; Guisset, 1968; Bergmans, 1970; Applegate & Burke, 1989; Taylor, Burke & Heywood, 1992).

1. Following *short trains of impulses*, a positive afterpotential (termed P_1) develops, and the axonal membrane becomes hyperpolarized (Gasser, 1935, 1958; Baker *et al.*, 1987). This results in a decrease in axonal excitability, termed H_1 by Bergmans (1970). The decrease in excitability becomes greater, both in depth and duration, as the number of impulses in the train and/or the frequency of stimulation are increased, but a plateau is reached after about 10–20 impulses at frequencies of about 200/s.

    This phenomenon can be demonstrated in human motor axons (Bergmans, 1970) and in human cutaneous afferents (Guisset, 1968; Taylor *et al.*, 1992), and is illustrated in Fig. 2.5 from experiments using similar techniques to those used for Fig. 2.4. Panel (*a*) shows the changes in excitability following conditioning by either a single supramaximal pulse or a train of ten supramaximal pulses delivered at 200/s. The difference between these recovery cycles is largely due to the H_1 phase of posttetanic depression of axonal excitability. In panel (*b*), the data in the top traces have been subtracted to eliminate the refractory and supernormal periods and to define more clearly the decrease in excitability that resulted from the train of ten impulses. Such a subtraction ignores the fact that there will be differences in the degree of supernormality with the two different conditioning stimuli, but the discrepancies are probably minor (Taylor *et al.*, 1992).

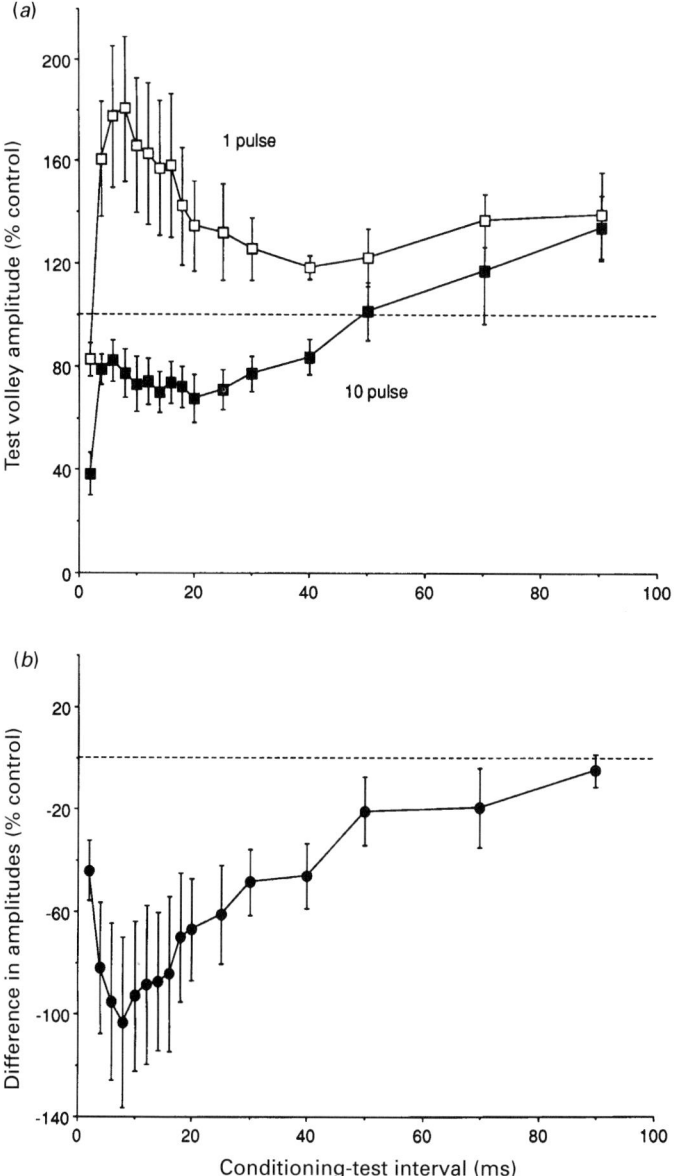

Fig. 2.5. Post-tetanic depression of excitability following short conditioning trains. Panel (*a*) shows the variations in amplitude of a test volley evoked by a constant stimulus when conditioned by a single supramaximal pulse (open squares) or a train of ten supramaximal pulses delivered at 200/s (filled squares), using conditioning-test intervals up to 90 ms. Data for six subjects (±SEM). In panel (*b*) the data in panel (*a*) have been subtracted to display more clearly the post-tetanic depression in excitability, on the assumption that refactoriness and supernormality are similar following conditioning with a single pulse and conditioning with ten pulses. Taylor et al. (1992) reproduced with permission.

Pathophysiological basis of paresthesiae 27

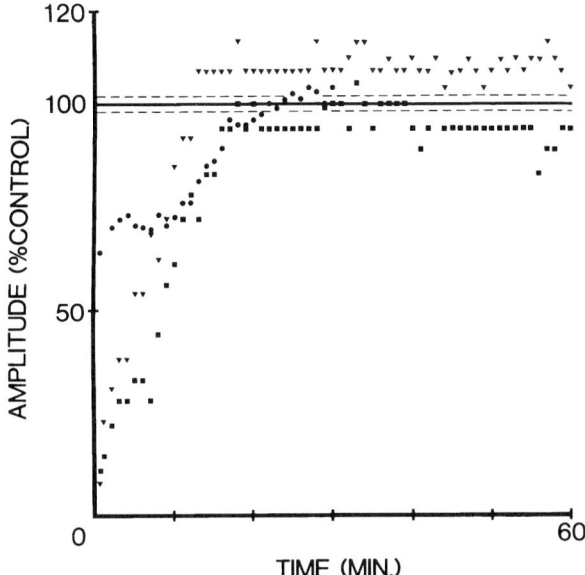

Fig. 2.6. The H2 phase of post-tetanic depression. Changes in amplitude of a submaximal test potential following repetitive stimulation at 200/s for 1 min. Data for three subjects, each represented by a different symbol. The dotted lines represent the variability of the test potential prior to tetanization, assessed for each subject using 8 consecutive averages of 16 responses. Excitability is depressed for 15–20 min following prolonged stimulation. Applegate & Burke (1989) reproduced with permission.

In animal experiments, it has been clearly established that this subexcitability is due to activation of slow potassium channels at the node of Ranvier (Baker *et al.*, 1987; Eng *et al.*, 1988). Because they mediate this subexcitability, slow potassium channels tend to produce accommodation to a sustained stimulus, and this will contribute to spike frequency adaptation: the decrease in discharge rate of an axon even though the stimulus to the receptor is maintained (Krylov & Makovsky, 1978; Awiszus, 1990). However, once a train of impulses is set up, the safety margin for neural transmission is quite high, and this slow-potassium-channel phenomenon will only limit the axonal discharge at sites of lowered safety margin, such as where the axon branches or at sites of demyelination.

2. Following *long trains of impulses*, a more profound subexcitability occurs. Such trains produce a large positive afterpotential (P_2 in the terminology of Gasser, 1935, 1958), associated with marked axonal hyperpolarization (Bostock & Grafe, 1985) and thereby a decrease in excitability (H_2 in the terminology of Bergmans, 1970). This decrease in excitability increases in depth and duration with the frequency and duration of the conditioning

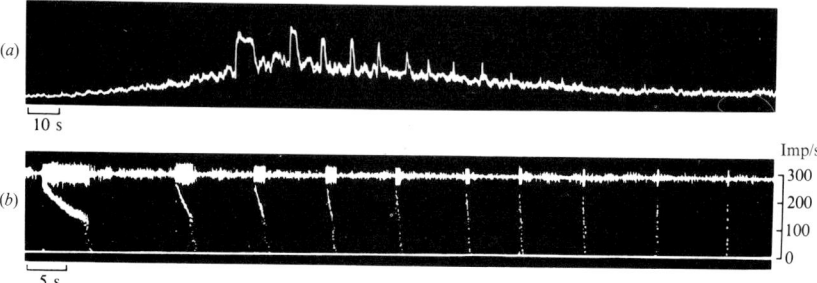

Fig. 2.7. 'Spontaneous' neural activity recorded in the ulnar nerve at elbow level in a subject who hyperventilated following release of an ischemic block produced using a sphygmomanometer around the upper arm. In panel (*a*), there is a gradual increase in the mean voltage neurogram interrupted by ten abrupt increases in activity, representing bursts of unitary discharges. These are better defined in the original neurogram in panel (*b*), which shows the ten bursts of high-frequency discharge of the single unit (upper trace). The instantaneous frequency plot in the lower trace shows a rapid onset of activity with a maximal discharge frequency of about 250 imp/s, with a fairly constant maximal frequency for successive bursts, but progressive steepness in the decay of impulse frequency in repeated bursts. Ochoa & Torebjörk (1980) reproduced with permission.

stimulus train, there being no plateau as with H_1 (Guisset, 1968; Bergmans, 1970; Applegate & Burke, 1989). Figure 2.6 shows, for three subjects, the slow recovery of excitability of human cutaneous afferents after stimulation at 200/s for 1 min. It took >10 min before excitability returned to normal. After stimulation of cutaneous afferents at 200/s for 10 min, it can take over an hour for normal excitability to be regained (see Fig. 2.8).

The H_2 phase of posttetanic subexcitability is not due to passive channel mechanisms. It is due to activation of the electrogenic sodium/potassium pump, and can be blocked by ouabain and by replacement of sodium ions by lithium (Bostock & Grafe, 1985).

Clinical implications

These normal homeostatic mechanisms open therapeutic options for patients in whom the safety margin for neural transmission has been impaired by demyelination, be it within the central nervous system or the peripheral nerve. In demyelinating diseases, loss of function results from loss of functioning axons, owing to a combination of conduction block in otherwise intact axons and axonal loss; it is not due to slowing of conduction. If impulses can be conducted across the lesion in a sufficient number of axons, there will be no clinically detectable functional deficit. Intact demyelinated axons may suffer complete conduction block affecting single impulses. However, even when able to conduct single impulses, affected axons may be unable to maintain conduction

Pathophysiological basis of paresthesiae 29

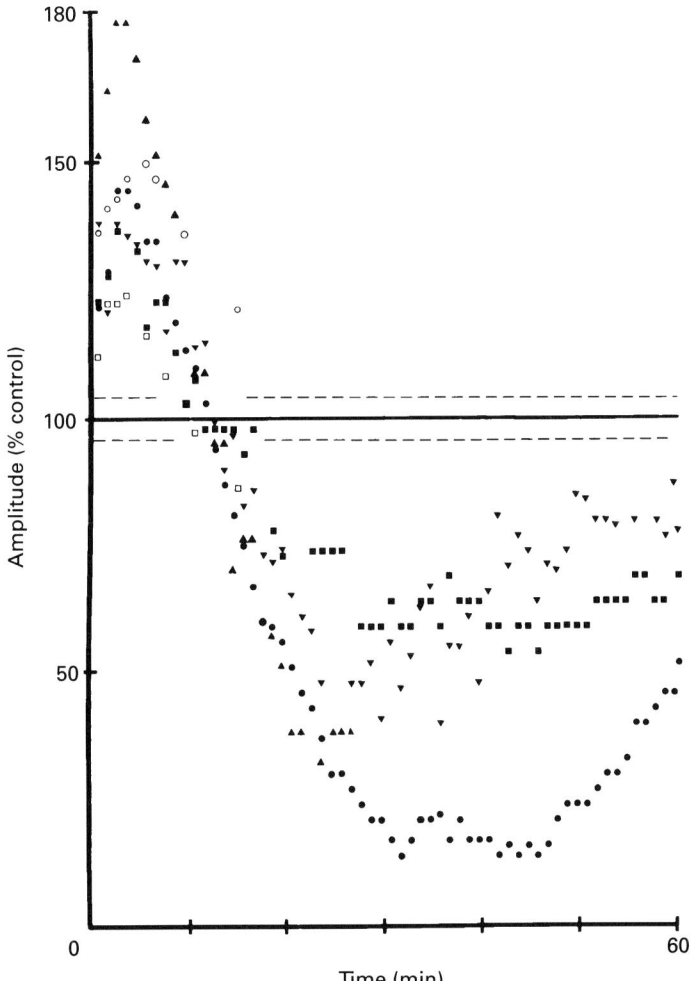

Fig. 2.8. Posttetanic increase in neural excitability. Changes in amplitude of a test potential following repetitive stimulation at 200/s for 10 min. Data for six subjects, three of whom were followed for one hour. The dotted lines illustrate the variability of the test potential prior to stimulation. Note the transient increase in excitability superimposed on the profound long-lasting posttetanic depression in excitability (H_2). Each subject experienced paraesthesiae in the distribution of the tetanized nerve during the phase of increased excitability. Applegate & Burke (1989) reproduced with permission.

of a physiologically meaningful impulse train. The deficit in these patients can be reduced by restoring the ability of demyelinated axons to conduct meaningful information, and this may be important in multiple sclerosis, in which remyelination is limited. Similarly, such measures could be useful in patients in the acute phase of Guillain–Barré syndrome.

Table 2.1 *Potential therapeutic strategies to overcome conduction block in demyelinated axons*

---

1. *Prolong the duration of the action potential by slowing the inactivation of Na^+ current*
   - scorpion venom
   - cooling
2. *Block fast K^+ channels*
   - 4 aminopyridine
   - 3,4 diaminopyridine
3. *Block slow K^+ channels*
   - tetraethyl ammonium
4. *Inhibit the electrogenic Na^+/K^+ pump*
   - digitalis glycosides
5. *Lower ionized Ca^{2+}*
   - hyperventilation
   - oral phosphates
   - Na_2EDTA
   - $NaHCO_3$

Such strategies have been tested in patients with multiple sclerosis (Table 2.1), with the exceptions of deliberate administration of scorpion venom (Bostock, Sherratt & Sears, 1978) or the administration of tetraethyl ammonium, an agent that was once used to treat peripheral vascular disease. Though no single strategy has received enthusiastic support, the underlying principle remains that functional deficits due to demyelination can be manipulated independently of the pathology (for example, Davis *et al.*, 1970; Davis & Jacobson, 1971; Becker, Michael & Davis, 1974; Schauf & Davis, 1974; Jones *et al.*, 1983; Davies, Carroll & Mastaglia, 1986; Davis, Stefoski & Rush, 1990; Hsueh, Toyoshima & Mayer, 1989; Kaji & Sumner, 1989; Kaji, Happel & Sumner, 1990).

Pathophysiological mechanisms of paresthesiae

From introspection, paresthesiae consist of multiple discrete nonpainful cutaneous sensations, which individually do not last long but which are recurrent. For ectopic impulse activity in individual cutaneous afferents to be directly responsible for this sensation, it is necessary for the activity in individual cutaneous afferents to evoke a percept. This seems to be the case for three of the four types of cutaneous mechanoreceptor in the nonhairy skin of the human hand. Thus the activity of single cutaneous sensory axons, particularly those in the pulp of the digits, may be perceived by a subject, whether that activity is set up mechanically by stimulation of the receptor, electrically by stimulation of the axon or spontaneously as an ectopic impulse train (Ochoa & Torebjörk, 1983; Vallbo et al., 1984; Macefield, Gandevia & Burke, 1990).

Figure 2.7 shows a recording made using a microelectrode within a cutaneous nerve fascicle of a subject experiencing paresthesiae (Ochoa & Torebjörk, 1980). The recording in (a) is of the mass discharge from the nerve, showing a number of abrupt transient increases in neural activity. These were due to repeated bursts of activity in a single myelinated sensory axon, as shown in panel (b). In each burst, the discharge reached ~250 imp/s, then gradually decayed to ~150 imp/s, and then stopped, pausing for a while before beginning again. These bizarre high-frequency discharges represent single paresthesiae. When this sort of activity occurs in multiple axons, discharging asynchronously in irregular bursts, the resultant perception is that of 'pins-and-needles'.

Why does this ectopic discharge start? Why does the afferent axon discharge at these high rates? Why does the burst stop and start again?

Techniques for producing paraesthesiae in normal subjects

To address the above questions, paresthesiae have been induced in normal subjects so that comparisons of normal and disturbed physiology could be made in the same subject in a single session under identical experimental conditions. Three techniques reliably produce paresthesiae: compression/ ischemia of the nerve (Ochoa & Torebjörk, 1980), long-lasting tetanization of cutaneous afferents (Bergmans, 1970; Applegate & Burke, 1989; Burke & Applegate, 1989; Macefield & Burke, 1991b), and hyperventilation (Macefield & Burke, 1991a). Each of these techniques can also produce fasciculation (Kugelberg, 1948; Bergmans, 1970; Macefield & Burke, 1991a; Bostock et al., 1991a; Bostock, Baker & Reid, 1991b), though less readily.

Postischemic states

During nerve ischemia there is commonly a stage of weak paresthesiae, associated with axonal depolarization and increased axonal excitability (Stöhr, 1981b), a phenomenon which is defective in diabetic patients (Strupp et al., 1990). Following ischemia, conventional excitability measurements using brief test pulses have usually reported that the excitability of both sensory and motor axons is depressed (Bergmans, 1970; Stöhr, 1981b; Bostock et al., 1991a), indicating that the membrane is hyperpolarized. However, accommodation breaks down such that the threshold to very long test pulses is decreased

(Kugelberg, 1948; Bostock *et al*., 1991*a*). If the duration of ischemia is sufficiently long, patients experience paresthesiae and fasciculation. It has recently been shown that human motor axons can exist in two stable equilibrium states in the postischemic phase, states of increased or decreased excitability (Bostock *et al*., 1991*b*), transition to the former presumably responsible for the generation of the ectopic motor discharges. Existence of axons in these two opposite states can be adequately explained by hyperpolarization due to activation of the electrogenic Na^+/K^+ pump in the presence of a high extracellular potassium concentration (Bostock *et al*., 1991*b*).

Prolonged tetanic stimulation of cutaneous afferents

It is perhaps paradoxical that prolonged tetanic stimulation can produce long-lasting paraesthesiae following cessation of the stimulus train, given that tetanic stimulation depresses axonal excitability (the H_2 phase of posttetanic depression, discussed above), and that paresthesiae presumably reflect axonal hyperexcitability. Nevertheless, prolonged tetanic stimulation of motor axons can induce fasciculation (Bergmans, 1970), and tetanic stimulation of cutaneous axons can cause single cutaneous afferents to become spontaneously active (Ochoa & Torebjörk, 1983) and can produce paresthesiae (Bergmans, 1970).

The explanation for this paradox is that, while the depth and duration of the H_2 phase of posttetanic depression of axonal excitability grows with the duration of the tetanic train (as discussed above), tetanic stimulation also produces a transient increase in excitability of at least some cutaneous afferents (Applegate & Burke, 1989). With trains at 200/s lasting 7–10 min, nerve excitability exceeds control values for some 5–10 min, subsequently declining dramatically and recovering slowly over up to one hour (Fig. 2.8). During this transient increase in excitability cutaneous afferents can become sufficiently excitable that they begin to discharge spontaneously (Ochoa & Torebjörk, 1983; Burke & Applegate, 1989) and, in parallel with this activity, subjects experience paresthesiae, starting some 20–30 s after cessation of the train and lasting some 5–10 min (see also Macefield & Burke, 1991*b*).

The mechanism responsible for the transient increase in excitability following very long tetanic trains is unknown, but it has been attributed to depolarization due to the accumulation of potassium ions extracellularly at the node of Ranvier, sufficient to offset the H_2 phase of post-tetanic depression (Applegate & Burke, 1989). There are some experimental data in support of this hypothesis. In experiments on isolated single axons, abolition of post-tetanic depression by ouabain (Schoepfle & Katholi, 1973) or replacement of sodium ions in

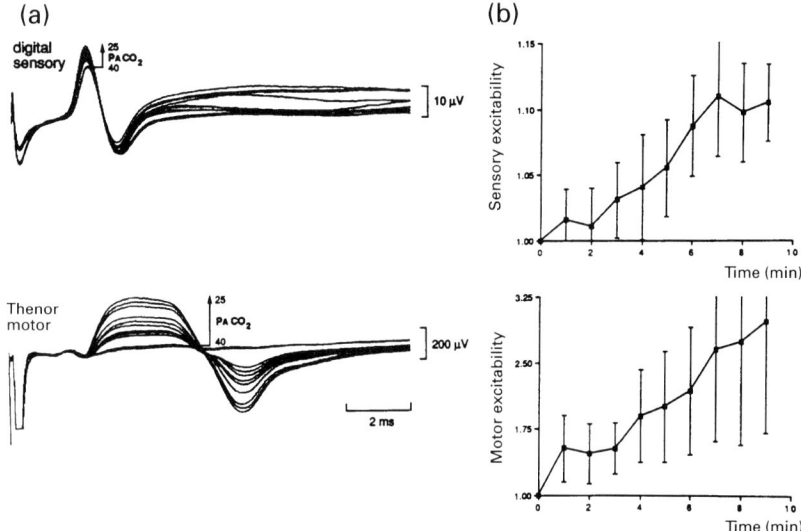

Fig. 2.9. Effects of hyperventilation on neural excitability. Panel (*a*) shows the antidromically conducted compound sensory action potential recorded from the digital nerves of the index finger (upper records), and the simultaneously evoked, orthodromically conducted compound muscle action potential recorded from the thenar muscles (lower records), produced by stimuli of 25.5 mA delivered to the median nerve at the wrist in one subject. Each set of traces consists of three control records and nine records obtained at 1-min intervals during hyperventilation. The arrows indicate the increases in sensory and motor excitability as alveolar pCO_2 declined from 40 to 25 mmHg. Panel (*b*) shows the increases in sensory and motor excitability as a function of the time from the beginning of hyperventilation; normalized data (mean ±SEM) for six subjects. Macefield & Burke (1991*a*) reproduced with permission.

the bathing medium by lithium ions (Bostock & Grafe, 1985) reveals a transient increase in excitability, attributed to extracellular potassium accumulation (Schoepfle & Katholi, 1972). Further support has come by analogy with the mechanisms responsible for postischemic fasciculation (see above).

Hyperventilation.

As end-tidal pCO_2 is reduced to around 20–25 mmHg by hyperventilation, normal subjects develop paresthesiae and fasciculations which gradually increase in severity until overt muscle spasm develops. Characteristically, subjects develop periumbilical and perioral paresthesiae in addition to limb paresthesiae. Indeed, hyperventilation is one of the few conditions that will produce paresthesiae in that distribution, and this can be a useful clinical clue in a patient who may have anxiety overbreathing.

As pCO_2 falls there is a progressive increase in excitability of cutaneous sensory axons (Fig. 2.9) and of motor axons (Kugelberg, 1948). This increase in excitability reflects progressive depolarization of axonal membrane, with spontaneous activity occurring when threshold is reached, first in sensory axons and shortly later in motor (Macefield & Burke, 1991a). The spontaneous activity continues as long as the membrane remains depolarized, and that depends on the balance between the driving force (which, here, is related to the decreased ionized calcium level) and accommodative responses (see later).

Involvement of the supernormal period in the ectopic discharge

Although axons become hyperexcitable during hyperventilation, there is no change in the degree of supernormality following an impulse (Macefield & Burke, 1991a). Thus, a change in supernormality does not provide a self-perpetuating driving force that would maintain the ectopic impulse train. Similarly, in the other models of paresthesiae, an abnormality of the supernormal period does not provide the driving force for the paresthesiae. Nevertheless, the supernormal period will still act as a pacemaker for the ectopic discharge, the driving force for which comes from membrane depolarization. A hyperexcitable sensory axon will be reexcited on the rising phase of the supernormal period following the previous discharge and, as the axon becomes less depolarized during the train, the discharge will occur progressively later on the rising phase of the supernormal period.

From the data in Fig. 2.4, cutaneous axons have a supernormal period that starts at 3.5 ms and peaks at 6–7 ms. This implies that axons will discharge at intervals of 3–4 ms early in the ectopic train, at frequencies of 250–300 imp/s. The interval between successive discharges will progressively lengthen to 6–7 ms and the discharge frequency will slow to about 150 imp/s as the train continues and then stops. This is precisely what is seen in recordings from single axons discharging spontaneously, such as Fig. 2.7 (Ochoa & Torebjörk, 1980). Hence, the supernormal period does not drive the ectopic discharge but it will act to entrain it.

Accommodation and the duration of the ectopic burst

Why does the train stop and then start again even though there is no obvious change in the driving stimulus? Two mechanisms that act to limit axonal firing were mentioned earlier: activation of slow potassium channels, and activation of the sodium/potassium pump. In addition, in depolarized axons, accommodation can be mediated by inactivation of sodium channels (Baker & Bostock, 1989). Hence there are at least three homeostatic mechanisms that will tend to curtail any one burst of impulses. When the ectopic discharge stops, the activity of the pump and of slow potassium channels will return to the basal level, and the axon will begin to discharge again if the increase in excitability has not dissipated.

Treatment of paresthesiae

Perhaps the commonest cause of paresthesiae is hyperventilation. While this may initially begin as an anxiety- or stress-related physiological response, it commonly continues as a learnt disorder or habit, in the absence of current anxiety or stress (Cluff, 1984). Accordingly the patient may deny any psychological distress and the physician may fail to recognize chronic hyperventilation as the cause of the patient's symptoms. The appropriate treatment is educational, with retraining under physiotherapeutic supervision.

To control paresthesiae pharmacologically in patients with peripheral nerve disease, the therapeutic strategies should be the opposite of those adopted to restore conduction in demyelinated axons. The most logical approach would be to interfere with the voltage-dependent sodium channels that are responsible for the action potential, but to do so selectively so that the interference was greater the more the axon was depolarized and greater the higher the axon's discharge rate. This is precisely what 'membrane-stabilizing drugs', such as phenytoin, are believed to do (see Mackenzie, this volume). Carbamazepine has the same action, as also does valproate.

Acknowledgements

The work reviewed here was supported by the National Health and Medical Research Council of Australia. The author acknowledges the role of his colleagues in the studies reviewed, and thanks Dr T. Gavranic for drawing attention to the paper by Cluff (1984).

References

Applegate, C. & Burke, D. (1989). Changes in excitability of human cutaneous afferents following prolonged high-frequency stimulation. *Brain*, **112**, 147–64.
Awiszus, F. (1990). Effects of a slow potassium permeability on repetitive activity of the frog node of Ranvier. *Biological Cybernetics*, **63**, 155–9.
Baker, M. & Bostock, H. (1989). Depolarization changes the mechanism of accommodation in rat and human motor axons. *Journal of Physiology (London)* **411**, 545–61.

Baker, M, Bostock, H., Grafe, P. & Martius, P. (1987). Function and distribution of three types of rectifying channel in rat spinal root myelinated axons. *Journal of Physiology (London)*, **383**, 45–67.

Barrett, E.F. & Barrett, J.N. (1982). Intracellular recording from vertebrate myelinated axons: mechanism of the depolarizing afterpotential. *Journal of Physiology (London)*, **323**, 117–44.

Becker, F.O., Michael, J.A. & Davis, F.A. (1974). Acute effects of oral phosphate on visual function in multiple sclerosis. *Neurology*, **24**, 601–7.

Bergmans, J. (1970). *The Physiology of Single Human Nerve Fibres.* Louvain: Vander.

Black, J.A., Felts, P., Smith, K.J., Kocsis, J.D. & Waxman, S.G. (1991). Distribution of sodium channels in chronically demyelinated spinal cord axons: immuno-ultrastructural localization and electrophysiological observations. *Brain Research*, **544**, 59–70.

Black, J.A., Kocsis, J.D. & Waxman, S.G. (1990). Ion channel organization of the myelinated fiber. *Trends in Neurosciences*, **13**, 48–54.

Blight, A.R. & Someya, S. (1985). Depolarizing afterpotentials in myelinated axons of mammalian spinal cord. *Neuroscience*, **15**, 1–12.

Bostock, H., Baker, M., Grafe, P. & Reid, G. (1991a). Changes in excitability and accommodation of human motor axons following brief periods of ischaemia. *Journal of Physiology*, **441**, 513–35.

Bostock, H., Baker, M. & Reid, G. (1991b). Changes in excitability of human axons underlying post-ischaemic fasciculations: evidence for two stable states. *Journal of Physiology (London)*, **441**, 537–57.

Bostock, H. & Grafe, P. (1985). Activity-dependent excitability changes in normal and demyelinated rat spinal root axons. *Journal of Physiology (London)*, **365**, 239–57.

Bostock, H., Sears, T.A. & Sherratt, R.M. (1981). The effects of 4-aminopyridine and tetraethylammonium ions on normal and demyelinated mammalian nerve fibres. *Journal of Physiology (London)*, **313**, 301–15.

Bostock, H., Sherratt, R.M. & Sears, T.A. (1978). Overcoming conduction failure in demyelinated nerve fibres by prolonging action potentials. *Nature*, **274**, 385–7.

Brismar, T. (1980). Potential clamp analysis of membrane currents in rat myelinated nerve fibres. *Journal of Physiology (London)*, **398**, 171–84.

Burke, D. & Applegate, C. (1989). Paraesthesiae and hypaesthesia following prolonged high-frequency stimulation of cutaneous afferents. *Brain*, **112**, 913–29.

Chiu, S.Y., Ritchie, J.M., Rogart, R.B. & Stagg, D. (1979). A quantitative description of membrane currents in rabbit myelinated nerve. *Journal of Physiology (London)*, **292**, 149–66.

Cluff, R.A. (1984). Chronic hyperventilation and its treatment by physiotherapy: discussion paper. *Journal of the Royal Society of Medicine*, **77**, 855–62.

Davies, H.D., Carroll, W.M. & Mastaglia, F. (1986). Effects of hyperventilation on pattern-reversal visual evoked potentials in patients with demyelination. *Journal of Neurology, Neurosurgery, and Psychiatry*, **49**, 1392–6.

Davis, F.A., Becker, F.O., Michael, J.A. & Sorensen, E. (1970). Effect of intravenous sodium bicarbonate, disodium edetate (Na_2EDTA), and hyperventilation on visual and oculomotor signs in multiple sclerosis. *Journal of Neurology, Neurosurgery, and Psychiatry*, **33**, 723–32.

Davis, F.A. & Jacobson, S. (1971). Altered thermal sensitivity in injured and demyelinated nerve. *Journal of Neurology, Neurosurgery, and Psychiatry*, **34**, 551–61.

Davis, F.A., Stefoski, D. & Rush, J. (1990). Orally administered 4-aminopyridine improves clinical signs in multiple sclerosis. *Annals of Neurology*, **27**, 186–92.

Eng, D.L., Gordon, T.R., Kocsis, J.D. & Waxman, S.G. (1988). Development of 4-AP and TEA sensitivities in mammalian myelinated nerve fibers. *Journal of Neurophysiology*, **60**, 2168–79.

Gasser, H.S. (1935). Changes in nerve-potentials produced by rapidly repeated stimuli and their relation to the responsiveness of nerve to stimulation. *American Journal of Physiology*, **111**, 35–50.

Gasser, H.S. (1958). The postspike positivity of unmedullated fibers of dorsal root origin. *Journal of General Physiology*, **41**, 613–32.

Gasser, H.S. & Grundfest, H. (1936). Action and excitability in mammalian A fibres. *American Journal of Physiology*, **117**, 113–33.

Gilliatt, R.W. & Willison, R.G. (1963). The refractory and supernormal periods of the human median nerve. *Journal of Neurology, Neurosurgery, and Psychiatry*, **26**, 136–47.

Graham, H.T. & Lorente de Nó, R. (1938). Recovery of blood-perfused mammalian nerves. *American Journal of Physiology*, **123**, 326–40.

Guisset, M. (1968). The recovery of excitability of human sensory nerve fibres following activity. *Archives Internationales de Physiologie et de Biochimie*, **76**, 139–41.

Hagbarth, K.E., Torebjörk, H.E. & Wallin, B.G. (1984). Microelectrode recordings from human skin and muscle nerves. In *Peripheral Neuropathy*, 2nd edn, vol. 1, eds. P.J. Dyck, P.K. Thomas, E.H. Lambert & R. Bunge, pp. 1016–29. Philadelphia: Saunders Co.

Hsueh, I.-H., Toyoshima, E. & Mayer, R.F. (1989). The effects of 4-aminopyridine on focal nerve conduction block. *Restorative Neurology and Neuroscience*, **1**, 39–46.

Jones, R.E., Heron, J.R., Foster, D.H., Snelgar, R.S. & Mason, R.J. (1983). Effects of 4-aminopyridine in patients with multiple sclerosis. *Journal of the Neurological Sciences*, **60**, 353–62.

Kaji, R., Happel, L. & Sumner, A.J. (1990). Effect of digitalis on clinical symptoms and conduction variables in patients with multiple sclerosis. *Annals of Neurology*, **28**, 582–4.

Kaji, R. & Sumner, A.J. (1989). Ouabain reverses conduction disturbances in single demyelinated nerve fibers. *Neurology*, **39**, 1364–8.

Kocsis, J.D., Bowe, C.M. & Waxman, S.G. (1986). Different effects of 4-aminopyridine on sensory and motor fibers: pathogenesis of paresthesias. *Neurology*, **36**, 117–20.

Krylov, B.V. & Makovsky, V.S. (1978). Spike frequency adaptation in amphibian sensory fibres is probably due to slow K channels. *Nature*, **275**, 549–51.

Kugelberg, E. (1948). Activation of human nerves by hyperventilation and hypocalcemia. *Archives of Neurology and Psychiatry*, **60**, 153–64.

Lance, J.W., Burke, D. & Pollard, J. (1979a). Hyperexcitability of motor and sensory neurons in neuromyotonia. *Annals of Neurology*, **5**, 523–32.

Lance, J.W., Burke, D. & Pollard, J. (1979b). Neuromyotonia in the spinal form of Charcot-Marie-Tooth disease. *Clinical and Experimental Neurology*, **16**, 49–56.

Macefield, G. & Burke, D. (1991a). Paraesthesiae and tetany induced by voluntary hyperventilation: increased excitability of human cutaneous and motor axons. *Brain*, **114**, 527–40.

Macefield, G. & Burke, D. (1991b). Long-lasting depression of central synaptic transmission following prolonged high-frequency stimulation of cutaneous afferents: a mechanism for post-vibratory hypaesthesia. *Electroencephalography and Clinical Neurophysiology*, **78**, 150–8.

Macefield, G., Gandevia, S.C. & Burke, D. (1990). Perceptual responses to microstimulation of single afferents innervating the joints, muscles and skin of the human hand. *Journal of Physiology (London)*, **429**, 113–29.

Meiri, H., Baum, Z. & Rosenthal, Y. (1989). Dynamic changes in sodium channels at demyelinated axons. *Progress in Neurobiology*, **32**, 159–79.

Ng, A., Burke, D. & Al-Shehab, A. (1987). Hyperexcitability of cutaneous afferents during the supernormal period: relevance to paraesthesiae. *Brain*, **110**, 1015–31.

Nordin, M., Nyström, B., Wallin, U. & Hagbarth, K.-E. (1984). Ectopic sensory discharges and paresthesiae in patients with disorders of peripheral nerves, dorsal roots and dorsal columns. *Pain*, **20**, 231–45.

Nyström, B. & Hagbarth, K.-E. (1981). Microelectrode recordings from transected nerves in amputees with phantom limb pain. *Neuroscience Letters*, **27**, 211–16.

Ochoa, J.L. & Torebjörk, H.E. (1980). Paraesthesiae from ectopic impulse generation in human sensory nerves. *Brain*, **103**, 835–53.

Ochoa, J. & Torebjörk, E. (1983). Sensations evoked by intraneural microstimulation of single mechanoreceptor units innervating the human hand. *Journal of Physiology (London)*, **342**, 633–54.

Oda, K., Araki, K., Totoki, T & Shibasaki, H. (1989). Nerve conduction study of human tetrodotoxication. *Neurology*, **39**, 743–5.

Röper, J. & Schwarz, J.R. (1989). Heterogeneous distribution of fast and slow potassium channels in myelinated rat nerve fibres. *Journal of Physiology (London)*, **416**, 93–110.

Schauf, C.L. & Davis, F.A. (1974). Impulse conduction in multiple sclerosis: a theoretical basis for modification by temperature and pharmacological agents. *Journal of Neurology, Neurosurgery, and Psychiatry*, **37**, 152–61.

Schoepfle, G.M. & Katholi, C.R. (1973). Posttetanic changes in membrane potential of single medullated nerve fibers. *American Journal of Physiology*, **225**, 1501–7.

Stöhr, M. (1981a). Activity-dependent variations in threshold and conduction velocity of human sensory fibers. *Journal of the Neurological Sciences*, **49**, 47–54.

Stöhr, M. (1981b). Modification of the recovery-cycle of human median nerve by ischemia. *Journal of the Neurological Sciences*, **51**, 171–80.

Strupp, M., Bostock, H., Weigl, P., Piwernetz, K., Renner, R. & Grafe, P. (1990). Is resistance to ischaemia of motor axons in diabetic subjects due to membrane depolarization? *Journal of the Neurological Sciences*, **99**, 271–80.

Stys, P.K. & Ashby, P. (1990). An automated technique for measuring the recovery cycle of human nerves. *Muscle & Nerve*, **13**, 750–8.

Taylor, J.L., Burke, D. & Heywood, J. (1992). Physiological evidence for a slow K^+ conductance in human cutaneous afferents. *Journal of Physiology (London)*, **453**, 575–89.

Vallbo, Å. B., Olsson, K.Å., Westberg, K.-G. & Clark, F.J. (1984). Microstimulation of single tactile afferents from the human hand. *Brain*, **107**, 727–49.

3
Spinal pain: backache and neck pain

N. BOGDUK

Faculty of Medicine,
University of Newcastle,
Callaghan NSW, Australia

Spinal pain is a complaint that does not fall within the province of any single branch of medicine. Alone, or in consultation, rheumatologists, orthopaedic surgeons, neurosurgeons, anesthetists, physical therapists or neurologists may be called upon to assess and manage patients with spinal pain. Conventional wisdom has been that nerve root compression is the cardinal, if not the most common, cause of spinal pain, but anatomical and clinical research over the past 15 years has shown this not to be correct. Spinal pain typically arises from the musculoskeletal elements of the vertebral column, and uncommonly it is the result of nerve root disorders. For the lumbar spine, clinical studies have shown that nerve root compression can be legitimately held to be the cause of pain in fewer than 30% of presentations (Horal, 1969) and as few as 5% (Friberg, 1954). Contemporary authorities place the figure perhaps as low as 1% (Mooney, 1987a).

Radiculopathy

Frank radiculopathy involves loss of nerve function as a result of compression, ischemia, inflammation or fibrosis of a nerve root or spinal nerve. Its manifestations are loss of conduction in sensory fibres (resulting in segmental numbness) or in motor fibres (resulting in segmental weakness). These features can be detected on careful clinical examination and confirmed by electrodiagnostic studies (Knuttson, 1961). However, the majority of patients with spinal pain do not exhibit features of nerve conduction loss, in which case radiculopathy cannot be held to be the cause of their pain (Friberg, 1954, Horal, 1969).

Radicular pain

There is no clinical or experimental evidence that nerve root compression causes pain. Compressing a lumbar nerve root with the balloon of a urinary catheter causes paraesthesiae, but not pain (MacNab, 1972). In experimental animals, squeezing a normal lumbar nerve root evokes only a momentary discharge in nociceptive afferents, but not activity consistent with prolonged or chronic pain (Howe, Loeser & Calvin, 1977; Howe, 1979). In humans, pulling or squeezing normal lumbar nerve roots does not evoke pain (Norlen, 1944; Smyth & Wright, 1959).

In experimental animals, prolonged discharges can be evoked from nerve roots only if the dorsal root ganglion is squeezed or if a previously damaged nerve root is squeezed (Howe *et al.*, 1977; Howe, 1979) but, even then, activity occurs in large diameter afferents as well as nociceptive afferents, and cannot be held to be purely nociceptive. In humans, squeezing or pulling previously damaged (ostensibly inflamed) nerve roots evokes pain (Norlen, 1944; Smyth & Wright 1959), and the same occurs when lumbar nerve roots are stimulated electrically (McCulloch & Waddell, 1980), but the pain is characteristic in quality, and only this type of pain can legitimately be held to be radicular.

Radicular pain is described as lancinating, and is perceived along bands not more than two inches wide reminiscent of, although not identical to, the bands of dermatomes (Norlen, 1944; Smyth & Wright, 1959). This quality and distribution is different from that of somatic pain.

Somatic pain

Somatic pain is pain that is evoked by nociceptive activity from the musculoskeletal elements of the body. In mechanism it differs from radicular pain in that somatic pain is evoked by the stimulation of nerve endings as opposed to the generation of ectopic impulses along the course of axons. In quality, somatic pain is described as dull and aching or pressure-like; it is perceived deeply and is constant or fixed in location, although the patient may find it hard to define the actual boundaries. In all studies that have sought to produce somatic pain experimentally in normal volunteers it is this quality that has been produced (Kellgren, 1938, 1939; Campbell & Parsons, 1944; Feinstein *et al.*, 1954; Hockaday & Whitty, 1967; Bogduk, 1980a). It is this type of pain that has been relieved whenever musculoskeletal elements have been anesthetized in patients suffering spinal pain.

Somatic referred pain

Somatic referred pain is pain perceived to arise in a location topographically distinct from the region of the actual source of pain. A stricter, and more accurate definition would be that, in neurological terms, referred pain is pain perceived to arise in a region innervated by nerves other than those that innervate the primary

Spinal pain 41

source of pain. Spinal pain may be referred to regions of the trunk wall (Kellgren, 1938, 1939; Feinstein *et al.*, 1954; Hockaday & Whitty, 1967), the limb girdles or the limbs themselves (Kellgren, 1938, 1939; Feinstein *et al.*, 1954; Bogduk, 1980*a*). Upper cervical spinal pain may be referred to the head (Cyriax, 1938; Campbell & Parsons, 1944; Feinstein *et al.*, 1954; Ehni & Benner, 1984; Bogduk & Marsland, 1985, 1986; Dwyer, Aprill & Bogduk, 1990).

Somatic pain postulates

For a given structure to be deemed a source of pain, certain criteria should be satisfied. First, the structure should be innervated so that it is capable of evoking nociceptive activity; a structure with no nerve supply cannot be a primary source of pain. Secondly, it is desirable that the structure has been shown experimentally to be able to generate pain, ideally in studies of normal volunteers. Finally, it is desirable that, in a given patient, the structure be shown to be affected by a lesion that can acceptably be presumed to be a cause of pain.

Recent research into spinal pain has provided much data on several elements of the vertebral column to satisfy the first two of these criteria. The innervation of the vertebral column has been thoroughly studied, and experiments have been conducted to produce spinal pain in normal volunteers. Research still lags with respect to the third criterion. The actual causes of spinal pain remain elusive, but understandably so. Spinal pain is not fatal so that postmortem studies have never been performed on patients with a recent history in whom the source of pain has been established. Meanwhile, modern imaging techniques can demonstrate a variety of changes in the vertebral column but studies in normal subjects have not been conducted to allow valid inferences to be drawn about whether or not a particular change is pathognomonic of a pain-producing lesion.

Anatomical studies

Microdissection and histological studies have demonstrated that virtually all elements of the cervical and lumbar vertebral columns are innervated. The dorsal rami of the spinal nerves innervate the back muscles, the ligamenta flava and the joints of the vertebral arches (Bogduk, 1982, 1983; Bogduk, Wilson & Tynan, *et al.*, 1982). In the lumbar region, the dorsal rami form lateral, intermediate and medial branches which innervate the iliocostalis, longissimus and multifidus muscles respectively (Bogduk *et al.*, 1982). In the neck, the lateral branches of the cervical dorsal rami innervate the more superficial, posterior neck muscles: splenius, iliocostalis and longissimus, while the medial branches innervate the deeper semispinalis muscles and multifidus (Bogduk,

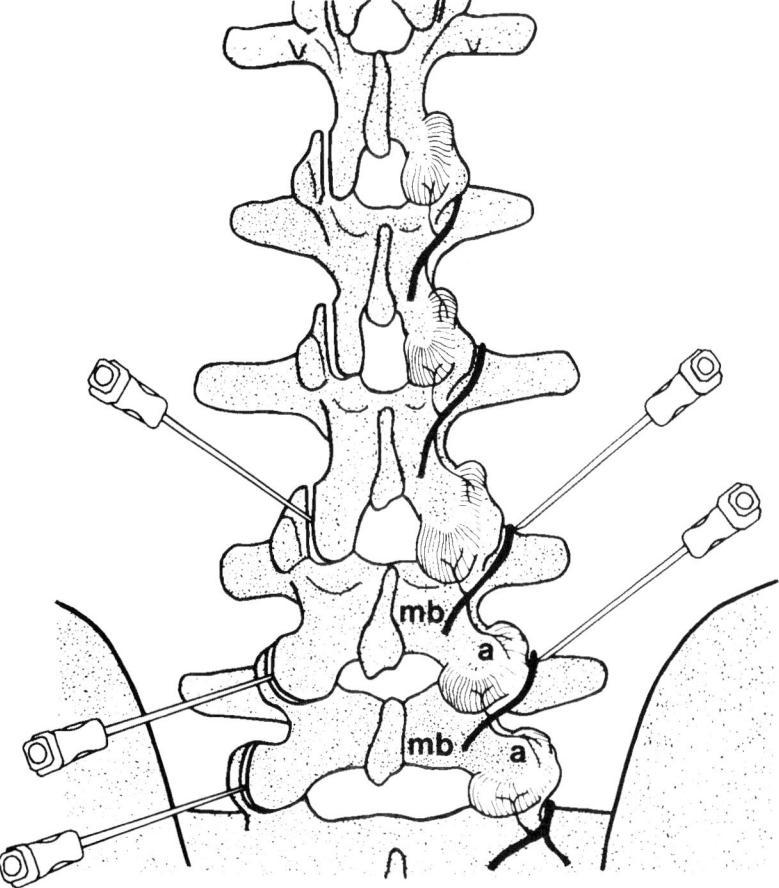

Fig. 3.1. A sketch of a posterior view of the lumbar spine showing, on the left, the courses of the medial branches (mb) of the lumbar dorsal rami and the distribution of articular branches (a) to the zygapophysial joints, with needles directed onto the medial branches of L3 and L4 where they cross the transverse processes, as they would be used to anaesthetise these nerves to block the L4–5 zygapophysial joint. On the right, needles are shown as they would be used for intra-articular blocks of the lower three zygapophysial joints. Bogduk (1988a) reproduced with permission.

1982). In both the neck and lumbar region, the multifidus is innervated segmentally, with distinct bands of muscle being innervated solely by one medial branch (Bogduk, 1982; Bogduk *et al.*, 1982). This provides a basis for accurate, segmental paraspinal electromyography (Macintosh *et al.*, 1986).

The medial branches of the lumbar and cervical dorsal rami follow constant courses in relation to bone. At lumbar levels each medial branch crosses the root of the transverse process before turning medially into the multifidus muscle

Spinal pain 43

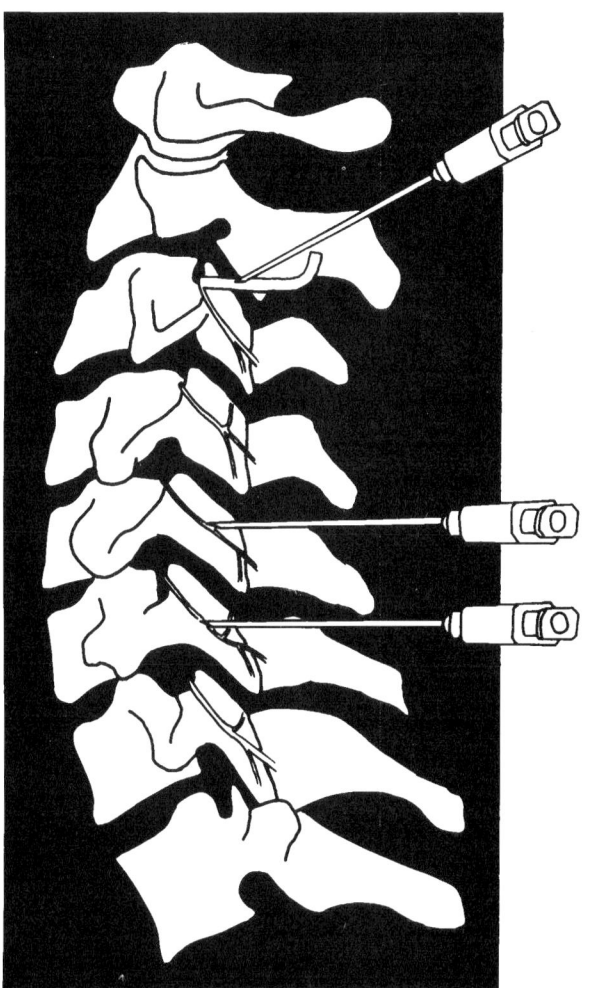

Fig. 3.2. A sketch of a lateral view of the cervical spine showing the courses of the medial branches of the typical cervical dorsal rami around the waists of the articular pillars, and the distribution of articular branches to the zygapophysial joints, and the course of the third occipital nerve around the C2–3 zygapophysial joint. Needles are illustrated in position as they would be used to block the third occipital nerve and the medial branches of C5 and C6 to anesthetise the C2–3 and C5–6 zygapophysial joints, respectively.

(Bogduk, 1983; Bogduk *et al.*, 1982). At typical cervical levels, each medial branch winds around the waist of the articular pillar (Bogduk, 1982). The medial branches of the dorsal rami innervate the zygapophysial joints. Typically, each joint receives articular branches from the medial branch above and below (Bogduk 1982, 1983; Bogduk *et al.*, 1982). The one exception is the C2–3 zygapophysial joint which is innervated by the third occipital nerve (the superfi-

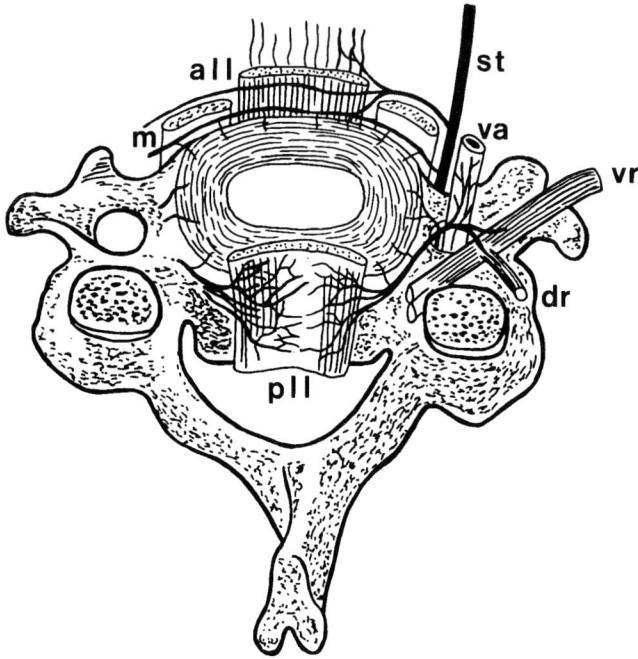

Fig. 3.3. A sketch of the plexuses surrounding and innervating a cervical intervertebral disc (based on Groen, Baljet & Drukker, 1990). The sinuvertebral nerves form a dense plexus accompanying the posterior longitudinal ligament (pll). Anteriorly, branches of the sympathetic trunk (st) supply the front of the disc and form a plexus accompanying the anterior longitudinal ligament (all). vr: ventral ramus, dr: dorsal ramus, va: vertebral artery, m: prevertebral muscles.

cial medial branch of the C3 dorsal ramus) which winds around the capsule of that joint and furnishes articular branches from its deep surface (Bogduk, 1982).

The constant relationship between the medial branches of the dorsal rami and bone allows the location of each nerve to be visualized radiographically so that each nerve can be selectively blocked with local anesthetic as a diagnostic test for pain mediated by these nerves and arising from the zygapophysial joints innervated by them (Bogduk, 1982, 1983, 1988*a*; Bogduk & Long, 1980; Bogduk *et al.*, 1982) (Figs. 3.1 and 3.2).

The anterior elements of the vertebral column receive an innervation from a variety of sources. The atlanto-occipital and lateral atlanto-axial joints are not zygapophysial joints; they lie ventral to the spinal nerves and are not innervated by dorsal rami. Instead they receive direct, articular branches from the C1 and C2 ventral rami respectively (Lazorthes & Gaubert, 1956; Bogduk, 1981). The other anterior elements of the vertebral column are innervated by extensive, microscopic plexuses of nerves derived from the sympathetic trunks

Spinal pain

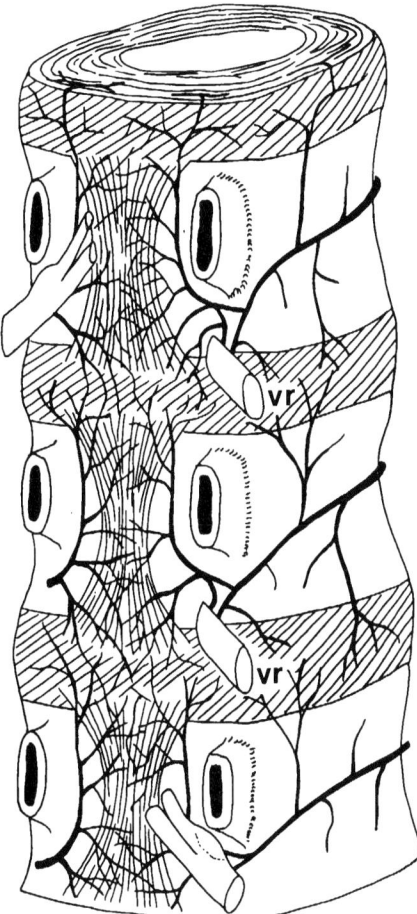

Fig. 3.4. A sketch of the plexuses of the lumbar intervertebral discs (based on Groen *et al.*, 1990). The sinuvertebral nerves generate a plexus accompanying the posterior longitudinal ligament from which the intervertebral discs are innervated. Laterally, the grey rami communicantes generate a fine plexus over the surfaces of the discs before reaching the ventral rami (vr).

and grey rami communicantes that accompany the anterior and posterior longitudinal ligaments (Groen, Baljet & Drukker, 1990).

Histological studies in human fetuses have shown that, along the entire length of the vertebral column, the anterior plexus furnishes branches to the anterior longitudinal ligament, the vertebral bodies and the anterior portion of the intervertebral discs (Groen *et al.*, 1990). The posterior plexus accompanies the posterior longitudinal ligament supplying the ligament, the vertebral bodies and intervertebral discs from behind (Figs. 3.3 and 3.4), and gives rise to a parallel plexus on the anterior surface of the dural sac (Groen *et al.*, 1990). The dural plexus supplies the anterior half of the dural sac and the nerve root sleeves, but no neural elements reach the posterior surface of dura which is devoid of nerve endings (Groen, Baljet & Drukker, 1988). At upper cervical levels, the posterior plexus supplies the ligaments of the atlanto-odontoid joint (Kimmel, 1960).

Larger elements within the posterior plexus constitute what had previously been recognized on dissection as the sinuvertebral nerves (Bogduk, 1983; Bogduk *et al.*, 1981*a*; Bogduk, Windsor & Inglis, 1988), but these large fibres are only a portion of what is now known to be a far more extensive, filamentous plexus within the floor of the vertebral canal (Groen *et al.*, 1990). In the cervical region, another plexus accompanies the vertebral artery; it is formed by grey rami communincantes from the stellate ganglion, and from the cervical sympathetic trunks, and has traditionally been known as the vertebral nerve (Bogduk *et al.*, 1981*b*).

It had previously been held that the intervertebral discs lacked a nerve supply (Wyke, 1980), but it is now quite evident that this is incorrect. Lumbar discs receive nerves from the anterior and posterior plexuses and from the grey rami communicantes over their lateral aspects (Bogduk *et al.*, 1981*b*; Groen *et al.*, 1990). What previously had been interpreted as direct branches to the discs from lumbar ventral rami (Bogduk *et al.*, 1981*a*) now appear to be proximal elements of a posterolateral plexus formed by the grey rami communicantes as they approach the ventral rami (Groen *et al.*, 1990). Cervical discs receive an innervation from the anterior and posterior plexuses and from the grey rami communicantes which accompany the vertebral artery (Bogduk *et al.*, 1988; Groen *et al.*, 1990).

Histological studies of human fetal material, neonatal, child and adult specimens obtained both from cadavers and at operation have demonstrated that within both lumbar and cervical intervertebral discs, nerve endings occur in the outer anulus fibrosus, extending definitely through the outer third of the anulus and sometimes as deeply as the outer half (Malinsky, 1959; Yoshizawa *et al.*, 1980; Bogduk *et al.*, 1988; Groen *et al.*, 1990). Experiments in animals have shown these endings to contain transmitter substances such as vasoactive intestinal polypeptide, substance P and calcitonin-gene-related peptide: a profile similar to that of nociceptive axons elsewhere in the body (Korkala *et al.*, 1985; Weinstein, Claverie & Gibson, 1988; Kottinen *et al.*, 1990).

Clinical studies

Several studies have been conducted to demonstrate the algogenic potential of many structures of the vertebral column. The classical studies of Kellgren

Spinal pain 47

(1938, 1939), Feinstein *et al.* (1954), Hockaday and Whitty (1967) and Campbell and Parsons (1944) supplemented by others (Bogduk, 1980*a*) demonstrated that spinal pain could be produced experimentally by injection of hypertonic saline into interspinous spaces or back muscles, and that this pain was accompanied by referred pain into the trunk wall or limbs, depending on which spinal segment was stimulated. These experiments established the principle that somatic referred pain could be produced by stimulating posterior spinal structures; they were not meant to imply that interspinous structures were a leading source of back pain and referred pain.

More contemporary studies have used similar methods to focus on particular structures that are more likely to be a natural source of spinal pain. Experimental stimulation of lumbar zygapophysial joints with injections of hypertonic saline in normal volunteers produces back pain and referred pain in the lower limbs (Mooney & Robertson, 1976; McCall, Park & O'Brien, 1979). Distention of cervical zygapophysial joints with contrast medium in normal volunteers produces neck pain and referred pain (Dwyer *et al.*, 1990). When lower cervical joints are stimulated, the referred pain extends into the upper limb girdle; when the C2–3 zygapophysial joint is stimulated, the referred pain occurs in the head (Dwyer *et al.*, 1990).

These experimental studies in normal volunteers complement the clinical observations that spinal pain can be relieved by anesthetising one or more zygapophysial joints. This applies to both patients presenting with low back pain (Mooney & Robertson, 1976; Fairbank *et al.*, 1981; Carrera & Williams, 1984; Lippit, 1984; Lewinnek & Warfield, 1986; Lynch & Taylor, 1986; Helbig & Lee, 1988; Murtagh, 1988) and patients with neck pain (Wedel & Wilson, 1985; Bogduk & Marsland, 1985, 1986, 1988; Aprill, Dwyer & Bogduk, 1990; Hove & Gyldensted; 1990). The atlanto-axial joints have not been studied in normal volunteers, but clinical studies have demonstrated that certain patients with headache or so-called occipital neuralgia can have their pain relieved by intraarticular (McCormick, 1987) or periarticular (Ehni & Benner, 1984) local anesthetic blocks of the lateral atlanto-axial joints (see also Anthony, this volume).

These studies are relevant in the assessment of headache, for it is now clearly evident that referred pain to the head can arise from the upper cervical synovial joints, notably the lateral atlanto-axial joints (McCormick, 1987; Ehni & Benner, 1984) and the C2–3 zygapophysial joints (Bogduk & Marsland, 1985, 1986, 1988). Anesthetic block of these joints in appropriate patients results in complete relief of their pain. Clinically, these patients resemble those with tension headache but their pain is more typically unilateral affecting the occipital region with radiation to the orbit or forehead. For these reasons the pain tends to be diagnosed as occipital neuralgia. However, the original anatomical basis for occipital neuralgia has been refuted (Bogduk, 1980*b*), and there is no firm evidence that this pain stems from lesions of the greater occipital nerve; contemporary evidence indicates that it constitutes referred pain from the upper cervical synovial joints (Ehni & Benner, 1984; Bogduk & Marsland, 1985, 1986, 1988; Bogduk, 1989, 1992*a*).

The dura mater can be a source of spinal pain and referred pain. Pulling on dural sleeves produces dull pain in the buttocks which is different in quality from the pain produced by pulling damaged nerve roots (Smyth & Wright, 1959). Injecting hypertonic saline around lumbar nerve root sleeves produces low back and referred pain (El Mahdi, Latif & Janko, 1981). Recent reports show that some forms of intractable, postsurgical back pain can be relieved by sectioning the nerves that supply the anterior dura mater (Cuatico *et al.*, 1988; Cuatico & Parker, 1989).

Pain from intervertebral discs has been studied using discography as the stimulus. However, discography is not painful in normal volunteers. Previous studies suggested the opposite (Holt, 1968), but these have now been discredited (Simmons *et al.*, 1988). Meticulous, stringently executed, modern studies have shown that pain is not produced when normal lumbar intervertebral discs are injected with contrast medium (Walsh *et al.*, 1990), reinforcing much older, but infrequently quoted studies which reported that pain was rarely produced by discography in normal volunteers (Massie & Stevens, 1967). Yet, in many patients with back pain or neck pain, provocation discography is frequently positive (Cloward, 1959; Collis & Gardner, 1962; Wiley, MacNab & Wortzman, 1968; Collins, 1975; Simmons & Segil, 1975; Brodsky & Binder, 1979; Park, 1980; Kikuchi, MacNab & Moreau, 1981; Colhoun *et al.*, 1988; McFadden, 1988; Bernard, 1990). This can be understood in the following way.

Provocation discography delivers a physical stimulus to the nucleus pulposus. There are no nerve endings in the nucleus that would be capable of sensing this stimulus. For a disc to be painful, the stimulus must stress the outer anulus fibrosus which is where the nerve endings of a disc are located. In an intact, normal disc, the outer anulus fibrosus is buffered from physical stimuli to the nucleus by the intact inner layers of the anulus whereupon, only extremely high pressures of injection might be transmitted to the outer anulus. In contrast, in abnormal discs in which the inner anulus had been disrupted, or in which radial fissures allow communication from the nucleus to the outer anulus, stimuli delivered to the nucleus are more readily transmitted to the outer, innervated anulus whereupon, such stimuli are more likely to be painful (Bogduk, 1991*b*).

This argument predicts that painful discs should be morphologically abnormal with communications between the nucleus and outer anulus, and indeed, this proves to be so. CT-discography provides a means of studying the internal

Spinal pain 49

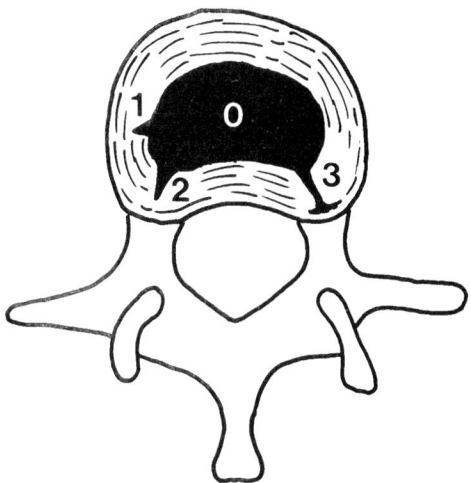

Fig. 3.5. The Dallas discogram description (Sachs *et al.*, 1987), showing the four grades of anular disruption evident in CT-discograms of discs with internal disc disruption. Bogduk (1991*b*) reproduced with permission.

morphology of discs that have undergone discography (Videman, Malmivaara & Mooney, 1987). A classification scheme has been devised to grade the extent of anular disruption in such discs (Sachs *et al.*, 1987). In discs with no disruption the contrast medium remains within the nucleus pulposus. In grade 1 disruption, the contrast medium reveals radial fissures extending into the inner third of the anulus. In grade 2 disruption, fissures reach the inner two-thirds. In grade 3 disruption, the fissures reach the outer third (Figs. 3.5 and 3.6).

Strong correlations exist between pain reproduction on discography, the grade of disruption and the density of innervation of lumbar discs (Vanharanta *et al.*, 1987). Grade 0 and grade 1 discs are rarely painful, but over 70% of grade 3 discs are associated with exact or similar reproduction of the patient's pain. Grade 2 disruptions are less often painful. These respective prevalences of pain reproduction are consistent with the nucleus and inner anulus having no nerve supply, the outer third regularly being innervated, and the middle third being variably innervated.

Leading sources of spinal pain

The spinal muscles are held, in some circles, to be the leading source of spinal pain, but few pathological models have been advanced to explain adequately how muscles can be sources of pain. For acute back strains or neck strain, muscle sprains are likely causes of pain. The pathology of acute muscle sprain has

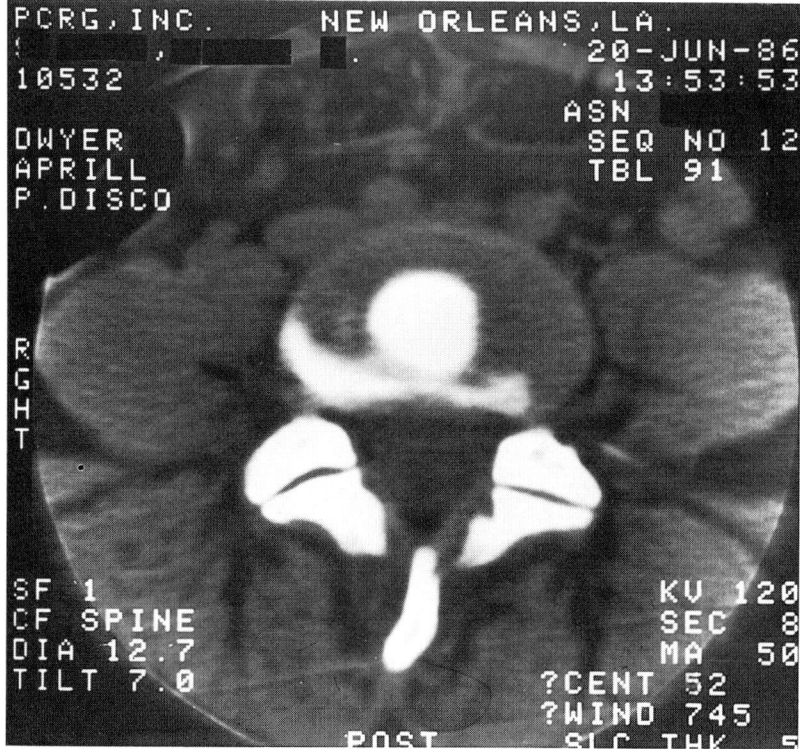

Fig. 3.6. An example of a CT-discogram of a painful L5-S1 intervertebral disc exhibiting grade 3 internal disruption. Bogduk & Twomey (1991) reproduced with permission.

been studied in experimental animals, and was found to consist of rupture of myotendinous junctions which attract an inflammatory repair response (Garrett et al., 1987, 1988; Nikolau et al., 1987). However, there are no laudable models to explain how muscles alone might be the source of chronic spinal pain.

'Trigger points' have been popularized as a source of chronic muscle pain (Simons, 1988), but there have been no clinical studies to show exactly how frequently trigger points in spinal muscles are the actual and sole source of spinal pain. 'Muscle spasm' has been enunciated as a cause of chronic spinal pain, but evidence is lacking as to what constitutes painful muscle spasm in a physiological sense or how it can be validly diagnosed (Roland, 1986; Andersson et al., 1989). For these reasons, the notion of chronic muscle pain has not found universal favour, and other models have been developed in which joints are believed to be the primary source of pain.

In many circles, a 'tissue diagnosis' is not pursued for patients with chronic spinal pain other than excluding nerve root compression by way of CT and MRI scans. However, for investigators who seek to establish the source of pain, the current, leading contenders are the zygapophysial joints and the intervertebral discs.

For the lumbar spine, internal disc disruption is the most common source of pain amongst patients presenting to tertiary referral units. These patients present with low back pain and referred pain but no neurological signs and exhibit essentially normal imaging studies; disc prolapse, foraminal or spinal stenosis are not evident on CT scans or MRI, and degenerative changes which may or may not be present are nonspecific, and essentially irrelevant to the diagnosis (Crock, 1986; Bogduk, 1991*b*). Conventional imaging studies are normal because the painful morphological changes are confined to the interior of the affected disc, and are not manifest by external bulges, protrusions or prolapse (Crock, 1986; Bogduk, 1991*b*). The diagnosis is established only on discography, with CT confirmation of the extent of disruption if required.

Throughout the 1970s and 1980s, lumbar zygapophysial joint pain, or 'facet syndrome' was popularized (Mooney & Robertson, 1976; Fairbank *et al.*, 1981; Carrera & Williams, 1984; Lippit, 1984; Lewinnek & Warfield, 1986; Lynch & Taylor, 1986; Mooney, 1987*b*; Helbig & Lee, 1988; Murtagh, 1988). It remains an entertainable differential diagnosis of low back pain, but its prevalence appears to be low. No proper prevalence studies have yet been conducted, but estimates place the prevalence of lumbar zygapophysial joint pain as low as 16 to 22% (Raymond & Dumas, 1984; Lynch & Taylor, 1986; Moran, O'Connell, & Walsh, 1988) and perhaps less than 8% (Jackson *et al.*, 1988).

The converse applies for the neck. As for the lumbar spine, discography was advocated as a cardinal tool for pinpointing the actual source of chronic neck pain, and cervical discography has been reported as a useful diagnostic exercise (Cloward, 1959; Kikuchi *et al.*, 1981). Cervical zygapophysial joint blocks, in contrast, have been explored only recently and by only a small number of groups around the world. However, prevalence studies have been conducted, and show that cervical zygpophysial joint pain is not uncommon (Aprill & Bogduk, 1992). In patients with chronic posttraumatic neck pain, the zygapophysial joints are the source of pain in at least 25% and possibly in as many as 64% (Aprill & Bogduk, 1992). Cervical zygapophysial joint pain can occur alone or in association with painful and presumably injured discs.

Pathology

Pathological studies have lagged conspicuously behind anatomical and clinical developments in the field of spinal pain. Much of what can be said of the pathology of disc pain and zygapophysial joint pain is conjectural or based only on circumstantial evidence. Internal disc disruption appears to be due to a biochemical degradation of the disc matrix following compression injury (Bogduk, 1991b). Some form of osteoarthrosis, subchondral fractures or capsular injuries seem the likely cause of lumbar zygapophysial joint pain (Bogduk, 1991a, 1992b). The causes of disc pain or zygapophysial joint pain in the neck have barely been explored, but contemporary evidence suggests that tears of the anulus fibrosus underlie cervical disc pain, and occult fractures may underlie cervical zygapophysial joint pain (Bogduk, 1988b). These entities still need to be investigated in a stringent manner.

Conclusions

Amongst contemporary investigators intent upon establishing the sources and causes of chronic spinal pain, attention has diverted from nerve root compression, and has been directed to investigating the articular elements of the vertebral column. Intervertebral discs and zygapophysial joints are known to be potential sources of pain, and these are the primary targets of interest.

Experience to date suggests that, for the lumbar spine, the discs are the leading sources of idiopathic pain, with internal disc disruption being the current favourite, pathological diagnosis. Lumbar zygapophysial joint pain can occur, but appears to be less common than originally intimated. Formal epidemiological studies are still required to establish how prevalent these conditions and others like muscle sprain and trigger point syndromes are in different classes of patients: those attending primary care with their first episode of pain, those suffering alleged industrial injury, and those with chronic pain.

The picture is somewhat different for the neck. Despite its past popularity, cervical disc pain may not be as prevalent as once believed. As more attention is paid to them, the cervical zygapophysial joints emerge as a common but hitherto unrecognised and overlooked source of chronic neck pain. The C2–3 zygapophysial joints and the lateral atlanto-axial joints can cause referred pain into the head that may be mistaken for occipital neuralgia.

References

Andersson, G., Bogduk, N., De Luca, C., Goldenberg, D., Mayer, T., Roy, S. & Smidt, G. (1989). Muscle: clinical perspectives. In *New Perspectives on Low Back Pain,* ed. J.W. Frymoyer & S.L. Gordon, pp. 293–334. Park Ridge, Illinois: American Academy of Orthopaedic Surgeons.

Aprill, C. & Bogduk, N. (1992). The prevalence of cervical zygapophysial joint pain: a first approximation. *Spine* **17**, 744–7.

Aprill, C., Dwyer, A. & Bogduk, N. (1990). Cervical zygapophysial joint pain patterns II: a clinical evaluation. *Spine,* **15**, 458–61.

Bernard, T. N. (1990). Lumbar discography followed by computed tomography: refining the diagnosis of low-back pain. *Spine*, **15**, 690–707.

Bogduk, N. (1980*a*). Lumbar dorsal ramus syndrome. *Medical Journal of Australia*, **2**, 537–41.

Bogduk, N. (1980*b*). The anatomy of occipital neuralgia. *Clinical and Experimental Neurology*, **17**, 167–84.

Bogduk, N. (1981). Local anaesthetic blocks of the second cervical ganglion: a technique with an application in occipital headache. *Cephalalgia*, **1**, 41–50.

Bogduk, N. (1982). The clinical anatomy of the cervical dorsal rami. *Spine*, **7**, 319–30.

Bogduk, N. (1983). The innervation of the lumbar spine. *Spine*, **8**, 286–93.

Bogduk, N. (1988*a*). Back pain: zygapophysial blocks and epidural steroids. In *Neural Blockade in Clinical Anaesthesia and Management of Pain*, 2nd edn, ed. M. J. Cousins & P. O. Bridenbaugh, pp. 935–54. Philadelphia: Lippincott.

Bogduk, N. (1988*b*). Neck pain: and update. *Australian Family Physician*, **17**, 75–80.

Bogduk, N. (1989). Greater occipital neuralgia. In *Current Therapy in Neurological Surgery*, 2nd edn, ed. D. M. Long, pp. 263–7. Philadelphia: Decker.

Bogduk, N. (1991*a*). Sources of low back pain. In T*he Lumbar Spine and Back Pain*, 4th edn, ed. M.I.V. Jayson, pp. 61–68, Edinburgh: Churchill Livingstone.

Bogduk, N. (1991*b*). The lumbar disc and low back pain. *Neurosurgical Clinics of North America*, **2**, 791–806.

Bogduk, N. (1992*a*). Cervical causes of headache and dizziness. In *Grieve's Modern Manual Therapy of the Vertebral Column*, 2nd edn, ed. J. D. Boyling & N. Palastanga. Edinburgh: Churchill Livingstone (in press).

Bogduk, N. (1992*b*). Lumbar dorsal ramus syndrome. In *Grieve's Modern Manual Therapy of the Vertebral Column*, 2nd edn, ed J. D. Boyling & N. Palastanga. Edinburgh: Churchill Livingstone (in press).

Bogduk, N., Lambert, G. & Duckworth, J. W. (1981a). The anatomy and physiology of the vertebral nerve in relation to cervical migraine. *Cephalalgia*, **1**, 11–24.

Bogduk, N. & Long, D. M. (1980). Percutaneous lumbar medial branch neurotomy. A modification of facet denervation. *Spine*, **5**, 193–200.

Bogduk, N. & Marsland, A. (1985). Third occipital headache. *Cephalalgia*, **5** Suppl 3, 310–11.

Bogduk, N. & Marsland, A. (1986). On the concept of third occipital headache. *Journal of Neurology Neurosurgery and Psychiatry*, **49**, 775–80.

Bogduk, N. & Marsland, A. (1988). The cervical zygapophysial joints as a source of neck pain. *Spine*, **13**, 610–17.

Bogduk, N., Tynan, W. & Wilson, A. S. (1981*b*). The nerve supply to the human lumbar intervertebral discs. *Journal of Anatomy*, **132**, 39–56.

Bogduk, N. & Twomey, L. T. (1991). *Clinical Anatomy of the Lumbar Spine*, 2nd edn. Melbourne: Churchill Livingstone.

Bogduk, N., Wilson, A.S. & Tynan, W. (1982). The human lumbar dorsal rami. *Journal of Anatomy*, **134**, 383–97.

Bogduk, N., Windsor, M. & Inglis, A. (1988). The innervation of the cervical intervertebral discs. *Spine*, **13**, 2–8.

Brodsky, A. E. & Binder, W. F. (1979). Lumbar discography. Its value in diagnosis and treatment of lumbar disc lesions. *Spine*, **4**, 110–20.

Campbell, D. G. & Parsons, C. M. (1944). Referred head pain and its concomitants. *Journal of Nervous and Mental Diseases*, **99**, 544–51.

Carrera, G. F. & Williams, A. L. (1984). Current concepts in evaluation of the lumbar facet joints. *CRC Critical Reviews in Diagnostic Imaging*, **21**, 85–104.

Cloward, R. B. (1959). Cervical diskography. A contribution to the aetiology and mechanism of neck, shoulder and arm pain. *Annals of Surgery*, **130**, 1052–64.

Colhoun, E., McCall, I. W. & Williams, L., Cassar Pullicino, V. N. (1988). Provocation discography as a guide to planning operations on the spine. *Journal of Bone and Joint Surgery*, **70B**, 267–71.

Collins, H. R. (1975). An evaluation of cervical and lumbar discography. *Clinical Orthopaedics and Related Research*, **107**, 133–8.

Collis, J. S. & Gardner, W. J. (1962). Lumbar discography – an analysis of 1,000 cases. *Journal of Neurosurgery*, **19**, 452–61.

Crock, H. V. (1986). Internal disc disruption: a challenge to disc prolapse fifty years on. *Spine*, **11**, 650–3.

Cuatico, W., Parker, J. C., Pappert, E. & Pilsi, S. (1988). An anatomical and clinical investigation of spinal meningeal nerves. *Acta Neurochirurgica*, **90**, 139–41.

Cuatico, W & Parker, J. C. (1989). Further observations on spinal meningeal nerves and their role in pain production. *Acta Neurochirurgica*, **101**, 126–8.

Cyriax, J. (1938). Rheumatic headache. *British Medical Journal*, **2**, 1367–8.

Dwyer, A., Aprill, C. & Bogduk, N. (1990). Cervical zygapophysial joint pain patterns I: a study in normal volunteers. *Spine*, **15**, 453–57.

Ehni, G. & Benner, B. (1984). Occipital neuralgia and the C1–2 arthrosis syndrome. *Journal of Neurosurgery*, **61**, 961–65.

El Mahdi, M. A., Latif, F. Y. A. & Janko, M. (1981). The spinal nerve root innervation, and a new concept of the clinico-pathological interrelations in back pain and sciatica. *Neurochirurgia*, **24**, 137–41.

Fairbank, J. C. T., Park, W. M., McCall, I. W. & O'Brien, J. P. (1981). Apophyseal injection of local anesthetic as a diagnostic aid in primary low-back pain syndromes. *Spine*, **6**, 598–605.

Feinberg, S. B. (1964). The place of diskography in radiology as based on 2,320 cases. *American Journal of Roentgenology*, **92**, 1275–81.

Feinstein, B., Langton, J. B. K., Jameson, R. M. & Schiller, F. (1954). Experiments on referred pain from deep somatic tissues. *Journal of Bone and Joint Surgery*, **36A**, 981–97.

Friberg, S. (1954). Lumbar disc degeneration in the problem of lumbago sciatica. *Bulletin of the Hospital for Joint Diseases*, **15**, 1–20.

Garrett, W. E., Nikolau, P. K., Ribbeck, B. M., Glisson, R. R. & Seaber, A. V. (1988). The effect of muscle architecture on the biomechanical failure properties of skeletal muscle under passive tension. *American Journal of Sports Medicine*, **16**, 7–12.

Garrett, W. E., Saffrean, M. R., Seaber, A. V., Glisson, R. R. & Ribbeck, B. M. (1987). Biomechanical comparison of stimulated and non-stimulated muscle pulled to failure. *American Journal of Sports Medicine*, **15**, 448–54.

Groen, G. J., Baljet, B. & Drukker, J. (1988). The innervation of the spinal dura mater: anatomy and clinical implications. *Acta Neurochirurgica*, **92**, 39–46.

Groen G. J., Baljet, B. & Drukker, J. (1990). Nerves and nerve plexuses of the human vertebral column. *American Journal of Anatomy*, **188**, 282–96.

Helbig, T. & Lee, C. K. (1988). The lumbar facet syndrome. *Spine*, **13**, 61–4.

Hockaday, J. M. & Whitty, C. W. M. (1967). Patterns of referred pain in the normal subject. *Brain*, **90**, 481–96.

Holt, E. P. (1968). The question of lumbar diskography. *Journal of Bone and Joint Surgery*, **50A**, 720–5.

Horal, J. (1969). The clinical appearance of low back disorders in the City of Gothenburg Sweden. *Acta Orthopaedica Scandinavica*, Suppl. 118.

Howe, J. F. (1979). A neurophysiological basis for the radicular pain of nerve root compression. In *Advances in Pain Research and Therapy*, vol. 3, ed. J. J. Bonica, J. C. Liebeskind & D. G. Albe-Fessard D G, pp. 647–57. New York: Raven Press.

Howe J. F., Loeser, J. D. & Calvin, W. H. (1977). Mechanosensitivity of dorsal root ganglia and chronically injured axons: a physiological basis for the radicular pain of nerve root compression. *Pain*, **3**, 25–41.

Hove, B. & Gyldensted, C. (1990). Cervical analgesic facet joint arthrography. *Neuroradiology*, **32**, 456–9.

Jackson, R. P., Jacobs, R. R. & Montesane, P. X. (1988). Facet joint injection in low back pain. *Spine*, **13**, 966–71.

Kellgren, J. H. (1938). Observations on referred pain arising from muscle. *Clinical Science*, **3**, 175–90.

Kellgren, J. H. (1939). On the distribution of referred pain arising from deep somatic structures with charts of segmental pain areas. *Clinical Science*, **4**, 35–46.

Kikuchi, S., MacNab, I. & Moreau, P. (1981). Localisation of the level of symptomatic cervical disc degeneration. *Journal of Bone and Joint Surgery*, **63B**, 272–7.

Kimmel, D. L. (1960). Innervation of spinal dura mater and dura mater of the posterior cranial fossa. *Neurology*, **10**, 800–9.

Knuttson, B. (1961). Comparative value of electromyography, myelography and clinico- neurological examination in diagnosis of lumbar root compression syndrome. *Acta Orthopaedica Scandinavica Supplementum* 49.

Kottinen, Y. T., Gronblad, M., Antti-Poika, I., Seitsulo, S., Santavirta, S., Hukanen, M. & Polak, J. M. (1990). Neuroimmunohistochemical analysis of peridiscal nociceptive neural elements. *Spine*, **15**, 383–6.

Korkala, O., Gronblad, M., Liesi, P. & Karaharju, E. (1985). Immunohistochemical demonstration of nociceptors in the ligamentous structures of the lumbar spine. *Spine*, **10**, 156–7.

Lazorthes, G. & Gaubert, J. (1956). L'innervation des articulations interapophysaire vertebrales. *Comptes Rendues de l'Association des Anatomistes*, pp. 488–94.

Lewinnek, G. E. & Warfield, C. A. (1986). Facet joint degeneration as a cause of low back pain. *Clinical Orthopaedics and Related Research*, **213**, 216–22.

Lippit, A. B. (1984). The facet joint and its role in spine pain: management with facet joint injections. *Spine*, **9**, 746–50.

Lynch M. C. & Taylor, J. F. (1986). Facet joint injection for low back pain. *Journal of Bone and Joint Surgery*, **68B**, 138–41.

Macintosh, J. E., Valencia, F., Bogduk, N. & Munro, R. R. (1986). The morphology of the lumbar multifidus. *Clinical Biomechanics*, **1**, 196–204.

MacNab, I. (1972). The mechanism of spondylogenic pain. In *Cervical Pain*, ed. C. Hirsch & Y. Zotterman, pp. 89–95. Oxford: Pergamon.

Malinsky, J. (1959). The ontogenetic development of nerve terminations in the intervertebral discs of man. *Acta Anatomica*, **38**, 96–113.

Massie, W. K., Stevens, D. B. (1967). A critical evaluation of discography. *Journal of Bone and Joint Surgery*, **49A**, 1243–4.

McCall, I. W., Park, W. M. & O'Brien, J. P. (1979). Induced pain referral from posterior lumbar elements in normal subjects. *Spine*, **4**, 441–6.

McCormick, C. C. (1987). Arthrography of the atlanto-axial (C1 – C2) joints: technique and results. *Journal of Interventional Radiology*, **2**, 9–13.

McCulloch, J. A., Waddell, G. (1980). Variation of the lumbosacral myotomes with bony segmental anomalies. *Journal of Bone and Joint Surgery*, **62B**, 475–80.

McCutcheon, M. E. (1986). CT scanning of lumbar discography: a useful diagnostic adjunct. *Spine*, **11**, 257–9.

McFadden, J. W. (1988). The stress lumbar discogram. *Spine*, **13**, 931–3.

Mooney, V. (1987*a*). Where is the pain coming from? *Spine*, **12**, 754–9.

Mooney, V. (1987*b*). Facet joint syndrome. In *The Lumbar Spine and Back Pain*, 3rd edn, ed. M.I.V. Jayson, pp. 370–82. Edinburgh: Churchill Livingstone.

Mooney, V. & Robertson, J. (1976). The facet syndrome. *Clinical Orthopaedics and Related Research*, **115**, 149–56.

Moran, R., O'Connell, D. & Walsh, M. G. (1988). The diagnostic value of facet joint injections. *Spine*, **12**, 1407–10.

Murtagh, F. R. (1988). Computed tomography and fluoroscopy guided anaesthesia and steroid injection in facet syndrome. *Spine*, **13**, 686–9.

Nikolau, P. K., MacDonald, B. L., Glisson, R. R., Seaber, A. V. & Garrett, W. E. (1987). Biomechanical and histological evaluation of muscle after controlled strain injury. *American Journal of Sports Medicine*, **15**, 9–14.

Norlen, G. (1944). On the value of the neurological symptoms in sciatica for the localisation of a lumbar disc herniation. *Acta Chirurgica Scandinavica*, Suppl 95, 1–96.

Park, W. (1980). The place of radiology in the investigation of low back pain. *Clinics in the Rheumatic Diseases*, **6**, 93–132.

Raymond, J. & Dumas, J-M. (1984). Intra-articular facet block: diagnostic test or therapeutic procedure? *Radiology*, **151**, 333–6.

Roland, M. O. (1986). A critical review of the evidence for a pain- spasm-pain cycle in spinal disorders. *Clinical Biomechanics*, **1**, 102–9.

Roth, D. A. (1976). Cervical analgesic discography. A new test for the definitive diagnosis of the painful-disk syndrome. *Journal of the American Medical Association*, **235**, 1713–14.

Sachs, B. L., Vanharanta, H., Spivey, M. A., Guyer, R. D., Videman, T., Rashbaum, R. F., Johnson, R. G., Hochschuller, S. H. & Mooney, V. (1987). Dallas discogram description: a new classification of CT/discography in low back disorders. *Spine*, **12**, 287–94.

Simmons, E. H. & Segil, C. M. (1975). An evaluation of discography in the localisation of symptomatic levels in discogenic disease of the spine. *Clinical Orthopaedics and Related Research*, **108**, 57–69.

Simmons, J. W., Aprill, C. N., Dwyer, A. P. & Brodsky, A. E. (1988). A reassessment of Holt's data on: 'the question of lumbar discography'. *Clinical Orthopaedics and Related Research*, **237**, 120–4.

Simons, D. G. (1988). Myofascial pain syndromes: Where are we? Where are we going? *Archives of Physical Medicine and Rehabilitation*, **69**, 207–12.

Smyth, M. J. & Wright, V. (1959). Sciatica and the intervertebral disc. An experimental study. *Journal of Bone and Joint Surgery*, **40A**, 1401–18.

Vanharanta, H., Sachs, B. L., Spivey, M. A., Guyer, R. D., Hochschuller, S. H., Rashbaum, R. F., Johnson, R. G., Ohrmeiss, D. & Mooney, V. (1987). The relationship of pain provocation to lumbar disc deterioration as seen by CT/discography. *Spine*, **12**, 295–8.

Videman, T., Malmivaara, A. & Mooney, V. (1987). The value of the axial view in assessing discograms: an experimental study with cadavers. *Spine*, **12**, 299–304.

Walsh, T. R., Weinstein, J. N., Spratt, K. F., Lehmann, T. R., Aprill, D. & Sayre, H. (1990). Lumbar discography in normal subjects. *Journal of Bone and Joint Surgery*, **72A**, 1081–8.

Wedel, D. J. & Wilson, P. R. (1985). Cervical facet arthrography. *Regional Anaesthesia*, **10**, 7–11.

Weinstein, J., Claverie, W. & Gibson, S. (1988). The pain of discography. *Spine*, **13**, 1344–8.

Wiley, J. J., MacNab, I. & Wortzman, G. (1968). Lumbar discography and its clinical applications. *Canadian Journal of Surgery*, **11**, 280–9.

Wyke, B. (1980). The neurology of low back pain. In *The Lumbar Spine and Back Pain*, 2nd edn, ed. M.I.V. Jayson, pp. 265–339. Turnbridge Wells: Pitman.

Yoshizawa, H., O'Brien, J. P., Thomas-Smith, W. & Trumper, M. (1980). The neuropathology of intervertebral discs removed for low-back pain. *Journal of Pathology*, **132**, 95–104.

Part II
Normal motor control and its disorders

4

Corticospinal control of movement

R. PORTER

Faculty of Medicine,
Monash University,
Clayton, Melbourne VIC, Australia

Introduction

Perusal of the published volumes of major journals such as the *Journal of Physiology* and the *Journal of Neurophysiology*, from about 40 years ago, will show that there was, at that time, a high level of interest in, and a very large number of observations being made on, the electrophysiology of the pyramidal or corticospinal tracts in mammals. The conduction velocities of the fibres in the pyramidal tract of the cat were shown to occupy a range from 7 to 70 m/s and to be grouped into two populations. The fast population of pyramidal tract fibres in the cat spans conduction velocities from 22 to 70 m/s and has a peak at 50 m/s. The slow population of pyramidal tract fibres ranges from 7 to 22 m/s with a peak at 14 m/s (Bishop, Jeremy & Lance, 1953; Lance & Manning, 1954; Lance, 1954). All pyramidal tract fibres appear to be myelinated. Two-thirds of these pyramidal tract fibres have their origins in the homologue of the Rolandic cortex, around the cruciate sulcus in the cat, and the vast majority of pyramidal tract fibres are derived from Brodmann's cyto-architectonic areas 4, 3, 1 and 5, with only a few slow fibres arising from area 6 (Lance & Manning, 1954).

At that time it was clear that there were two populations of corticospinal fibres in the macaque monkey. Brookhart (1952) had identified these populations and attempted differential stimulation of them. He recorded activation of different muscle groups with different patterns of stimulation. He concluded from his experiments that the rapidly conducting pyramidal tract fibres in the monkey terminated functionally in relation to motoneurones controlling facial and digital muscles. He thought that these fast fibres were responsible for the phasic elements of pyramidal control of movement. In contrast, the slow pyramidal tract fibres seemed to provide the principal control over proximal limb muscles and could be responsible for the tonic elements of motor function.

Although suggestive evidence had been obtained long before from experiments in Sherrington's laboratory, it was also in the 1950s that Bernhard, Bohm & Petersén (1953) and Bernhard & Bohm (1954) described the work they had done earlier to establish that rapidly conducting axons in the monkey's pyramidal tract made direct monosynaptic connections with spinal motoneurones. They named this component of the pyramidal tract, which exists in human and subhuman primates, the cortico-motoneuronal system. Landgren, Phillips & Porter (1962) used intracellular recording in spinal motoneurones to characterize the monosynaptic nature of the excitatory postsynaptic potentials (EPSPs) generated in motoneurones by activation of corticomotoneuronal axons. Then, Lawrence & Kuypers (1968) demonstrated that integrity of corticomotoneuronal connections was an essential requirement for independent movements of the digits of monkeys, refined control of dextrous manipulations, and the 'precision grip'.

We are witnessing an increasing use of electrical and magnetic methods of stimulation of the human brain, and especially of the human corticospinal tract, in clinical investigation. This should make it timely to assess our present state of knowledge about the nature of corticospinal neurones, and their role in the control of movement (Gandevia, this volume). We need to seek new evidence about the processes by which particular corticospinal cells are selected and recruited to action during movement performance and about the details of the distribution of their influences within the motor apparatus of the spinal cord. For this we will need to rely heavily on experimental observations made in monkeys, and their correlation with human studies where these are available.

Cells of origin of the corticospinal tract

Jane, Yashon, de Meyer and Bucy (1967) examined the cerebral cortex of human subjects and concluded that 60% of all pyramidal tract fibres arose from area 4 and the remainder derived from cortex in the immediate vicinity of area 4. This result was consistent with the view of Nathan and Smith (1955) that most of the cells of origin of the corticospinal tract in man were situated in the precentral gyrus, mainly in its upper two-thirds, and in the paracentral lobule. This region of origin appears more restricted in the human brain than in that of the cat and monkey. In these other animals, the identification of cortical neurones which have transported labels such as horseradish peroxidase (HRP) retrogradely from their terminals in the spinal cord, reveals that corticospinal fibres arise from areas 6, 4, 3a, 1, 2 and 5 and from SII (the second somatic sensory area) (Coulter, Ewing & Carter, 1976). The cells of origin of all corticospinal fibres are exclusively located in the deeper parts of lamina V (Jones & Wise, 1977).

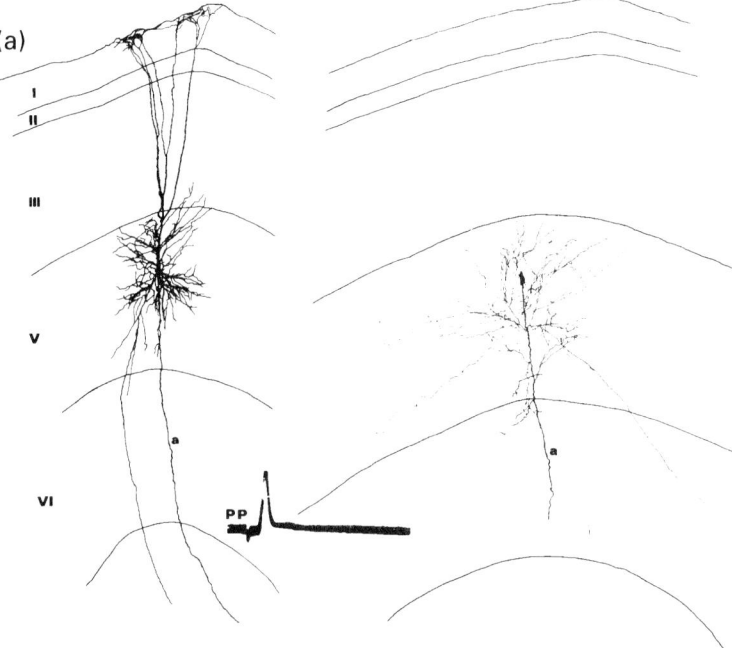

Fig. 4.1. *Left-hand panel*: Camera lucida reconstruction of the soma, dendrites and axon (a) of a pyramidal tract neurone with a slowly conducting axon. The antidromic response recorded with the intracellular electrode during the injection of HRP is indicated at PP. *Right-hand panel*: Soma, axon (a) and axon collaterals of the same cell. This cell was located in the anterior sigmoid gyrus of the cat. The illustration is oriented as a parasaggital section, anterior to the left and posterior to the right. Ghosh *et al.* (1988) reproduced with permission.

Recently, Ghosh, Fyffe and Porter (1988) and Ghosh and Porter (1988a) studied the morphology of identified pyramidal tract neurones in area 4 of the cat and monkey cortex using the technique of intracellular recording from these cells in barbiturate anesthetized animals. The intracellular electrode contained a solution of horseradish peroxidase (HRP) which could be injected into identified cells to reveal, in subsequent histological preparations of the cortex, the full dendritic and axon collateral architecture of the cells. Figure 4.1 illustrates a camera lucida reconstruction (on the left) of the soma, dendrites and axon (a) of a slow pyramidal tract neurone in the anterior sigmoid gyrus of the cat's cortex. The right hand panel illustrates the soma, axon and intracortical axon collateral arborization of the same cell. The completely reconstructed and fully examined slow pyramidal tract neurones in the cat (of which this is a representative example), had axonal conduction velocities in the range 6.7 to 19.2

Table 4.1 *Morphology of pyramidal tract neurones in the cat*

	Fast PTN	Slow PTN
Number studied	3	7
Axonal conduction velocity (m/s)	27.5 – 33.3	6.7 – 19.2
Soma size (μm)	18–24 x 30–42	19–24 x 32–73
Mean number of apical dendrites	9.7	6.4
Range	(7–14)	(5–9)
Mean extent of dendritic arbor (μm)	782(A-P) x 640(M-L)	733(A-P) x 742(M-L)
Range (μm)	(664–900) (480–720)	(536–973) (560–1280)
Mean number of axon collaterals	4.7	3.7
Range	(4–5)	(2–5)

Notes: A-P = anteroposterior.
M-L = mediolateral.

m/s and morphological dimensions in the ranges indicated in Table 4.1. These cells (and the fast pyramidal tract neurones) exhibited intracortical axon collaterals which were limited in their distribution to lamina V with only minimal extensions to lamina VI. This distribution of axon collaterals of pyramidal tract neurones may be contrasted with the much more variable and more widespread distribution of the intracortical collaterals of lamina III pyramidal neurones.

Figure 4.2 illustrates a camera lucida reconstruction of a fast pyramidal tract neurone (conduction velocity 24.5 m/s) recorded from area 4 of the 'forearm' representation in the precentral gyrus of a monkey. Although the conduction velocity of the axon of this cell was at the low end of the fast range, the soma dimensions (25.0 x 40.0 μm), spread of apical dendrites (over a cortical zone with a diameter of the order of 500 μm) and more extensive distribution of lateral and oblique dendrites (covering 650–700 μm) and basal dendrites (880–900 μm) were characteristic of this class of neurone. The intracortical axon collateral arbors (right-hand side of Fig. 4.2) extended for 1.5 to 2.2 mm, were made up of short and long collaterals and were limited in their extent to laminae V and VI. Again, in the monkey the intracortical axon collaterals of lamina III neurones were more variable, more extensive and could project widely into regions of all laminae (Porter, Ghosh & Fyffe, 1988).

The range of diversity, and the variations in complexity, exhibited by these identified corticospinal neurones were examined using fractal analysis (Porter *et al.*, 1991). The mathematical concepts of Mandelbrot's fractal geometry have allowed quantitative measurements of the complexity of the borders or

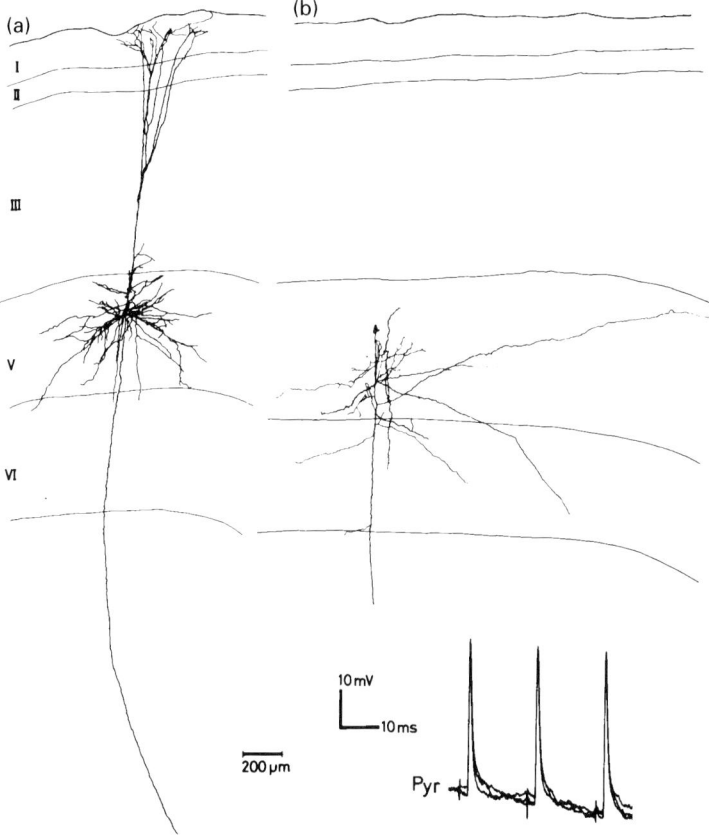

Fig. 4.2. Camera lucida reconstruction of a fast pyramidal tract neurone situated in the anterior lip of the central sulcus in monkey motor cortex. The soma, dendrites and axon are illustrated on the left (*a*) while the soma, axon and intracortical axon collaterals are illustrated on the right (*b*). Antidromic responses to repetitive stimulation of the pyramidal tract are illustrated (Pyr). Ghosh and Porter (1988a) reproduced with permission.

geometrical outlines of the projections of neurones and glia in an unbiased and objective manner (Mandelbrot, 1982; Smith *et al.*, 1989). In the cat, lamina V neurones (which include the pyramidal tract neurones) were revealed to be more complex in their geometry than lamina III neurones. The reverse, however, was true for the monkey, in which the lamina V pyramidal neurones were found to be less complex than those in the cat and also less complex than monkey lamina III neurones. The significance of these differences is still to be investigated. The high level of complexity in a cat lamina V pyramidal tract neurone may signify a large number of computational operations to be accommodated on its many complex, dendritic branches. In the larger, more extensive motor cortex of the monkey, with many more pyramidal tract neurones, less densely packed, some of these computational operations may be subserved by different cells rather than by different branches of the same cells, allowing segregation of different functions to be performed by separate neurones, each with lesser requirements for morphological complexity.

Spinal terminations of corticomotoneuronal fibres in the monkey

Histological demonstration, at the light microscopic level, of the collateral arborization of corticomotoneuronal fibres which innervate the motoneurones of the intrinsic muscles of the hand of the monkey was provided by the work of Lawrence, Porter and Redman (1985). They employed intraaxonal and intracellular recordings with HRP filled electrodes to study identified corticospinal axons and 'hand' motoneurones located close together mostly within the C8-T1 segments of the spinal cord. Shinoda, Yokota and Futami (1981) had used intraaxonal injection of HRP into corticospinal fibres in the monkey to visualize the preterminal and terminal axonal arborizations of these axons. They found, in transverse sections of the spinal cord, that the arborizations ramified in laminae V, VI and VII of the spinal cord and some collaterals extended into lamina IX where the cell bodies and proximal dendrites of motoneurones innervating different forelimb muscles had been retrogradely labelled with HRP. By intracellular labelling of a few intrinsic hand muscle motoneurones in the same segment of the spinal cord as the intraaxonal injection of the corticospinal axon, Lawrence et al. (1985) were able to identify fibres which did, indeed, make monosynaptic contact with the surface of a motoneurone, to describe these contacts and their locations, and then to describe the complete distribution of the arborization of this particular branch of the corticomotoneuronal axon. The reconstructions were made from serial parasagittal sections of the spinal cord so that they exhibit the longitudinal projection of the corticomotoneuronal arborization, albeit condensed into a single plane.

Figure 4.3 illustrates the camera lucida reconstruction of a main collateral of a corticomotoneuronal fibre assembled from superimposition of 14 parasagittal 100 μm thick sections of the C8-TI segment of the monkey's spinal cord. In the same diagram are depicted the somata and proximal dendrites (black) of two ulnar motoneurones innervating intrinsic muscles of the hand. Each of these motoneurones, separated by almost 2 mm, was completely reconstructed: their dendrites extended to cover the territories included in the dotted outlines which overlap near the middle of the collateral's most dense branching. Each motoneurone received a synapse from a branch of this main collateral

Corticospinal control of movement 67

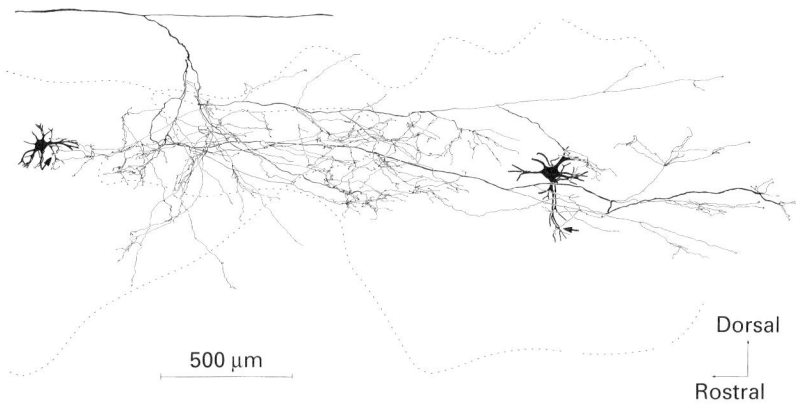

Fig. 4.3. Camera lucida reconstruction of the arborization of a main collateral arising from a corticomotoneuronal axon running in the lateral funiculus of the monkey's spinal cord. The arborization occupies a length of about 2 mm in the central region of the C8 segment of the cord between two ulnar motoneurones innervating intrinsic muscles of the hand. The outermost limits of the dendritic trees of these two motoneurones are indicated by interrupted lines. The thick arrows indicate the sites of a single synaptic contact upon each motoneurone. Lawrence *et al.* (1985) reproduced with permission.

(arrowed). Hence the parent fibre, running horizontally in the lateral white matter of the cord (top left of the figure) must be defined as a corticomotoneuronal axon. The corticomotoneuronal stem axon, which could be followed for 1.25 mm in the lateral funiculus, gave rise to a main collateral in the C8 segment which ran caudally and ventromedially for approximately 0.5 mm before entering lamina IX. In lamina IX, the collateral branched extensively and the majority of the branches occupied a longitudinal cylinder 0.5 mm in diameter and 1.5 mm in length. More distant branches spread well beyond this cylinder, however. Longitudinal extension was provided by two thick collaterals which arose from proximal branches of the main collateral and ran caudally within lamina IX. Many terminal and en passant synapses were evident on all of the branches and some of the most extensive projections continued beyond the limits of these sections. This extensive divergence of the collateral branches of corticomotoneuronal fibres is characteristic of the very small sample of identified axons that has been studied. Whenever a stem axon issuing from the 'hand' motor area of the cerebral cortex was successfully filled with HRP and exhibited a collateral in the vicinity of 'hand' motoneurones, revealed by intracellular filling with HRP, one or a few synaptic contacts on each of these motoneurones was discovered at the light microscopic level. Most often, only one synaptic knob was contributed to the cell by the corticomotoneuronal fibre. Frequently the synapse was small: they ranged in size from 0.6 x 3.0 µm to 2.4

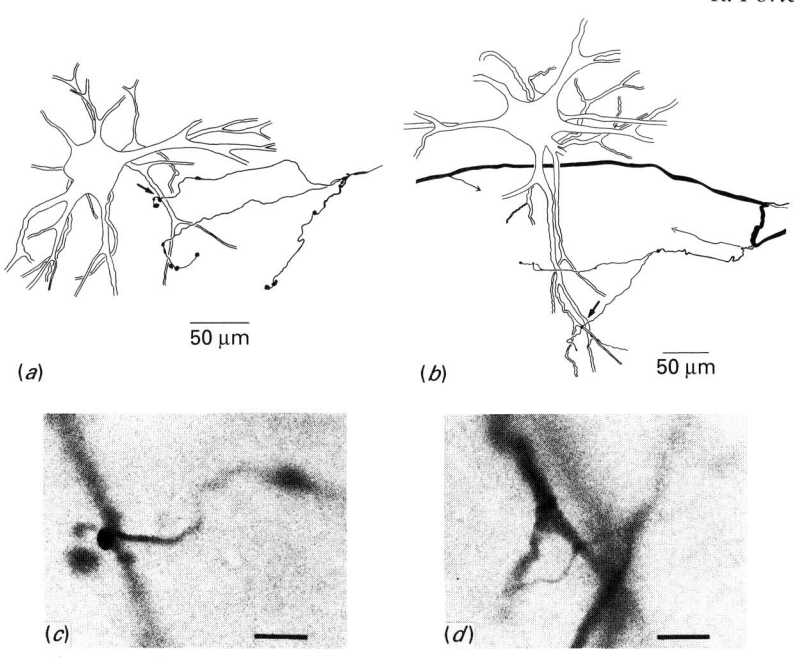

Fig. 4.4. High power illustration of the reconstruction of the rostral (*a*) and caudal (*b*) motoneurones shown in Fig. 4.3 to indicate the sites of synaptic contacts (arrows). Photomicrographs (*c*) and (*d*) show the plane of these synaptic sites as revealed by photography of the thick (100 μm) sections. Calibration bars = 10 μm. Lawrence *et al.* (1985) reproduced with permission.

x 3.6 μm. Often they were located on dendrites and some could be distant from the cell soma (Fig. 4.4).

These results indicate that a single corticomotoneuronal axon may exhibit great divergence of its influence within the spinal cord. In addition, this small sample illustrates that whenever the collateral of a corticomotoneuronal axon destined to innervate 'hand' motoneurones branched in the vicinity of appropriate motoneurones, it contacted each of these. Hence the probability of each corticomotoneuronal fibre contacting all members of a selected class of motoneurones (in this case those supplying the intrinsic muscles of the hand) must be very high and there must be a very high degree of convergence of influences from multiple corticomotoneuronal axons on to each motoneurone. In the one instance in which two corticomotoneuronal axons and two adjacent motoneurones in the same region of the cord were successfully revealed for full reconstruction, collaterals of both axons made contacts with the dendrites of both motoneurones. Part of one of these collaterals contacting (at a and b) a dendrite of each of the motoneurones is illustrated in Fig. 4.5.

Corticospinal control of movement 69

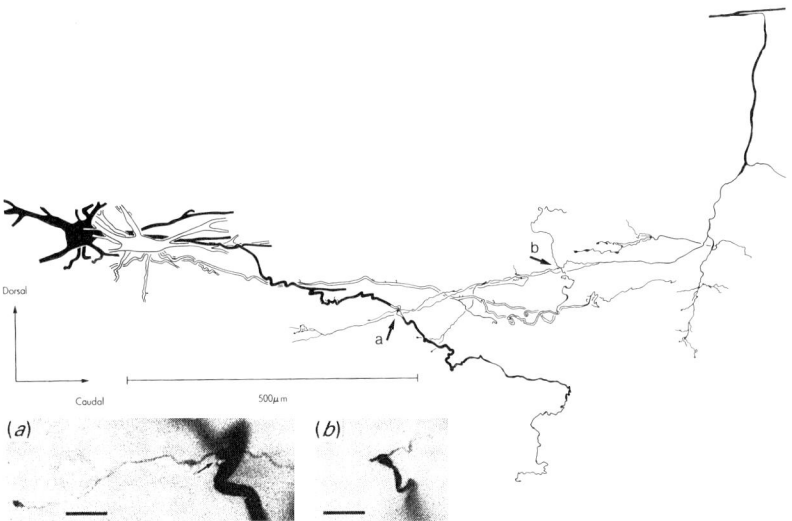

Fig. 4.5. Reconstruction of the segments of the collateral branches of a corticomotoneuronal axon which establish one synaptic contact (arrows) with a dendrite of each of two ulnar motoneurones located in the T1 segment of the spinal cord and innervating intrinsic muscles of the hand. The photomicrographs of these single contacts (a and b) are illustrated in the insets. Note that a is probably an axospinous synapse. Calibration bars = 10 μm. Lawrence *et al.* (1985) reproduced with permission.

Discharges of individual corticomotoneuronal neurones during natural movement

Fetz and Cheney (1980) and Muir and Lemon (1983) used cross-correlation methods to extend the meaningfulness of the technique of recording natural discharges of cortical pyramidal cells during movement performance (Evarts, 1968) and to allow specific examination of functionally identified corticomotoneuronal cells. If the discharge of a cortical cell was followed by a postspike facilitation of the electromyogram (EMG) in a target muscle at an appropriately short latency, the cortical cell could be defined functionally as a corticomotoneuronal element including that muscle in its 'muscle field'. Postspike facilitation of muscles in the monkey's forelimb had a mean onset latency for the small muscles of the hand which was about 3 ms later than those of forearm muscles (Lemon, Mantel & Muir, 1986).

Cheney and Fetz (1980) showed that corticomotoneuronal cells, as a special class of corticospinal neurones, discharged in advance of the onset of EMGs in their target muscles. They also demonstrated changes in firing rate in association with changes in force required to execute natural movements. In spite of this, Muir and Lemon (1983) found that some motor cortical cells, which were particularly active during a precision grip, fired at much higher frequencies when their target muscle (e.g. first dorsal interosseous) was engaged in this controlled pincer movement of the tips of the thumb and index finger than they did when the same muscle was generating a greater force during a power grip of the hand around a rod. Then, the same corticomotoneuronal cell could be almost silent.

The muscle field to which a given corticomotoneuronal cell was found to project could involve several muscles normally coactivated in a particular movement in a given direction about the wrist joint (Fetz & Cheney, 1978). In contrast, some corticomotoneuronal cells, particularly those producing post-spike facilitation of the intrinsic muscles of the hand or muscles acting about the distal joints of the fingers, may have a much more restricted muscle field. Buys and colleagues (1986) found that many of the corticomotoneuronal cells which were active in association with precision grip movements had a restricted connectivity which was relatively specific to a small number of muscles with synergistic functions, or to only one muscle.

Significance

The extensive dendritic territories of single corticospinal neurones present an expanded, complex receptive surface onto which a vast multitude of synaptic terminals, derived from collaterals of cells in this region or in other more remote parts of the brain, may project. These wide territories may contribute to the selection of the specific combinations of outputs from the motor cortex needed for voluntary movement. Ghosh, Brinkman and Porter (1987) described the origins of the cortico-cortical projections to a restricted part of the hand area of motor cortex in the monkey. They found that cortico-cortical inputs from cells in the premotor and supplementary motor areas were quantitatively most numerous and, potentially, most significant. Ghosh and Porter (1988*b*) demonstrated that short-latency excitatory postsynaptic potentials (EPSPs) were produced in pyramidal tract neurones by activation of the postarcuate premotor cortex in the monkey. It is well known that both premotor and supplementary motor areas influence movement performance. Furthermore, the latter is involved in man in 'internal programming' of movement sequences and in memorizing the queue of 'time-ordered commands' that are needed for the performance of complex voluntary movements (Roland *et al.*, 1980). Moreover, there are important indications that, in man and monkey, each supplementary motor area is able to have an impact on the motor output to both the ipsilateral and the contralateral hand, and that the supplementary motor area is important in the performance of tasks requiring bimanual co-ordination

(Brinkman & Porter, 1979; Roland et al., 1980; Brinkman, 1981; Schell, Hodge & Cacayorin, 1986; Porter, 1990).

The precise selection of the corticospinal outputs which need to be recruited in specific order in the organization and direction of a complex voluntary movement will depend on the interactions between synaptic inputs derived from premotor, supplementary motor and many other brain regions. The dendritic surface membrane of the pyramidal cells of origin of these corticospinal fibres is not only extremely expanded to receive this multitude of inputs but its complexity (which can be quantified mathematically) varies from cell to cell and from one species to another. In all cases, the complexity is contributed to, not only by the profuse branching of the dendrites, but by the distribution of a myriad of spines over their surfaces. The spines are highly specialized synaptic protrusions of these surfaces. We speculate that, in the monkey, where separate aspects of the complex task of computing the necessary outputs may be shared by a number of corticospinal neurones, the demands for complexity in the organization of the neurone's geometry may be less than it is in the cat where many associated functions may have to be served by the same cell.

The relatively limited distribution of the intracortical axon collaterals of the corticospinal neurones, essentially to lamina V and parts of lamina VI, indicates that corticospinal neurones are able to make axo-somatic or axo-dendritic contacts only with pyramidal cells in these deep layers, although they may also contact other neurones such as stellate or basket cells. It has been found consistently that activation of these intracortical axon collaterals of pyramidal tract neurones is capable of producing excitatory or inhibitory actions (EPSPs or IPSPs) on other pyramidal tract neurones. Thus, the outputs of the motor cortex, through particular corticospinal fibres, are able to influence the probabilities of involvement of neighbouring corticospinal neurones (within a few mm) and thereby shape the membership of the population of output projections involved in a motor task. Ghosh et al. (1988) found that, when they examined two closely associated nearby pyramidal tract neurones in the cat, many 'close crossings' of apical dendritic shafts, basal dendrites and lateral dendrites of the two neurones occurred. It is not known whether these close crossings allow functional interactions between neighbouring cells. Two particular neurones were separated by about 100 μm in the cortex. Intracortical axon collaterals of the two were examined in detail with high power microscopy. A terminal bouton of the axon collateral of the slow pyramidal tract neurone was found closely apposed to the shaft of a basal dendrite of the fast pyramidal tract neurone. This form of connection clearly provides an anatomical basis for the excitation of fast pyramidal tract neurones by action potentials in slow pyramidal tract axons which is consistently described in the cat's motor cortex (Armstrong, 1965; Takahashi, Kubota & Uno, 1967).

The small size of the synaptic boutons provided to the surface of a 'hand' motoneurone by the spinal collaterals of a corticomotoneuronal fibre, together with the dendritic location of many of these contacts, and the fact that only one or a few contacts are provided by each fibre, may be consistent with the small contribution made by each corticomotoneuronal axon to the depolarization of the receiving motoneurone (Asanuma et al., 1979). Yet the extensive divergence of even a single spinal collateral of a corticomotoneuronal fibre destined to innervate motoneurones of the intrinsic hand muscles, coupled with the fact that the probability of synaptic contact being made with all members of a population of synergistic motoneurones must be very high, could explain why the cortex appears to deliver the largest quantities of excitation to motoneurones of muscles acting distally (Phillips & Porter, 1964, 1977; Clough, Kernell & Phillips, 1968). It is clear, then, that this potentially large excitatory influence, which must be generated by the convergence onto individual motoneurones of a very large number of very small individual contributions, could provide the substrate for precise selection, at a spinal level, of the exact motoneurones which will be recruited to a given task when their thresholds are exceeded. It will determine the fractionated use of muscles which is essential for independent digit movements and the precision grip. This capacity for selection of motor units and fractionation of use of distally acting muscles is lost when the pyramidal tracts are interrupted.

References

Armstrong, D.M. (1965). Synaptic excitation and inhibition of Betz cells by antidromic pyramidal volleys. *Journal of Physiology (London)*, **178**, 37–8P.

Asanuma, H., Zarzecki, P., Jankowska, E., Hongo, T. & Marcus, S. (1979). Projection of individual pyramidal tract neurons to lumbar motor nuclei of the monkey. *Experimental Brain Research*, **34**, 73–89.

Bernhard, C.G., Bohm, E. & Petersén, I. (1953). Investigations on the organization of the cortico-spinal system in monkeys (*Macaca mulatta*). *Acta Physiologica Scandinavica*, **29**, (Supplement 106), 79–105.

Bernhard, C.G. & Bohm, E. (1954). Cortical representation and functional significance of the cortico-motoneuronal system. *Acta Neurologica Psychiatrica*, **72**, 473–502.

Bishop, P.O., Jeremy, D. & Lance, J.W. (1953). Properties of pyramidal tract. *Journal of Neurophysiology*, **16**, 537–50.

Brinkman, C., & Porter, R. (1979). Supplementary motor area in the monkey. Activity of neurons during the performance of a learned motor task. *Journal of Neurophysiology*, **42**, 681–709.

Brinkman, C. (1981). Lesions in the supplementary motor area interfere with a monkey's performance of a bimanual co-ordination task. *Neuroscience Letters*, **27**, 267–70.

Brookhart, J.M. (1952). A study of corticospinal activation of motoneurons. *Research Publications, Association for Research in Nervous and Mental Diseases*, **30**, 157–73.

Buys, E.J., Lemon, R.N., Mantel, G.W.H. & Muir, R.R. (1986). Selective facilitation of different hand muscles by single corticospinal neurones in the conscious monkey. *Journal of Physiology (London)*, **381**, 329–49.

Cheney, P.D. & Fetz, E.E. (1980). Functional classes of primate cortico-motoneuronal cells and their relation to active force. *Journal of Neurophysiology*, **44**, 773–91.

Clough, J.F.M., Kernell, D. & Phillips, C.G. (1968). The distribution of monosynaptic excitation from the pyramidal tract and from primary spindle afferents to motoneurones of the baboon's hand and forearm. *Journal of Physiology (London)*, **216**, 257–79.

Coulter, J.D., Ewing, L. & Carter, C. (1976). Origin of primary sensorimotor cortical projections to lumbar spinal cord of cat and monkey. *Brain Research*, **103**, 366–72.

Evarts, E.V. (1968). Relation of pyramidal tract activity to force exerted during voluntary movement. *Journal of Neurophysiology*, **31**, 14–27.

Fetz, E.E. & Cheney, P.D. (1978). Muscle fields of primate cortico-motoneuronal cells. *Journal de Physiologie (Paris)*, **74**, 239–45.

Fetz, E.E. & Cheney, P.D. (1980). Postspike facilitation of forelimb muscle activity by primate cortico-motoneuronal cells. *Journal of Neurophysiology*, **44**, 751–72.

Ghosh, S., Brinkman, C. & Porter, R. (1987). A quantitative study of the distribution of neurons projecting to the precentral motor cortex in the monkey (*M. fascicularis*). *Journal of Comparative Neurology*, **259**, 424–44.

Ghosh, S., Fyffe, R.E.W. & Porter, R. (1988). Morphology of neurons in area 4 of the cat's cortex studied with intracellular injection of HRP. *Journal of Comparative Neurology*, **269**, 290–312.

Ghosh, S. & Porter, R. (1988a). Morphology of pyramidal neurones in monkey motor cortex and the synaptic actions of their intracortical axon collaterals. *Journal of Physiology (London)*, **400**, 593–615.

Ghosh, S. & Porter, R. (1988b). Corticocortical synaptic influences on morphologically identified pyramidal neurones in the motor cortex of the monkey. *Journal of Physiology (London)*, **400**, 617–29.

Jane, J.A., Yashon, D., de Meyer, W. & Bucy, P.C. (1967). The contribution of the precentral gyrus to the pyramidal tract of man. *Journal of Neurosurgery*, **26**, 244–8.

Jones, E.G. & Wise, S.P. (1977). Size, laminar and columnar distribution of efferent cells in the sensory-motor cortex of monkeys. *Journal of Comparative Neurology*, **175**, 391–438.

Lance, J.W. (1954). Pyramidal tract in spinal cord of cat. *Journal of Neurophysiology*, **17**, 253–70.

Lance, J.W. & Manning, R.L. (1954). Origin of the pyramidal tract in the cat. *Journal of Physiology (London)*, **124**, 385–99.

Landgren, S., Phillips, C.G. & Porter, R. (1962). Minimal synaptic actions of pyramidal impulses on some alpha motoneurones of the baboon's hand and forearm. *Journal of Physiology (London)*, **161**, 91–111.

Lawrence, D.G. & Kuypers, H.G.J.M. (1968). The functional organization of the motor system. I. The effects of bilateral pyramidal lesions. *Brain*, **91**, 1–14.

Lawrence, D.G., Porter, R. & Redman, S.J. (1985). Corticomotoneuronal synapses in the monkey: light microscopic localisation upon motoneurons of intrinsic muscles of the hand. *Journal of Comparative Neurology*, **232**, 499–510.

Lemon, R.N., Mantel, G.W.H. & Muir, R.B. (1986). Corticospinal facilitation of hand muscles during voluntary movement in the conscious monkey. *Journal of Physiology (London)*, **381**, 497–527.

Mandelbrot, B. (1982). *The Fractal Geometry of Nature*. New York: W.H. Freeman.

Muir, R.B. & Lemon, R.N. (1983). Corticospinal neurons with a special role in precision grip. *Brain Research*, **261**, 312–6.

Nathan, P.W. & Smith, M.C. (1955). Long descending tracts in man. I. Review of present knowledge. *Brain*, **78**, 248–303.

Phillips, C.G. & Porter, R. (1964). The pyramidal projection to motoneurones of some muscle groups of the baboon's forelimb. In *Physiology of Spinal Neurons* (ed. J.C. Eccles & J.P. Schadé) Progress in Brain Research, vol. 12, pp. 222–45 Amsterdam: Elsevier.

Phillips, C.G. & Porter, R. (1977). *Corticospinal Neurones: Their Role in Movement.* New York: Academic Press.

Porter, R. (1990). The Kugelberg Lecture. Brain mechanisms of voluntary motor commands – a review. *Electroencephalography and Clinical Neurophysiology*, **76**, 282–93.

Porter, R., Ghosh, S. & Fyffe, R.E.W. (1988). Axon-collaterals of pyramidal cells in laminae III and V of area 4 of monkey's cortex, revealed by intracellular HRP. *Society for Neuroscience,* **14**, 820.

Porter, R., Ghosh, S., Lange, G.D. & Smith, T.G. Jnr. (1991). A fractal analysis of pyramidal neurons in mammalian motor cortex. *Neuroscience Letters,* **130**, 112–6.

Roland, P.E., Skinhøj, E., Larsen, N.A. & Lassen, B. (1980). Different cortical areas in man in organization of voluntary movements in extra-personal space. *Journal of Neurophysiology*, **43**, 137–50.

Schell, G., Hodge, C.J. & Cacayorin, E. (1986). Transient neurological deficit after therapeutic embolization of the arteries supplying the medial wall of the hemisphere including the supplementary motor area. *Neurosurgery*, **18**, 353–6.

Shinoda, Y., Yokota, J. & Futami, T. (1981). Divergent projections of individual corticospinal axons to motoneurons of multiple muscles in the monkey. *Neuroscience Letters*, **23**, 7–12.

Smith, T.G. Jnr., Marks, W.B., Lange, G.D., Sheriff, W.H. & Neale, E.A. (1989). A fractal analysis of cell images. *Journal of Neuroscience Methods*, **27**, 173–80.

Takahashi, K., Kubota, K. & Uno, M. (1967). Recurrent facilitation in cat pyramidal tract cells. *Journal of Neurophysiology*, **30**, 22–34.

5
Assessment of corticofugal output: strength testing and transcranial stimulation of the motor cortex

S. C. GANDEVIA

Department of Clinical Neurophysiology,
The Prince Henry and Prince of Wales Hospitals
and
Prince of Wales Medical Research Institute, University of New South Wales,
Sydney NSW, Australia

Lesions which damage the motor cortex or its output result in the classical upper motoneurone syndrome: this is characterized by muscle weakness, usually with heightened tendon reflexes and abnormalities of cutaneous reflexes such as the Babinski response. Not all muscles are equally weak in a hemiplegic patient. Thus, upper facial muscles, masticatory muscles and those of the trunk are relatively spared (e.g. Hughlings Jackson, 1865; Bastian, 1886; Gowers, 1893; Willoughby & Anderson, 1984), presumably due to their bilateral cortical representation. Furthermore, in both upper and lower limbs the distal musculature appears more severely affected than proximal musculature; and the extensors in the upper limb and the flexors in the lower limb are believed to be more affected than their antagonists (e.g. Wernicke, 1889).

However, while the above picture has become ingrained in the clinical literature, there have been few attempts to confirm this distribution of weakness in patients with upper motoneurone lesions using quantitative measurements of muscle strength. This chapter describes some quantitative measurements in patients with upper motoneurone lesions and discusses some relevant results of transcranial motor cortical stimulation in human subjects. Both approaches are, however, indirect ways to determine the functional output of the upper motoneurones.

Distribution of strength in upper motoneurone lesions

The distal predominance in the distribution of weakness following upper motor neurone lesions is clinically evident in the frequent presentation of patients with minimal movement of the fingers but reasonable movement at more proximal joints such as the shoulder. Furthermore, in the recovery from stroke the distal musculature seems the last to recover both to clinical strength testing and to

Table 5.1 *Muscle actions affected in hemiparesis*

Upper limb	Lower limb
most affected:	most affected:
finger flexion	hallux plantarflexion
wrist flexion	hallux dorsiflexion
thumb extension	ankle plantarflexion
thumb flexion	ankle dorsiflexion
least affected:	least affected:
shoulder adduction	hip flexion
shoulder abduction	hip extension

Data from Colebatch et al. (1989) and Adams et al. (1990).

assessment of the functional use of the limb. During recovery, the flexors of the elbow are thought to recover earlier than the extensors, although, as argued below this may be fortuitous. We have measured the strength of muscle groups in the upper (Colebatch, Gandevia & Spira, 1986; Colebatch & Gandevia, 1989) and lower limbs (Adams, Gandevia & Skuse, 1990) in patients with unilateral lesions. Studies were conducted both in patients with hemiparesis in whom some strength was present in most muscle groups on the affected side, and in patients with hemiplegia in whom strength was absent in at least four, usually distal, muscle groups. Subjects contracted against isometric myographs fashioned for each muscle group with restraints provided to minimize the contribution of remote muscle groups to the recorded torques. Results in the patient groups were compared with those from control subjects to determine the extent of any strength deficit on the 'unaffected' side. Strength was defined as the peak force produced in any of three brief 'maximal' efforts; during these the subjects received verbal encouragement and had access to a visual signal of their performance. Strength was expressed as an isometric torque (rather than a force), and torques necessary to oppose gravity were taken into account (e.g. the weight of the leg for assessment of hip flexion). Twelve muscle groups were assessed in the upper and eight in the lower limb. Although it was not possible to mimic the positions used in all clinical examinations, the same testing postures applied to hemiparetic and normal subjects.

For hemiparetic subjects, several features of the results were common for both the upper and lower limb. First, the proximal muscles on the affected side were relatively stronger than distal muscles, when relative strength was expressed as the ratio of strength on the affected to the unaffected side. It is this side-to-side comparison which forms part of the clinical examination. Thus strength at the shoulder and hip was better preserved than strength at distal joints. Data for relative strength are given in Table 5.1.

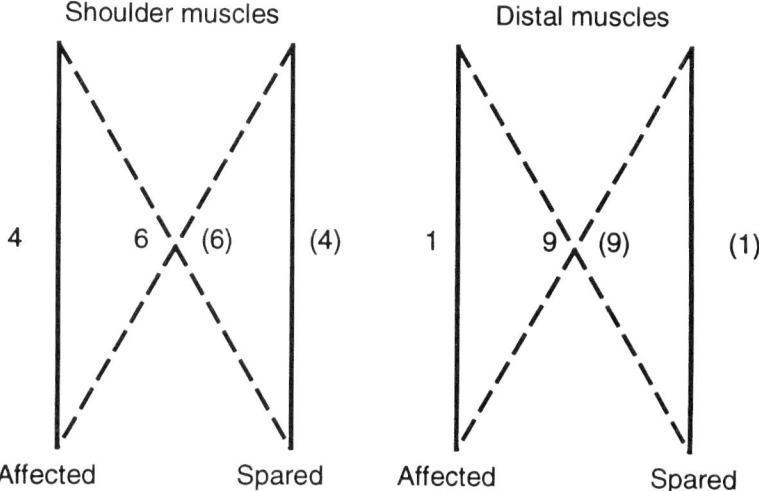

Fig. 5.1. Schematic representation, assuming the presence of bilateral projections, showing possible corticofugal output for shoulder adductors (left) and distal muscles (right). The strength of ipsilateral and contralateral projections is given by a number and the lesion affects the right hemisphere (dashed lines). Normal power is 10 for both muscle groups. On the weaker ('affected') side, distal muscles are more severely affected (1) and shoulder adductors relatively spared (4). By contrast, on the stronger ('spared') side, shoulder adduction is more impaired. This simplified scheme shows that the relative strength of the ipsilateral (bilateral) projection may determine the pattern of weakness that follows a unilateral cerebral lesion. Colebatch & Gandevia (1989) reproduced with permission.

Secondly, the distribution of weakness did not show a striking difference between the flexor and extensor muscle groups as might have been predicted. In the study of multiple muscle groups in the upper limb, flexors were on average more severely affected than extensors, but these differences were not statistically significant (Colebatch & Gandevia, 1989). One contributory factor is that the flexors have a greater cross-sectional area than their corresponding extensors and thus exert more torque at the elbow, wrist and metacarpophalangeal joints. This makes it more difficult for the examiner to overcome the flexor than the extensor torque. Strength in the elbow flexors and extensors has been separately studied and, in this study, flexor strength was slightly but significantly more affected than extensor strength (Colebatch et al., 1986). This suggestion for the elbow has been corroborated in some but not all of the studies by Bohannon and colleagues (e.g. Bohannon & Smith, 1987; Bohannon, 1987; Bohannon & Andrews, 1987). No study of strength has yet supported the obvious impression on clinical examination that hemiparetic elbow extensors are preferentially weakened.

Thirdly, variability between patients in the apparent distribution of strength was marked. However, patients who were weak in one muscle group tended to be weak in its antagonist (see also Bohannon, 1990). This presumably reflects the fact that corticofugal fibres destined to affect an agonist–antagonist pair are likely to be close together at cortical and subcortical sites. Furthermore, in patients with capsular infarcts, the likelihood of affecting the bulk of fibres for an entire limb is high. The descending fibres for trunk muscles are dispersed more widely, at least at the level of the cerebral peduncle (Warabi *et al.*, 1990).

Fourthly, while the assumption during the neurological examination is that the muscle groups on the 'unaffected' side have normal strength, this was not borne out when strength was measured quantitatively. Strength of muscles in the arm and leg on the 'ipsilateral' side was less than expected when compared with that in control subjects matched for age and sex. This ipsilateral deficit was relatively uniform in the leg, but affected the shoulder most and the distal muscles in the upper limb least. The deficit can presumably be explained by the pattern of 'ipsilateral' projections emanating from the motor cortex. Such projections have recently been studied electrophysiologically in monkeys (Tanji, Okano & Sato, 1988). Figure 5.1 shows diagrammatically how the variable distribution of corticospinal output can produce changes in absolute and relative strength on the ipsilateral and contralateral sides. Other ipsilateral motor deficits have been established following stroke (for review, see Jones, Donaldson & Perkin, 1989).

The distribution of weakness revealed in the studies above has implications for the routine rehabilitation of patients following stroke. In particular, no two patients are likely to have the same distribution of strength and hence of disability. Unless the strength is measured, the strength distribution in a patient with a more extreme pattern will go unrecognized. Strength is affected on the 'normal' side, particularly at the shoulder, so that it is likely that some benefit will ensue from direction of the rehabilitation effort to both sides.

Use of transcranial motor stimulation

Since the advent of the electrical (Merton & Morton, 1980) and electromagnetic methods (Barker, Jalinous & Freeston, 1985) to activate the human motor output, many studies have reported, not surprisingly, abnormalities in patients with upper motoneurone lesions. Responses in hemiplegic and hemiparetic muscles may be absent, delayed or normal (e.g. Macdonell, Donnan & Bladin, 1989; Dominkus, Grisold & Jelinek, 1990). To determine whether the corticofugal output to particular limb muscles is potentially functional, it is better to examine the projection when the patient is attempting to contract the weakened

Assessment of corticofugal output

muscle, and if it is paralysed on electromyographic examination, then an attempt must be made to activate the relevant motoneurone pools with attempted bilateral efforts or reinforcement manoeuvres. Failure to provide effective 'facilitation' may mean that an evoked descending corticofugal volley remains subthreshold for the motoneurones and hence for a motor response. A review of the appropriate methods to determine the 'central' motor conduction time is beyond the current scope, but it is notable that, while the abnormalities in patients with lesions affecting upper motoneurones such as stroke and multiple sclerosis may be obvious (on clinical or electrical testing, or both), the changes in central conduction in patients with motoneurone disease may be much more subtle. Thus repeated testing, use of electrical as well as magnetic transcranial stimulation, and accurate measurement of peripheral conduction delays to a number of muscles can be crucial in establishing of a definite abnormality (e.g. Schriefer *et al.*, 1989; Eisen *et al.*, 1990).

Studies using transcranial stimulation have provided some insight into the distribution of strength following unilateral upper motoneurone lesions. Based on the size of electromyographic potentials in muscles such as first dorsal interosseous and more proximal muscles such as elbow flexors, it appears that the initial contralateral excitation produced by cortical stimulation is larger for distal musculature (Rothwell *et al.*, 1987). The conduction velocity of the descending presumed corticospinal excitation is approximately 65 m/s to proximal, distal and truncal musculature (e.g. Boyd *et al.*, 1986; Gandevia & Plassman, 1988). Furthermore, transcranial stimulation increases the probability of discharge of single motor units, and can give a measure of the time course and amplitude of the excitatory postsynaptic potentials due to the descending motor cortical volleys (e.g. Day *et al.*, 1987, 1989; Plassman & Gandevia, 1989; Colebatch *et al.*, 1990). Comparable changes in the probability of firing of single motor units have been observed in monkeys performing precise tasks with finger muscles in which the discharge of motor units is plotted in relation to a discharge in a single motor cortical cell (Fetz & Cheney, 1980; Lemon, Mantel & Muir, 1986).

Given that changes in muscle strength on the ipsilateral side have been observed in stroke, it is of interest to know whether the corticofugal volleys evoked by transcranial stimulation produce bilateral electromyographic responses. In a recent study, the discharge of motor units in shoulder muscles was examined following electrical and magnetic transcranial stimulation. This revealed powerful excitation of contralateral deltoid (similar to that of intrinsic hand muscles), with smaller excitation of the antagonist pectoralis (Colebatch *et al.*, 1990; see also Gandevia & Rothwell, 1987; *cf.* Palmer & Ashby, 1992). In addition, after the initial short-latency excitation, a 'middle' latency excitation occurred on both the contralateral and ipsilateral side. This indicates that the cortical output evoked by transcranial stimulation has access to bilateral descending pathways, possibly corticoreticulo-spinal pathways. Bilateral responses to motor cortical stimulation have also been produced in facial and some neck muscles (Benecke *et al.*, 1988; Gandevia & Applegate, 1988).

While the cortical output to truncal muscles has usually been considered to lack a powerful monosynaptic component (e.g. Kuypers, 1987), indirect evidence against this view is accumulating. Transcranial electromagnetic stimulation produces short-latency contralateral responses in truncal muscles, including the diaphragm (Gandevia & Rothwell, 1987; Gandevia & Plassman, 1988; Plassman & Gandevia, 1989; Murphy *et al.*, 1990). This output is rapidly conducting and appears able to access the motoneurone pool as readily as the cortical output to intrinsic muscles of the hand. While transcranial stimulation does not reveal whether this output is used to produce a specific movement, other data support the likely role of the motor cortex in control of truncal and respiratory muscles. First, an appropriately located premovement potential is observed following voluntary inspiratory tasks, but not during unstimulated breathing (Macefield & Gandevia, 1991). Secondly, voluntary activation of the phrenic motoneurone pool can be as complete in a voluntary task as that of limb muscles (for discussion, see Gandevia, McKenzie & Plassman, 1990). Thirdly, positron emission tomography reveals increased activity in the appropriate motor cortical region when voluntary targeted breathing is performed (Colebatch *et al.*, 1991).

There are some uncertainties about which cortical elements are excited by the different forms of electromagnetic transcranial stimulation and the exact site of excitation. Based on the slightly longer latency to magnetic stimulation, the effects on the discharge of single motoneurones and theoretical considerations, it seems that the predominant excitation of corticofugal elements is via trans-synaptic activation, whereas the activation produced by electrical stimulation also involves direct activation of corticofugal axons (for discussion see Amassian *et al.*, 1987; Rothwell *et al.*, 1991). Certainly, these assumptions have allowed the convenient interpretation of much data on the responses to transcranial motor cortical stimulation in patients with real or suspected upper motoneurone abnormalities (e.g. Thompson *et al.*, 1987). Transcranial motor cortical stimulation has already found a significant place in the clinical armamentarium for the investigation of putative upper motoneurone lesions, possible multiple sclerosis and suspected functional weakness, and for monitoring spinal cord function during neurosurgical procedures. Furthermore, the technique may lend itself to assessment of levels of 'excitability' of the cortex and spinal cord. Already, a generalized disturbance of motor cortical excitability has been detected in patients with intractable temporal lobe epilepsy (Hufnagel *et al.*, 1990).

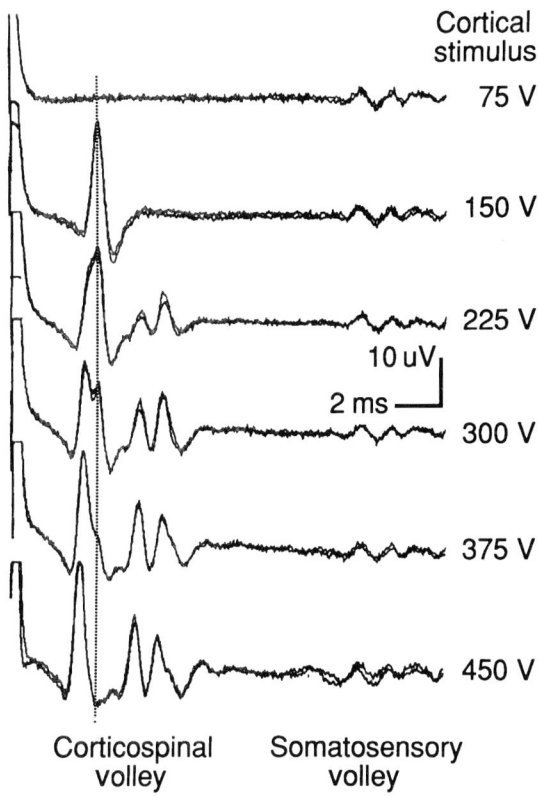

Fig. 5.2. The effect of increasing stimulus intensity (time constant 50 ms) on the corticospinal volley recorded at the high-thoracic site. Electrode configuration: anode at vertex, cathode 7 cm lateral to vertex. Duplicate averages of 25 sweeps are superimposed for each stimulus intensity. The low-amplitude polyphasic deflection to the right of each trace represents the ascending somatosensory volley set up by bilateral stimulation of the tibial nerves in the popliteal fossae. The cortical and somatosensory stimuli were delivered simultaneously, at the onset of the sweeps. The dotted vertical line indicates the D wave. Note that this peak is completely replaced at 450 V by an earlier peak that began to appear at 225 V. Burke et al. (1990) reproduced with permission.

Problems with minimal latencies

At least with the electrical stimulator, some caution is warranted in interpreting *minimal* latencies to cortical stimulation because there is evidence from epidural recordings from the human thoracic spinal cord that the site of activation which produces the initial descending volley (the direct or 'D' wave) is not simply the axon hillock of the corticospinal cells (Burke, Hicks & Stephen, 1990). The latency of this volley can diminish by 0.8 or 1.7 ms with only modest increases in stimulus intensity, thus suggesting that the site of activation of the D-wave moves to the internal capsule, cerebral peduncle or even lower. An

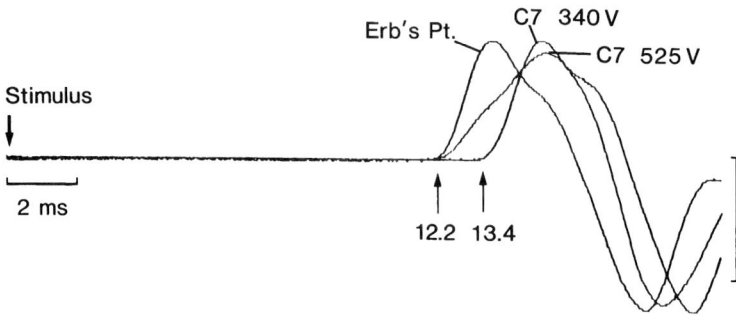

Fig. 5.3. Single trial EMG responses recorded in the thenar muscles following cathodal stimulation over the C7 vertebral process (anode: 6 cm cephalad) at two stimulus intensities (340 and 525 V) and stimulation at Erb's point. In all subjects, as the stimulus intensity was increased sufficiently, the latency to vertebral stimulation decreased. In the case illustrated, stimulation over C7 at a supramaximal level activated a population of rapidly conducting motor axons near Erb's point. Response latencies are indicated below the responses. Vertical calibration: 2 mV. Plassman & Gandevia (1989) reproduced with permission.

example of this is shown in Fig. 5.2. A similar change in activation site is evident in records published by Inghilleri *et al*. (1989). Use of high-voltage stimulation at the cortex may spuriously produce values for the latency of the EMG responses (and central conduction time) which are within the normal range. Supportive evidence for a shift in the site of activation with increased intensity comes from recordings of pyramidal tract cells in the monkey (Edgley *et al*., 1990). This evidence, plus the observed production of D-waves with the magnetic stimulator in monkeys provides an alternative explanation for the latency difference observed in electromyographic responses to magnetic and electrical stimulation. Magnetic stimulation may activate the corticofugal output directly with the electrical stimulation activating the axons subcortically.

When the spinal cord is stimulated electrically to measure the peripheral latency a related complicating factor can arise in assessment of central motor conduction time: distal shifts in the site of activation along the motor axons occur with increased stimulus intensity (Fig. 5.3; Plassman & Gandevia, 1989; see also Schmid *et al*., 1990). If such phenomena occur at both sites of stimulation, i.e. motor cortex and spinal cord, a fortuitous cancellation of errors may result in a 'normal' value for the 'central' conduction time. Pathology at a cortical level may be missed altogether, while proximal nerve root lesions may give rise to an apparent 'central' delay in the motor pathway.

However, despite these qualifying remarks, these techniques of noninvasive stimulation have the potential to reveal other dramatic changes in the behaviour of motor output such as the reorganization within human motor pathways.

Thus, ipsilateral responses to cortical stimulation have been observed in patients with long standing hemispherectomy (Benecke, Meyer & Freund, 1991). The motor cortical map is apparently distorted in congenital amputation but not when amputation is acquired in recent trauma (Hall *et al.*, 1990). In addition, cerebral palsy changes the pattern of innervation revealed by motor cortical stimulation such that ankle plantar flexors receive a more prominent facilitation (Brouwer & Ashby, 1991).

Voluntary drive to muscles assessed in maximal voluntary efforts and 'fatigue'

Excessive fatigue and a feeling of heaviness in the limb are not only common symptoms in patients with overt neuromuscular disease, but occur in patients with upper motoneurone lesions. Indeed, Samuel Johnson comments on the apparent heaviness of his watering pots after a 'paralytick stroke'. One mechanism for this is that the patient is reporting the excessive motor commands or sensations of innervation, rather than the actual (reduced) muscle force (Gandevia & McCloskey, 1977). Thus despite increased effort directed through damaged corticofugal and other motor pathways, insufficient excitation occurs at the level of the motoneurone pool and a weaker contraction than expected results.

The extent to which the descending corticofugal output can recruit and increase the frequency of firing of spinal motoneurones in a voluntary effort has long been controversial. Is there considerable reserve in the excitation reaching motoneurone pools in an attempted maximal voluntary contraction such that motoneurones are driven at supratetanic frequencies? Such an arrangement would allow the production of the truly maximal force output by a muscle, but at the expense of the ability to modulate force accurately by changing the firing frequency of motor units. Alternatively, does the central nervous system only just achieve optimal driving of motor units to produce 'maximal' force in an attempted maximal effort? Introspection suggests the latter, at least for the subelite athlete, and some experimental evidence supports this view.

The technique of twitch interpolation was introduced by Merton (1954) as a way to determine whether muscle fatigue occurred peripherally, i.e. within the muscle, or in the central nervous system during maximal isometric contractions. He found that the force produced by adductor pollicis in the strongest voluntary efforts could not be incremented by a supramaximal stimulus to the ulnar nerve innervating the muscle and concluded that all motor units in adductor pollicis were recruited in the voluntary effort and that their discharge frequencies were high enough for fusion of their twitches. Because this still

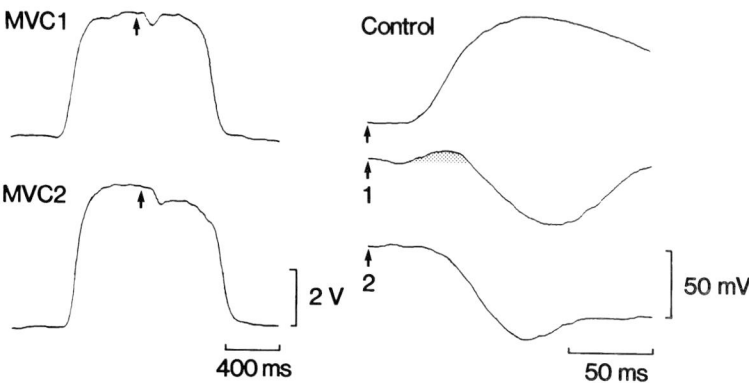

Fig. 5.4. MVC1 and MVC2 depict the forces generated in two attempted maximal voluntary contractions. The right panel shows the responses to supramaximal nerve stimulation for a control twitch obtained at rest (upper trace), during MVC1 (middle trace) and MVC2 (lower trace). Arrows mark the time of stimulation. Shaded region along MVC1 shows a small increase in force at the appropriate latency after the stimulation for a superimposed twitch. This indicates that the stimulated muscle may not have been maximally activated at the time of stimulus delivery. MVC2 shows no increase in force at the appropriate latency. Hales & Gandevia (1988) reproduced with permission.

occurred when the muscular force was diminished by fatigue, it seemed that voluntary drive was always sufficient to fully activate the motoneurone pool. However, more recent studies using sensitive versions of the test indicate that normal subjects, even when provided with verbal and visual feedback of their performance, are capable of achieving the full output from the muscle in *some* efforts, but that usually force is about 5–10% below maximum for elbow flexor muscles (e.g. Lloyd, Gandevia & Hales, 1991). It is even further below maximum for the diaphragm (e.g. McKenzie *et al.*, 1992) and, at the ankle, it is more difficult to drive the ankle plantarflexor maximally than dorsiflexor muscles (Belanger & McComas, 1981). An example of two consecutive attempted maximal efforts is shown in Fig. 5.4: only in the first was the contracting muscle being maximally driven when the stimulus was delivered to the elbow flexors.

Another relevant observation is that the discharge frequencies of motor units during brief sustained maximal efforts are not well above those predicted to produce twitch fusion (e.g. Bigland-Ritchie *et al.*, 1983; Gandevia *et al.*, 1990). Furthermore, during submaximal or maximal exercise which produces peripheral fatigue, there is a concomitant decline in voluntary activation in limb muscles (e.g. Thomas *et al.*, 1990; Lloyd *et al.*, 1991; McKenzie & Gandevia, 1991). Thus, some of the force reduction during exercise reflects a diminution in the ability of the central nervous system to recruit and drive motoneurones.

These observations have several implications for clinical practice. First, it must be appreciated that a muscle may not be driven optimally by 'maximal' voluntary effort, particularly when the muscles or joints are painful (e.g. Stokes & Young, 1984; Rutherford *et al.*, 1986; Gandevia & McKenzie, 1988). Furthermore, the degree of voluntary drive (and force produced) can be quantified if required. Secondly, not all processes relevant to muscle fatigue (measured as a reduction in maximal voluntary force) occur in the muscle. Perhaps mechanisms exist to diminish the effective central drive to motoneurones in parallel with peripheral reductions in muscular force. Such mechanisms would be efficient and may exert a 'protective' effect on diseased or damaged muscles and joints.

Acknowledgements

This work was supported by the National Health and Medical Research Council of Australia. The author is most grateful to the collaborators whose work is cited here.

References

Adams, R. W., Gandevia, S. C. & Skuse, N.F. (1990). The distribution of muscle weakness in upper motor lesions affecting the lower limb. *Brain*, **113**, 1459–76.

Amassian, V. E., Stewart, M., Quirk, G. J. & Rosenthal. J. L. (1987). Physiological basis of motor effects of a transient stimulus to cerebral cortex. *Neurosurgery*, **20**, 74–93.

Bard, G. & Hirschberg, G. G. (1965). Recovery of voluntary motion in upper extremity following hemiplegia. *Archives of Physical Medicine & Rehabilitation*, **45**, 567–72.

Barker, A. T., Jalinous, R. & Freeston, I. L. (1985). Non-invasive stimulation of human motor cortex. *Lancet*, **i**, 1106–7.

Bastian, H. C. (1886). *Paralyses: Cerebral, Bulbar and Spinal.* London: K. H. Lewis.

Belanger, A. Y. & McComas, A. J. (1981). Extent of motor unit actvation during effort. *Journal of Applied Physiology*, **51**, 1131–5.

Benecke, R., Meyer, B.-U., Schonle, P. & Conrad, B. (1988). Transcranial magnetic stimulation of the human brain – responses in muscles supplied by cranial nerves. *Experimental Brain Research*, **71**, 623–32.

Benecke, R., Meyer, B.-U. & Freund, H.-J. (1991). Reorganisation of descending motor pathways in patients after hemispherectomy and severe hemispheric lesions demonstrated by magnetic brain stimulation. *Experimental Brain Research*, **83**, 419–26.

Bigland-Ritchie, B., Johansson, R., Lippold, O. C. J., Smith, S. & Woods, J. J. (1983). Changes in motoneurone firing rates during sustained maximal voluntary contractions. *Journal of Physiology (London)*, **340**, 335–46.

Bohannon, R. W. (1987). Relationship between static strength and various other measures in hemiparetic stroke patients. *International Rehabilitation Medicine*, **8**, 125–8.

Bohannon, R. W. (1990). Significant relationships exist between muscle group strengths following stroke. *Clinical Rehabilitation*, **4**, 27–31.

Bohannon, R. W. & Andrews, A. W. (1987). Relative strength of seven upper extremity muscle groups in hemiparetic stroke patients. *Journal of Neurological Rehabilitation*, **1**, 161–5.

Bohannon, R. W. & Smith, M. B. (1987). Assessment of strength deficits in eight paretic upper extremity muscle groups of stroke patients with hemiplegia. *Physical Therapy*, **67**, 522–5.

Boyd, S. G., Rothwell, J. C., Cowan, J. M. A., Webb, P. J., Morley, T., Asselman, P. & Marsden, C. D. (1986). A method of monitoring function in corticospinal pathways during scoliosis surgery with a note on motor conduction velocities. *Journal of Neurology, Neurosurgery and Psychiatry*, **49**, 251–7.

Brouwer, B. & Ashby, P. (1991). Altered corticospinal projections to lower limb motoneurons in subjects with cerebral palsy. *Brain*, **114**, 1395–407.

Burke, D., Hicks, R. G. & Stephen, J. P. H. (1990). Corticospinal volleys evoked by anodal and cathodal stimulation of the human motor cortex. *Journal of Physiology (London)*, **425**, 283–99.

Colebatch, J. G. & Gandevia, S. C. (1989). The distribution of muscular weakness in upper motor neuron lesions affecting the arm. *Brain*, **112**, 749–63.

Colebatch, J. G., Gandevia, S. C. & Spira, P. J. (1986). Voluntary muscle strength in hemiparesis: distribution of weakness at the elbow. *Journal of Neurology, Neurosurgery and Psychiatry*, **49**, 1019–24.

Colebatch, J. G., Adams, L., Murphy, K., Martin, A. J., Lammertsma, A. A., Tochon-Danguy, H. J., Clark, J. C., Friston, K. J. & Guz, A. (1991). Regional cerebral blood flow during volitional breathing in man. *Journal of Physiology (London)*, **443**, 91–103.

Colebatch, J. G., Rothwell, J. C., Day, B. L., Thompson, P. D. & Marsden, C. D. (1990). Cortical outflow to proximal arm muscles in man. *Brain*, **113**, 1843–56.

Day, B. L., Dressler, D., Maertens de Noordhout, A., Marsden, C. D., Nakashima, K., Rothwell, J. C. & Thompson, P. D. (1989). Electric and magnetic stimulation of human motor cortex: surface EMG and single motor unit responses. *Journal of Physiology (London)*, **412**, 449–73.

Day, B. L., Rothwell, J. C., Thompson, P. D., Dick, J. P. R., Cowan, J. M. A., Berardelli, A. & Marsden, C. D. (1987). Motor cortex stimulation in intact man. 2. Multiple descending volleys. *Brain*, **110**, 1191–209.

Dominkus, M., Grisold, W. & Jelinek, V. (1990). Transcranial electrical motor evoked potentials as a prognostic indicator for motor recovery in stroke patients. *Journal of Neurology, Neurosurgery and Psychiatry*, **53**, 745–8.

Edgley, S. A., Eyre, J. A., Lemon, R. N. & Miller, S. (1990). Excitation of the corticospinal tract by electromagnetic and electric stimulation of the scalp in the macaque monkey. *Journal of Physiology (London)*, **425**, 301–20.

Eisen, A. A. & Shtybel, W. (1990). Clinical experience with transcranial magnetic stimulation - AAEM minimonograph 35. *Muscle & Nerve*, **13**, 995–1011.

Eisen, A. A., Shytbel, W., Murphy, K. & Hoirch, M. (1990). Cortical magnetic stimulation in amyotrophic lateral sclerosis. *Muscle & Nerve*, **13**, 146–51.

Fetz, E. E. & Cheney, P. D. (1980). Postspike facilitation of forelimb muscle activity by prmate corticomotoneuronal cells. *Journal of Neurophysiology*, **44**, 751–72.

Gandevia, S. C. (1989). Human motor cortical output studied with transcranial stimulation. *Current Opinion in Neurology and Neurosurgery*, **2**, 782–7.

Gandevia, S. C. & Applegate, C. (1988). Activation of neck muscles from the human motor cortex. *Brain*, **111**, 801–13.

Gandevia, S. C., Macefield, G., Burke, D. & McKenzie, D. K. (1990). Voluntary activation of human motor axons in the absence of muscle afferent feedback: the control of the deafferented hand. *Brain*, **113**, 1563–81.

Gandevia, S. C. & McCloskey, D. I. (1977). Sensations of heaviness. *Brain*, **100**, 345–54.
Gandevia, S. C. & McKenzie, D. K. (1988). Activation of human muscles at short muscle lengths during maximal static efforts. *Journal of Physiology (London)*, **407**, 599–613.
Gandevia, S. C., McKenzie, D. K. & Plassman, B. L. (1990). Activation of human respiratory muscles during different voluntary manoeuvres. *Journal of Physiology (London)*, **428**, 387–403.
Gandevia, S. C. & Plassman, B. L. (1988). Responses in human intercostal and truncal muscles to motor cortical and spinal stimulation. *Respiration Physiology*, **73**, 325–38.
Gandevia, S. C. & Rothwell, J. C. (1987). Activation of the human diaphragm from the motor cortex. *Journal of Physiology (London)*, **384**, 109–18.
Gowers, W. R. (1893). *A Manual of Diseases of the Nervous System.* Second edition. London: J. and A. Churchill. Reprinted 1970. Darien, C. N.: Hafner.
Hales, J. P. & Gandevia, S. C. (1988). Assessment of maximal voluntary contraction with twitch interpolation: an instrument to measure twitch responses. *Journal of Neuroscience Methods*, **25**, 97–102.
Hall, E. J., Flament, D., Fraser, C. & Lemon, R. N. (1990). Non-invasive brain stimulation reveals reorganised cortical outputs in amputees. *Neuroscience Letters*, **116**, 379–86.
Hufnagel, A., Elger, C. E., Marx, W. & Ising, A. (1990). Magnetic motor-evoked potentials in epilepsy: effects of the disease and of anticonvulsant medication. *Annals of Neurology*, **28**, 680–6.
Hughlings Jackson, H. J. (1865). Lectures on hemiplegia. *Clinical Lectures and Reports by the Medical and Surgical Staff of the London Hospital*, **2**, 297–332.
Inghilleri, M., Berardelli, A., Cruccu, G., Priori, A. & Manfredi, M. (1989). Corticospinal potentials after transcranial stimulation in humans. *Journal of Neurology, Neurosurgery, and Psychiatry*, **52**, 970–4.
Jones, R. D., Donaldson, I. M. & Perkin, P. J. (1989). Impairment and recovery of ipsilateral sensory-motor function following unilateral cerebral infarction. *Brain*, **112**, 113–32.
Kuypers, H. G. J. M. (1987). Some aspects of the organization of the output of the motor cortex. In *Motor Areas of the Cerebral Cortex*. Ciba Foundation Symposium 132, ed. G. Bock, M. O'Connor & J. Marsh, pp. 63–82. Chichester: John Wiley.
Lemon, R. N., Mantel, G. W. H. & Muir, R. B. (1986). Corticospinal facilitation of hand muscles during voluntary movement in the conscious monkey. *Journal of Physiology (London)*, **381**, 497–527.
Lloyd, A. R., Gandevia, S. C. & Hales, J. P. (1991). Muscle performance, voluntary activation, twitch properties and perceived effort in normal subjects and patients with chronic fatigue syndrome. *Brain*, **114**, 85–98.
McKenzie, D. K., Bigland-Ritchie, B., Gorman, R. B. & Gandevia, S. C. (1992). Central and peripheral fatigue of human diaphragm and limb muscles assessed by twitch interpolation. *Journal of Physiology (London)*, **454**, 643–56.
McKenzie, D. K. & Gandevia, S. C. (1991). Recovery from fatigue of human diaphragm and limb muscles. *Respiration Physiology*, **84**, 49–60.
Macdonell, R. A. L., Donnan, G. A. & Bladin, P. F. (1989). A comparison of somatosensory evoked and motor evoked potentials in stroke. *Annals of Neurology*, **25**, 68–73.
Macefield, G. & Gandevia, S. C. (1991). The cortical drive to human respiratory muscles in the awake state assessed by premotor cerebral potentials. *Journal of Physiology (London)*, **439**, 545–8.

Merton, P. A. (1954). Voluntary strength and fatigue. *Journal of Physiology (London)*, **123**, 553–64.

Merton, P. A. & Morton, H. B. (1980). Stimulation of the intact cerebral cortex in the intact human subject. *Nature*, **285**, 227.

Murphy, K., Mier, A., Adams, L. & Guz, A. (1990). Putative cerebral cortical role in the ventilatory response to inhales CO_2 in conscious man. *Journal of Physiology (London)*, **420**, 1–18.

Palmer, E. & Ashby, P. (1992). Corticospinal projections to upper limb motoneurones in humans. *Journal of Physiology (London)*, **448**, 397–412.

Plassman, B. L. & Gandevia, S. C. (1989). Comparison of human motor cortical projections to abdominal muscles and intrinsic muscles of the hand. *Experimental Brain Research*, **78**, 301–8.

Rothwell, J. C., Thompson, P. D., Day, B. L., Boyd, S. & Marsden, C. D. (1991). Stimulation of the human motor cortex through the scalp. *Experimental Physiology*, **72**, 159–95.

Rothwell, J. C., Thompson, P. D., Day, B. L., Dick, J. P. R., Kachi, T., Cowan, J. M. A. & Marsden, C. D. (1987). Motor cortex stimulation in intact man. 1. General characteristics of EMG responses in different muscles. *Brain*, **110**, 1173–90.

Rutherford, O. M, Jones, D. A. & Newham, D. J. (1986). Clinical and experimental application of the percutaneous twitch superimposition technique for the study of human muscle activation. *Journal of Neurology, Neurosurgery, and Psychiatry*, **49**, 1288–94.

Schmid, U. D., Walker, G., Hess, C. W. & Schmid, J. (1990). Magnetic and electrical stimulation of cervical motor roots: technique, site and mechanisms of excitation. *Journal of Neurology, Neurosurgery, and Psychiatry*, **53**, 770–7.

Schriefer, T. N., Hess, C. W., Mills, K. R. & Murray, N. M. F. (1989). Central motor conduction studies in motor neurone disease using magnetic brain stimulation. *Electroencephalogaphy and Clinical Neurophysiology*, **74**, 431–7.

Stokes, M. & Young, A. (1984). The contribution of reflex inhibition to arthrogenous muscle weakness. *Clinical Science*, **67**, 7–14.

Tanji, J., Okano, K. & Sato, K. C. (1988). Neuronal activity in cortical motor areas related to ipsilateral, contralateral, and bilateral digit movements of the monkey. *Journal of Neurophysiology*, **60**, 325–43.

Thomas, C. K. Woods, J. J. & Bigland-Ritchie, B. (1989). Impulse propagation and muscle activation in long maximal voluntary contractions. *Journal of Applied Physiology*, **67**, 1835–42.

Thompson, P. D., Dick, J. P. R., Asselman, P., Griffin, G. B., Day, B. L., Rothwell, J. C., Sheehy, M. P. & Marsden, C. D. (1987). Examination of motor function in lesions of the spinal cord by stimulation of the motor cortex. *Annals of Neurology*, **21**, 389–96.

Warabi, T., Inoue, K., Noda, H. & Murakami, S. (1990). Recovery of voluntary movement in hemiplegic patients. *Brain*, **113**, 177–89.

Wernicke, C. (1889). Zur Kenntniss der cerebralen Hemiplegie. *Berliner Klinische Wochenschrift,* **26**, 969–70.

Willoughby, E. W. & Anderson, N. E. (1984). Lower cranial nerve motor function in unilateral vascular lesions of the cerebral hemisphere. *British Medical Journal*, **289**, 791–4.

6
Muscle spindles, muscle tone and the fusimotor system

D. BURKE AND S. C. GANDEVIA

Prince of Wales Medical Research Institute,
and University of New South Wales,
Sydney NSW, Australia

In their response to external stimuli, muscle spindles are no more privileged than other mechanoreceptors. However, the ability of the nervous system to accelerate the discharge of a muscle spindle ending by activating fusimotor neurones innervating the spindle sets this receptor apart. This 'internal' stimulus has intrigued physiologists and clinicians alike. Fusimotor innervation has been considered central to normal muscle tone, and disturbed fusimotor control of spindle endings has been postulated to underlie most if not all disorders of muscle tone, such as cerebellar hypotonia, spasticity, rigidity and stiff-man syndrome (Rushworth, 1960; Gordon, Januszko & Kaufman, 1967; Gilman, 1969; Dietrichson, 1971).

This chapter examines some of the evidence on the role of the fusimotor system in the control of movement, stretch reflexes and muscle tone in human subjects. Are muscle spindles and the fusimotor system essential for the initiation and maintenance of movement? Is intact fusimotor innervation of spindle endings necessary for normal tendon jerks? Is the tendon jerk an exclusively monosynaptic reflex, dependent on an afferent volley that comes only from primary spindle endings in the percussed muscle? Is normal muscle tone dependent on intact fusimotor innervation? Are abnormalities of muscle tone, hypotonia and hypertonia, due to disturbed fusimotor activity?

Initiation of movement

Voluntary contractions are initiated by activation of α-motoneurones by corticospinal and other descending volleys (see Porter, this volume; Gandevia, this volume), not indirectly through the fusimotor system, as postulated by Merton (1953) in his original formulation of the servo theory. In all voluntary contractions studied using microneurography in human subjects (fast or slow, isometric or isotonic, practised or unlearnt, self-paced or cued), electromyographic

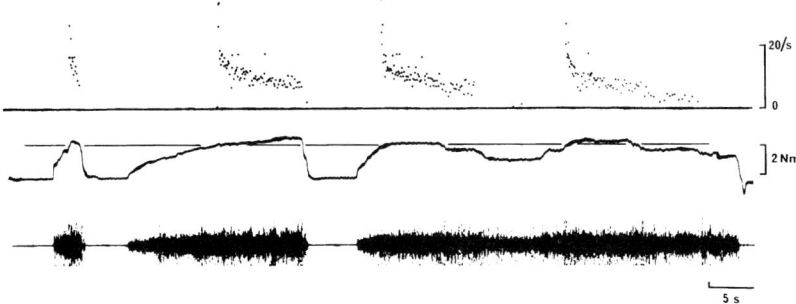

Fig. 6.1. Recording from a spindle ending in tibialis anterior during isometric voluntary contractions with different speeds of onset. Traces are: instantaneous frequency of discharge of the ending, contraction force (expressed as torque at the ankle), and EMG of tibialis anterior. Note that the afferent was activated after the onset of the contraction, in this instance when the force of contraction exceeded 2.5 Nm. Note also that the discharge frequency declines steadily when the contraction is maintained. Burke et al. (1978) reproduced with permission.

activity (EMG) appears in the contracting muscle before the discharge of muscle spindle endings accelerates (e.g., Vallbo, 1971; Vallbo et al., 1979; Burke, 1981; Vallbo & Al-Falahe, 1990; see Fig. 6.1). Furthermore, subjects acutely deprived of feedback from the target muscle by a local anesthetic nerve block can activate motoneurones innervating the paralysed muscle and can grade the activation (Gandevia et al., 1990).

Theoretically, an intense synchronized discharge in fusimotor neurones could produce a sufficiently intense discharge from spindle endings that a reflex contraction would result. However, whether the normal nervous system ever operates in this manner is debatable. For example, a volley in fusimotor (and skeletofusimotor) neurones can be produced by transcranial stimulation of the human motor cortex, but the effects on muscle spindle discharge are quite subtle (Rothwell, Gandevia & Burke, 1990). In patients with cortical myoclonus, the latencies of the jerk after the causative cerebral potential are relatively short, 20–25 ms for the upper limb; 30–35 ms for the lower limb (see Marsden, Obeso & Rothwell, 1983). This suggests that the descending corticofugal volley initiates the involuntary contraction by a direct action on the α-motoneurones, rather than indirectly by an action on fusimotor neurones (for a full discussion of latencies, see Rothwell et al., 1990).

Maintenance of contraction

The fusimotor innervation of a muscle is activated whenever the muscle contracts voluntarily, and this can result in increased afferent feedback from spindle endings in the muscle (Vallbo et al., 1979; Burke, 1981). The intensity of the feedback varies directly with the strength of contraction and, if the contracting muscle is allowed to shorten, inversely with the speed of shortening (see also Al-Falahe, Nagaoka & Vallbo, 1990). Through spinal reflex pathways, the heightened spindle discharge will result in supportive excitation of the active motoneurone pool, but this feedback is not essential for subjects to be able to maintain the contraction.

When deprived of muscle spindle feedback by an acute nerve block, subjects can keep motoneurones firing with no greater variability than in the intact situation (Gandevia et al., 1990; Macefield et al., 1991b; cf. Fukushima et al., 1976). However, in maximal voluntary efforts, the firing rates of motoneurones only reach about two-thirds of the rates achievable in the intact state. Afferent feedback seems critical to the generation of normal maximal firing rates, and it is likely that there is a similar contribution to motoneurone discharge in submaximal efforts. If this is so, the reflex mechanism makes a substantial contribution to contraction strength, even if contraction is possible in its absence (see Hagbarth et al., 1986).

The spindle support available to the contracting motor pool does not remain constant throughout a prolonged contraction, whether the contraction is strong or weak (Bongiovanni & Hagbarth, 1990; Macefield et al., 1991a). Spindle discharge may reach its maximum and begin to decline even before peak force is reached, and in sustained contractions the discharge rates decrease to, on average, 60–70% of the peak rate by 30 s (Figs. 6.1 and 6.2(a),(b)). Such prolonged submaximal contractions produce fatigue, as evidenced by increasing EMG (Fig. 6.2(d)) to maintain a constant force (Fig. 6.2(c)). Thus, during fatigue, the greater voluntary effort required to maintain a desired force level is not translated into a greater (or even maintained) spindle discharge. This finding is unexpected given the servo theory (Merton, 1953) and its subsequent modifications (Matthews, 1972): muscle fatigue should result in enhanced fusimotor drive and greater spindle feedback, and this should drive the motoneurone pool harder to compensate for the fatigue.

The tendon jerk

One of few objective signs in clinical neurology is the tendon jerk. It is usually thought that abrupt percussion on the appropriate tendon (i) directly stretches the relevant muscle, (ii) activates primary spindle endings in that muscle, and (iii) produces a synchronized volley in group Ia afferents from the muscle, and that (iv) this afferent input produces a tendon jerk through a monosynaptic pathway. It was also once taught that, (v) in the absence of background fusimotor drive,

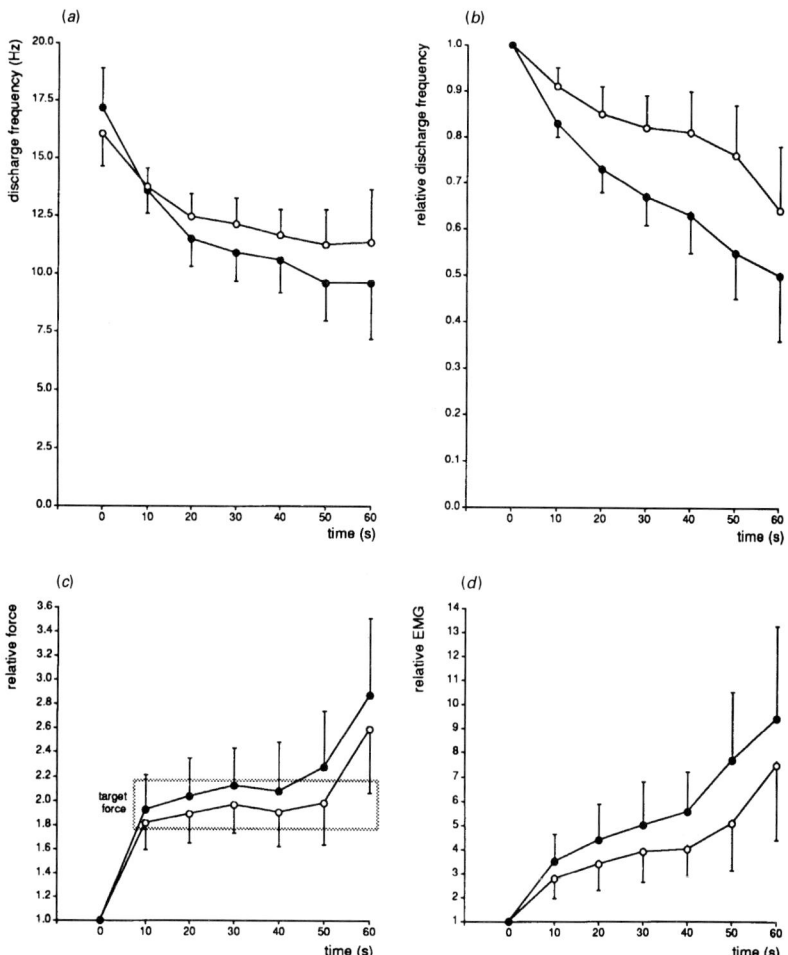

Fig. 6.2. Decline in muscle spindle activity during sustained voluntary contractions (pooled data, mean ±SEM). (a), discharge frequency (in Hz) of muscle spindle endings during sustained contractions in which the subjects attempted to maintain force constant. (b), relative discharge frequency for data in panel (a). (c), force during the contractions. (d), amplitude of integrated EMG during the contractions. Open circles, data from 16 spindle endings; and filled circles, data from 11 of the 16 endings with which there was a progressive decline in discharge frequency. In all panels time 0 is the time at which the peak spindle discharge was reached. The outline in panel (c) indicates the time over which the target force was maintained. During this time EMG increased progressively, but spindle discharge rate fell. Macefield et al. (1991) reproduced with permission.

spindle endings would be insufficiently sensitive for tendon percussion to evoke a reflex response. None of these views can be accepted without qualification.

(i) Percussion-induced direct muscle stretch is not necessary to evoke a tendon jerk (Lance & de Gail, 1965). Indeed, when percussion is delivered so that

Muscle spindles, muscle tone, fusimotor system 93

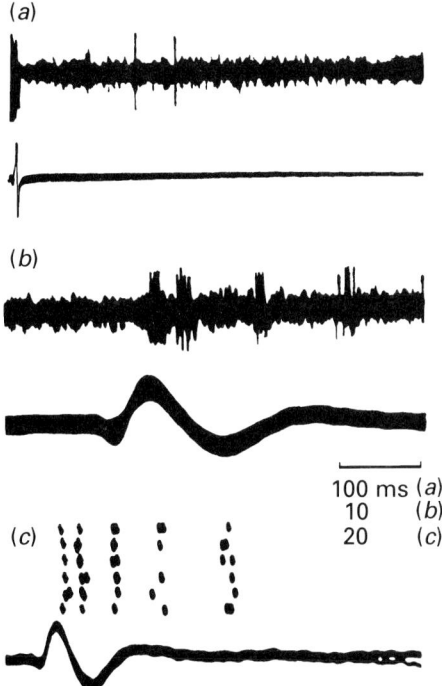

Fig. 6.3. The response of a primary spindle ending in medial gastrocnemius to weak tendon percussion, insufficient to produce a tendon jerk. (*a*), electrically induced twitch test, the afferent discharging twice at 170 ms and 220 ms, latencies appropriate to the falling phase of twitch force (not illustrated). Upper trace: raw neural activity; lower trace: EMG of triceps surae. (*b*), the response to tendon percussion subthreshold for Achilles tendon jerk. Upper trace: raw neural activity; lower trace: output of accelerometer on Achilles tendon. Multiple sweeps have been superimposed. The fifth discharge from the afferent (see (*c*)) is not seen on this time base. (*c*), Z-modulated raster of the responses to individual tendon taps, with up to five spike discharges in response to any one tap. Lower trace: superimposed accelerometer tracings for these taps. Burke *et al.* (1983) reproduced with permission.

it shortens the relevant muscle and tendon, a tendon jerk still occurs. In subjects with hyperactive tendon jerks, percussion on a nearby bony prominence is often sufficient to evoke the reflex contraction, and in patients with pathological hyperreflexia percussion on the clavicle can produce reflex contractions in upper-limb muscles, and percussion on the pubis can produce reflex contractions in lower-limb muscles, particularly the adductors of the thigh. 'Irradiation of reflexes' in patients with spasticity is due not to intraspinal spread of an afferent volley from the percussed muscle but to spread of the percussion-induced vibration wave through bone to spindle endings in remote muscles (Lance & de Gail, 1965). In other words, the reflex contraction of muscles remote from the percussion site occurs because the percussion wave stimulates spindle endings in these remote muscles. Direct support for these conclusions came from experiments in which afferent volleys were recorded using microneurography (Burke, Gandevia & McKeon, 1983), as described below.

(ii) In the cat, abrupt muscle stretch can be applied directly through a tendon in such a way that it activates selectively primary spindle endings in the stretched muscle (Lundberg & Winsbury, 1960; Stuart, Willis & Reinking, 1971). However, tendon percussion involves a transversely applied indirect stimulus that is difficult to quantify let alone control. Percussion on the Achilles tendon is an effective stimulus for primary spindle endings in triceps surae, and responsive endings may discharge multiple times in any one percussion-induced volley (Fig. 6.3). Activation of different spindle endings depends on transmission of the percussion wave to spindles dispersed throughout the muscle belly, and therefore does not occur at precisely the same moment for each ending. As a result, the group Ia volley from spindles in triceps surae is quite dispersed even at the popliteal-fossa level, taking some 10 ms to reach its peak and lasting some 30–40 ms (Burke *et al*., 1983). Tendon organs and secondary spindle endings may also respond to percussion, the former usually with only a single discharge, the latter commonly as a transient increase in background discharge frequency. In addition, the total afferent input produced by tendon percussion will involve group Ia activity from other calf muscles and even from the pretibial and peroneal muscles, innervated by the peroneal nerve. Percussion insufficiently strong to produce an ankle jerk activates muscle and cutaneous afferents from the foot (Burke *et al*., 1983). Afferent volleys from this source take a few milliseconds longer to reach the popliteal fossa, but they can still contribute to the rising phase of the percussion-induced afferent volley. It can be concluded that tendon percussion is not a clean stimulus: it is virtually impossible to produce a volley only in group Ia afferents from the percussed muscle, especially if the intensity of percussion is sufficient to evoke an ankle jerk.

(iii) As discussed above, the percussion-evoked group Ia volley from triceps surae is highly dispersed at popliteal-fossa level (Burke *et al*., 1983). In the human thigh, the conduction velocity of fast group I muscle afferents is approximately 60 m/s (Macefield, Gandevia & Burke, 1989). If the slowest group Ia afferents travel at 40 m/s, even a pure Ia volley evoked electrically by stimulation of the tibial nerve in the popliteal fossa would have a dispersion of approximately 5 ms by the time it reached the motoneurone pool. A percussion-evoked volley will be much more desynchronized because it is already dispersed at popliteal-fossa level.

Muscle spindles, muscle tone, fusimotor system 95

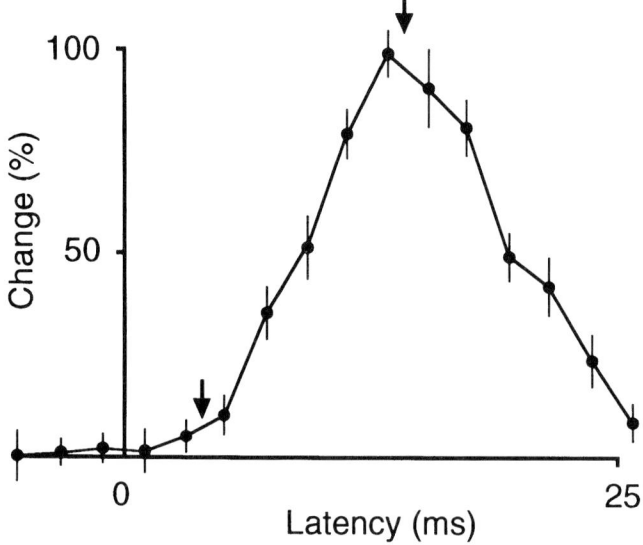

Fig. 6.4. The time course of the excitability changes set-up in the soleus motoneurone pool of one subject produced by tendon percussion, with motoneurone excitability tested by the soleus H reflex. Mean ±SEM are shown for the change at each conditioning test interval. The time between the arrows represents the rise time of the excitability change, equivalent to the rise time of the compound EPSP in the motoneurone pool. Adapted from Burke *et al.* (1984), with permission.

(iv) The monosynaptic pathway would be the sole spinal pathway involved in the tendon jerk, only if other inputs failed to reach the motoneurone pool before the target motoneurones discharge. This condition cannot be fulfilled. Using two separate techniques (H reflexes conditioned by subthreshold tendon percussion; and poststimulus time histograms of voluntarily active motor units in soleus in response to tendon percussion), it has been found that the rise-time of the compound excitatory postsynaptic potential (EPSP) produced by tendon percussion lasts >10 ms (Fig. 6.4), even when the intensity of percussion is too weak to evoke an ankle jerk (Burke, Gandevia & McKeon, 1984). Motoneurones which are only just recruited into the reflex discharge will be activated some 10 ms after the onset of monosynaptic excitation. There is adequate time for interneuronal transmission from homonymous, heteronymous and antagonist Ia afferents and from Ib and cutaneous afferents, and there is sufficient time for second impulses in the fastest Ia afferents to reach late-recruited motoneurones before they discharge. In addition, 10 ms is sufficient for recurrent inhibition from low-threshold motoneurones to affect the discharge of higher-threshold motoneurones, and for the earliest group I impulses to evoke

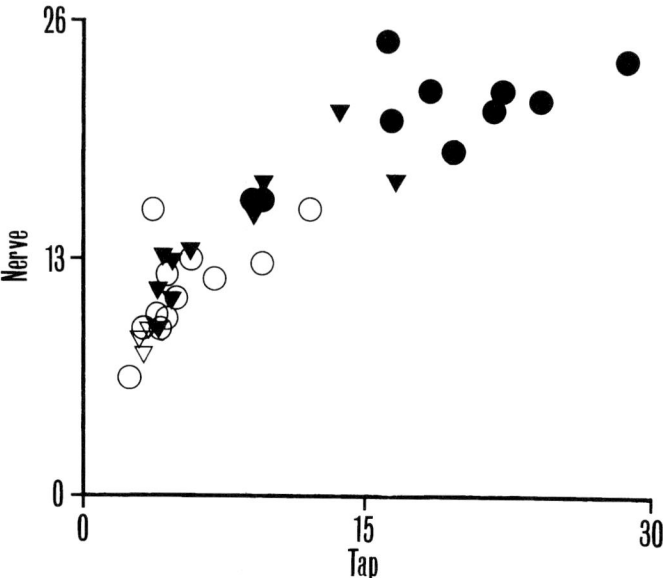

Fig. 6.5. The relationship between the intensity of tendon percussion (in arbitrary units and the amplitude of the rectified and smoothed muscle afferent volley from soleus produced by tendon percussion (also in arbitrary units) for one subject. Circles represent sequences when the subject was relaxed, triangles, those for which the subject performed Jendrassik reinforcement manoeuvres. With the solid symbols, tendon percussion produced a tendon jerk; with the open symbols, tendon percussion failed to produce a reflex response. Note that when the subject was relaxed, the intensity of percussion had to be >10 units and produce a nerve volley of >15 units for a tendon jerk to occur. When the subject performed the reinforcement manoeuvre, the intensity of the tap necessary to produce a tendon jerk was <5 units and the required neural volley was reduced to 9 units. These data clearly show that performance of the manoeuvre effectively reinforced the reflex, lowering its threshold when expressed in both mechanical and neural terms, but note that the relationship between nerve volley and tendon percussion is the same whether the subject is at rest or performing the reinforcement manoeuvre. The decrease in threshold for the reflex response is not due to a change in sensitivity of muscle afferents to tendon percussion. Burke *et al.* (1981a) reproduced with permission.

presynaptic inhibition of slow Ia afferents or of the second impulses in fast Ia afferents. The situation may be even more complex. The only certainty is that pathways other than the monosynaptic pathway will be active, and that these pathways will affect the recruitment of some motoneurones.

In a strong reflex contraction, the first-recruited motoneurones may discharge early on the rising phase of the compound EPSP, and so might be activated exclusively by monosynaptic inputs. However, in relaxed subjects, much of the compound EPSP must be spent raising the motoneurone pool to firing threshold (Burke *et al.*, 1984), so that, even with large rapidly rising EPSPs, there will be some delay before low-threshold motoneurones reach threshold. Moreover, an increase in tendon jerk amplitude is produced by recruiting motoneurones that had previously not been recruited; inhibition of the tendon jerk will affect preferentially those motoneurones that had only just been recruited. In both cases, this will occur at the peak of the compound EPSP >10 ms after its monosynaptic onset. It is of interest that oligosynaptic Ia excitation of homonymous and heteronymous motoneurones has been documented for lower- and upper-limb muscles of human subjects (Fournier et al., 1986; Malmgren & Pierrot-Deseilligny, 1988).

(v) Tendon jerks can be elicited in relaxed normal subjects. There is considerable evidence that the level of fusimotor drive directed to spindle endings in relaxed human muscles is quite low, possibly negligible (Vallbo et al., 1979; Burke, 1981). Recently, evidence has been found for background fusimotor activity directed to spindle endings in EMG-silent pretibial muscles of subjects who were standing upright (Ribot, Roll & Vedel, 1986; Aniss et al., 1990), but this does not alter the generality that, for muscles not engaged in any task, there is little background fusimotor drive. This view is supported by nerve block experiments, in which the percussion-evoked volley from soleus was recorded from the tibial nerve in the popliteal fossa while the sciatic nerve was blocked completely by concentrated lignocaine (Burke, McKeon & Skuse, 1981b). There was no change in the evoked afferent volley when all fusimotor influences were eliminated. The tendon jerk can be potentiated by a number of different manoeuvres including voluntary contraction of a remote muscle (Fig. 6.5), contraction of the test muscle, discomfort, arousal, and a decrease in percussion rate. None of these manoeuvres enhances the muscle afferent response to percussion (Burke et al., 1981a). This indicates that changes in amplitude of the tendon jerk may be produced centrally by altering reflex transmission within the spinal cord, rather than peripherally by altering the muscle afferent input to the spinal cord. Similarly, in spasticity, the site of the reflex enhancement appears to be central rather than peripheral (see later).

The tonic vibration reflex and normal muscle tone

By analogy with the stretch reflex of the decerebrate cat, it is widely held that normal muscle tone is dependent on the 'tonic stretch reflex', and for this reason the description in 1966 that muscle vibration could evoke a tonic reflex contraction in both cat (Matthews, 1966) and man (de Gail, Lance & Neilson, 1966; Hagbarth & Eklund, 1966) met with wide interest. Hitherto, only phasic reflexes such as the tendon jerk and the H reflex could be studied in human subjects: a tonic reflex contraction in response to muscle stretch is not nor-

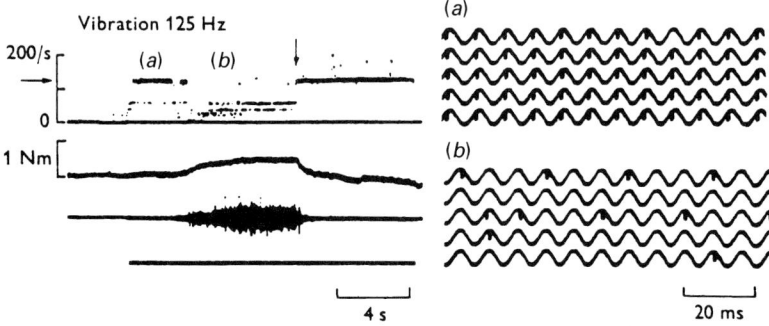

Fig. 6.6. The tonic vibration reflex and the effect of its development on the response to vibration of a primary spindle ending in tibialis anterior. The tibialis anterior tendon was vibrated for the period indicated by the horizontal bar. There was an artefactual thickening of the EMG trace, but EMG activity and a reflex increase in muscle force did not develop for some 3 s. The reflex contraction slowly increased to a plateau over some 2–4 s. As indicated in the instantaneous frequency plot, the muscle spindle ending responded to vibration 1:1, until the reflex contraction developed and unloaded the spindle ending. At the vertical arrow, the subject voluntarily suppressed the reflex contraction, and the spindle response reverted to 1:1. In the panels on the right, the spindle discharge has been superimposed on the vibration wave, panels (a) and (b) coming from the indicated sites on the instantaneous frequency plot, to the left. Burke et al. (1976b) reproduced with permission.

mally elicitable unless the subject performs a reinforcement manoeuvre or deliberately contracts the muscle being stretched (Neilson, this volume). This elusiveness of the tonic stretch reflex in relaxed subjects raises the question whether reflex mechanisms normally contribute to muscle tone.

The characteristics of the tonic vibration reflex (TVR) support the view that stretch reflex mechanisms have little to do with normal muscle tone. Tendon vibration can create an intense afferent barrage (Fig. 6.6), predominantly though not exclusively in group Ia afferents, sustained for as long as the vibration is sustained (Burke et al., 1976a) unless a TVR develops and unloads muscle spindles (Burke et al., 1976b). The intensity of the afferent input exceeds that when non-contracting muscles are passively stretched. Despite the intensity of the input to the spinal cord, the TVR develops slowly, if at all, generates little force and is readily suppressed by the subject (de Gail et al., 1966; Hagbarth & Eklund, 1966). Thus, in relaxed subjects, the 'gain' of the reflex pathways is initially quite low, though it then increases slowly. By contrast, if the excitability of the reflex pathways is enhanced by asking the subject to perform a voluntary contraction or if the subject suffers from spasticity, the reflex response to vibration appears more rapidly and reaches a plateau more quickly (Hagbarth & Eklund, 1968; Burke, Andrews & Lance, 1972). These

Muscle spindles, muscle tone, fusimotor system

characteristics suggest that reflex mechanisms are quite sluggish in relaxed man and can make little contribution to muscle tone. Accordingly, limb stiffness is not altered when neurologically normal subjects are rendered flaccid by general anesthesia, in order to remove neural contributions to tone (Lakie *et al.*, 1980) and, as all electromyographers can confirm, resting muscles are electrically silent unless the patient tenses them.

What then determines normal muscle tone? To a clinician, muscle tone is the resistance to stretch felt by an examiner when he passively moves a joint of a patient who is trying to be relaxed. Some subjects find it impossible to relax completely or will insist that they are relaxed when they are clearly not, particularly with proximal muscles. When a subject is fully relaxed, muscles are EMG-silent, and all but the most abrupt passive movements fail to evoke a reflex contraction. In such subjects, any resistance to stretch is due to the passive viscoelastic (and thixotropic) properties of muscle, noncontractile tissues and joints (Lakie, Walsh & Wright, 1984). The stretch reflex then makes no contribution to 'tone', whatever the level of fusimotor drive. In patients who fail to relax, passive stretch can evoke a reflex contraction, and in these patients the resistance to stretch will be determined by a combination of nonneural factors, the unintentional 'voluntary' contraction and the evoked reflex activity. In these patients, fusimotor drive could contribute to muscle tone. However, underactivity of the fusimotor system would not be detected because, at most, it could only remove something that is not effective in fully relaxed subjects with normal muscle tone.

Abnormalities of muscle tone

Hypotonia

Hypotonia is said to occur in lower motoneurone lesions, in cerebellar disease and in spinal or cerebral shock. In none of these changes is fusimotor underactivity responsible for the 'disordered tone'.

The reflex arc can be interrupted in lower motoneurone lesions, but hypoactive tendon jerks are found particularly when there is also involvement of the afferent side of the reflex arc. A peripheral lesion that affects only motor units cannot abolish tendon jerks unless it involves all low-threshold reflexly accessible motoneurones. If the motor involvement is random (or preferentially affects large motoneurones and large axons, as compressive lesions usually do), absence of the tendon jerk in a muscle that can still contract betokens a sensory disturbance. However, even when the tendon jerk is depressed, a decreased resistance to stretch is not due to absent reflex activity but to a soft-tissue change.

In *cerebellar disorders* and *shock syndromes*, the low level of muscle tone is accompanied by other evidence of abnormality: dysmetria, dysdiadokokinesis, intention tremor and rebound in cerebellar disorders; flaccidity with tendon jerk areflexia, with or without loss of cutaneous reflexes in shock syndromes. By itself, and in the absence of these collateral signs, muscle tone is unremarkable, not different to that seen in normal subjects who relax fully (Van der Meché & Van Gijn, 1986). In chronic cerebellar disturbances, soft-tissue changes can occur, leading to enhanced compliance of muscle and joint, such that joints in a relaxed limb can adopt abnormally lax positions (see Lance & McLeod, 1981). Presumably such changes occur because the patients have lost the "shock absorber" action of the stretch reflex during voluntary activities, and accordingly the hypotonia at rest could be attributed to an indirect effect of diminished reflex responsiveness.

Spasticity and rigidity

In spasticity and rigidity, muscle tone is enhanced, and this is largely due to enhanced reflex activity. In the spastic patient, the characteristic feature of the increase in muscle tone is its velocity dependence, with greater resistance the faster the passive joint movement (Burke & Lance, 1973). The stiffness of spastic muscle contributes to the increased resistance, even in the absence of overt contracture (e.g. Dietz, Quintern & Berger, 1981; Tang & Rymer, 1981; Lee, Boughton & Rymer, 1987; Dietz, Trippel & Berger, 1991), particularly during voluntary movement. However, there seems little doubt that 'stretch reflex' mechanisms within the spinal cord are abnormally active (see Ashby, this volume) and that much or most of the resistance is generated through these pathways, at least when patients are quiescent (Burke & Lance, 1973; Thilmann, Fellows & Garms, 1992).

Over the last decade, emphasis has been placed on the central nervous system as the prime site for changes in stretch reflex 'gain', at the expense of fusimotor-induced changes in the afferent volley entering the spinal cord (for example, Burke, 1981, 1988; Prochazka, 1989). There are abnormalities in afferent transmission in spastic patients, but there is, as yet, no evidence that, in spastic man or in appropriate animal models of spasticity, a fusimotor-induced exaggeration of the spindle response to stretch contributes to the abnormal reflex enhancement (for reviews, see Burke, 1983; Ashby, this volume). Few comparable data are available for Parkinsonian rigidity, but again there is no convincing evidence of fusimotor overactivity as the prime cause of rigidity (Wallin *et al.*, 1973; Burke, Hagbarth & Wallin, 1977).

Whether central nervous system lesions ever produce fusimotor overactivity in human patients remains unresolved. Given that selective stimulation of, or discrete lesions in, specific brain structures can produce selective overactivity of components of the fusimotor system in the cat (see Matthews, 1972; Hulliger, 1984), it is reasonable to presume that comparable lesions would produce fusimotor overactivity in man. In, for example, a spinal patient with pressure areas or bladder/bowel disturbance, disturbed cutaneous and visceral afferent volleys might produce enhanced background fusimotor drive, given that appropriate reflex pathways exist in man (Gandevia *et al.*, 1986; Aniss *et al.*, 1990), much as in the cat (Hulliger, 1984). Nevertheless, it is the thesis of this chapter that any fusimotor overactivity is unlikely to play a major role in increasing reflex gain in patients at rest, although in patients who are not quiescent or whose condition is unstable, abnormal or inappropriate fusimotor activity could disturb function and potentiate spasms.

Conclusions

It has been argued above that the fusimotor system probably plays little role in modulating the gain of stretch reflex pathways. There are powerful mechanisms for adjusting reflex gain centrally, without need to invoke an additional peripherally acting mechanism. The striking finding from all studies that have used microneurography to record spindle afferent activity in human subjects has been the contrast between rest and activity. The fusimotor system is a motor system, relatively quiescent at rest but called into activity during movement. It is unfortunate that clinicians are compelled to test muscle tone in subjects who are attempting to be relaxed, because this is when fusimotor drive and the excitability of reflex pathways are lowest. The major contribution of fusimotor-induced muscle spindle assistance to the α-motoneurone pool is likely to be during movement, and that is when defective fusimotor control is likely to create or contribute to a patient's deficit. As yet, we lack the necessary clinical skills to assess reflex activity adequately during movement, but already significant insights have come from correlations of EMG and muscle stretch performed on patients and normal subjects during sustained contractions (see Neilson, this volume).

Acknowledgements

The authors are grateful for collaboration of numerous colleagues and for the inspiration and support from Professor J.W. Lance. Their studies have been supported by the National Health and Medical Research Council of Australia.

References

Al-Falahe, N.A., Nagaoka, M. & Vallbo, Å.B. (1990). Response profiles of human muscle afferents during active finger movements. *Brain,* **113**, 325–46.

Aniss, A.M., Diener, C., Hore, J., Burke, D. & Gandevia, S.C. (1990). Reflex influences on muscle spindles in human pretibial muscles during standing. *Journal of Neurophysiology,* **64**, 671–9.

Bongiovanni, L.G. & Hagbarth, K.-E. (1990). Tonic vibration reflexes elicited during fatigue from maximal voluntary contractions in man. *Journal of Physiology (London),* **423**, 1–14.

Burke, D. (1981). The activity of human muscle spindle endings in normal behavior. In *International Review of Physiology. Neurophysiology IV,* ed. R. Porter, pp. 91–126. University Park Press: Baltimore.

Burke, D. (1983). Critical examination of the case for or against fusimotor involvement in disorders of muscle tone. In *Motor Control Mechanisms in Health and Disease, Advances in Neurology,* vol. 39, ed. J.E. Desmedt, pp. 133–50. New York: Raven Press.

Burke, D. (1988). Spasticity as an adaptation to pyramidal tract injury. In *Functional Recovery in Neurological Disease, Advances in Neurology,* vol. 47, ed. S.G. Waxman, pp. 401–23. New York: Raven Press.

Burke, D., Andrews, C.J. & Lance, J.W. (1972). Tonic vibration reflex in spasticity, Parkinson's disease, and normal subjects. *Journal of Neurology, Neurosurgery, and Psychiatry,* **35**, 477–86.

Burke, D., Gandevia, S.C. & McKeon, B. (1983). The afferent volleys responsible for spinal proprioceptive reflexes in man. *Journal of Physiology (London),* **339**, 535–52.

Burke, D., Gandevia, S.C. & McKeon, B. (1984). Monosynaptic and oligosynaptic contributions to the human ankle jerk and H reflex. *Journal of Neurophysiology,* **52**, 435–48.

Burke, D., Hagbarth, K.-E., Löfstedt, L. & Wallin, B.G. (1976a). The responses of human muscle spindle endings to vibration of non-contracting muscles. *Journal of Physiology (London),* **261**, 673–93.

Burke, D., Hagbarth, K.-E., Löfstedt, L. & Wallin, B.G. (1976b). The responses of human muscle spindle endings to vibration during isometric contraction. *Journal of Physiology (London)* **261**, 695–711.

Burke, D., Hagbarth, K.-E. & Skuse, N.F. (1978). Recruitment order of human spindle endings in isometric voluntary contractions. *Journal of Physiology (London)* **285**, 101–12.

Burke, D., Hagbarth, K.-E. & Wallin, B.G. (1977). Reflex mechanisms in Parkinsonian rigidity. *Scandinavian Journal of Rehabilitation Medicine,* **9**, 15–23.

Burke, D. & Lance, J.W. (1973). Studies of the reflex effects of primary and secondary spindle endings in spasticity. In *New Developments in Electromyography and Clinical Neurophysiology,* vol. 3, ed. J.E. Desmedt, pp. 475–95. Basel: Karger.

Burke, D., McKeon, B.B. & Skuse, N.F. (1981a). Dependence of the Achilles tendon reflex on the excitability of spinal reflex pathways. *Annals of Neurology,* **10**, 551–6.

Burke, D., McKeon, B.B. & Skuse, N.F. (1981b). The irrelevance of fusimotor activity to the Achilles tendon jerk of relaxed humans. *Annals of Neurology,* **10**, 547–50.

de Gail, P., Lance, J.W. & Neilson, P.D. (1966). Differential effects on tonic and phasic reflex mechanisms produced by vibration of muscles in man. *Journal of Neurology, Neurosurgery, and Psychiatry,* **29**, 1–11.

Dietrichson, P. (1971). Phasic ankle reflex in spasticity and Parkinsonian rigidity. *Acta Neurologica Scandinavica*, **47**, 22–51.

Dietz, V., Quintern, J. & Berger, W. (1981). Electrophysiological studies of gait in spasticity and rigidity. Evidence that altered mechanical properties of muscle contribute to hypertonia. *Brain*, **104**, 431–49.

Dietz, V., Trippel, M. & Berger, W. (1991). Reflex activity and muscle tone during elbow movements of patients with spastic paresis. *Annals of Neurology*, in press.

Fournier, E., Meunier, S., Pierrot-Deseilligny, E. & Shindo, M. (1986). Evidence for interneuronally mediated Ia excitatory effects to human quadriceps motoneurones. *Journal of Physiology (London)*, **377**, 143–69.

Fukushima, K., Taniguchi, K., Kamishima, Y. & Kato, M. (1976). Peripheral factors contributing to the volitional control of firing rates of the human motor units. *Neuroscience Letters*, **3**, 33–6.

Gandevia, S.C., Macefield, G., Burke, D. & McKenzie, D.K. (1990). Voluntary activation of human motor axons in the absence of muscle afferent feedback: the control of the deafferented hand. *Brain*, **113**, 1563–81.

Gandevia, S.C., Miller, S., Aniss, A.M. & Burke, D. (1986). Reflex influences on muscle spindle activity in relaxed human leg muscles. *Journal of Neurophysiology*, **56**, 159–70.

Gilman, S. (1969). The mechanism of cerebellar hypotonia. An experimental study in the monkey. *Brain*, **92**, 621–38.

Gordon, E.E., Januszko, D.M. & Kaufman, L. (1967). A critical survey of stiff-man syndrome. *American Journal of Medicine*, **42**, 582–99.

Hagbarth, K.-E. & Eklund, G. (1966). Motor effects of vibratory stimuli in man. In *Nobel Symposium 1, Muscular Afferents and Motor Control*, ed. R. Granit, pp. 177–86. Stockholm: Almqvist & Wiksell.

Hagbarth, K.-E. & Eklund, G. (1968). The effects of muscle vibration in spasticity, rigidity and cerebellar disorders. *Journal of Neurology, Neurosurgery, and Psychiatry*, **31**, 207–13.

Hagbarth, K.-E., Kunesch, E.J., Nordin, M., Schmidt, R. & Wallin, E.U. (1986). Gamma loop contributing to maximal voluntary contractions in man. *Journal of Physiology (London)*, **380**, 575–91.

Hulliger, M. (1984). The mammalian muscle spindle and its central control. *Reviews of Physiology, Biochemistry and Pharmacology*, **101**, 1–110.

Lakie, M., Tsementzis, S.T., Walsh, E.G. & Wright, G. (1980). Anaesthesia does not (and cannot) reduce muscle tone? *Journal of Physiology (London)*, **301**, 23P.

Lakie, M., Walsh, E.G. & Wright, G.W. (1984). Resonance at the wrist demonstrated by the use of a torque motor: an instrumental analysis of muscle tone in man. *Journal of Physiology (London)*, **353**, 265–85.

Lance, J.W. & de Gail, P. (1965). Spread of phasic muscle reflexes in normal and spastic subjects. *Journal of Neurology, Neurosurgery, and Psychiatry*, **28**, 328–34.

Lance, J.W., de Gail, P. & Neilson, P.D. (1966). Tonic and phasic spinal cord mechanisms in man. *Journal of Neurology, Neurosurgery, and Psychiatry*, **29**, 535–44.

Lance, J.W. & McLeod, J.G. (1981). *A Physiological Approach to Clinical Neurology*, 3rd edn. London: Butterworths.

Lee, W.A., Boughton, A. & Rymer, W.Z. (1987). Absence of stretch reflex gain enhancement in voluntarily active spastic muscle. *Experimental Neurology*, **98**, 317–35.

Lundberg, A. & Winsbury, G. (1960). Selective adequate activation of large afferents from muscle spindles and Golgi tendon organs. *Acta Physiologica Scandinavica*, **49**, 155–64.

Macefield, G., Gandevia, S.C. & Burke, D. (1989). Conduction velocities of muscle and cutaneous afferents in the upper and lower limbs of human subjects. *Brain*, **112**, 1519–32.

Macefield, G., Hagbarth, K.-E., Gorman, R.B., Gandevia, S.C. & Burke, D. (1991a). Decline in spindle support to α-motoneurones during sustained voluntary contractions. *Journal of Physiology (London)*, **440**, 497–512.

Macefield, G., Gandevia, S.C., Bigland-Ritchie, B., Gorman, R. & Burke, D. (1991b). The discharge rate of human motoneurones innervating ankle dorsiflexors in the absence of muscle afferent feedback. *Journal of Physiology (London)*, **438**, 219P.

Malmgren, K. & Pierrot-Deseilligny, E. (1988). Evidence for non-monosynaptic Ia excitation of human wrist flexor motoneurones, possibly via propriospinal neurones. *Journal of Physiology (London)*, **405**, 747–64.

Marsden, C.D., Obeso, J.A. & Rothwell, J.C. (1983). Clinical neurophysiology of muscle jerks: myoclonus, chorea, and tics. In *Motor Control Mechanisms in Health and Disease, Advances in Neurology*, vol. 39, ed. Desmedt, J.E., pp. 865–81. New York: Raven.

Matthews, P.B.C. (1966). The reflex excitation of the soleus muscle of the decerebrate cat caused by vibration applied to its tendon. *Journal of Physiology (London)*, **184**, 450–72.

Matthews, P.B.C. (1972). *Mammalian Muscle Receptors and Their Central Actions*. London: Arnold.

Merton, P.A. (1953). Speculations on the servo-control of movement. In *The Spinal Cord, Ciba Foundation Symposium, London*, ed. Malcolm, J.L., Gray, J.A.B. and Wolstenholme, G.E.W., pp. 247–55. London: Churchill.

Prochazka, A. (1989). Sensorimotor gain control; a basic strategy of motor systems? *Progress in Neurobiology*, **33**, 281–307.

Ribot, E., Roll, J.-P. & Vedel, J.-P. (1976). Efferent discharges recorded from single skeletomotor and fusimotor fibres in man. *Journal of Physiology (London)*, **375**, 251–68.

Rothwell, J.C., Gandevia, S.C. & Burke, D. (1990). Activation of fusimotor neurones by motor cortical stimulation in human subjects. *Journal of Physiology (London)*, **430**, 105–17.

Rushworth, G. (1960). Spasticity and rigidity: an experimental study and review. *Journal of Neurology, Neurosurgery, and Psychiatry*, **23**, 99–117.

Stuart, D.G., Willis, W.D. & Reinking, R.M. (1971). Stretch-evoked excitatory postsynaptic potentials in motoneurons. *Brain Research*, **33**, 115–25.

Tang, A. & Rymer, W.Z. (1981). Abnormal force-EMG relations in paretic limbs of hemiparetic human subjects. *Journal of Neurology, Neurosurgery, and Psychiatry*, **44**, 690–8.

Thilmann, A.F., Fellows, S.J. & Garms, E. (1992). The mechanism of spastic muscle hypertonia: variation in reflex gain over the time course of spasticity. *Brain*, in press.

Vallbo, Å.B. (1971). Muscle spindle responses at the onset of isometric voluntary contractions in man. Time difference between fusimotor and skeletomotor effects. *Journal of Physiology (London)*, **318**, 405–31.

Vallbo, Å.B. & Al-Falahe, N.A. (1990). Human muscle spindle response in a motor learning task. *Journal of Physiology (London)*, **421**, 553–68.

Vallbo, Å.B., Hagbarth, K.-E., Torebjörk, H.E. & Wallin, B.G. (1979). Somatosensory, proprioceptive, and sympathetic activity in human peripheral nerves. *Physiological Reviews*, **59**, 919–57.

Van Der Meché, F.G.A. & Van Gijn, J. (1986). Hypotonia: an erroneous clinical concept? *Brain*, **109**, 1169–78.

Wallin, B.G., Hongell, A. & Hagbarth, K.-E. (1973). Recordings from muscle afferents in Parkinsonian rigidity. In *New Developments in Electromyography and Clinical Neurophysiology*, vol. 3, ed. Desmedt, J.E., pp. 263–72. Basel: Karger.

7
The neurophysiology of human spinal spasticity

P. ASHBY

Playfair Neuroscience Unit,
The Toronto Hospital (Western Division)
University of Toronto,
CANADA M5T 2S8

Introduction

A well-known sequence of events follows transection of the spinal cord in man (Riddoch, 1917; Kuhn, 1950). Initially, the muscles below the level of the lesion are flaccid and the tendon jerks are absent. After 1–2 weeks, noxious stimulation of the foot may elicit an extensor plantar response and contraction of the flexor muscles of the leg (flexion reflex). After about 1 month, the tendon jerks reappear and become exaggerated. Passive displacement of a joint, especially if it is made rapidly, elicits a contraction of the stretched muscles (stretch reflex). By about six months, prolonged contractions of muscles (spasms) can be elicited by stretching muscles or touching the skin. Passive stretch of certain muscles may result in a series of stretch reflexes producing a rhythmic oscillation at the joint (clonus). The stretch reflex elicited in the quadriceps is often maximum with the knee extended diminishing as the knee is flexed (clasp-knife phenomenon). After 6–12 months, the resistance to passive stretch felt by the examiner may be increased even in the absence of obvious reflex activity, and the extensibility of the muscles may be reduced (contracture). The muscles below a long-standing, complete spinal lesion often become wasted. The term spasticity is used loosely to describe any combination of these clinical features but in this review, according to Lance's (1980) suggestion, it will connote a subgroup of these signs: increased tendon jerks and increased, velocity-dependent, stretch reflexes. What happens to the spinal circuitry and muscle to give rise to these clinical features? This question has been the topic of several other recent reviews (Pierrot-Deseilligny & Mazières, 1985; Ashby & McCrea, 1987; Burke, 1988; Katz & Rymer, 1989; Ashby, 1990). This chapter will provide a brief summary.

Deductions form clinical observation

All the clinical features listed above can be found below a complete spinal lesion. It follows that the abnormalities giving rise to these signs must be in the spinal cord below the level of the lesion. The spinal reflexes are first depressed and then exaggerated. This cannot be explained by the loss of some single facilitatory or inhibitory tonic drive from above. There must be some 'plastic change' below the level of the lesion.

Some of the clinical signs can occur independently. In Friedreich's ataxia the flexion reflex is exaggerated but tendon jerks and stretch reflexes are absent. Similarly, the tendon jerks and stretch reflexes do not become exaggerated when a transverse lesion of the spinal cord complicates tabes dorsalis or the Holmes-Adie syndrome (Swash & Earl, 1975). Thus there are at least two independent reflex abnormalities that follow a spinal lesion in man: the release of the flexion reflex and the increase in tendon jerks and stretch reflexes. The latter depend on the integrity of large-diameter afferents.

Certain reflex abnormalities seen after spinal lesions (such as the extensor plantar response) can, in other circumstances, be of virtually instantaneous onset and remission (e.g. during transient ischemia of the brain), whereas others, such as extensor spasms, only develop weeks or months after the lesion. The mechanisms underlying such immediate and delayed signs are likely to be different.

Motor circuitry of the normal mammalian spinal cord

The circuitry of the mammalian spinal cord is now known in considerable detail in the cat (Baldissera, Hultborn & Illert, 1981) and to a lesser extent in man (e.g. Pierrot-Deseilligny *et al.*, 1981; Bayoumi & Ashby, 1989; see Fig. 7.1). The Ia afferents from the velocity-sensitive primary spindle endings, which can be influenced by dynamic gamma motoneurones (gd), project monosynaptically to alpha motoneurones (a). The powerful monosynaptic facilitation from these afferents contributes to the tendon jerk. Transmission in the presynaptic terminals of Ia afferents can be modified by neurones mediating presynaptic inhibition (p). Ia afferents also project to the Ia inhibitory interneurones (Ia$^-$) responsible for reciprocal inhibition of antagonist motoneurones. These interneurones receive convergent input from many sources (not shown in Fig. 7.1). The activation of a motoneurone pool by many fibre systems is accomplished by simultaneous excitation of alpha, gamma and Ia inhibitory interneurones thus ensuring by three routes, the certain activation of agonist and inhibition of antagonist motoneurones. Renshaw cells, which inhibit these three elements, have the opposite action.

Ib afferents from the tension-sensitive Golgi tendon organs project to another set of interneurones which, in turn, project to the motoneurones of many limb muscles and are capable of producing inhibition of extensor motoneurones. These Ib interneurones also receive convergent input from many sources including joint and cutaneous afferents (Fig. 7.1).

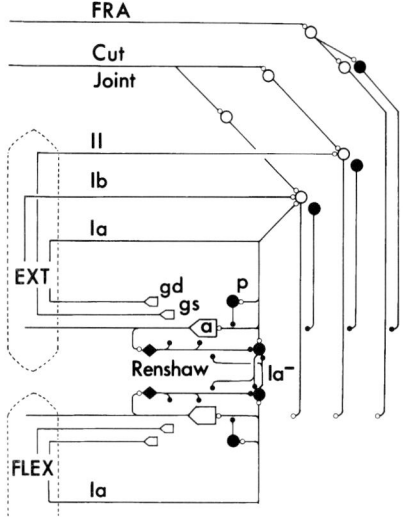

Fig. 7.1. Simplified diagram of some of the segmental reflex pathways controlling extensor (EXT) and flexor (FLEX) muscles. Filled neurones inhibitory. Open neurones excitatory. The Ia afferents from the primary spindle endings, which are controlled by dynamic gamma motoneurones (gd), project monosynaptically to alpha motoneurones (a). Transmission in the presynaptic terminals of Ia afferents can be modulated by neurones (p) mediating presynaptic inhibition. Ia afferents also project to Ia inhibitory interneurones (Ia⁻) producing reciprocal inhibition of antagonist motoneurones. Renshaw cells inhibit alpha and gamma motoneurones and Ia inhibitory neurones. The Ib afferents from Golgi tendon organs can produce facilitation or inhibition of motoneurones via Ib interneurones which also receive convergence from Ia and cutaneous afferents. Group II afferents from secondary spindle endings, which are controlled by static gamma motoneurones (gs), project to motoneurones through interneurones receiving convergent input from joint and cutaneous (Cut) afferents. Small muscle and cutaneous afferents project to interneurones producing facilitation of flexor motoneurones and inhibition of extensor motoneurones. These afferents have been called the flexor reflex afferents (FRA). The second-order interneurones in this reflex pathway are inhibited by supraspinal centers. Ashby (1990) reproduced with permission.

The group II muscle afferents, from the length-sensitive secondary spindle endings, which are influenced by static gamma motoneurones (gs), have strong projections to interneurones producing facilitation or inhibition of subsets of motoneurones. These group II interneurones also receive facilitation from cutaneous, joint and group III afferents. Lundberg, Malmgrem and Schomberg (1987) have postulated that movements could be executed by tonic facilitation of static gamma motoneurones (which would cause the group II endings to signal until the muscle shortened). A transient facilitation of the group II interneurones would then be sufficient to cause them to excite motoneurones and initiate the movement. The activation of joint and cutaneous afferents by the

Table 7.1. *Reasons for the exaggeration of the tendon jerks and short-latency stretch reflex*

1. Increased receptor discharge
   a) Changes in muscle spindle properties
   b) Changes in fusimotor drive

2. A normal synaptic input brings more motoneurones to threshold
   a) The motoneurones are depolarized
   b) The biophysical properties of motoneurones are altered

3. A normal receptor discharge results in increased synaptic input to motoneurones
   a) Presynaptic modulation of the monosynaptic pathway
   b) Altered transmission through oligosynaptic pathways

movement would then take over the support of group II interneurones until the muscle shortened and the spindle activity died away. The gain of such a system would have to be low enough that motoneurones were not activated during passive movements and also carefully controlled to prevent positive feedback from the movement causing continuous excitation of motoneurones.

Certain small diameter afferents project to interneurones facilitating flexor and inhibiting extensor motoneurones, thus resulting in withdrawal of the limb. These are termed the flexor reflex afferents (FRA) and their interneurones are inhibited from the brainstem.

Why are the tendon jerks and stretch reflexes increased?

The tendon jerk is often considered to be a purely monosynaptic reflex but the afferent volley produced by a tendon tap (Burke, Gandevia & McKeon, 1983) and the rise times of the composite EPSPs produced in motoneurones are long enough (Birnbaum & Ashby, 1982) that, as Burke, Gandevia and McKeon (1984) have pointed out, there is ample time for oligosynaptic pathways to contribute to the net motoneurone output. Thus, when seeking an explanation for exaggerated tendon jerks, alterations in transmission through oligosynaptic as well as monosynaptic pathways have to be considered. The possible reasons for the exaggeration of tendon jerks are given in Table 7.1. Each of these possibilities is considered in turn.

The discharge from the spindle is increased

The increased tendon jerks and stretch reflexes could be accounted for if the spindle discharge in response to a given stretch were exaggerated. The evidence is to the contrary.

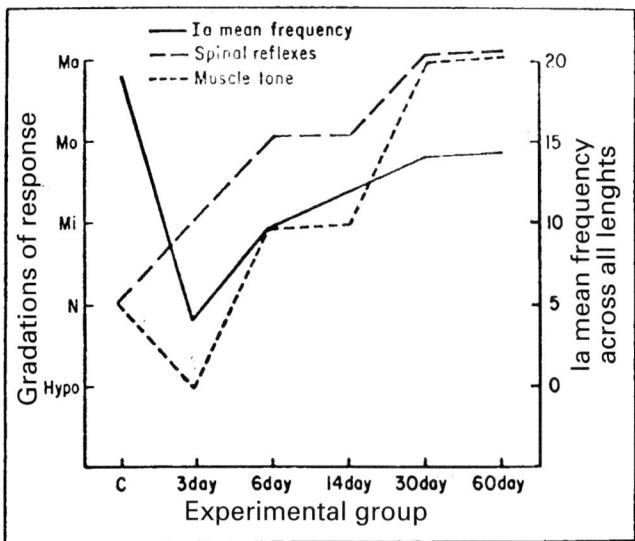

Fig. 7.2. Comparison of mean spindle firing frequency, spinal reflexes and muscle tone of control cats (C) and cats at intervals following spinal cord transection. The right ordinate indicates the mean firing frequency, averaged across all muscle lengths, of the primary spindle endings in each experimental group. The left ordinate indicates the levels of depression or exaggeration of spinal reflexes and of muscle tone. (N = normal, Mi = mildy increased, Mo = moderately increased, Ma = markedly increased). Bailey et al. (1980) reproduced with permission.

Bailey and colleagues (1980) examined the responses of Ia afferents from medial gastrocnemius to static stretch in normal cats and in cats 3–60 days after a complete spinal transection at L1. Spindle firing rates were depressed following the spinal transection and remained so 30 and 60 days later even though the animals were clinically hypertonic and hyperreflexic (Fig. 7.2). Hagbarth, Wallin and Löfstedt (1973) used microelectrodes to record the firing rates of Ia afferents in the medial gastrocnemius nerve in eight normal subjects and in two patients with upper motoneurone lesions (one with cervical spondylosis). There was no obvious difference in the static or dynamic responses of the primary endings to dorsiflexion of the foot in 20 successive 1.8 degree increments. (Fig. 7.3) The results of earlier experiments, originally interpreted as indicating increased spindle receptor sensitivity (in which electrical and mechanical stimuli were compared or in which dilute procaine was used to block fusimotor fibres), have other explanations (Burke, 1983). At present, there is no positive evidence that an increased spindle receptor discharge causes the hyperreflexia following a chronic spinal lesion.

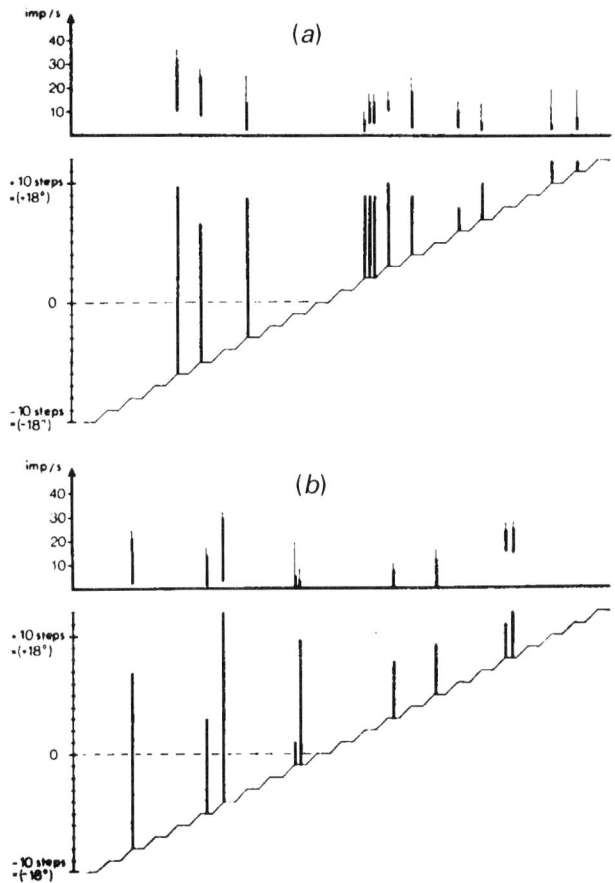

Fig. 7.3. Data from eight normal subjects (*a*) and two patients with spasticity (*b*). Diagrams illustrating the static (thick line) and dynamic (thin line) firing rates of spindle afferents (top traces) in response to increments of muscle stretch (bottom traces). The vertical bars in the lower traces indicate the range over which the discharge could be recorded. Hagbarth *et al.* (1973) reproduced with permission.

A normal synaptic input brings more motoneurones to threshold

In the cat, electrical stimulation of muscle nerves results in larger monosynaptic (and polysynaptic) reflexes on the side below a chronic spinal hemisection (Hultborn & Malmstem, 1983*a*) indicating a change in central excitability. Is it that motoneurones are more readily brought to threshold? This could occur if the motoneurones were depolarized by other neurones or if an alteration in their membrane properties allowed them to be discharged more readily by a given synaptic current. These possibilities have been carefully examined (most

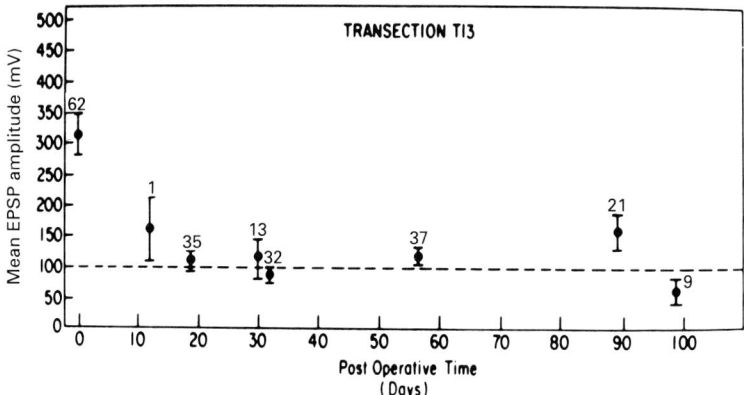

Fig. 7.4. Time course of changes in single fibre Ia EPSP amplitude following spinal transection in the cat at T13. Each point represents the mean EPSP amplitude (± 1 SEM) obtained that number of days after transection. The number of motoneurones tested is given above the point. The dotted line gives the mean EPSP amplitude for intact cats. Nelson and Mendell (1979) reproduced with permission.

recently by Hochman & McCrea, 1993*a,b*) with the conclusion that the minor changes in the biophysical properties of motoneurones are unlikely to be the principal cause of the increased motoneurone activity following spinal section. It is impossible to test the membrane properties of motoneurones in man but there is evidence that motoneurones can be brought to threshold more easily following an upper motoneurone lesion (Schiller & Stålberg, 1978; Katz & Rymer, 1989). This, of course, might mean that they receive more synaptic activity producing depolarization.

A normal receptor discharge results in an increased synaptic input to motoneurones

Altered transmission through the monosynaptic pathway to motoneurones: Ia EPSPs

Within hours of spinal cord section in the cat (when the hind limbs would be flaccid) the amplitude of the synaptic potentials produced by Ia afferents in motoneurones increases by up to 300% in some species of neurone. Paradoxically these acute changes, which are, in part, due to extraneural factors, last only 12 days in cats with T13 lesions (Fig. 7.4) and the Ia EPSPs are subsequently normal even though the animals by then would be expected to have brisk tendon jerks and exaggerated stretch reflexes (Nelson & Mendell,

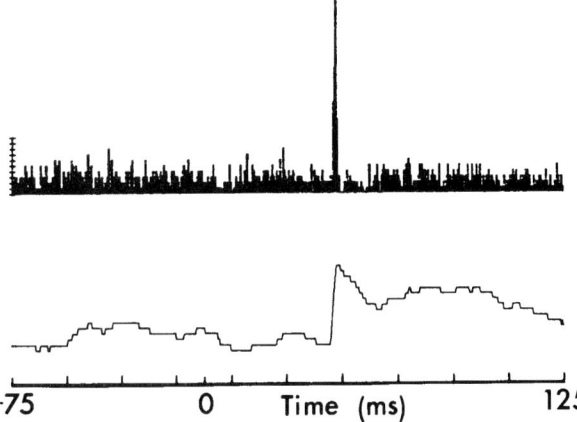

Fig. 7.5. Poststimulus time histogram of a human soleus motor unit showing the changes in firing probability produced by a group I volleys delivered at time zero (upper trace). The profile of the EPSP can be obtained by integration (lower trace).

1979). Other studies have shown shortening of the rise time of EPSPs after spinal transection and the authors have postulated that Ia fibres might make new synapses on more proximal parts of the motoneurone (e.g. Hochman & McCrea, 1993*b*).

The characteristics of postsynaptic potentials in single human motoneurones can be derived from changes in the firing probability of single motor units (Fig. 7.5) (Ashby & Zilm, 1982*a,b*; Fetz & Gustafsson, 1983; Cope, Fetz & Matsumura, 1987; Midroni & Ashby, 1989). Mailis and Ashby (1990) used this technique to examine composite homonymous Ia EPSPs in soleus motoneurones in patients with spinal lesions. In patients with lesions less than 8 weeks in duration, the EPSPs were larger. However, in the whole group (with lesions up to 4 years in duration), the Ia EPSPs were not detectably different from normal (Fig. 7.6). Thus, in animals and perhaps man, Ia EPSPS are enlarged immediately after a spinal lesion but when hyperreflexia is established, after several months, Ia EPSPs are normal in amplitude.

Dynamics of transmission at the Ia-motoneurone synapse: presynaptic inhibition

Could there be changes in the way that bursts of impulses as opposed to single impulses in Ia afferents are processed in the spinal cord?

Transmission at the central terminals of a primary afferent fibre may be altered by the passage of a prior impulse in that fibre. These alterations have

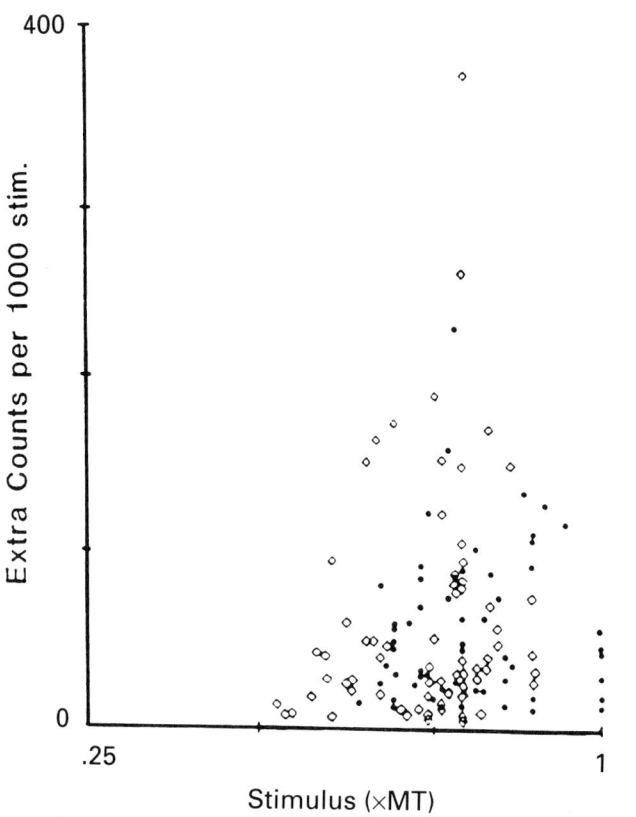

Fig. 7.6. Magnitude of the changes in firing probability produced in single soleus motoneurones by group I volleys in the posterior tibial nerve in the normal subjects (solid circles) and patients with chronic spinal lesions (open diamonds). The intensity of stimulation (expressed in terms of the threshold (MT) of the alpha motoneurone axons of soleus) is on the abcissa. The change in firing probability (in extra counts), which is proportional to the amplitude of the composite Ia EPSP, is on the ordinate. There are no statistically significant differences between the two groups. Mailis and Ashby (1990) reproduced with permission.

been attributed to the dynamics of transmitter release (Curtis & Eccles, 1960). Synaptic transmission can also be modulated by activity in the same or other afferents by a process known as presynaptic inhibition (Schmidt, 1971; Burke & Rudomin, 1977; Baldissera *et al.*, 1981). Presynaptic inhibition is produced by interneurones with synapses on the terminals of primary afferents whose depolarizing action ('primary afferent depolarization') reduces the amount of transmitter released by an impulse in the primary afferent. Primary afferent

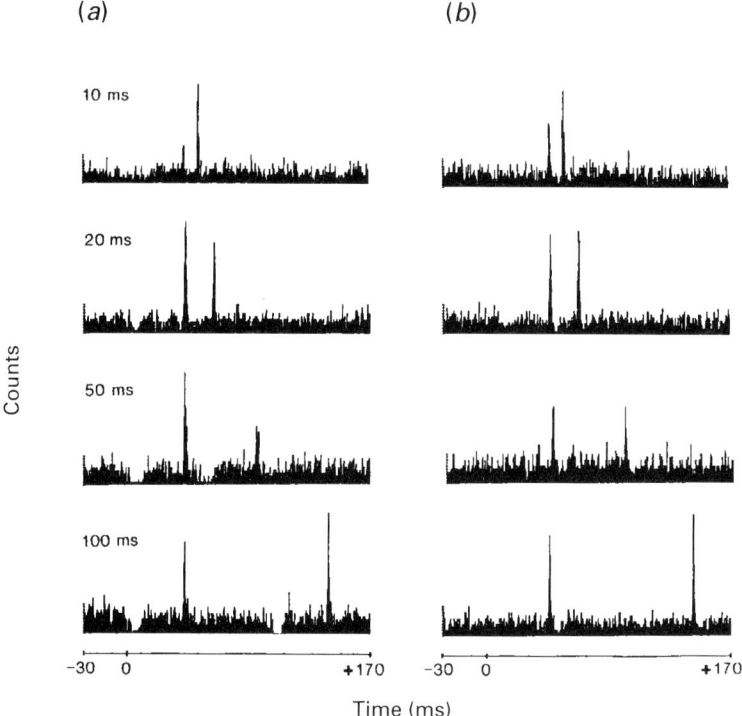

Fig. 7.7. Peristimulus time histograms of a soleus motor unit in a normal subject (left) and a patient with quadriplegia 16 weeks in duration (right) using paired stimuli at the intervals indicated. Note the changes in the relative magnitude of the two periods of facilitation in the normal subject (e.g. depression at 50 ms interval). This is not present in the patient with the spinal lesion. Mailis and Ashby (1990) reproduced with permission.

depolarization can be detected with recording electrodes over the cord dorsum as a 'dorsal root potential'. Results of studies of presynaptic inhibition in chronic spinal animals have been conflicting (Naftchi, Schlosser & Horst, 1980; Hultborn & Malmsten, 1983b).

Presynaptic inhibition of Ia afferents can be demonstrated in man using conditioning and test volleys in different afferents (Ashby *et al.*, 1987, Hultborn *et al.*, 1987). The 'homosynaptic depression' produced when the conditioning and test volleys travel in the same fibres (Katz, Morin & Pierrot-Deseilligny, 1987; Mailis & Ashby, 1990) probably includes both presynaptic inhibition and transmitter depletion. The depression of the H reflex by mechanical vibration applied to the limb (DeGail, Lance & Neilson, 1966) probably includes these and other mechanisms.

Vibration produces greater depression of the H reflex in patients with recent

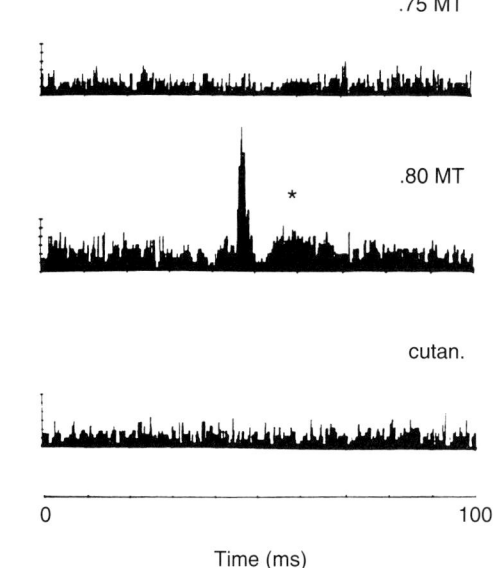

Fig. 7.8. Evidence for polysynaptic facilitation of soleus motoneurones from group I afferents in the tibial nerve in a patient with marked spasticity. Stimuli just above the threshold of group I afferents (0.8 of the threshold of the alpha motoneurone axons of soleus) resulted in a large, short-latency facilitation (reflecting the Ia EPSP) followed by a second, broader peak of increased firing probability reflecting polysynaptic facilitation from these afferents. This was not seen in normal subjects. The later peak could not be produced by cutaneous stimulation (bottom trace). Mailis and Ashby (1990) reproduced with permission.

spinal lesions (Ashby & Verrier, 1975) and, in general, less depression in patients with chronic spinal lesions (Burke & Ashby, 1972; Delwaide, 1973; Taylor, Ashby & Verrier, 1984). These changes have been attributed to modulations of presynaptic inhibition. Furthermore, when two group I volleys are delivered in quick succession, the second of the two EPSPs evoked in individual soleus motoneurones is depressed at intervals between 20 and 100 ms. This is caused by a presynaptic process, partly due to changes in the excitability of peripheral nerves, and partly due to changes presumed to occur at the Ia motoneurone synapse. In patients with chronic spinal lesions (Fig. 7.7), this depression is less (Mailis & Ashby, 1990).

With recording electrodes placed over the lumbosacral cord, it is possible to record potentials in response to stimulation of the sciatic nerve. The evoked potentials include an early negative peak and a late slow positive wave which has been equated with the dorsal root potential and thus presynaptic inhibition.

Delwaide, Schoenen and DePasqua (1985) reported that the area of this positive wave (relative to the area of the initial peak) was reduced in patients with hyperreflexia from multiple sclerosis. Thus there is some evidence, albeit still conflicting, that one of the abnormalities associated with hyperreflexia is a reduction in presynaptic inhibition (or 'homosynaptic depression') of Ia afferents.

Altered transmission through oligosynaptic pathways to motoneurones

Polysynaptic Ia actions

In addition to their monosynaptic action on motoneurones, Ia afferents in the cat have facilitatory actions on motoneurones through polysynaptic pathways (Baldissera et al., 1981) (Fig. 7.1). Certain polysynaptic projections which have been demonstrated from low-threshold muscle afferents to human upper limb motoneurones, are believed to be analogous to these (Malmgren & Pierrot-Deseilligny, 1988). Transmission in these pathways has not been specifically examined in chronic spinal cats but Mailis and Ashby (1990) found a later facilitation of soleus motoneurones from group I volleys in those patients with spinal lesions who had the greatest increase in stretch reflexes (Fig. 7.8). Thus there may be increased transmission in polysynaptic pathways from group I afferents to motoneurones in some patients with hyperreflexia.

Ib actions

There have been no studies of Ib effects in chronic spinal animals. Effects attributed to Ib actions (e.g. inhibition of soleus motoneurones from medial gastrocnemius group I volleys) have been demonstrated in man (Pierrot-Deseilligny et al., 1981; Mao et al., 1984). So far there have been no studies in spinal man, but Delwaide and Oliver (1988) have shown with hemiplegia that group I volleys in the medial gastrocnemius nerve inhibit soleus motoneurones on the normal side through a presumed Ib action, but facilitate soleus motoneurones on the hemiplegic side (Fig. 7.9). Thus, there may be alterations in Ib pathways following an upper motoneurone lesion.

Reciprocal inhibition

The cocontraction of agonists and antagonists that may be seen during attempted voluntary movements in patients with hyperreflexia led to speculation that disynaptic reciprocal inhibition might be impaired (McLellan, 1977). This does not seem to be the case. In the chronic hemisected cat, reciprocal

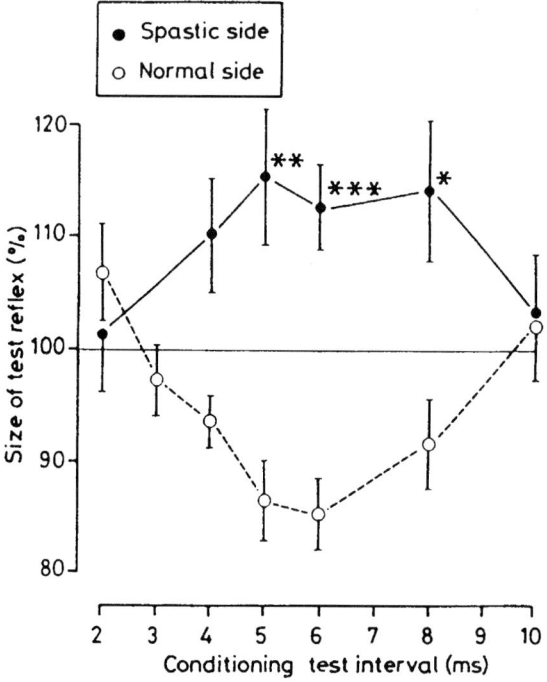

Fig. 7.9. Effects of conditioning the H reflex of soleus with a group I volley in the nerve from medial gastrocnemius in 6 patients with hemiplegia. The reduction of the H reflex on the normal side (open circles), atributed to Ib inhibition, is not present on the spastic side (filled circles) The vertical bars indicate SEM. Asterisks indicate significant differences between the two sides. Delwaide and Oliver (1988) reproduced with permission.

inhibition was increased on the side below the hemisection (Hultborn & Malmsten, 1983b).

In human patients with spinal lesions, the short-latency (presumed disynaptic) inhibition from extensors to flexors (Ashby & Wiens, 1989) (Fig. 7.10), and from flexors to extensors (Boorman et al., 1991) is greater than normal. In man, the presumed disynaptic reciprocal inhibition is followed by a later, more powerful, and prolonged inhibition (Ashby & Wiens, 1989). This is lost in patients with spinal lesions (El-Tohamy & Sedgwick, 1982; Ashby & Wiens, 1989) (Fig. 7.10). In summary, disynaptic reciprocal inhibition is enhanced following a spinal lesion but a later, more prolonged inhibition is absent. By analogy with reciprocal inhibition between forearm muscles (Berardelli et al., 1987), the long-latency phase of inhibition could be due to presynaptic inhibition, and its depression in spinal patients is further evidence for a defect in presynaptic inhibition.

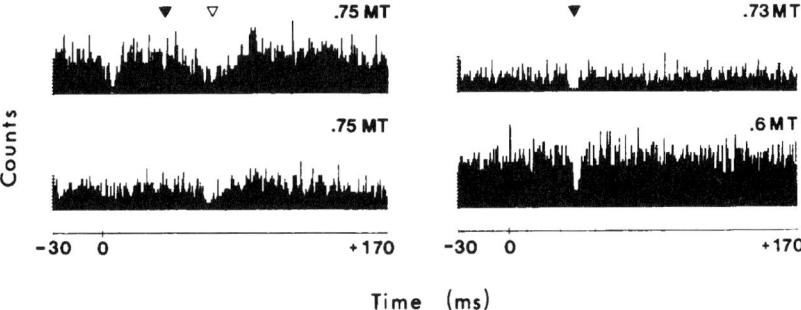

Fig. 7.10. Histograms showing the changes of firing probability of tibialis anterior motoneurones in two normal subjects (left) and two patients with spinal cord lesions (right) following group I volleys in the tibial nerve delivered at time zero. In the normal subject (top left) there is a short-latency decrease in firing probability (filled triangle) representing disynaptic reciprocal inhibition. This is followed by a more prominent later inhibition (hollow triangle) which is sometimes the only inhibition seen (bottom left). In the patients with spinal cord lesions (right) the early inhibition increased, the later inhibition is absent. Ashby and Wiens (1989) reproduced with permission.

Renshaw cells

In chronic spinal hemisected cats, Hultborn and Malmstem (1983b) found greater Renshaw inhibition on the hemisected side. This is in keeping with earlier observations of increased Renshaw cell spiking following ventral root stimulation in chronic spinal cats (Goldfarb, 1976). Renshaw cell inhibition of human soleus can be assessed in man by an ingenious technique devised by Bussel and Pierrot-Deseilligny (1979). This inhibition is less during voluntary contractions of soleus (Hultborn & Pierrot-Deseilligny, 1979) and greater during voluntary contractions of tibialis anterior (Katz & Pierrot-Deseilligny, 1984). Katz and Pierrot-Deseilligny (1982) found that Renshaw cell inhibition was increased at rest in half of the patients with upper motoneurone lesions, and that the inhibition was no longer modulated during voluntary movements.

Thus, Renshaw cell inhibition is increased following a spinal lesion. This abnormality is unlikely to contribute to the hyperreflexia although the loss of ability to modulate Renshaw cell function could adversely affect the control of muscle contractions and of reciprocal inhibition.

Reflex pathways from cutaneous afferents

Stimulation of digital nerves results in short-latency facilitation of motoneurones of hand muscles in patients with upper motoneurone lesions (Jenner & Stephens, 1982) indicating release of a short-latency pathway from cutaneous afferents to motoneurones not 'open' in normal adults. Thus transmission from cutaneous afferents to motoneurones may be exaggerated.

Reflex pathways from Group II muscle afferents

Reflex effects attributed to the Group II afferents have been demonstrated in man (Burke, Andrews & Ashby, 1971), but as yet there have been no comparisons between normal subjects and patients with spinal lesions.

Are spasms due to increased interneurone excitability?

Noting that there was convergence from cutaneous, joint and group II afferents onto reflex pathways to motoneurones (Fig. 7.1), Lundberg and colleagues (1987) postulated that, once initiated, movements may be sustained by the inevitable discharges in these afferents (which they called the 'general reflex afferents') during the course of the movement. In the relaxed state, the gain in this reflex pathway would have to be low so that passive movements did not result in reflex activation of motoneurones. An abnormally high gain in this system would result in excessive driving of motoneurones from any movement or cutaneous input. Pathways of this sort may contribute to the prolonged 'spasms' which occur in some patients with chronic spinal lesions or contribute to the exaggeration of 'tonic' rather than 'phasic' reflexes (see Neilson, this volume).

The clasp-knife phenomenon

When an attempt is made to flex the knee in a patient with a chronic spinal lesion, a stretch reflex occurs in the quadriceps which then fades away again as the movement is continued. This is known as the 'clasp-knife phenomenon'. The clasp-knife phenomenon in the human quadriceps is due to active inhibition. It is impossible to elicit a stretch reflex in this muscle with the knee flexed (Burke, Gillies & Lance, 1970). This inhibition has been attributed to the central effects of group II spindle afferents (Burke *et al.*, 1970), or to muscle group III and IV afferents (Rymer, Houk & Crago, 1979) which converge on the interneurones of the flexor reflex afferent pathway (see below) normally inhibited by a dorsolateral reticulospinal pathway (Baldissera *et al.*, 1981). The loss of Ib inhibition in hemiplegic subjects documented by Delwaide and Oliver (1988) provides a further argument against the classical explanation of the clasp-knife phenomenon as an autogenetic inhibition caused by mobilization of Golgi tendon organs.

The apparent clasp-knife phenomena in the hamstrings, biceps or triceps are not associated with length-dependent inhibition of stretch reflexes and can be explained by the fact that the stretching movement carried out by the examiner is not linear (Ashby & Burke, 1971).

Clonus

A sudden dorsiflexion of the ankle in a patient with a chronic spinal lesion may result in rhythmic oscillations at that joint known as clonus. During clonus spindle afferent discharges occur only during the period of relaxation between EMG bursts and are not coincident with EMG as in centrally generated movements (Hagbarth *et al.*, 1975). Thus clonus results from recurrent activation of the stretch reflex.

The flexion reflex

In addition to their own individual reflex pathways, small-diameter (high-threshold) muscle and cutaneous afferents converge on a common reflex pathway which results in flexion of the limb (Fig. 7.1). The interneurones in this pathway from the 'flexor reflex afferents' (FRA) are strongly inhibited from the reticular formation through systems descending in the dorsolateral columns (Baldissera *et al.*, 1981). Spinal transection results in an immediate release of the flexion reflex.

The immediate release of this pathway doubtless accounts, in part, for the exaggerated flexion reflexes seen in spinal animals but the actions of the FRA are greater if there has been a preceding spinal hemisection indicating that additional alterations in this pathway occur below a chronic spinal lesion (Hultborn & Malmsten, 1983*b*). Baker and Chandler (1987) also observed alterations in the amplitude of EPSPs and IPSPs in triceps surae motoneurones of chronic cats in response to sural nerve stimulation.

Exaggerated flexion reflexes and spontaneous flexor spasms frequently follow a spinal lesion in man. Meinck, Benecke and Conrad (1985) recorded the flexion reflex in the human lower limbs by stimulating the medial plantar nerve and recording EMG activity from limb muscles. In normal subjects, the flexor muscles showed short latency (70 ms) facilitation and inhibition followed by much later (150–200 ms) facilitation. In patients with upper motoneurone lesions (including some with lesions of the spinal cord), the early facilitation and inhibition were not observed. Weak stimuli produced an enhanced and prolonged late (150 ms) facilitation which could occur as early as 70 ms if the stimulus intensity was increased or the interstimulus interval shortened

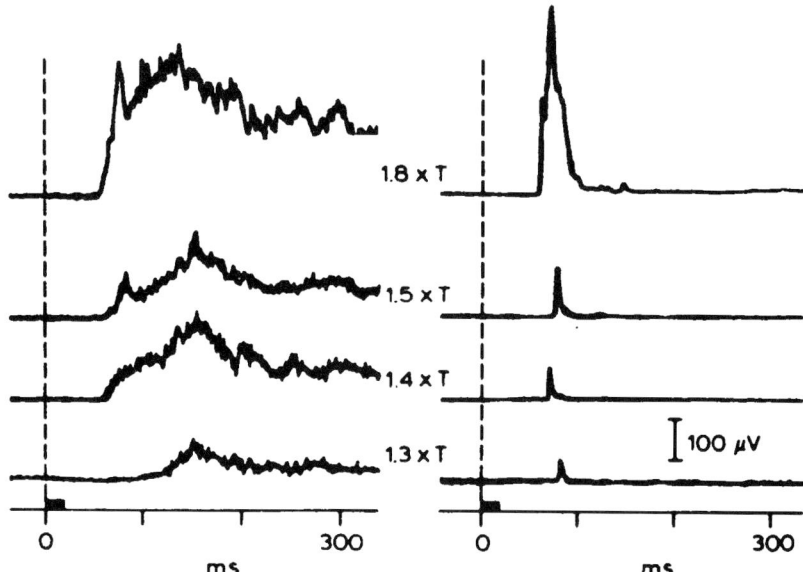

Fig. 7.11. Flexion reflex recorded with surface electrodes over the tibialis anterior in response to stimulation of the medial plantar nerve at various intensities above the threshold of the alpha motoneurone axons of flexor hallicis brevis in a patient with hemiplegia. On the normal side (right) there is a brief, short-latency response. On the abnormal side there is a delayed and prolonged response. Meinck *et al.* (1985) reproduced with permission.

(Fig. 7.11). These changes could occur immediately after an acute lesion (see also Shahani & Young, 1971). Flexor spasms have the same physiological properties as the evoked flexion reflex in the same patient and can be considered flexor reflexes for which the triggering stimulus is unrecognized (Shahani & Young, 1973).

Secondary changes in motor units below a chronic spinal lesion

Degeneration of motoneurones

Patients with longstanding spinal lesions often have wasting of muscles below the level of the lesion. This is due not just to atrophy of muscle fibres but to a loss of motor units. Hunter and Ashby (1984), for example, found a reduction in the number of motor units in the abductor hallucis muscle in 11 of 23 patients with spinal lesions, a finding which could not be explained by a coincidental peripheral nerve lesion. Fibrillation potentials are common in muscles below the level of a spinal cord lesion (Nyboer & Johnson, 1971) and histological changes in muscle indicative of denervation have been described (Reske-Nielsen, Harmsen & Ovesen, 1971). Van Alphen, Lammers and Walder (1962) found morphological changes in motoneurones within a few days of a spinal cord injury and Reske-Nielsen *et al.* (1971) found loss of anterior horn cells far below the level of the lesion in patients dying 1–28 months after spinal injury.

Changes in muscle fibres

In cats with chronic spinal transection there is atrophy of the paralysed muscles associated with decreased force output. The proportion of slow twitch (type S) and resistant fast twitch (type FR) units diminish while the proportion of fast fatiguable (type FF) and intermediate units (FI) increase. All motor unit types have shorter contraction times (Mayer *et al.*, 1984; Munson *et al.*, 1986, Cope *et al.*, 1986).

In biopsies of human muscles affected by upper motoneurone lesions (mostly hemiplegia), the type II ('fast') fibres show more atrophy than type I ('slow') fibres (Edström, 1970; Dietz *et al.*, 1986). The decline in tension with repetitive activation of muscles and the decline in muscle phosphocreatine is greater in spasticity suggesting that the excessive fatiguability has a biochemical basis (Miller *et al.*, 1990).

Changes in the rheological properties of muscles

There is a difference in the resistance to slow passive movements of the limbs in patients with spinal shock and those with a longstanding spinal cord lesion. In such movements, which are insufficient to provoke reflex EMG, the resistance felt by the examiner is largely due to the mechanical properties of the muscle and connective tissues. These changes in the mechanical properties of muscles make a major contribution to the resistance to passive stretch in patients with chronic upper motoneurone lesions (Herman, 1970; Dietz, Quintern & Berger, 1981). Allum & Mauritz (1984) have described a method for dissecting out the inertial, the passive elastic and viscous and the reflex contributions to the resistance to muscle stretch (components which will depend, to some extent, on the immediate past history of movement, see Hagbarth *et al.*, 1985). Hufschmidt and Mauritz (1985) examined the resistance to displacement in patients with cerebral or spinal lesions. Where the lesion had been present for less than one year, only the reflex component was increased. After one year there was, in addition, an increase in the mechanical stiffness of the muscle which they attributed to structural alterations within muscle fibres. A reduction in the number of sarcomeres has been described in muscles that have been actively shortened by repeated stimulation (Tabary *et al.*, 1981). The importance of this change is also considered by Neilson (this volume).

There are times when the scrutiny of individual reflex pathways in the human spinal cord may seem a somewhat academic exercise, but such studies have revealed the humbling complexity of the spinal machinery, and, in the final analysis, these circuits must be understood if drug therapy (Davidoff, 1985), physical therapy or spinal transplants (Tessler, 1991) are to be effectively employed.

Summary and conclusions

The exaggeration of spinal reflexes below the level of a chronic spinal lesion is not due to increased muscle spindle discharge or to alteration in the biophysical properties of the motoneurones. The synaptic potentials produced in motoneurones from multiple spindle afferents are not larger, although there are changes in the way that multiple volleys in these afferents are transmitted to motoneurones. This is possibly related to changes in presynaptic inhibition. The greatest abnormalities are found in polysynaptic spinal reflex pathways. Polysynaptic facilitation from Ia afferents to motoneurones is increased and polysynaptic Ia inhibition of antagonists is decreased. Disynaptic reciprocal inhibition and Renshaw cell inhibition are increased. Transmission from cutaneous afferents to motoneurones is enhanced. There are also important secondary consequences of a spinal cord lesion including loss of motoneurones and alterations in the mechanical properties of muscle fibres which contribute to the clinical picture.

Acknowledgements

Recent work carried out in our laboratory was funded by the Canadian Medical Research Council grant #MA 6727 and The National Institutes of Health grant #RO1 NS 27873-02.

References

Allum, J. H. J. & Mauritz, K.-H. (1984). Compensation for intrinsic muscle stiffness by short latency reflexes in human triceps surae muscles. *Journal of Neurophysiology*, **52**, 797–818.

Ashby, P. (1990). Neurophysiological effects of chronic spinal lesions in animals and man. In *The Origin and Treatment of Spasticity*. eds. R. Benecke, M. Emre & R.A. Davidoff, pp. 29–61. Carnforth: Parthenon.

Ashby, P. & Burke D. (1971). Stretch reflexes in the upper limb of spastic man. *Journal of Neurology, Neurosurgery, and Psychiatry*, **34**, 765–71.
Ashby, P. & McCrea D. (1987). Neurophysiology of spinal spasticity. In *Handbook of the Spinal Cord*, ed. R.A. Davidoff, pp. 119–43. New York: Marcel Dekker.
Ashby, P., Stålberg E., Winkler, T. & Hunter, J. P. (1987). Further observations on the depression of group Ia facilitation of motoneurones by vibration in man. *Experimental Brain Research*, **69**, 1–6.
Ashby, P. & Verrier, M. (1975). Neurophysiological changes following spinal cord lesions in man. *Canadian Journal of Neurological Science*, **2**, 91–100.
Ashby, P. & Wiens, M. (1989). Reciprocal inhibition following lesions of the spinal cord in man. *Journal of Physiology (London)*, **414**, 145–57.
Ashby, P. & Zilm, D. (1982a). Relationship between EPSP shape and cross-correlation profile explored by computer simulation for studies on human motoneurons. *Experimental Brain Research*, **47**, 33–40.
Ashby, P. & Zilm, D. (1982b). Characteristics of postsynaptic potentials in single human motoneurons by homonymous group I volleys. *Experimental Brain Research*, **47**, 41–8.
Bailey, C. S., Lieberman J.S. & Kitchell, R. L. (1980). Response of muscle spindle primary endings to static stretch in acute and chronic spinal cats. *American Journal of Veterinary Research.*, **41**, 2030–6.
Baker, L. L. & Chandler, S. H. (1987). Characterization of post-synaptic potentials evoked by sural nerve stimulation in hindlimb motoneurons from acute and chronic spinal cats. *Brain Research*, **420**, 340–50.
Baldissera, F., Hultborn, H. & Illert, M. (1981). Integration in spinal neuronal systems. In *Handbook of Physiology, Section I, The Nervous System*, Vol 2, part 1, ed. J.M. Brookhart & V.B. Mountcastle, pp. 509–95. Maryland: American Physiological Society.
Bayoumi, A. & Ashby P. (1989). Projections of group Ia afferents to motoneurons of thigh muscles in man. *Experimental Brain Research,* **76**, 223–8.
Berardelli, A., Day, B. L., Marsden, C. D. & Rothwell, J. C. (1987). Evidence favouring presynaptic inhibition between antagonist muscle afferents in the human forearm. *Journal of Physiology (London)*, **391**, 71–83.
Birnbaum, A. & Ashby, P. (1982). Post synaptic potentials in individual soleus motoneurons in man produced by Achilles tendon taps and electrical stimulation of tibial nerve. *Electroencephalography and Clinical Neurophysiol*ogy, **54**, 46–9.
Boorman, G., Hulliger M., Lee R. G., Tako, K. & Tanaka, R. (1991). Reciprocal Ia inhibition in patients with spinal spasticity. *Neuroscience Letters*, **127**, 57–60.
Burke, D. (1983). Critical examination of the case for and against fusimotor involvement in disorders of muscle tone. In *Progress in Clinical Neurophysiology, 10, Advances in Neurology*, vol 39, ed. J. E. Desmedt, pp. 133–50. New York: Raven Press.
Burke, D. (1988). Spasticity as an adaptation to pyramidal tract injury. In *Advances in Neurology, vol 47, Functional Recovery in Neurological Disease*, ed. S. G. Waxman pp. 401–23. New York: Raven Press.
Burke, D., Andrews, C. & Ashby, P. (1971). Autogenic effects of static muscle stretch in spastic man. *Archives of Neurolology*, **25**, 367–72.
Burke, D. & Ashby, P. (1972). Are spinal 'presynaptic' inhibitory mechanisms suppressed in spasticity? *Journal of Neurological Sciences*, **15**, 321–6.
Burke, D., Gandevia, S. C & McKeon, B. (1983). The afferent volleys responsible for spinal proprioceptive reflexes in man. *Journal of Physiology (London)*, **339**, 535–52.

Burke, D., Gandevia, S. C. & McKeon, B. (1984). Monosynaptic and oligosynaptic contributions to human ankle jerk and H reflex. *Journal of Neurophysiology*, **52**, 435–48.

Burke, D., Gillies, J. D. & Lance J. W. (1970). The quadriceps stretch reflex in human spasticity. *Journal of Neurology, Neurosurgery, and Psychiatry*, **33**, 216–23.

Burke, R. E. & Rudomin, P. (1977). Spinal neurons and synapses. In *Handbook of Physiology, Section 1, The Nervous System*, vol. 1, eds. J. M. Brookhart & V. B. Mountcastle, pp. 877–944. Baltimore: Wilkins.

Bussel, B. & Pierrot-Deseilligny, E. (1977). Inhibition of human motoneurons, probably of Renshaw origin, elicited by an orthodromic motor discharge. *Journal of Physiology (London)*, **269**, 319–39.

Cope, T. C., Bodine, S. C., Fournier, M., & Edgerton, V.R. (1986). Soleus motor units in chronic spinal transected cats: physiological and morphological alterations. *Journal of Neurophysiology*, **55**, 1202–20.

Cope, T. C., Fetz, E. E. & Matsumura, M. (1987). Cross-correlation assessment of synaptic strength of single Ia fiber connections with triceps surae motoneurons in cats. *Journal of Physiology (London)*, **390**, 161–88.

Curtis, D. R. & Eccles J.C. (1960). Synaptic action during and after repetitive stimulation. *Journal of Physiology (London)*, **150**, 374–98.

Davidoff, R. A. (1985). Antispastic drugs: mechanisms of action. *Annals of Neurology*, **17**, 107–16.

DeGail, P., Lance, J. W. & Neilson, P. D. (1966). Differential effects on tonic and phasic reflex mechanisms produced by vibration of muscles in man. *Journal of Neurology, Neurosurgery, and Psychiatry*, **29**, 1–11.

Delwaide, P. J. (1973). Human monosynaptic reflexes and presynaptic inhibition: an interpretation of spastic hyperreflexia. In *New Developments in Electromyography and Clinical Neurophysiology*, vol. 3, ed. J. E. Desmedt, pp. 505–22. Basel: Karger.

Delwaide, P. J. & Oliver, E. (1988). Short latency autogenic inhibition (Ib inhibition) in human spasticity. *Journal of Neurology, Neurosurgery, and Psychiatry*, **51**, 1546–50.

Delwaide, P. J., Schoenen, J. & DePasqua, V. (1985). Lumbosacral spinal evoked potentials with multiple sclerosis. *Neurology*, **35**, 174–9.

Dietz, V., Quintern, J. & Berger, W. (1981). Electrophysiological studies of gait in spasticity and rigidity. Evidence that altered mechanical properties of muscles contribute to hypertonia. *Brain*, **104**, 431–49.

Dietz, V., Ketelsen, V.-P., Berger W. & Quintem, J. (1986). Motor unit involvement in spastic paresis. Relationship between leg muscle activation and histochemistry. *Journal of the Neurological Sciences*, **75**, 89–103.

Edström, L. (1970). Selective changes in the sizes of red and white muscle fibers in upper motor lesions and Parkinsonism. *Journal of the Neurological Sciences*, **11**, 537–50.

El-Tohamy, A. & Sedgwick, E. M. (1982). Spinal inhibitory mechanisms in spasticity. *Electroencephalography and Clinical Neurophysiology*, **53**, 3–4P.

Fetz, E. E. & Gustafsson, B. (1983). Relation between shapes of post-synaptic potentials and changes in firing probability of cat motoneurons. *Journal of Physiology (London)*, **341**, 387–410.

Goldfarb, I. (1976). Excitation of Renshaw cells via motor neuron collaterals in acute and chronic spinal cats. *Brain Research*, **106**, 176–83.

Hagbarth, K.-E., Hägglund, J. V., Nordin, M. & Wallin, E. V. (1985). Thixotropic behaviour of human finger flexor muscles with accompanying changes in spindle and reflex response to stretch. *Journal of Physiology (London)*, **368**, 323–42.

Hagbarth, K.-E., Wallin, G. & Löfstedt, L. (1973). Muscle spindle responses to stretch in normal and spastic subjects. *Scandinavian Journal of Rehabilitation Medicine*, **5**, 156–9.

Hagbarth, K.-E., Wallin, G., Löfstedt, L. & Aquilonius, S. M. (1975). Muscle spindle activity in alternating tremor of Parkinsonism and in clonus. *Journal of Neurology, Neurosurgery, and Psychiatry*, **38**, 636–41.

Herman, R. (1970). The myotatic reflex. Clinico-physiological aspects of spasticity and contracture. *Brain*, **93**, 273–312.

Hochman, S. & McCrea, D. A. (1993*a*). Changes in composite Ia EPSPs in triceps surae and plantaris motoneurons following chronic spinalization. *Journal of Neurophysiology*, (in press).

Hochman, S. & McCrea, D. A. (1993*b*). Electrical properties of ankle extensor motoneurons following chronic spinalization. *Journal of Neurophysiology*, (in press).

Hufschmidt, A. & Mauritz, K.-H. (1985). Chronic transformation of muscle in spasticity: a peripheral contribution to increased tone. *Journal of Neurology, Neurosurgery, and Psychiatry*, **48**, 676–85.

Hultborn, H. & Malmstem, J. (1983*a*). Changes in segmental reflexes following chronic spinal cord hemisection in the cat. I Increased monosynaptic and polysynaptic ventral root discharges. *Acta Physiologica Scandinavica*, **119**, 405–22.

Hultborn, H. & Malmstem, J. (1983*b*). Changes in segemental reflexes following chronic spinal cord hemisection in the cat. II Conditioned monosynaptic test reflexes. *Acta Physiologica Scandinavica*, **119**, 423–33.

Hultborn, H., Meunier, S., Morin C. & Pierrot-Deseilligny, E. (1987). Assessing changes in presynaptic inhibition of Ia fibres: a study in man and in the cat. *Journal of Physiology (London)*, **389**, 729–56.

Hultborn, H. & Pierrot-Deseilligny, E. (1979). Changes in recurrent inhibition during voluntary soleus contractions in man studied by an H-reflex technique. *Journal of Physiology (London)*, **297**, 229–51.

Hunter, J. & Ashby, P. (1984). Secondary changes in segmental neurons below a spinal cord lesion in man. *Archives of Physical Medicine and Rehabilitation*, **65**, 702–5.

Jenner, J. R. & Stephens, J. A. (1982). Cutaneous reflex responses and their central nervous pathways studied in man. *Journal of Physiology (London)*, **333**, 405–9.

Katz, R., Morin, C. & Pierrot-Deseilligny, E. (1987). Conditioning of H reflex by a preceding subthreshold tendon reflex stimulus. *Journal of Neurology, Neurosurgery, and Psychiatry*, **40**, 575–80.

Katz, R. & Pierrot-Deseilligny, E. (1982). Recurrent inhibition of alpha motoneurons in patients with upper motor neuron lesions. *Brain*, **105**, 103–24.

Katz, R. & Pierrot-Deseilligny, E. (1984). Facilitation of soleus-coupled Renshaw cells during voluntary contraction of pretibial flexor muscles in man. *Journal of Physiology (London)*, **355**, 587–603.

Katz, R. T. & Rymer W. Z. (1989). Spastic hypertonia: mechanisms and measurement. *Archives of Physical Medicine and Rehabilitation*, **70**, 144–55.

Kuhn, R. A. (1950). Functional capacity of the isolated human spinal cord. *Brain*, **73**, 1–51.

Lance, J. W. (1980). Symposium synopsis. In *Spasticity: Disordered Motor Control*, eds. R. G. Feldman, R. R. Young & W. P. Koella, pp. 485–95. Chicago: Year Book Medical Publishers.

Lundberg, A., Malmgren K. & Schomburg, E. D. (1987). Reflex pathways from group II muscle afferents. 3 secondary spindle afferents and the FRA: a new hypothesis. *Experimental Brain Research*, **65**, 294–306.

Mailis, A. & Ashby P. (1990). Alterations in group Ia projections to motoneurons following spinal lesions in man. *Journal of Neurophysiology,* **64**, 637–47.

Malmgren, K. & Pierrot-Deseilligny, E. (1988). Evidence for non-monosynaptic Ia excitation of human wrist flexor motoneurones, possibly via propriospinal neurones. *Journal of Physiology (London),* **405**, 747–64.

Mao, C. C., Ashby, P., Wang, M. & McCrea, D. (1984). Synaptic connections from large muscle afferents to the motoneurones of various leg muscles in man. *Experimental Brain Research,* **56**, 341–50.

Mayer, R. F., Burke, R. E., Toop, J., Walmsley, B. & Hodgson, J. A. (1984). The effect of spinal cord transection on motor units in cat medial gastrocnemius muscles. *Muscle and Nerve,* **7**, 23–31.

McLellan, D. L. (1977). Co-contraction and stretch reflexes in spasticity during treatment with baclofen. *Journal of Neurology, Neurosurgery, and Psychiatry,* **40**, 30–8.

Meinck, H.-M., Benecke R. & Conrad, B. (1985). Spasticity and the flexion reflex. In *Clinical Neurophysiology in Spasticity,* eds. P. J. Delwaide & R. R. Young, pp. 41–54. Amsterdam: Elsevier.

Mendell, L. M. (1984). Modifiability of spinal synapses. *Physiological Reviews,* **64**, 260–324.

Midroni, G. & Ashby, P. (1989). How synaptic noise may affect cross-correlations. *Journal of Neuroscience Methods,* **27**, 1–12.

Miller, R. G., Green, A. T., Moussavi, R. S., Carson, P. J. & Weiner M. W. (1990). Excessive muscular fatigue in patients with spastic paraparesis. *Neurology,* **40**, 1271–4.

Munson, J. B., Foehring, R. C., Lofton, S. A., Zengel, J. E. & Sypert, G. W. (1986). Plasticity of medial gastrocnemius motor units following cordotomy in the cat. *Journal of Neurophysiology,* **55**, 619–34.

Naftchi, N. E. Schlosser, W. & Horst, W. D. (1980). Correlation of changes in the GABA-ergic system with the development of spasticity in paraplegic cats. In *GABA – Biochemistry and CNS functions,* eds. P. Mandel & F. V. DeFeudis, pp. 431–50. New York: Plenum Press.

Nelson, S. G. & Mendell, L. M. (1979). Enhancement in Ia-motoneuron synaptic transmission caudal to chronic spinal cord transection. *Journal of Neurophysiology,* **42**, 642–54.

Nyboer, V. J. & Johnson, H. E. (1971). Electromyographic findings in lower extremities of patients with traumatic quadriplegia. *Archives of Physical Medicine and Rehabilitation,* **52**, 256–9.

Pierrot-Deseilligny, E. & Mazières, L. (1985). Spinal mechanisms underlying spasticity. In *Clinical Neurophysiology of Spasticity,* eds. P. J. Delwaide & R. R. Young, pp. 63–76. Amsterdam: Elsevier.

Pierrot-Deseilligny, E., Morin, C., Bergego, C. & Tankov, N. (1981). Pattern of group I fibre projections from ankle flexor and extensor muscles in man. *Experimental Brain Research,* **42**, 337–50.

Reske-Nielsen, E., Harmsen, A. & Ovesen, N. (1971). Pathological study of muscle biopsies from the legs in patients with fractures of the cervical spine. *Actualités de Pathologie Neuromusculaire* pp. 509–21. Paris: Expansion Scientifique.

Riddoch, G. (1917). The reflex functions of the completely divided spinal cord in man compared to those associated with less severe lesions. *Brain,* **40**, 264–404.

Rymer, W. Z., Houk, J. C. & Crago, P. E. (1979). Mechanisms of the clasp knife reflex studied in an animal model. *Experimental Brain Research,* **37**, 93–113.

Schiller, H. H. & Stålberg, E. (1978). F response studied with single fiber EMG in normal subjects and spastic patients. *Journal of Neurology, Neurosurgery, and Psychiatry* **41**, 45–53.

Schmidt, R. F. (1971). Presynaptic inhibition in the vertebrate central nervous system. *Ergeben Physiology*, **63**, 20–101.

Shahani, B. T. & Young, R. R. (1971). Human flexor reflexes. *Journal of Neurology, Neurosurgery, and Psychiatry*, **34**, 616–27.

Shahani, B. T. & Young, R. R. (1973). Human flexor spasms. In *New Developments in Electromyography and Clinical Neurophysiology*, vol. 3, ed. J. E. Desmedt, pp. 734–43. Basel: Karger.

Swash, M. & Earl C. J. (1975). Flaccid paraplegia: a feature of spinal cord lesions in Holmes Adie syndrome and tables dorsalis. *Journal of Neurology, Neurosurgery, and Psychiatry*, **38**, 317–21.

Tabary, J.-C., Tardieu, C., Tardieu, G. & Tabary, C. (1981). Experimental and rapid sarcomere loss with concomitant hypoextensibility. *Muscle and Nerve*, **4**, 198–203.

Taylor, S., Ashby, P. & Verrier, M. (1984). Neurophysiological changes following spinal lesions in man. *Journal of Neurology, Neurosurgery, and Psychiatry*, **47**, 1102–8.

Tessler, A. (1991). Intraspinal transplants. *Annals of Neurology*, **29**, 115–23.

Van Alphen, H. A. M., Lammers, H. J. & Walder, H. A. D. (1962). Sur une réactoin remarquable des motoneurons de la région lombo-sacrée après une transection cervicale traumatique chez l'homme. *Neurochirurgie*, **8**, 328–30.

8

Future strategies for the treatment of Parkinson's disease

S. FAHN

Department of Neurology,
Columbia University College of Physicians & Surgeons
and
The Neurological Institute of New York,
Presbyterian Hospital, New York NY, USA

It is hazardous to read a 'crystal ball' and predict the future. Any new, unexpected development in understanding the pathophysiology and biochemistry of Parkinson's disease (PD) could lead to an entirely new therapeutic approach that is unpredictable at present. In trying to predict the future of new therapy for PD, I think there will be at least five separate directions: (i) attempts to slow the progression of the disease, (ii) attempts to overcome complications from current therapies, (iii) attempts to treat other neurotransmitter and chemical defects, (iv) attempts to control symptoms by noncentral approaches, and (v) attempts to control symptoms and progression by surgical approaches.

Attempts to slow the progression of the disease

The attempts to slow progression of PD will be on two different paths. There will be investigations of trophic factors such as brain derived neurotrophic factor (BDNF) and epidermal growth factor (EGF). There will be new attempts to prevent oxidative stress, such as using new MAO-B and MAO-A inhibitors, free radical scavengers, and interfering with the iron-induced catalysis in forming free radicals. Whether any will be successful is uncertain. These approaches would be particularly useful in the presymptomatic stage of the disease, and there will be attempts to detect the disorder preclinically (Langston, 1990; Calne & Snow, 1992) so that protective measures can be applied as soon as possible.

Investigating trophic factors

Three trophic factors have been implicated with the dopaminergic nigrostriatal system: brain derived neurotrophic factor (BDNF) (Hyman *et al.*, 1991), epidermal growth factor (EGF) (Hadjiconstantinou *et al.*, 1991; Pezzoli *et al.*,

1991), and acidic and basic fibroblast growth factors (Bean et al., 1991; Cintra et al., 1991). The first two have already been studied in models of PD, both in vitro and in vivo and will be discussed here.

Hyman et al. (1991) tested BDNF in dissociated rat ventral mesencephalic cell cultures. These cultures normally show a 75% loss of dopaminergic neurons after 8 days. In the presence of BDNF, there was a loss of only one-third as much. If the cultures are exposed to MPP+, the toxic metabolite of MPTP, over a 2-day period, a loss of 75% of cells are seen but, when grown in the presence of BDNF, there was a loss of only 32% of cells. These studies suggest that BDNF in vitro is a trophic factor for dopaminergic neurons. It needs to be determined if BDNF is also effective with in vivo systems.

Hadjiconstantinou and her colleagues (1991) evaluated EGF both in vivo and in vitro. In the former, they showed that the striatal content of dopamine, the acid metabolite of dopamine, and tyrosine hydroxylase activity increased after intracerebroventricular infusion of EGF in mice with MPTP-induced loss of striatal dopamine. In the in vitro experiments they exposed embryonic mesencephalon cultures to MPTP, which interferes with dopamine uptake. When the cultures were then treated with EGF, the uptake of dopamine increased.

Pezzoli and his colleagues (1991) tested EGF in a rat model of parkinsonism. The nigrostriatal tract of rats were sectioned with a knife cut, and the dopamine pathway was allowed to degenerate. Animals were tested for the effectiveness of the lesion by testing rotation to amphetamine stimulation. Experimental animals were treated with intracerebroventricular infusion of EGF beginning 36 days after the lesion was made. Control animals received either saline or denatured EGF. The infusions lasted 28 days, using two infusion solutions, lasting 14 days each. The EGF solution was tested for viability at the end of each 14-day period, and found to be potent. Three sets of rotations were carried out over the next 130 days after the infusions had ceased. The EGF-infused animals had a significant reduction in the number of rotations compared to control animals. After the animals were sacrificed, immunohistochemical staining for tyrosine hydroxylase showed increased staining in nigra and striatum in the EGF-treated animals. These studies indicate that EGF can partially reverse Parkinsonism induced by transection of the nigrostriatal pathway, even after EGF infusions had been discontinued.

Antioxidants to prevent oxidant stress

Oxidant stress has been proposed as a mechanism for dopaminergic neuronal death in PD (Graham, 1978; Cohen, 1983; Mann & Yates, 1983; Fahn, 1989; Halliwell, 1989; Olanow, 1990; Jenner, 1991). Normally a balance between oxidative events and antioxidative forces maintains the status quo within living cells. A variety of enzymes help to maintain cells in a reduced state despite the presence of an aerobic environment. Thus major cellular reducing agents, such as ascorbate, glutathione and tocopherol, are present predominantly in their reduced states (rather than their oxidized forms). In addition, a number of enzymes scavenge and remove reactive chemical species such as hydrogen peroxide or the superoxide radical. When the normal balance is upset, either by loss of reducing agents or protective enzymes, or by increased production of oxidizing species (such as hydrogen peroxide and the hydroxyl radical), or by both events simultaneously, the tissue is considered to be under 'oxidant stress'.

Oxidant stress has been implicated in PD because of the coalition of four biochemical features at, or around, the major site of cell death, namely the monoaminergic neurons, particularly the dopaminergic neurons, and especially the pigmented ones, in the substantia nigra (Hirsch, Graybiel & Agid, 1988). The four unique biochemical features of the substantia nigra's dopaminergic cells and processes are monoamine oxidase activity, autoxidation, accumulation of iron, and presence of neuromelanin.

The oxidant stress hypothesis in the etiology of PD was a major factor leading to the decision to test deprenyl as a possible protective agent. The initial studies showed that deprenyl delays the need for levodopa in otherwise untreated patients with PD (Tetrud & Langston, 1989; Parkinson Study Group, 1989; Teräväinen, 1990; Myllyla et al., 1992). Because deprenyl has mild symptomatic effects, there is some doubt whether the delay in initiating levodopa was due to protective or symptomatic effect, or both. Other MAO-B inhibitors will be tested, and this may help provide an answer. Already lazabemide, a reversible MAO-B inhibitor without active metabolites, has begun to be investigated (Parkinson Study Group, 1993).

In addition to testing new MAO-B inhibitors, there will be tests on reversible MAO-A inhibitors because this enzyme acts intraneuronally while MAO-B acts extraneuronally (Westlund et al., 1985; Konradi et al., 1989). Other free radical scavengers, besides α-tocopherol and ascorbate (Fahn, 1992a) are being developed, particularly drugs known as lazeroids and agents to chelate free iron. Like other proposed drugs for protective therapy, these approaches would be particularly useful in the presymptomatic stage of the disease.

Attempts to overcome complications from current therapies

Much of today's problems in treating the patient with PD stems from the complications of long-term levodopa therapy (Fahn, 1992b). The major difficulty is that of clinical response fluctuations. For this problem, longer-acting dopamine agonists, such as ropinerol, and dopa-releasers, such as Sinemet CR and Madopar HBS, will be studied, along with drugs to extend the biological half-life of dopamine, such as catechol-o-methyltransferase (COMT) and monoamine oxidase (MAO) inhibitors. There will be investigations of new delivery systems, such as subcutaneous and enteral infusions, and skin patches containing new dopamine agonists.

Surgical approaches to extend duration of benefit from dopamine are being devised. Fetal mesencephalic implants have received the most publicity (Freed *et al.*, 1990; Lindvall *et al.*, 1992; Sawle *et al.*, 1992), but studies are under way stereotactically to place dopamine producing cells inside semipermeable membranes in the striatum. This approach has been successful in animals (Aebischer *et al.*, 1991), and tests on humans should soon begin.

Psychosis from levodopa therapy is already being counteracted by adding the atypical neuroleptic, clozapine (Pfeiffer *et al.*, 1990; Friedman & Lannon, 1990). Because of its potential adverse effect of agranulocytosis in 2% of the patients, a search is on for other atypical neuroleptics. These will also be tried on patients with PD with psychiatric complications from the medications.

Drugs targeted at nondopamine receptors

Drugs acting at other neurotransmitter sites

Although most of the motor problems in PD are due to striatal dopamine deficiency, some motor problems, such as freezing (motor blocks) and loss of postural reflexes, as well as nonmotor problems, such as bradyphrenia, depression and dementia, appear not to respond to dopaminergic agents. Therefore, neurotransmitter deficits other than dopamine may be involved in this disorder. Unfortunately, drugs able to affect most of the other transmitters, be they agonists or antagonists, are not particularly active, compared to drugs that affect the dopamine system. However, with time, there should be drugs available that act at other neurotransmitter sites, e.g. norepinephrine, serotonin, GABA, and peptides. When available, these drugs will be tested against the less tractable symptoms of PD.

Drugs inhibiting NMDA and other excitatory receptors

Glutamate is an excitatory neurotransmitter, and the glutamate receptors are described by the chemical agent that drives them to continuous excitatory activity, leading to neuronal death by exhaustion. The N-methyl-D-aspartate (NMDA) receptor is one of these.

Greenamyre and O'Brien (1991) listed several reasons why antagonists of the glutamate receptors might be effective in PD. Then Klockgether and his colleagues (1991) tested the concept in a primate model of Parkinsonism. The subthalamic nucleus is overactive in PD, and its output to the medial segment of the globus pallidus, and to the substantia nigra reticulata, is glutaminergicatergic. By treating with a glutamate antagonist, one might be able to lessen the symptoms and signs of PD. These investigators tested a quisqualate receptor antagonist in two aged Rhesus monkeys with bilateral, moderate, stable Parkinsonism after bilateral carotid artery injections of MPTP. They found that this provided anti-Parkinsonian effects, and was also synergistic with subclinical doses of levodopa. Thus one can predict that further investigations along these lines will be conducted, and eventually tested in patients.

Attempts to control symptoms by noncentral approaches

Jankovic and Schwartz (1991) showed that botulinum toxin injected into forearm muscles can reduce the severity of essential tremor. The same principle can be applied to the treatment Parkinsonian tremor. If successful, this has the potential to treat a symptom without having to use centrally active drugs, with all their possible adverse effects.

Surgical approaches

In the 1950s and 1960s stereotaxic thalamotomy was a common procedure in patients with PD. It was particularly effective for controlling tremor, somewhat effective for rigidity, and less so for bradykinesia. It probably had no effect on loss of postural reflexes. Bilateral surgery would not uncommonly produce an unwanted disabling dysarthria. The procedure did not slow progression of the disease, and elderly patients were prone to develop complications of stroke. So the procedure was largely abandoned after levodopa therapy became established as a more effective treatment.

However, time has shown that many patients on long-term levodopa therapy have complications of fluctuations and dyskinesias, and levodopa does not control tremor is a large minority of patients. Thus, thalamotomy is making a comeback to a small extent, particularly in patients with intractable tremor (Kelly et al., 1987; Fox, Ahlskog & Kelly, 1991). In some places, it may also be applied in patients to control dopa-induced dyskinesias and to treat the 'off' states (Narabayashi, Yokochi & Nakajima, 1984).

Another surgical approach is the stimulation of the same target used for ablation. Electrical stimulation of the thalamus can effectively control contralateral tremor and may be safer, particularly in the dominant hemisphere or when bilateral procedures are needed (Blond & Siegfried, 1991; Benabid *et al.*, 1991).

Stereotaxic ablative pallidotomy, namely in the ventroposterolateral portion of the medial segment of the pallidum, is also being revived (Laitinen, Bergenheim & Hariz, 1992). The results suggest that lesions in this area not only control tremor, but may also offer benefit for bradykinesia. This area of the medial segment of the globus pallidus receives the subthalamo-pallidal pathway. Thus a lesion in this region would be analogous to a lesion in the subthalamic nucleus.

Two groups had shown that surgical lesions of the subthalamic nucleus (corpus Luysi) can correct Parkinsonism in MPTP primates (Bergman, Wichmann & DeLong, 1990; Aziz *et al.*, 1991). The principle is that the subthalamic nucleus is overly active, and a lesion can restore the balance on the target nuclei, the medial globus pallidus and the substantia nigra reticulata. An experiment of nature has recently been reported in which a patient with PD had a stroke affecting the subthalamic nucleus; the result was an improvement of the Parkinsonism contralaterally (Sellal *et al.*, 1992). Thus it seems that there will be further investigations regarding luysiectomy and pallidotomy in the future.

Fetal transplantation technology was discussed briefly above. Benefits from transplanting dopaminergic neurons may be largely a form of replacement therapy of dopamine at the specific target site. This form of surgery is still investigational; too many questions remain as to the exact methodology and the benefit from the procedure. However, this is a fast moving field, and a number of centres worldwide are now involved. Adrenal medulla transplantation is still being conducted at some centres. One approach is to inject tissue that produces nerve growth factor along with the medullary tissue. Preliminary results (Olson *et al.*, 1991) indicate that there might be some merit to this approach.

Another transplantation concept is to engineer cells of the host so that these cells, such as fibroblasts or glia, contain the enzymes tyrosine hydroxylase and dopa decarboxylase, then the implanted cells would not be rejected by the host, and they would produce dopamine. Preliminary studies in animals indicate that this idea is feasible (Chen *et al.*, 1991).

Conclusion

All these strategies are reasonable, and any could lead to improvements in the patient with PD. This survey covered only the major aspects of the possibilities. Extensive research on PD continues and is ever more exciting. Emphasis is still on improving current therapy, but more and more attention is being paid to protective therapy, trophic factors, and surgical approaches. Some of these will eventually become an accepted treatment form in the future, but which of these will succeed best remains to be answered.

References

Aebischer, P., Tresco, P. A., Sagen, J. & Winn, S. R. (1991). Transplantation of microencapsulated bovine chromaffin cells reduces lesion-induced rotational asymmetry in rats. *Brain Research*, **560**, 43–9.

Aziz, T. Z., Peggs, D., Sambrook, M. A. & Crossman, A. R. (1991). Lesion of the subthalamic nucleus for the alleviation of 1-methyl-4-phenyl-1,2,3,6-tetrahydropyridine (MPTP)-induced parkinsonism in the primate. *Movement Disorders*, **6**, 288–92.

Bean, A. J., Elde, R., Cao, Y. H., Oellig, C., Tamminga, C., Goldstein, M., Pettersson, R. F. & Hökfeldt, T. (1991). Expression of acidic and basic fibroblast growth factors in the substantia-nigra of rat, monkey, and human. *Proceedings of the National Academy of Sciences, USA*, **88**, 10237–41.

Benabid, A. L., Pollak, P., Gervason, C., Hoffmann, D., Gao, D. M., Hommel, M., Perret, J.E. & de Rougemont, J. (1991). Long-term suppression of tremor by chronic stimulation of the ventral intermediate thalamic nucleus. *Lancet*, **337**, 403–6.

Bergman, H., Wichmann, T. & DeLong, M. R. (1990). Reversal of experimental parkinsonism by lesions of the subthalamic nucleus. *Science*, **249**, 1436–8.

Blond, S. & Siegfried, J. (1991). Thalamic stimulation for the treatment of tremor and other movement disorders. *Acta Neurochirurgica*, **52**, 109–11.

Calne, D. B. & Snow, B.J. (1992). Early markers of idiopathic parkinsonism. In *The Scientific Basis for the Treatment of Parkinson's Disease*, eds. C. W. Olanow & A. N. Lieberman, pp. 3–11. Carnforth, England: Parthenon Publishing Group.

Chen, L. S., Ray, J., Fisher, L. J., Kawaja, M. D., Schinstine, M., Kang, U. J. & Gage, F. H. (1991). Cellular replacement therapy for neurologic disorders: potential of genetically engineered cells. *Journal of Cell Biochemistry*, **45**, 252–7.

Cintra, A., Cao, Y. H., Oellig, C., Tinner, B., Bortolotti, F., Goldstein, M., Pettersson, R. F. & Fuxe, K. (1991). Basic FGF is present in dopaminergic neurons of the ventral midbrain of the rat. *Neuroreport*, **2**, 597–600.

Cohen, G. (1983). The pathobiology of Parkinson's disease: biochemical aspects of dopamine neuron senescence. *Journal of Neural Transmission*, Suppl. 19, 89–103.

Fahn, S. (1989). The endogenous toxin hypothesis of the etiology of Parkinson's disease and a pilot trial of high dosage antioxidants in an attempt to slow the progression of the illness. *Annals of the New York Academy of Science*, **570**, 186–96.

Fahn, S. (1992*a*). A pilot trial of alpha-tocopherol and ascorbate in early Parkinson's disease. *Annals of Neurology*, **32** (Suppl.), S 128–32.

Fahn, S. (1992*b*). Adverse effects of levodopa. In *The Scientific Basis for the Treatment of Parkinson's Disease*, eds. C. W. Olanow & A. N. Lieberman, pp. 89–112. Carnforth, England: Parthenon.

Fox, M. W., Ahlskog, J. E. & Kelly, P. J. (1991). Stereotactic ventrolateralis thalamotomy for medically refractory tremor in post-levodopa era Parkinson's disease patients. *Journal of Neurosurgery*, **75**, 723–30.

Freed, C. R., Breeze, R. E., Rosenberg, N. L., Schneck, S. A., Wells, T. H., Barrett, J. N., Grafton, S. T., Huang, S. C., Eidelberg, D. & Rottenberg, D. A. (1990). Transplantation of human fetal dopamine cells for Parkinson's disease: results at 1 year. *Archives of Neurology*, **47**, 505–12.

Friedman, J. H. & Lannon, M. C. (1990). Clozapine in idiopathic Parkinson's disease. *Neurology*, **40**, 1151–2.

Graham, D. G. (1978). Oxidative pathways for catecholamines in the genesis of neuromelanin and cytotoxic quinones. *Molecular Pharmcacology*, **14,** 633–43.

Greenamyre, J. T. & O'Brien, C. F. (1991). N-methyl-D-aspartate antagonists in the treatment of Parkinson's disease. *Archives of Neurology*, **48**, 977–81.

Hadjiconstantinou, M., Fitkin, J. G., Dalia, A. & Neff, N. H. (1991). Epidermal growth factor enhances striatal dopaminergic parameters in the 1-methyl-4-phenyl-1,2,3,6- tetrahydropyridine-treated mouse. *Journal of Neurochemistry*, **57**, 479–82.

Halliwell, B. (1989). Oxidants and the central nervous system: some fundamental questions. Is oxidant damage relevant to Parkinson's disease, Alzheimer's disease, traumatic injury or stroke? *Acta Neurologica Scandinavica*, **80** (Suppl. 126), 23–33.

Hirsch, E., Graybiel, A. M. & Agid, Y. (1988). Melanized dopaminergic neurons are differentially susceptible to degeneration in Parkinson's disease. *Nature*, **334**, 345–8.

Hyman, C., Hofer, M., Barde, Y. A., Juhasz, M., Yancopoulos, G. D., Squinto, S. P. & Lindsay, R. M. (1991). BDNF is a neurotrophic factor for dopaminergic neurons of the substantia nigra. *Nature*, **350**, 230–2.

Jankovic, J. & Schwartz, K. (1991). Botulinum toxin treatment of tremors. *Neurology*, **41**, 1185–8.

Jenner, P. (1991). Oxidative stress as a cause of Parkinson's disease. *Acta Neurologica Scandinavica*, **84**, 6–15.

Kelly, P. J., Ahlskog, J. E., Goerss, S. J., Daube, J. R., Duffy J. R. & Kall, B. A. (1987). Computer-assisted stereotactic ventralis lateralis thalamotomy with microelectrode recording control in patients with Parkinson's disease. *Mayo Clinic Proceedings*, **62**, 655–64.

Klockgether, T., Turski, L., Honore, T., Zhang, Z. M., Gash, D. M., Kurlan, R. & Greenamyre, J. T. (1991). The AMPA receptor antagonist NBQX has antiparkinsonian effects in monoamine-depleted rats and MPTP-treated monkeys. *Annals of Neurology*, **30**, 717–23.

Konradi, C., Kornhuber, J., Froelich, L., Fritze, J., Heinsen, H., Beckmann, H., Schulz, E. & Riederer, P. (1989). Demonstration of monoamine oxidase-A and -B in the human brainstem by a histochemical technique. *Neuroscience*, **33**, 383–400.

Laitinen, L. V., Bergenheim, A. T. & Hariz, M. I. (1992). Leksell's posteroventral pallidotomy in the treatment of Parkinson's disease. *Journal of Neurosurgery*, **76**, 53–61.

Langston, J. W. (1990). Predicting Parkinson's disease. *Neurology*, **40** (Suppl. 3), 70–4.

Lindvall, O., Widner, H., Rehncrona, S., Brundin, P., Odin, P., Gustavii, B., Frackowiak, R., Leenders, K. L., Sawle, G., Rothwell, J. C., Björklund, A. & Marsden, C. D. (1992). Transplantation of fetal dopamine neurons in Parkinson's disease: One-year clinical and neurophysiological observations in two patients with putaminal implants. *Annals of Neurology*, **31**, 155–65.

Mann, D. M. A. & Yates, P. O. (1983). Possible role of neuromelanin in the pathothogenesis of Parkinson's disease. *Mechanism of Ageing Development*, **21**, 193–203.

Myllyla, V. V., Sotaniemi, K. A., Vuorinen, J. A. & Heinonen, E. H. (1992). Selegiline as initial treatment in de novo parkinsonian patients. *Neurology*, **42**, 339–43.

Narabayashi, H., Yokochi, F. & Nakajima, Y. (1984). Levodopa-induced dyskinesia and thalamotomy. *Journal of Neurology, Neurosurgery, and Psychiatry*, **47**, 831–9.

Olanow, C. W. (1990). Oxidation reactions in Parkinson's disease. *Neurology*, **40** (Suppl. 3), 32–7.

Olson, L., Bäcklund, E. O., Ebendal, T., Freedman, R., Hamberger, B., Hansson, P., Hoffer, B., Lindblom, U., Meyerson, B., Strömberg, I., Sydow, O., Seiger, A. (1991). Intraputaminal infusion of nerve growth factor to support adrenal medullary autografts in Parkinsons disease: one-year follow- up of first clinical trial. *Archives of Neurology*, **48**, 373–81.

Parkinson Study Group. (1989). Effect of deprenyl on the progression of disability in early Parkinson's disease. *New England Journal of Medicine*, **321**, 1364–71.

Parkinson Study Group. (1993). Short-term assessment of lazabemide (RO19-6327) in untreated Parkinson's disease. *Movement Disorders*, in press.

Pezzoli, G., Zecchinelli, A., Ricciardi, S., Burke, R. E., Fahn, S., Scarlato, G. & Carenzi, A. (1991). Intraventricular infusion of epidermal growth factor restores dopaminergic pathway in hemiparkinsonian rats. *Movement Disorders*, **6**, 281–7.

Pfeiffer, R. F., Kang, J., Graber, B., Hofman, R. & Wilson, J. (1990). Clozapine for psychosis in Parkinson's disease. *Movement Disorders*, **5**, 239–42.

Sawle, G. V., Bloomfield, P. M., Björklund, A., Brooks, D. J., Brundin, P., Leenders, K. L., Lindvall, O., Marsden, C. D., Rehncrona, S., Widner, H. & Frackowiak, R. S. J. (1992). Transplantation of fetal dopamine neurons in Parkinson's disease: PET (F-18)6-L-fluorodopa studies in two patients with putaminal implants. *Annals of Neurology*, **31**, 166–73.

Sellal, F., Hirsch, E., Lisovoski, F., Mutschler, V., Collard, M. & Marescaux, C. (1992). Contralateral disappearance of parkinsonian signs after subthalamic hematoma. *Neurology*, **42**, 255–6.

Teräväinen, H. (1990). Selegiline in Parkinson's disease. *Acta Neurologica Scandinavica*, **81**, 333–6.

Tetrud, J. W. & Langston, J. W. (1989). The effect of deprenyl (selegiline) on the natural history of Parkinson's disease. *Science*, **245**, 519–22.

Westlund, K. N., Denney, R. M., Korchersperger, L. M., Rose, R. M. & Abell, C. W. (1985). Distinct monoamine oxidase A and B populations in primaimaimate brain. *Science*, **230**, 181–3.

9

Understanding the genetics of classical idiopathic torsion dystonia

S. FAHN

Dystonia Clinical Research Center, Department of Neurology,
Columbia University College of Physicians & Surgeons
and
The Neurological Institute of New York,
Columbia-Presbyterian Medical Center,
New York, NY, USA

Definition and classification of dystonia

Torsion dystonia refers to a group of neurologic disorders characterized by sustained muscle contractions, frequently causing twisting and repetitive movements, or abnormal postures (Fahn, 1988). Dystonia is classified etiologically into idiopathic (or primary) and symptomatic (or secondary) groups. The latter may be caused by other genetic disorders with metabolic disturbances (e.g. Wilson disease, GM1 and GM2 gangliosidoses, Hallervorden–Spatz disease), or environmental factors (e.g. stroke, trauma, birth injury, encephalitis or exposure to neuroleptics). Approximately a quarter of patients have symptomatic dystonia and three-quarters have idiopathic torsion dystonia (ITD). There is no marker (biochemical, radiological, or pathological) of idiopathic dystonia, and the diagnosis is based on clinical criteria (Fahn, Marsden & Calne, 1987) which can be highly variable (Fahn, 1984). Torsion dystonia is also classified by age at onset (childhood-, adolescent-, or adult-onset), and by distribution of involvement of body parts (focal, segmental, multifocal, generalized, and hemidystonia).

Aside from age at onset and body regions affected, other distinct features have been used to divide idiopathic torsion dystonia (ITD) into subgroups based on clinical, ethnic, pharmacologic and genetic differences (for reviews, see Eldridge, 1970; Fahn *et al.*, 1987; Fahn, 1989). Classical ITD is based on the clinical features recognized by the early neurologists who described this disorder (Oppenheim, 1911; Flatau & Sterling, 1911), with periodic updates by subsequent investigators (Herz, 1944; Fahn, 1984). The presence of myoclonus (myoclonic dystonia), the presence of fluctuations of symptoms (i.e. dystonia with diurnal variation and paroxysmal dystonic choreoathetosis), the response to therapy (e.g. dopa-responsive dystonia and alcohol-responsive dystonia), ethnic background (e.g. Ashkenazi, Philipino), and the presence of other affected family members (familial versus sporadic dystonia) have all been utilized as criteria to distinguish subgroups or variants of classical ITD (Fahn, 1989).

Historical aspects of heredity of dystonia

An inherited basis for classical ITD has long been suspected because some affected individuals have affected family members and because the disorder is more frequent among Ashkenazi Jews (Zeman & Dyken, 1967). The first description of familial ITD was a report of three affected Jewish siblings (Schwalbe, 1908; Truong & Fahn, 1988). This now famous Lewin family was subsequently reported with follow-up information in several publications (Ziehen, 1911; Jankowska, 1934; Vogt & Vogt, 1937; Regensburg, 1930) and contained, aside from the affected siblings, a mother and maternal grandfather with tremor and two affected children of one of the sibs. Over the years, dystonia in this family has been characterized as hysterical or tics (Schwalbe, 1908; Truong & Fahn, 1988), dominantly inherited (Zeman & Dyken, 1967), and of uncertain genetic pattern (Eldridge, 1970). Indeed, discord over categorization of this family has been representative of a larger debate: the importance and precise nature of genetic mechanisms in dystonia.

While Flatau and Sterling (1911) emphasized the possible genetic aspects, Oppenheim (1911), Mendel (1919), and Herz (1944) discounted or minimized the importance of heredity in ITD, although Herz was aware of previous reports of dystonia afflicting siblings (Bernstein, 1912; Abrahamson, 1920) and members of successive generations of Jewish families (Wechsler & Brock, 1922; Mankowsky & Czerny, 1929; Regensburg, 1930). The report by Wechsler and Brock (1922) had two siblings, a paternal uncle and a paternal first cousin through an unaffected aunt affected with dystonia. The reports of Mankowsky and Czerny (1929) and Regensburg (1930) each showed a parent and offspring affected with dystonia.

Subsequently, evidence for dominant inheritance of this disorder in several non-Jewish families was provided by reports of Zeman, Kaelbling and Rasamanick (1959), Johnson, Schwartz and Barbeau (1962), Larsson and Sjögren (1966), and Hoefnagel, Allen and Falk (1970). The study of Larsson and Sjögren of a large Swedish kindred, however, may not represent classical ITD because they included individuals with fibrillary tongue twitching, tremor, deforming arthropathy and muscle atrophy; also the average age of onset was 35–40 years. The reports of Hoefnagel *et al.* (1970) and Johnson *et al.* (1962) describe a total of three kindreds with more typical clinical features. Although dystonia varied in severity and age of onset, nondystonic signs were not considered part of the phenotype. Unaffected obligate gene carriers occurred in all three families and the authors stressed the incomplete penetrance and variable expression. The family described by Zeman et al. (1959) documented affected cases in three successive generations. The authors concluded that there is high or complete penetrance in this family although to do so they included nonspecific hyperkinesias, clubfeet, minor speech defects and mental retardation as expressions of the trait.

Zeman and Dyken (1967) concluded that the mode of inheritance in their non-Jewish families was autosomal dominant with incomplete penetrance. They emphasized, that in many instances, affected family members were identified only by personal examination and that only by such careful examination could the familial pattern be established. They described the clinical heterogeneity of this disorder and proposed criteria for the diagnosis of 'formes frustes'. They included focal dystonia, hyperkinesias, dystonia with action only, postural disturbances such as scoliosis, writer's cramp, restricted muscle spasm, dystonia that does not progress, and fluctuating hyper- and hypotonus upon passive movement as evidence that an individual carries the mutant gene, provided another family member displays 'classical' dystonia.

Using their criteria, Zeman and Dyken (1967) reviewed the pedigrees of 217 cases reported in the literature and 36 of their own cases. They categorized 148, coming from 31 families, as familial (19 families were Jewish and 12 non-Jewish) and 105 as sporadic. Assuming autosomal dominant inheritance in these 31 families, they found 74 obligate carriers, 39 of whom were affected, giving an estimated penetrance of 0.52. Based on 127 American-born Jewish cases reported during a 60-year period, they estimated the gene frequency in American Jews to be .000026 (1/38 000). This was fivefold higher than the gene frequency of .000005 (1/200 000) that they estimated for the general population in the state of Indiana. Although they commented on the higher gene frequency of dystonia in Jews and considered it an indication of the inherited nature of the disorder, they did not construe a distinct disease gene or mechanism of inheritance for this population.

Controversy on the mode of inheritance of ITD in the Ashkenazi Jewish population

In contrast to Zeman and Dyken (1967), Eldridge (1970) and Eldridge and Gottlieb (1976), evaluating the familial pattern in dystonic subjects who had undergone stereotaxic thalamotomy by Irving Cooper, proposed at least two modes of inheritance: autosomal recessive seen most commonly among Ashkenazi Jews and autosomal dominant in non-Jews. In a detailed report of his own 156 patients representing 96 families, and additional families from a review of the literature, Eldridge (1970) proposed 12 subcategories and found most cases fell into one of three categories: (i) Jewish, without parent and child affected; (ii) non-Jewish, without parent and child affected; and (iii) non-Jewish, parent and child affected. The first subcategory was thought to represent autosomal recessive inheritance; the second, new mutations of a dominant trait; and the third, autosomal dominant inheritance. For those cases that were Jewish but had parent–child affected, 'pseudodominance' was proposed (that is, the union of a homozygous affected individual with a heterozygous carrier would have a 50% chance of producing affected offspring). The argument for considering autosomal recessive inheritance among Jews was based on: (i) the high frequency of the disorder in this population, specifically among Ashkenazim from eastern Europe, (ii) the close agreement between the number of observed and expected siblings affected for this mode of inheritance, and (iii) clinical differences between Jews and non-Jews (i.e. Jews were more likely to have onset in childhood, limb rather than axial involvement and greater disability). Eldridge (1970) estimated the recessive gene frequency in Ashkenazi Jews to be 0.0077 with a disease frequency of 1:17 000.

However, Eldridge did not examine many first-degree relatives in his pedigrees. Of 32 Jewish families with proposed recessive inheritance, 23/55 living parents were not examined. The need to examine parents personally rather than rely on historical information is particularly important as parents are less likely to be severely affected due to reduced fertility among those severely affected. Moreover, we and others (Zeman & Dyken, 1967; Burke *et al.*, 1986) have observed that some mildly affected individuals are not aware that they manifest symptoms of dystonia. Also, a recessive gene frequency of .0077 corresponds to a risk of .0077 for parents and children of affected cases, which appears small compared to the number of parent-child pairs reported. Furthermore, a reanalysis of Eldridge's data by Burke *et al.* (1986) shows no significant difference between Jews and non-Jews in their ages and sites of onset.

Eldridge's study has also been criticized by Korczyn *et al.* (1981) for its lack of quantitative genetic analysis. They point out that, assuming single ascertainment rather than truncate ascertainment, one obtains a recurrence risk to siblings significantly lower than the 25% claimed by Eldridge. They also argue that Eldridge's data show too low a consanguinity rate and too many instances of parent-child affected for autosomal recessive inheritance unless a much higher disease allele frequency is postulated.

Zilber *et al.* (1984) challenged the notion of recessive inheritance among Ashkenazim. They presented a quantitative analysis of a nationwide survey of primary torsion dystonia in Israel which produced 47 affected individuals (40 of European origin). Probands were drawn from hospital discharge diagnoses and did not include individuals with focal dystonias. They calculated the frequency of ITD, based on 11 Israeli cases born between 1949 and 1959 and 455 000 live births during that period, to be 1:15 000 among Jews of Eastern Europe, 1:23 000 among Jews of combined Western and Eastern European origin, and 1:117 000 among Afro-Asian Jews. These data confirmed the high frequency of the disorder among Eastern European Jews. However, they found that a recessive model of inheritance did not fit their data unless a very high disease allele frequency were postulated. Consanguinity was too infrequent, the observed number of affected parents, offspring, aunts, uncles and cousins was too high, and the observed number of affected siblings of probands from presumed heterozygous parents was significantly lower than expected. However, an autosomal dominant pattern with a penetrance of 51% could explain their familial cases. If cases of tremor, blepharospasm, and stuttering were designated as affected, penetrance was raised to 68%. Further, mean paternal age at birth of isolated cases was found to be increased, suggesting that some isolated cases were the result of new mutations. Although this study provided more detailed demographic data, no information regarding the completeness of examination of family members was given. Clinical data are not provided and probands with focal dystonia such as graphospasm and torticollis were not included. Furthermore, the perplexing finding of a lower than expected number of affected siblings remained unexplained under recessive or dominant inheritance.

In a study in England, Bundey, Harrison and Marsden (1975) discerned two main clinical groups: a childhood-onset group of 17 patients who had dystonia involving the legs and arms, and an adult-onset group of 12 patients with torticollis or upper limb dystonia. There were two Ashkenazi probands among their childhood-onset cases. They had no cases of parent–child affected, but three probands had affected relatives (brothers in two non-Jewish cases, and a first cousin in a Jewish case). The authors concluded that a minority of childhood-onset cases (Jewish and non-Jewish) are autosomal recessive and the remainder are new mutations of a dominant trait; to support this latter conclusion they found an increased mean paternal age of isolated cases with childhood-onset. In contrast to Eldridge's results, Bundey *et al.* found that these two forms (recessive and dominant) were clinically indistinguishable. They also proposed that most adult-onset cases have a nongenetic disorder due to exogenous undefined causes.

A more recent study in England by Fletcher, Harding and Marsden (1990) investigated 100 families containing 107 index cases with generalized, multifocal or segmental dystonia. They found 79 cases of dystonia in relatives of 58 index cases. Jewish cases comprised 10.3% which is greater than expected, but they did not differ from non-Jewish cases either clinically or genetically. In contrast to the previous study in England (Bundey et al., 1975), these investigators concluded that, in the United Kingdom, 85% of ITD are due to autosomal dominant inheritance with about 40% penetrance and highly variable expression. They found no evidence for the existence of autosomal recessive or X-linked forms of ITD. A recent survey in England of non-Jewish families with probands with focal dystonia has confirmed that this is inherited as an autosomal dominant disorder as well (Waddy et al., 1991).

Resolution of the controversy of the mode of inheritance of ITD in the Ashkenazi Jewish population

Because of disagreement regarding the inheritance pattern of ITD in the Ashkenazi Jewish population, we designed a protocol to come to a definitive conclusion about the mode of inheritance of torsion dystonia in this ethnic group. We conducted a prospective study in which we systematically examined first- and second-degree relatives of Ashkenazi Jewish individuals with ITD.

Because age at onset for ITD ranges from 5 years upwards, with a bimodal distribution, consisting of modes at 9 and 55 years of age and with a nadir at age 27 years (Fahn, 1986), we elected to investigate just the families in which the probands had onset of symptoms before the age of 28 years. Moreover, it is not certain that older onset patients have the same disorder as those of young onset. For example, younger onset individuals (age at onset at 21 years or younger) are likely to have generalized dystonia (58%), less likely to have segmental dystonia (30%), and least likely to have focal dystonia (12%) (Fahn, 1986). For individuals with onset greater than 21 years, the opposite is found (2% generalized, 36% segmental, and 62% focal). Therefore, in order to study a clinically homogeneous population, we included as probands only those with onset of dystonia at 27 years or younger.

Field trips to the family members of 43 probands were made by neurologists trained in the field of movement disorders, and the family members were videotaped according to a standardized protocol, including evaluations of the face, jaw, tongue, voice, neck, trunk and limbs, at rest and with action (Bressman et al., 1989). We assessed rapid successive movements of the limbs, finger-to-nose, handwriting with each hand, speaking and walking. On-site examinations and videotapes were obtained on 215 (88.8%) of 242 relatives, on-site examinations only for 15 relatives (6.2%), and videotapes only on 12 relatives (5.0%). The videotaped examinations were assessed for evidence of dystonia by at least two independent neurologists who were blinded to the subject's identity and to any possible relationship to any other patients or subjects. Diagnoses were assigned on the basis of all available information, which included both on-site examination and videotape review. Definite dystonia was defined by twisting movements or postures that were apparent to all examiners, including focal, segmental, multifocal and generalized dystonia. Probable dystonia was diagnosed if there were twisting movements or postures that were apparent to some but not all examiners. Possible dystonia was diagnosed for suggestive abnormalities that were not diagnostic of dystonia. Scoliosis alone was not considered diagnostic of dystonia unless there was also dystonia of another body part. The presence of coexistent medical factors such as birth injury, use of neuroleptic drugs, and Parkinsonism led to the diagnosis of unrateable.

To determine compatibility with different modes of inheritance, we calculated rates of illness among classes of first- and second-degree relatives. Because symptoms may start in childhood or middle age, we calculated age-adjusted rates or lifetime risks with the Kaplan–Meier product-limit life table estimator and standard errors (Kaplan & Meier, 1958; Thompson & Weissman 1981) Dominant and recessive models were fitted to the lifetime risks by estimating penetrances and gene frequencies by maximum likelihood. The pedigree data were subjected to segregation analysis using the computer programs POINTER (Lalouel & Morton, 1981) and MENDEL (Lange, Weeks & Boehnke, 1988). To control for variable age at onset, five liability classes were defined on the basis of age that were determined by the cumulative age-at-onset distribution observed among the affected relatives in the present study, with the final lifetime risk assumed to be 1:15 000 (Zilber et al., 1984).

We found 19 definitely affected and two probably affected relatives (Bressman et al., 1989). Among the 19 definitely affected nonproband relatives, nine were first diagnosed as having dystonia as a result of this study. All were symptomatic; seven were sufficiently affected to seek medical opinion or treatment. There were no significant sex differences in rates of illness in the different classes of relatives. Of the 21 relatives definitely or probably affected, 10 were male and 11 female. Among the seven affected parents, six were mothers and one a father.

Age-adjusted lifetime risks (LTR) to age 45 years were calculated for each type of relative. The LTR for all first-degree relatives combined was 15.5 ± 3.4%. The LTR for all second-degree relatives was 6.5 ± 2.6%, slightly less than one-half the value for first-degree relatives. This finding is consistent with any single-locus pattern of inheritance. Discrimination between dominant and recessive inheritance was made by comparing the risks to parents and offspring (combined LTR 14.2 ± 4.2%) with that to siblings (17.2%). The similarity of these values argues strongly in favor of dominant inheritance and against recessive inheritance, because the latter would lead to a much higher risk to siblings than to parents and offspring.

We found an approximately equal frequency of dystonia among the parents, siblings, and offspring of our probands (Bressman et al., 1989), as would be expected with autosomal dominant inheritance. With autosomal recessive inheritance, one would expect a 25% risk to siblings, with virtual absence of the disorder in parents and offsprings. Our data strongly reject recessive inheritance and are also inconsistent with multifactorial inheritance. We found that the risk to first-degree relatives was twice that to second-degree relatives, a pattern consistent with a single locus disease rather than a multifactorial disease.

Our data are consistent with autosomal dominant transmission with reduced penetrance. Based on the relatives with definite dystonia, the penetrance rate was estimated to be 29.4%; the rate increases to 32.2% if probably affected relatives were included. Our results also indicate that all probands with onset before age 28 years were gene carriers. These conclusions were confirmed by the results of segregation analysis (Risch et al., 1990). With the analysis using POINTER, the recessive model was strongly rejected ($P<10^{-11}$, chi square) as was the polygenic model ($P<10^{-9}$, chi square). With the analysis using MENDEL, the generalized single-locus model also converged to the dominant model. The recessive model was rejected ($P<10^{-14}$, chi square).

Later, when Eldridge's original data were analysed by powerful statistical methods, the inheritance pattern of his Jewish cases were found to be autosomal dominant, rather than recessive as originally thought by Eldridge (Pauls & Korczyn, 1990).

Linkage analysis of ITD

Genetic linkage analysis, using DNA polymorphisms and protein markers, can be used to determine the chromosomal location of an abnormal gene. We investigated a large non-Jewish family with ITD inherited as an autosomal dominant and found the responsible gene to be located on chromosome 9 in the q32–34 region (Ozelius et al., 1989). Based on this information, we searched this region in Ashkenazi Jewish families with ITD, and discovered that their abnormal gene is located in the same region (Kramer et al., 1990).

Linkage between the dystonia locus and chromosome 9q markers showed that the highest lod scores obtained with two-point analysis was with a probe for the arginosuccinate synthetase locus, ASS. At a theta of 0.03 the lod score

Genetics of classical idiopathic torsion dystonia

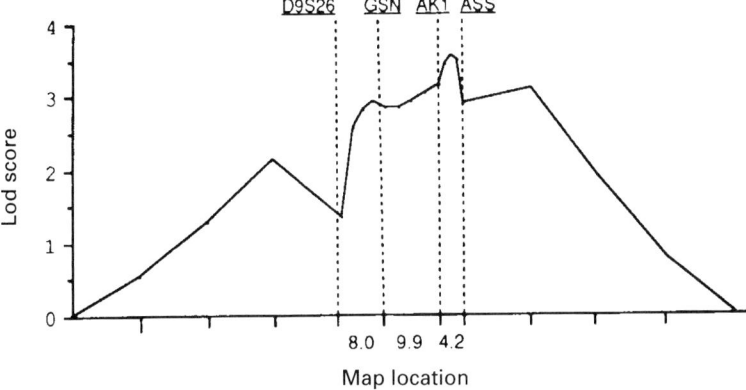

Fig. 9.1. Multipoint analysis of idiopathic torsion dystonia in 12 Ashkenazi Jewish families and chromosome 9q markers. Genetic distances between markers are given in centiMorgans (cM) and are calculated on the basis of sex-average recombination estimates. Names of the markers are given at top of Figure. Kramer et al. (1990) reproduced with permission.

for this locus was 3.49. Multipoint analysis with the dystonia locus were carried out with the probes for loci D9S26, GSN, AK1, and ASS because these gave suggestive evidence for linkage on two-point analysis and because these are in the region for the DYT1 locus in a large non-Jewish family. Results showed the highest lod score of 3.54 at a point midway between AK1 and ASS (Fig. 9.1).

We have evaluated two candidate genes located in the 9q34 region, namely the genes for gelsolin and for dopamine beta hydroxylase (DBH). We excluded linkage to both of them (Kwiatkowski et al., 1991a; Schuback et al., 1991). Not only was the DBH gene excluded in the large non-Jewish family and the Ashkenazi Jewish family, it was also excluded in a large family with dopa-responsive dystonia and myoclonic dystonia. By using GT repeat polymorphisms, the DYT1 gene was localized to a narrow region between two flanking markers, AK1 and D9S10, a distance of about 11 cM (Kwiatkowski et al., 1991a). The only remaining known genes in this region are the ABL oncogene and the argininosuccinate synthetase (ASS) gene, neither of which appears a likely candidate gene for this neurologic disorder.

Allelic association in the Ashkenazi Jewish population

Allelic association, also known as linkage disequilibrium, is the occurrence of particular alleles at nearby markers being found together more often than would be predicted from their frequency in the general population. Allelic association indicates tight linkage and that there is only one or a few origins for the disease mutations in the population being studied, i.e. a founder effect. Testing for several polymorphisms and GT repeats at the extended haplotype ABL/ASS, we found that 69% of Ashkenazi Jewish affected individuals had an association with haplotype 4A12, whereas only 0.75% of control chromosomes had this haplotype ($P<10^{-19}$) (Ozelius *et al.*, 1992). This finding narrows the DYT1 gene in the Ashkenazi Jewish population to the flanking markers AK1 and ABL/ASS, a distance of less than 2 cM.

We evaluated Ashkenazi Jewish patients with dystonia without a family history of the disorder (i.e. so-called sporadic dystonia) and found that 62% (8 of 13 cases) could carry the A12 allele on one of their two chromosomes No. 9 (Ozelius *et al.*, 1992). This finding indicates that most sporadic cases of dystonia in the Ashkenazi Jewish population are probably hereditary.

The allelic association found in the Ashkenazi Jewish population suggests that the mutation for dystonia may be due in part to the relatively recent occurrence of the mutation for dystonia. A search for the haplotype in several non-Jewish populations with dystonia reveal that they do not share the 4A12 haplotype, suggesting that different mutations at the same gene are responsible for DYT1 in the Jewish and non-Jewish populations.

Carrier detection in the Ashkenazi Jewish population

The goals of linkage research are (i) to locate the abnormal gene(s) to specific chromosomes, (ii) to allow the characterization of variability of expression in linked families, (iii) to determine factors allowing low penetrance, (iv) to permit carrier detection including prenatal detection, and (v) to identify gene product(s). Because the DYT1 gene lies between the flanking markers of AK1 and ABL/ASS, and because there is allelic association to the 4A12 haplotype in the Ashkenazi Jewish population, it is now possible to detect carriers of the DYT1 gene in this population. Such carrier detection has now been successfully carried out in one extended family (de Leon *et al.*, 1991).

Other genetic forms of torsion dystonia

Dopa-responsive dystonia is inherited as an autosomal dominant disorder (de Yebenes *et al.*, 1988; Nygaard, 1989; Nygaard *et al.*, 1990). It is usually inherited in childhood or adolescence, often beginning with dystonia affecting the legs and walking. It can occur in the neonatal period and resemble cerebral palsy. Parkinsonian signs are often present, such as bradykinesia and loss of postural stability. The absolute criterion for diagnosis is a dramatic response to

Genetics of classical idiopathic torsion dystonia 149

Linkage map of human 9q

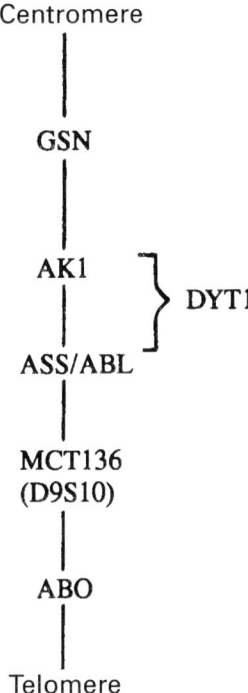

Fig. 9.2. Map of the human chromosome 9 showing physical assignments in the q34 region. The gene DYT1 represents the gene causing idiopathic torsion dystonia in the Ashkenazi Jewish population and in at least one non-Jewish family.

low dosage levodopa. The gene causing dopa-responsive dystonia has been excluded from 9q34 region (Kwiatkowski *et al.*, 1991*b*). The search for the linkage markers for this disorder continues.

X-linked recessive inheritance of dystonia in a Philippine population from the island of Panay was concluded based on a report of 28 cases from 25 families of males affected with torsion dystonia, with a mean age of onset of 31.5 years (Lee *et al.*, 1976). Six families were found to have more than one affected member. Four of these families each had one affected brother. The two other families had affected relatives in two generations, suggesting either sex-linked recessive inheritance or autosomal dominant inheritance with reduced penetrance. A large family of males-only affected from this Philippine island was recently described, confirming the X-linked hereditary pattern, but in this family those most affected also developed features of Parkinsonism (Fahn & Moskowitz, 1988). Lee *et al.* (1991) confirmed the presence of both

Parkinsonism and dystonia in a large survey of the clinical features of the disorder. X-linked recessive dystonia-parkinsonism (referred to as 'lubag' by the affected population) has been found to be due to a gene in the pericentromeric region of the X chromosome (Kupke *et al.*, 1990; Wilhelmsen *et al.*, 1991).

Summary

The controversy regarding the mode of inheritance of idiopathic torsion dystonia in the Ashkenazi Jewish population has been resolved. At one time it was believed to be inherited as an autosomal recessive disorder. But recent studies, including a prospective, systematic, blinded analysis of the first- and second-degree relatives of 43 probands with age at onset less than 28 years found the disorder to be inherited in an autosomal dominant manner with a penetrance of approximately 0.30. Linkage analysis of Ashkenazi Jewish families with multiple affected members revealed that the gene for dystonia in this population is located in the q34 region of chromosome 9. This is the same region found to encode the dominant DYT1 gene for dystonia in a large non-Jewish family with a penetrance of about 0.70. It is likely that the disorder in these two ethnic populations may be caused by the same locus, and that the difference in penetrance may reflect different mutations operating in these two populations. We have found no evidence for genetic heterogeneity in the Ashkenazi Jewish families studied for linkage analysis, but there is at least one non-Jewish family with idiopathic torsion dystonia that is not linked to this region. Allelic association in 9q34 in the Ashkenazi Jewish population has narrowed the dystonia gene to a region of less than 2 cM.

Acknowledgements

Studies on the genetics of dystonia were supported in part by the Dystonia Medical Research Foundation and by grants from the US Public Health Service.

References

Abrahamson, I. (1920). Presentation of cases of familial dystonia musculorum of Oppenheim. *Journal of Nervous and Mental Disease*, **51**, 451–4.

Bernstein, S. (1912). Ein Fall von Torsionskrampf. *Wiener Klinische Wochenschrift*, **25**, 1567–71.

Bressman, S. B., de Leon, D., Brin, M. F., Risch, N., Burke, R. E., Greene, P. E., Shale, H. & Fahn, S. (1989). Idiopathic torsion dystonia among Ashkenazi Jews: Evidence for autosomal dominant inheritance. *Annals of Neurology*, **26**, 612–20.

Bundey, S., Harrison, M. J. G. & Marsden, C. D. (1975). A genetic study of torsion dystonia. *Journal of Medical Genetics*, **12**, 12–9.

Burke, R. E., Brin, M. F., Fahn, S., Bressman, S. B. & Moskowitz, C. B. (1986). Analysis of the clinical course of non-Jewish autosomal dominant torsion dystonia. *Movement Disorders*, **1**, 163–78.

de Leon, D., Brin, M. F., Murphy, P., Bressman, S. B., Ozelius, L., Cardon, N., Reich, S., Breakefield, X. O & Fahn, S. (1991). Genetic counseling for idiopathic torsion dystonia: First use of DNA based carrier detection in Ashkenazic Jews. *Movement Disorders*, **6**, 273–4.

de Yebenes, J. G., Moskowitz, C., Fahn, S. & Saint-Hilaire, M. H. (1988). Long-term treatment with levodopa in a family with autosomal dominant torsion dystonia. *Advances in Neurology*, **50**, 101–11.

Eldridge, R. (1970). The torsion dystonias: literature review and genetic and clinical studies. *Neurology*, **20**, 1–78.

Eldridge, R. & Gottlieb, R. (1976) The primary hereditary dystonias: Genetic classification of 768 families and revised estimate of gene frequency, autosomal recessive form, and selected bibliography. *Advances in Neurology*, **14**, 457–74.

Fahn, S. (1984). The varied clinical expressions of dystonia. *Neurological Clinics of North America*, **2**, 541–54.

Fahn, S. (1986). Generalized dystonia: concept and treatment. *Clinical Neuropharmacology*, **9**, Suppl. 2, S37–48.

Fahn, S. (1988). Concept and classification of dystonia. *Advances in Neurology*, **50**, 1–8.

Fahn, S. (1989). Clinical variants of idiopathic torsion dystonia. *Journal of Neurology Neurosurgery, and Psychiatry, Special Supplement*, 96–100.

Fahn, S., Marsden, C. D. & Calne, D. B. (1987). Classification and investigation of dystonia. In *Movement Disorders 2*, eds. C. D. Marsden & S. Fahn, pp. 332–58. London: Butterworths.

Fahn, S. & Moskowitz, C. (1988). X-Linked recessive dystonia and parkinsonism in Filipino males. *Annals of Neurology*, **24**, 179.

Flatau, E. & Sterling, W. (1911). Progressiver Torsionspasms bie Kindern. *Zeitschrift Gesamte Neurologische und Psychiatrie*, **7**, 586–612.

Fletcher, N. A., Harding, A. E. & Marsden, C. D. (1990). A genetic study of idiopathic torsion dystonia in the United Kingdom. *Brain*, **113**, 379–95.

Herz, E. (1944). Dystonia. II. Clinical classification. *Archives of Neurology and Psychiatry*, **51**, 319–55.

Hoefnagel, D., Allen, F. H. Jr & Falk, C. (1970). Hereditary dystonia musculorum deformans. *Journal of Clinical Genetics*, **1**, 258–62.

Jankowska, H. (1934). Beitrag zur Hereditat der Torsionsdystonie. *Neurologie Polanska*, **16–17**, 258–64.

Johnson, W., Schwartz, G. & Barbeau, A. (1962). Studies on dystonia musculorum deformans. *Archives of Neurology*, **7**, 301–13.

Kaplan, E. L. & Meier, P. (1958). Nonparametric estimation from incomplete observations. *Journal of the American Statistical Association*, **53**, 457–81.

Korczyn, A. D., Zilber, N., Kahana, E. & Alter, M. (1981). Inheritance of torsion dystonia. Reply. *Annals of Neurology*, **10**, 204–5.

Kramer, P. L., Ozelius, L., de Leon, D., Risch, N., Brin, M. F., Bressman, S. B., Burke, R.E., Kwiatkowski, D. J., Schuback, D. E., Shale, H., Gusella, J. F., Breakefield, X. O. & Fahn, S. (1990). Dystonia gene in Ashkenazi Jewish population located on chromosome 9q32–34. *Annals of Neurology*, **27**, 114–20.

Kupke, K. G., Lee, L. V. & Muller, U. (1990). Assignment of the X-linked torsion dystonia gene to Xq21 by linkage analysis. *Neurology*, **40**, 1438–42.

Kwiatkowski, D. J., Nygaard, T. G., Schuback, D. E., Perman, S., Trugman, J. M., Bressman, S. B., Burke, R. E., Brin, M. F., Ozelius, L., Breakefield, X. O., Fahn, S. & Kramer, P. L. (1991b). Identification of a highly polymorphic microsatellite VNTR within the argininosuccinate synthetase locus: exclusion of the dystonia gene on 9q32–34 as the cause of dopa-responsive dystonia in a large kindred. *American Journal of Human Genetics*, **48**, 121–8.

Kwiatkowski, D. J., Ozelius, L., Kramer, P., Perman, S., Schuback, D. E., Gusella, J. F., Fahn, S. & Breakefield, X. O. (1991a). Torsion dystonia genes in two populations defined to a small region on chromosome 9q32–34. *American Journal of Human Genetics*, **49**, 366–71.

Lalouel, J.-M. & Morton, N. E. (1981). Complex segregation analysis with pointers. *Human Heredity*, **31**, 312–21.

Lange, K. L, Weeks, D. & Boehnke, M. (1988). Programs for pedigree analysis: Mendel, Fisher, and Dgene. *Genetics and Epidemiology*, **5**, 471–2.

Larsson, T. & Sjögren, T. (1966). Dystonia musculorum deformans. A genetic and clinical population study of 121 cases. *Acta Neurologica Scandinavia*, **42** (Suppl. 17), 1–235.

Lee, L. V., Kupke, K. G., Caballar-Gonzaga, F., Hebron-Ortiz, M. & Muller, U. (1991). The phenotype of the X-linked dystonia-parkinsonism syndrome: an assessment of 42 cases in the Philippines. *Medicine*, **70**, 179–87.

Lee, L. V., Pascasio, F. M., Fuentes, F. D. & Viterbo, G. H. (1976). Torsion dystonia in Panay, Philippines. *Advances in Neurology*, **14**, 137–52.

Mankowsky, B. N. & Czerny, L. I. (1929). Zur Frage Uber die Hereditat der Torsionsdystonie. *Monatschritt von Psychiatrie und Neurologie*, **72**, 165–79.

Mendel, K. (1919). Torsionsdystonie (Dystonia musculorum deformans, Torsionsspasmus). *Monatschritt von Psychiatrie und Neurologie*, **46**, 309–61.

Nygaard, T. G. (1989). Dopa-responsive dystonia: 20 years into the L-dopa era. In *Disorders of Movement: Clinical Pharmacological and Physiological Aspects*, eds. N. P. Quinn & P.G. Jenner, pp. 323–37. London: Academic Press.

Nygaard, T. G., Trugman, J. M., de Yebenes, J. G. & Fahn, S. (1990). Dopa-responsive dystonia: The spectrum of clinical manifestations in a large North American family. *Neurology*, **40**, 66–9.

Oppenheim, H. (1911). Uber eine eigenartige Krampfkrankheit des kindlichen und jugendlichen Alters (Dysbasia lordotica progressiva, Dystonia musculorum deformans). *Neurologische Centrablatt*, **30**, 1090–107.

Ozelius, L., Kramer, P. L., Moskowitz, C. B., Kwiatkowski, D. J., Brin, M. F., Schuback, D. E., Falk, C. T., Haines, J., Bressman, S. B., DeLeon, D., Burke, R. E., Gusella, J. F., Fahn, S. & Breakefield, X. O. (1989). Human gene for torsion dystonia located on chromosome 9q32–34. *Neuron*, **2**, 1427–34.

Ozelius, L. J., Kramer, P. L., de Leon, D., Risch, N., Bressman, S. B., Schuback, D. E., Brin, M. F., Kwiatkowski, D. J., Burke, R. E., Gusella, J. F., Fahn, S. & Breakefield, X. O. (1992). Strong allelic association between the torsion dystonia gene (DYT1) and loci on chromosome 9q34 in Ashkenazi Jews. *American Journal of Human Genetics*, **50**, 619–28.

Pauls, D. L. & Korczyn, A. D. (1990). Complex segregation analysis of dystonia pedigrees suggests autosomal dominant inheritance. *Neurology*, **40**, 1107–10.

Regensburg, I. (1930). Zur Klinik des hereditaren torsionsdystonischen Symptomkomplexes. *Monatschritt von Psychiatrie und Neurologie*, **75**, 323–45.

Risch, N., Bressman, S. B., deLeon, D., Brin, M. F., Burke, R. E, Greene, P. E., Shale, H., Claus, E. B., Cupples, L. A. & Fahn, S. (1990). Segregation analysis of idiopathic torsion dystonia in Ashkenazi Jews suggests autosomal dominant inheritance. *American Journal of Human Genetics*, **46**, 433–8.

Schuback, D. E., Kramer, P. L., Ozelius, L., Holmgren, G., Kyllerman, M., Wahlstrom, J., Forsgren, L., Kyllerman, M., Wahlstrom, J., Craft, C. M., Nygaard, T., Brin, M., DeLeon, D., Bressman, S., Moskowitz, C. B., Burke, R. E., Sanner, G., Drugge, U., Gusella, J. F., Fahn, S. & Breakefield, X. O. (1991). Dopamine beta-hydroxylase gene excluded in four subtypes of hereditary dystonia. *Human Genetics*, **87**, 311–16.

Schwalbe, W. (1908). Eine eigentumliche tonische Krampfform mit hysterischen Symptomen. *Inaugural Dissertations*, Berlin: G. Schade.

Thompson, W. D. & Weissman, M. M. (1981). Quantifying lifetime risk of psychiatric disorder. *Journal of Psychiatric Research*, **16**, 113–26.

Truong, D. D. & Fahn, S. (1988). An early description of dystonia: translation of Schwalbe's thesis and information on his life. *Advances in Neurology*, **50**, 651–64.

Vogt, C. & Vogt, O. (1937). Sitz und Wesen der krankheiten im Lichte der topistischen Hirnforschung und des Bariierens der Tiere. *Journal von Psychologie und Neurologie (Leipzig)*, **47**, 237–457.

Waddy, H. M., Fletcher, N. A., Harding, A. E. & Marsden, C. D. (1991). A genetic study of idiopathic focal dystonias. *Annals of Neurology*, **29**, 320–4.

Wechsler, I. S. & Brock, S. (1922). Dystonia Musculorum Deformans with especial reference to a myostatic form and the occurrence of decerebrate rigidity phenomena. *Archives of Neurology and Psychiatry*, **8**, 538–52.

Wilhelmsen, K. C., Weeks, D. E., Nygaard, T. G., Moskowitz, C. B., Rosales, R. L., dela Paz, D. C., Sobrevega, E. E. & Fahn, S. (1991). Genetic mapping of 'lubag' (X-linked dystonia-parkinsonism) in a Filipino kindred to the pericentromeric region of the X chromosome. *Annals of Neurology*, **29**, 124–31.

Zeman, W. & Dyken, P. (1967). Dystonia musculorum deformans; clinical, genetic and pathoanatomical studies. *Psychiatrie Neurologie Neurochirugia*, **70**, 77–121.

Zeman, W., Kaelbling, R. & Rasamanick, B. (1959). Idiopathic dystonia musculorum deformans I. The hereditary pattern. *American Journal of Human Genetics*, **11**, 188–202.

Ziehen, T. H. (1911). Ein Fall von tonischer Torsions Neurose. *Neurologische Zentralblatt*, **30**, 109–10.

Zilber, N., Korczyn, A. D., Kahana, E., Fried, K. & Alter, M. (1984). Inheritance of idiopathic torsion dystonia among Jews. *Journal of Medical Genetics*, **21**, 13–20.

10

Functional imaging in motor disorders

J. G. COLEBATCH

Department of Neurology,
Institute of Neurological Sciences,
The Prince Henry and Prince of Wales Hospitals,
Sydney NSW, Australia

Advances in medical imaging have had a major impact on the clinical practice of neurology. These advances in turn have depended upon improvements in computer hardware which have allowed more sophisticated data collection and image reconstruction. Both computed tomography (CT) and magnetic resonance imaging (MRI) produce high resolution images of brain structure that are now used widely in routine clinical practice. Complementing these anatomical images, new radioisotope methods allow images to be generated of brain physiology, i.e. images of function rather than structure. The two methods commonly in use to produce functional images of the brain are SPECT (single photon emission computed tomography) and PET (positron emission tomography).

Technical considerations

Both PET and SPECT depend upon the collection of emitted radiation by multiple (or rotating) detectors and back-projection techniques of image reconstruction. SPECT can be performed using a conventional gamma camera and technetium 99-based compounds, and is therefore quite widely available. It is not possible at present to estimate the degree of tissue attenuation of the emitted radiation so that SPECT images of count density cannot be converted to absolute tissue concentrations of the tracer. PET, on the other hand, uses positron-emitting isotopes produced from a cyclotron. The short halflives of many of the useful isotopes (e.g. 15 oxygen, half-life 123 s) require the camera to be sited near the source of their production. Because positron annihilation generates two photons travelling in opposite directions, calculation of tissue attenuation and thus absolute quantitation is possible. Further, positron-emitting isotopes can be substituted for the native element in a wide range of organic compounds of biological interest.

Functional imaging in motor disorders 155

Intrinsic limitations in spatial resolution of all current methods of functional imaging mean that small structures (e.g. the subthalamic nucleus) are likely to remain difficult or impossible to study as will inhomogeneities of function occurring within a single structure such as the putamen (Penney & Young, 1986). These considerations are particularly pertinent to studies of disordered motor function when assessing basal ganglia function: here current theories suggest that changes within the two adjacent subdivisions of the globus pallidus should have very different effects on movement (Albin, Young & Penney, 1989). Temporal resolution is also limited and all commonly used methods of generating functional images represent time averages of metabolic activity. This means that caution must be exercised in concluding whether a particular metabolic change is a cause or a consequence of the movement under study. Tracer tissue fixation time ultimately determines temporal resolution and thereby the minimum time for which the subject should remain at rest or perform a specific task. This time varies from less than a minute for some 15O methods of generating CBF images (Raichle *et al.*, 1983), to about 5 minutes for 99Tc-HMPAO (Ebmeier *et al.*, 1991) to 30–45 minutes for 18F-deoxyglucose (FDG). Tracer washout determines how soon after tracer administration the scan must be acquired and influences how long the subject remains within the camera. Both FDG and 99Tc-SPECT images may be collected well after initial tracer injection, an advantage when studying unpredictable, spontaneous events or when using agitated subjects who may require sedation during image acquisition.

Types of functional images

'Static' metabolic images
These images, for example of cerebral blood flow (CBF), oxygen consumption (CMRO) or brain glucose uptake, are usually obtained with the subject in a single, resting state. The studies are commonly used to characterize the patterns of disordered blood flow or tissue metabolism peculiar to specific disease processes. Studies of the same subjects can be repeated after clinical interventions or spontaneous improvement to define the brain regions at which the critical changes occurred. Affected subjects can be compared with control subjects studied under similar conditions.

'Neural activation' images
Repeated images, usually of blood flow, are collected of the one subject during a single study session under different physiological conditions or after pharma-

cological 'activation'. Subjects thus act as their own control but must be willing and able to cooperate fully with the task under study. Interpretation of the images relies upon the principle of local coupling of neural activity to blood flow and metabolism (Roy & Sherrington, 1890) although PET studies have suggested that dynamic increases in blood flow associated with neural activation are greater than increases in oxygen metabolism (Fox & Raichle, 1986). Brain regions of increased neural activation are deduced from the sites at which increases in blood flow or metabolism occur in association with the task under study, when compared to the resting state. Although conceptually simple, interpreting these changes in physiological terms is not straightforward. The nerve terminal appears to be the main site of change in metabolism (Kadekaro et al., 1987) so that greater postsynaptic inhibition may be associated with increased metabolism (e.g. Ackermann et al., 1984) and increments in metabolism within one brain region may actually be a consequence of increased discharges in projections to that area from other brain regions.

'Activation-type' studies are designed to detect local changes and absolute values are of secondary interest. Recently, a 'split-dose' technique has been developed for SPECT studies which can be used to detect sites of increased activation in a single study session (Ebmeier et al., 1991). Theoretically, because the activated image is effectively superimposed on the resting one, smaller increases would be expected than those obtained with PET images of flow.

Images of specific radioligands

These images reflect the uptake of particular organic compounds of biological interest, such as neurotransmitter precursors, drugs or receptor labels. Tracers are available to study the dopaminergic system which has particular relevance to disordered movement (Fig. 10.1). Presynaptic function can be assessed with 18F-dopa which is taken up and stored by dopaminergic neurons (Martin et al., 1989; see below). Postsynaptically, D2 receptor labels have been used both with PET (e.g. raclopride, spiperone) and, recently, with SPECT imaging (Brücke et al. 1991).

Data analysis

The resolution of both types of functional image is inadequate to allow recognition of particular sulci or gyri. Anatomical localization can be improved considerably if the anterior–posterior commissural (AC–PC) line can be determined (Fox, Perlmutter & Raichle, 1985) and reference made to a stereotactic atlas.

Functional imaging in motor disorders 157

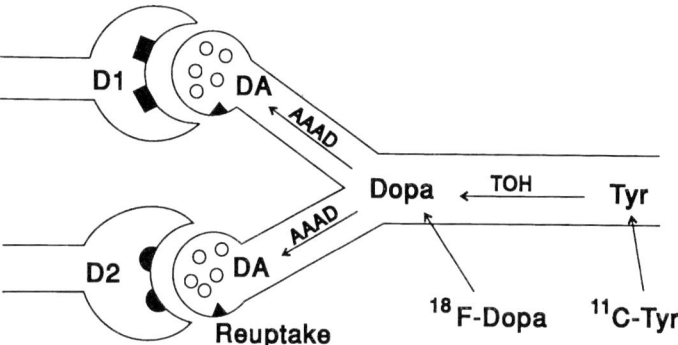

Fig. 10.1. Diagrammatic summary of dopamine transmitter metabolism demonstrating sites at which radiolabelled compounds have been used. TOH = tyrosine hydroxylase, AAAD = L-aromatic amino acid decarboxylase, DA = dopamine; D1, D2 = post synaptic dopamine receptors, which appear to be on separate populations of neurons within the striatum (Trugman & Wooten, 1990). Compounds that have been developed allow the study of precursor uptake and storage (11C-Tyrosine, 18F-Dopa), reuptake sites and postsynaptic receptor function.

The simplest way to analyse functional images consists of calculation of average values for regions of interest, preferably placed on the image in an objective fashion. This method may, however, 'dilute' focal increases if the regions of interest are too large, whereas, if too small, the risk of false positive results becomes unacceptable. Sophisticated methods of analysis that rely upon computer-processing of the images have been developed for 'activation' paradigms to localize sites of increase (Fox *et al.*, 1988; Friston *et al.*, 1991). The analysis of images of biological tracer uptake depends in part upon whether the behaviour of the compound can be modelled successfully. The simplest, semiquantitative method of analysis compares uptake in a region of interest at a given time after tracer administration with that of a reference region, known or assumed to have only nonspecific uptake.

Application of functional imaging techniques to disorders of voluntary movement

Human movement disorders may be subdivided into deficits of movement (akinesia), excess movements (hyperkinesia) and dystonia. Functional imaging has provided important insights into the pathophysiology of all three. The findings are summarized in Table 10.1.

Table 10.1.

	Metabolic	Striatal Dopaminergic
Akinetic-rigid syndromes:		
PD	Increased: pallidum?	FD: reduced (selective)
	Reduced: frontal cortex	D2: normal
PSP	Reduced: frontal cortex	FD: reduced
		D2: reduced
MSA	Reduced: striatum, cerebellum, brainstem	FD: reduced
		D2: reduced
CBD	Reduced: association cortex	FD: reduced (asymmetrical)
Dyskinesias:		
HC	Reduced: caudate, multiple cortical sites	FD: normal
		D2: reduced
NA	Reduced: striatum	FD: reduced (selective)
		D2: reduced
ET	Increased: olive, cerebellum	FD: normal
Dystonia:		
ID	Reduced: basal ganglia?	FD: reduced?
DRD	?	FD: reduced (mild)
WD	Reduced: multiple cortical sites	FD: reduced

Notes: Simplified summary of reported findings for metabolic imaging (FDG, blood flow or CMRO) and investigations of the integrity of striatal dopaminergic pathways in various types of movement disorder. Not all findings necessarily apply to individual patients. Some results have not been confirmed by all investigators and the pattern of metabolic changes may depend upon the stage of the illness. FD = 18F-dopa uptake, D2 - dopamine D2 receptor studies.

PD = Parkinson's disease, PSP = progressive supranuclear palsy, MSA = multiple system atrophy, CBD = corticobasal degeneration, HC = Huntington's chorea, NA = neuroacanthocytosis, ET = essential tremor, ID = idiopathic dystonia, DRD = dopa-responsive dystonia, WD = Wilson's disease.

Akinetic-rigid syndromes

Parkinsonism of all types is characterized by loss of dopamine content within the striatum. It appears that over 85% of total striatal dopamine content has been lost even in cases with mild clinical deficits (Hornykiewicz, 1981). PET images following the administration of 18F-dopa (which, like L-dopa, is taken up, decarboxylated and stored in vesicles within dopaminergic nerve terminals in the striatum; Martin et al., 1989), dramatically illustrate the pathology of Parkinson's disease (Fig. 10.2). Patients with 'on-off' fluctuations in response to treatment have more severely reduced uptake than those with shorter duration disease (Leenders et al., 1986b). An influx constant (Ki) can be calculated

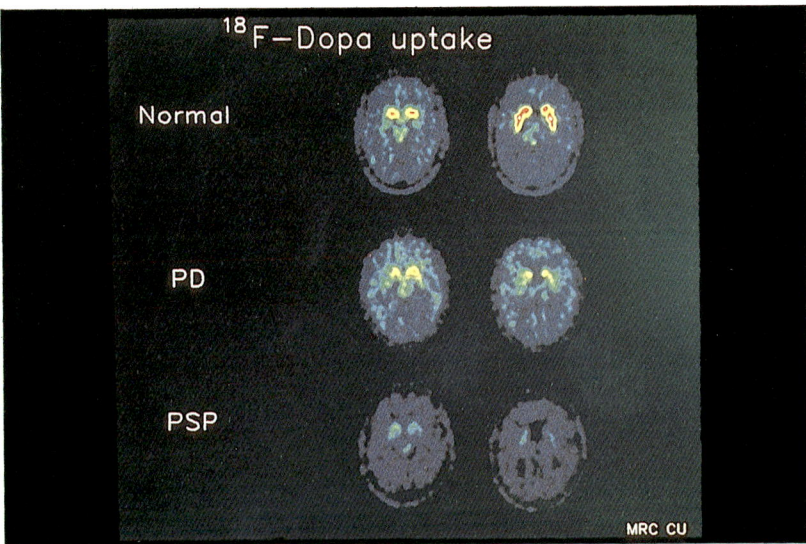

Fig. 10.2. Two axial views of 18F-dopa uptake in three different conditions. In normal subjects (uppermost) uptake is symmetrical and the caudate nuclei can be seen distinct from the putamen on each side. Subjects with Parkinson's disease (PD, middle of figure) show reduced 18F-dopa uptake, with relative sparing of the caudate nuclei. Patients with PSP (lowermost part of figure) have a generalized reduction in striatal 18F-dopa uptake. Brooks *et al.* (1990) reproduced with permission.

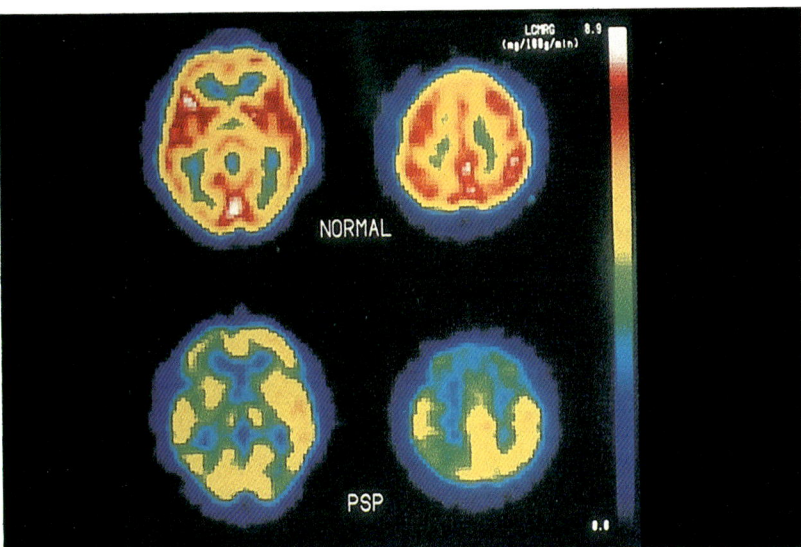

Fig. 10.3. Two transverse brain sections from a normal volunteer and a patient with progressive supranuclear palsy (PSP, lower half) showing regional metabolism of glucose. The patient with PSP shows a generalized reduction in cortical metabolism, most marked frontally. Foster *et al.* (1988) reproduced with permission.

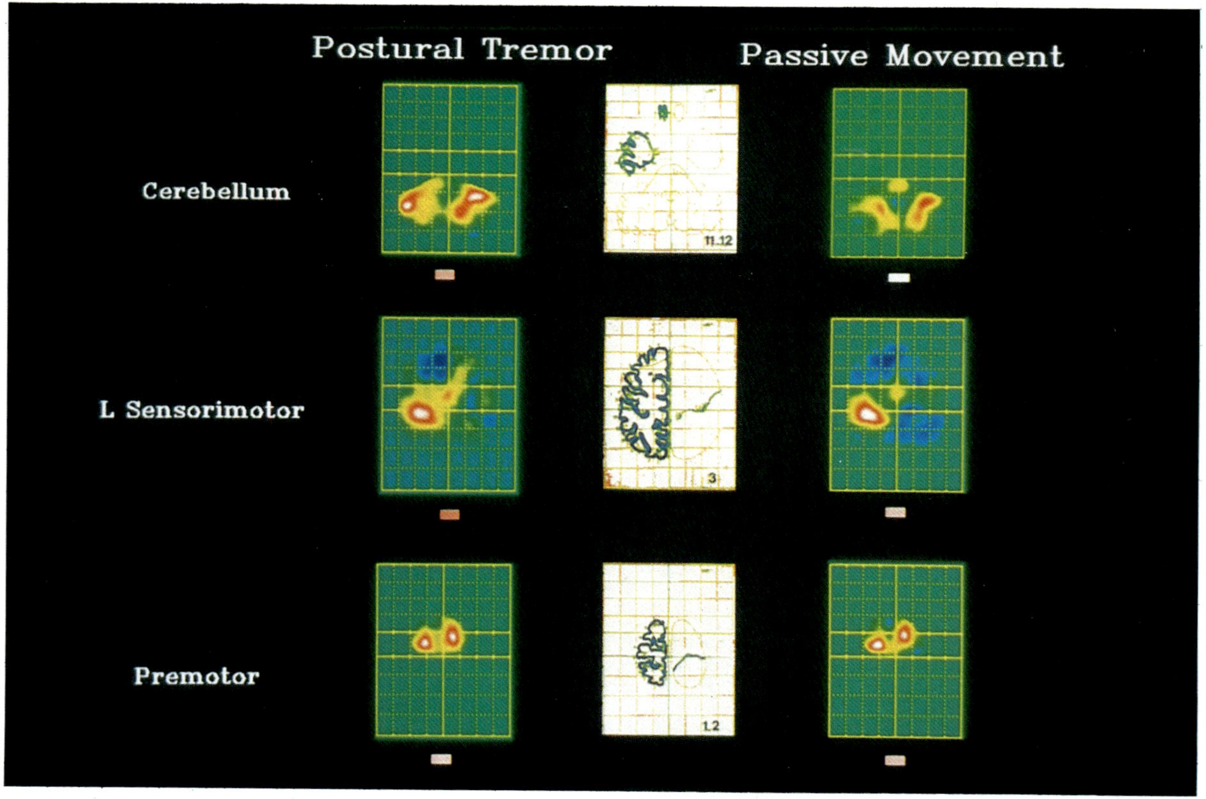

Fig. 10.4. Changes in cerebral blood flow in association with essential tremor (ET). An 'activation-type' paradigm was used. The images on each side are slices at three different transverse levels showing the significance of any increase in pixel flow compared with that in the resting state. All areas shown in white were sites of significant (corrected $p<0.05$) increases in blood flow. The three images on the left show the increases in flow in association with unilateral postural tremor of the right hand and those on the right show the changes in the same areas that occurred in response to passively imposed movement of the patient's right wrist. Only the flow increase occurring within the cerebellum was specifically associated with the presence of postural tremor. Central part of Figure shows slices at the same levels from the stereotaxic atlas of Talairach and Tournoux (1988), reproduced with permission. (Modified from Fig. 1 of Colebatch et al., 1990.)

from the PET data which reflects the rate of decarboxylation and synaptic storage (Martin *et al.*, 1989) and is negatively correlated with clinical disability (Brooks *et al.*, 1990). Ki values for 18F-dopa uptake by the putamen, which were halved on average, completely separated 16 patients with Parkinson's disease from 30 normal volunteers (Brooks *et al.*, 1990). In addition, the PET images have shown more severe loss of dopaminergic uptake in the putamen than in the caudate nucleus (Fig. 10.2), a pattern also found for dopamine depletion at post-mortem (Hornykiewicz, 1981).

The use of 18F-dopa with PET is capable of detecting asymptomatic reductions of nigrostriatal dopamine uptake (Calne *et al.*, 1985), a finding of considerable clinical importance given that specific therapy may delay the progression of Parkinson's disease (Shoulson *et al.*, 1989). Normal ageing has been reported to cause a fall in the striatal uptake of 18F-dopa in one (Martin *et al.*, 1989) but not in another study (Sawle *et al.*, 1991*b*), observations relevant to theories as to the cause and progression of Parkinson's disease. Even in established disease, the decline of nigrostriatal function as measured by 18F-dopa uptake appears to be slow (Bhatt *et al.*, 1991). D1 and D2 dopamine receptor numbers in the striatum are preserved in postmortem brains from patients with Parkinson's disease (Cortes *et al.*, 1989). PET estimates of D2 receptors in the striatum have shown little alteration from normal controls, while early in disease, receptor density may actually increase (Hägglund *et al.*, 1987).

The metabolic changes seen in resting images of Parkinsonian patients have been reviewed by Martin (1989). Experimentally produced lesions of the nigrostriatal tract increase metabolism in the ipsilateral globus pallidus, presumably a consequence of disinhibition causing increased activity in striatopallidal pathways (Wooten & Collins, 1981; Crossman, 1987). Wolfson *et al.* (1985) reported that CBF was increased in the contralateral basal ganglia in patients with clinically unilateral Parkinson's disease, but Perlmutter and Raichle (1985), who also studied unilaterally affected patients, could not confirm this finding. The latter authors speculated that their findings reflected clinical differences between the patients as to the presence or absence of tremor. Both studies reported a decrease in blood flow to the frontal lobes, which may be due in part to loss of mesocortical dopamine projections. Poor activation of medial frontal structures have been shown with neural activation paradigms (Passingham *et al.*, 1991). Metabolic changes consistent with Alzheimer's disease may be seen in demented parkinsonian patients (Martin, 1989).

Progressive supranuclear palsy (PSP) or Steele–Richardson–Olszewski syndrome may present as an akinetic-rigid syndrome prior to the development of a downgaze palsy. Neuropathologically, it is distinct from Parkinson's disease and is characterized by neurofibrillary tangles affecting brainstem nuclei.

Patients often show 'frontal lobe' signs and display defects on neuropsychological tests sensitive to frontal lobe function, even though the cerebral neocortex is usually normal on histological examination (Lees, 1987). It is particularly notable therefore that functional imaging demonstrates frontal lobe abnormalities, in addition to abnormalities of subcortical structures (Fig. 10.3). Glucose metabolism in the frontal lobes is depressed, most markedly so in the superior half (D'Antona et al., 1985; Foster et al., 1988; Goffinet et al., 1989), and oxygen uptake and local blood flow are reduced (Leenders, Frackowiak & Lees, 1988a). Similar abnormalities have been shown with SPECT blood flow markers (Neary et al., 1987). Although the mechanism of the frontal abnormalities is unknown, Laplane et al. (1989) have shown that lesions localized to the basal ganglia can cause 'frontal-lobe' behavioural changes associated with hypometabolism of prefrontal cortex. Postmortem studies have demonstrated loss of both D2 receptor sites and reductions in dopamine in the striatum of patients with PSP (Bokoboza et al., 1984). PET studies have confirmed the reduction in dopa uptake but have also shown that, unlike Parkinson's disease, the caudate uptake is as severely reduced as that for the putamen (Brooks et al., 1990; Fig. 10.2). A reduction in D2 receptors has been demonstrated by PET *in vivo* (Baron et al., 1985). Brooks et al. (1991b) observe that the changes in the dopaminergic pathway are not more severe than are seen in long-standing Parkinson's disease, so that the characteristic failure of patients with PSP to respond to L-dopa must be due to additional pathology. Degeneration of the internal segment of the globus pallidus, which is almost always present (Lees, 1987), might contribute to this refractoriness and, by virtue of projections via the thalamus, to the alterations in frontal metabolism.

Up to 10% of patients diagnosed on clinical grounds as having Parkinson's disease are found at postmortem to have striatonigral degeneration (SND) in which there is atrophy and gliosis affecting both the striatum and the substantia nigra (Adams, van Bogaert & Vander Eecken, 1964). This condition and others including the Shy Dräger syndrome and sporadic cases of olivopontocerebellar degeneration (OPCA) arguably are different manifestations of a common pathological process and the broader term 'multiple system atrophy' (MSA) has been proposed (Oppenheimer, 1984; Quinn, 1989). Glucose metabolism in such patients has been shown to be reduced in the regions known to be affected from pathological findings. Thus, hypometabolism of glucose occurs in the striatum of patients with SND (De Volder et al., 1989) and MSA (Dubinsky et al., 1988) and in the brainstem and cerebellum in OPCA (Gilman et al., 1988). Less predictable, however, have been these authors' findings of reduced frontal lobe metabolism in SND and of increased metabolism in the pallidum occurring in MSA. 18F-dopa uptake in the putamen of patients with MSA is reduced to a degree similar to that found in Parkinson's disease and is associated with a variable degree of reduction in uptake by the caudate (Brooks *et al.*, 1990). Putaminal D2 receptor binding sites are reduced (Brooks *et al.*, 1991*b*).

Corticobasal degeneration is an uncommon cause of an asymmetrical akinetic-rigid syndrome, often combined with other features (Gibb, Luthert & Marsden, 1989). Sawle *et al.* (1991*a*) studied six such patients and found multiple areas of reduced oxygen metabolism affecting most severely the temporal, parietal and occipital association cortices. 18F-dopa uptake in the striatum was reduced opposite the clinically affected side.

Dyskinesias

Chorea is characterized by frequent muscle jerks, irregularly timed and of variable duration. Patients with Huntington's disease, in which choreiform movements are usually a prominent feature, have reduced glucose utilization affecting both the caudate and the putamen (Kuhl *et al.*, 1982). Reductions in caudate metabolism precede structural abnormalities (Mazziotta *et al.*, 1987). More advanced stages of the disease show reductions in cortical metabolism (Kuwert *et al.*, 1990). Reduced caudate metabolism is not unique to Huntington's disease, being present also in some cases of benign hereditary chorea (Suchowersky *et al.*, 1986) and neuroacanthocytosis (Hosokawa *et al.*, 1987: see below). Leenders *et al.* (1986*a*) reported normal 18F-dopa uptake with reduced D2 receptor numbers in the striatum in a subject with Huntington's disease and Hägglund *et al.* (1987) also found reduced D2 receptor numbers in the striatum of a patient with Huntington's disease. Brücke *et al.* (1991), who used a SPECT D2 ligand, also demonstrated a marked reduction in receptor density in unmedicated patients with Huntington's disease.

In neuroacanthocytosis, a condition in which chorea and other movement disorders are often prominent features (for review, see Hardie *et al.*, 1991), hypometabolism of both the caudate and putamen has been reported to occur prior to evidence of atrophy of either structure (Dubinski *et al.*, 1989). Brooks *et al.* (1991*a*) found reduced 18F-dopa uptake selectively affecting the posterior part of the putamen coupled with evidence of reduced D2 receptor sites in both the caudate and putamen.

Tremor may be a prominent feature of Parkinson's disease but just as frequently occurs without other disease, sometimes in a familial form (essential tremor [ET]: Marsden, 1986). 18F-dopa scanning may be capable of identifying patients presenting with isolated tremor who are at risk for the subsequent development of Parkinson's disease (Brooks *et al.*, 1991*b*). No specific pathological change has been identified in ET (Herskovitz & Blackwood, 1969). Current hypotheses as to the cause of ET postulate either a peripheral/segmental abnormality such as dysfunction of the stretch reflex, or a central 'oscillator' located in either the thalamus or olivary nucleus (Elbe & Koller, 1990). Dubinsky and Hallett (1987) reported hypermetabolism of glucose in the inferior olive ipsilateral to the tremor in patients with ET. Colebatch *et al.* (1990) used a neural activation paradigm in patients with ET and deduced that the presence of tremor was specifically associated with increased blood flow within the cerebellum (Fig. 10.4). This is consistent with evidence of neural oscillation within cerebello-olivary pathways in experimental models of ET (Lamarre, 1984). Cerebellar projections to the motor cortex relay in the thalamus and the observations of Colebatch *et al.* may explain the effectiveness of thalamotomy as a treatment for ET. Thalamotomy is also an effective treatment of Parkinsonian tremor (Elbe & Koller, 1990) and Parker *et al.* (1991) have reported reduced blood flow in the cerebellum when they abolished Parkinsonian tremor by intrathalamic stimulation. Olivo-cerebellar oscillations may be common to several different types of tremor.

Dystonia

Dystonia, a syndrome of sustained muscle contractions producing abnormal postures, may be symptomatic or idiopathic and each may be generalized, segmental or focal (Fahn, 1988). The genetics of idiopathic dystonia are specifically considered elsewhere (Fahn, this volume). The primary dystonic syndromes include classical idiopathic torsion dystonia (dystonia musculorum deformans) as well as several other syndromes, including the paroxysmal dystonias and dopa-responsive dystonia (Marsden, 1987). Although symptomatic dystonia may follow lesions of the thalamus or neostriatum (Marsden *et al.*, 1985), no consistent structural abnormality has been described in association with idiopathic torsion dystonia. Abnormalities of monoamine transmitters, particularly noradrenaline, were reported in postmortem study of two cases of idiopathic torsion dystonia by Hornykiewicz *et al.* (1986). Electrophysiological studies have shown abnormalities of brainstem (Beradelli *et al.*, 1985) and segmental (Nakashima *et al.*, 1989) reflex pathways but these presumably result from neural dysfunction lying higher 'upstream'. PET studies have provided supporting evidence for a functional abnormality affecting higher levels of the neuraxis in association with dystonia.

Perlmutter and Raichle (1984) studied a case of 'paroxysmal hemidystonia' in whom no structural changes were evident, and showed a decrease in CMRO and an increase in CBF in the region of the basal ganglia contralateral to the movements. Similar abnormalities of glucose metabolism in the basal ganglia have since been shown to affect only a minority of dystonics (e.g. Gilman *et al.*, 1988; Lang *et al.*, 1988; *cf.* Chase, Tamminga & Burrows, 1988). In 16 subjects with torticollis, Stoessl *et al.* (1986) found no absolute abnormality in metabolic rates for glucose in the basal ganglia or thalamus but reported that the normal ratio between these values was disturbed. Lesions of the putamen causing symptomatic dystonia are associated with reduced glucose metabolism (Lang *et al.*, 1988). Temple and Perlmutter (1990) found that the increase in blood flow in the sensorimotor cortex in response to cutaneous vibration was less in dystonic patients than in normal controls, a finding which they argued might be relevant to the underlying pathophysiology.

Leenders *et al.* (1988*b*) studied 18F-dopa uptake in four subjects with hemidystonia, three of whom had associated hemiparkinsonism and two subjects with torticollis. All the subjects, including the two with torticollis, had reduced 18F-dopa uptake into the striatum compared to their six controls. In the hemidystonics, this reduction was more severe on the side contralateral to the clinical signs. D2 receptor levels were normal except in one subject in whom they were increased contralateral to the affected side. Sawle *et al.* (1991*c*) studied a clinically uniform series of patients all of whom were diagnosed to have dopa-responsive dystonia (DRD). This commonly familial condition begins in childhood or adolescence and is dramatically responsive to modest doses of L-Dopa for many years (Nygaard, Marsden & Fahn, 1991). Despite up to 17 years of L-dopa therapy, six of the seven patients studied by Sawle *et al.* (1991*c*) had only mild degrees of reduced 18F-dopa uptake in the caudate and putamen, while the remaining subject's uptake was in the range seen in Parkinson's disease. Clearly, the nigrostriatal pathway is well preserved in the majority of cases of DRD.

Wilson's disease is an important cause of a number of movement disorders, including dystonia and tremor. Glucose metabolism is widely depressed in the later stages of the disease, both cortically and subcortically. The lenticular nuclei are relatively severely affected and the thalmus is spared (Hawkins, Mazziotta & Phelps, 1987). Striatal 18F-dopa uptake is reduced in most cases (Snow *et al.*, 1991).

Conclusion

Functional imaging provides in vivo images of brain metabolism in humans. It shows potential for being an important research tool, although some of the studies may be criticised as having merely confirmed what is already known from post-mortem findings. However, previously unsuspected abnormalities have been demonstrated and future studies should allow detailed investigations of the evolution of different illnesses and correlation with clinical features. 'Activation'-type studies can be applied to investigate the neural basis of both normal and disordered brain function. As a clinical tool in movement disorders, the role of functional imaging in the individual patient is still to be fully defined: this role may not be primarily one of diagnosis, as the patterns demonstrated are rarely unique to a single pathological condition. Despite this, important contributions can be made with currently available techniques in the detection of presymptomatic changes in individuals at risk of Parkinson's disease or Huntington's disease and in monitoring patient responses to current and experimental treatments.

References

Ackermann, R. F., Finch, D. M., Babb, T. L. & Engel, J. (1984). Increased glucose metabolism during long-duration recurrent inhibition of hippocampal pyramidal cells. *Journal of Neuroscience*, **4**, 251–64.

Adams, R. D., van Bogaert, L. & Vander Eecken, H. (1964). Striato-nigral degeneration. *Journal of Neuropathology and Experimental Neurology*, **23**, 584–608.

Albin, R. L., Young, A. B. & Penney, J. B. (1989). The functional anatomy of basal ganglia disorders. *Trends in Neurosciences*, **12**, 366–75.

Baron, J. C., Maziere, B., Loc'h, C., Sgouropoulos, P., Bonnet, A. M. & Agid, Y. (1985). Progressive supranuclear palsy: loss of striatal dopamine receptors demonstrated *in vivo* by positron tomography. *Lancet*, **i**, 1163–4.

Berardelli, A., Rothwell, J. C., Day, B. L. & Marsden, C. D. (1985). Pathophysiology of blepharospasm and oromandibular dystonia. *Brain*, **108**, 593–608.

Bhatt, M. H., Snow, B. J., Martin, W. R. W., Pate, B. D. & Calne, D. B. (1991). Positron emission tomography suggests that the rate of progression of idiopathic parkinsonism is slow. *Annals of Neurology*, **29**, 673–7.

Bokoboza, B., Ruberg, M., Scatton, B., Javoy-Agid, F. & Agid, Y. (1984). [^{3}H]Spiperone binding, dopamine and HVA concentrations in Parkinson's disease and supranuclear palsy. *European Journal of Pharmacology*, **99**, 167–75.

Brooks, D. J., Ibanez, V., Sawle, G. V., Quinn, N., Lees, A. J., Mathias, C. J., Bannister, R., Marsden, C. D. & Frackowiak, R. S. J. (1990). Differing patterns of striatal 18F-dopa uptake in Parkinson's disease, multiple system atrophy, and progressive supranuclear palsy. *Annals of Neurology*, **28**, 547–55.

Brooks, D., Ibanez, V., Playford, E. D., Sawle, G. V., Leigh, P. N., Kocen, R. S., Harding, A. E. & Marsden, C. D. (1991a) Presynaptic and postsynaptic striatal dopaminergic function in neuroacanthocytosis: a positron emission tomographic study. *Annals of Neurology*, **30**, 166–71.

Brooks, D., Ibanez, V., Sawle, G. V., Playford, E. D., Quinn, N., Lees, A. J, Marsden, C. D. & Frackowiak, R. S. J. (1991b). Striatal D2 receptor density in Parkinson's disease, striatonigral degeneration, and progressive supranuclear palsy, measured with 11C-Raclopride and PET. *Journal of Cerebral Blood Flow and Metabolism*, **11** (Suppl. 2), S229.

Brooks, D., Playford, E. D., Thompson, P. D., Findley, L. J. & Marsden, C. D. (1991c). The relationship between nigral pathology and isolated tremor: an 18F-dopa study. *Journal of Cerebral Blood Flow and Metabolism*, **11** (Suppl. 2), S817.

Brücke, T., Podreka, I., Angelberger, P., Wenger, S., Topitz, A., Kufferle, B., Muller, C. & Deecke, L. (1991). Dopamine D2 receptor imaging with SPECT: studies in

different neuropsychiatric disorders. *Journal of Cerebral Blood Flow and Metabolism*, **11**, 220–8.
Calne, D. B., Langston, J. W., Martin, W. R. W., Stoessl, A. J., Ruth, T. J., Adam, M. J., Pate, B. D. & Schulzer, M. (1985). Positron emission tomography after MPTP: observations relating to the cause of Parkinson's disease. *Nature*, **317**, 246–8.
Chase, T. N., Tamminga, C. N. & Burrows, H. (1988). Positron emission tomographic studies of regional cerebral glucose metabolism in idiopathic dystonia. In *Dystonia 2: Advances in Neurology*, vol 50, ed. S. Fahn, C. D. Marsden & D. B. Calne, pp. 237–41. New York: Raven Press.
Colebatch, J. G., Findley, L. J., Frackowiak, R. S. J., Marsden, C. D. & Brooks, D. J. (1990). Activation of the cerebellum in essential tremor. *Lancet*, **336**, 1028–30.
Cortes, R., Camps, M., Gueye, B., Probst, A. & Palacios, J. M. (1989). Dopamine receptors in human brain: autoradiographic distribution of D1 and D2 sites in Parkinson syndrome of different etiology. *Brain Research*, **483**, 30–8.
Crossman, A. R. (1987). Primate models of dyskinesia: the experimental approach to the study of basal ganglia-related involuntary movement disorders. *Neuroscience*, **21**, 1–40.
D'Antona, R., Baron, J. C., Samson, Y., Serdaru, M., Vader, F., Agid Y. & Cambier J. (1985). Subcortical dementia: frontal cortex hypometabolism detected by positron tomography in patients with progressive supranuclear palsy. *Brain* **108**, 785–99.
De Volder, A. G., Francart, J., Laterre, C., Dooms, G., Bol, A., Michel, C. & Goffinet, A. M. (1989). Decreased glucose utilization in the striatum and frontal lobe in probable striatonigral degeneration. *Annals of Neurology*, **26**, 239–47.
Dubinsky, R. M., Brown, R. T., Polinsky, R. J., Di Chiro, G. D. & Hallett, M. (1988). Regional brain glucose hypometabolism in multiple system atrophy. *Neurology*, **38** (Suppl. 1), 330.
Dubinsky, R. M. & Hallett, M. (1987). Glucose hypermetabolism of the inferior olive in patients with essential tremor. *Annals of Neurology*, **22**, 118.
Dubinsky, R. M., Hallett, M., Levey, R. & Di Chiro, G. (1989). Regional brain metabolism in neuroacanthocytosis. *Neurology*, **39**, 1253–5.
Ebmeier, K. P., Dougall, N. J., Austin, M.-P. V., Murray, C. L., Curran, S. M., O'Carroll, R., Moffoot, A. P. R., Hannan, J. & Goodwin, G. M. (1991). The split-dose technique for the study of psychological and pharmacological activation with the cerebral blood flow marker exametazime and single photon emission computed tomography (SPECT): reproducibility and rater reliability. *International Journal of Methods in Psychiatric Research*, **1**, 27–38.
Elbe, R. J. & Koller, W. C. (1990). *Tremor*. Baltimore: Johns Hopkins.
Fahn, S. (1988). Concept and classification of dystonia. In *Dystonia 2. Advances in Neurology*, vol 50, ed. S. Fahn, C. D. Marsden & D. B. Calne, pp. 1–8. New York: Raven Press.
Foster, N. L., Gilman, S., Berent, S., Morin, E. M., Brown, M. B. & Koeppe, R. A. (1988). Cerebral hypometabolism in progressive supranuclear palsy studied with positron emission tomography. *Annals of Neurology*, **24**, 399–406.
Fox, P. T., Mintun, M. A., Reiman, E. M. & Raichle, M. E. (1988). Enhanced detection of focal brain responses using intersubject averaging and change-distribution analysis of subtracted PET images. *Journal of Cerebral Blood Flow and Metabolism*, **8**, 642–53.
Fox, P. T., Perlmutter, J. S. & Raichle, M. E. (1985). A stereotactic method of anatomical localization for Positron Emission Tomography. *Journal of Computer Assisted Tomography*, **9**, 141–53.

Fox, P. T. & Raichle, M. E. (1986). Focal physiological uncoupling of cerebral blood flow and oxidative metabolism during somatosensory stimulation in human subjects. *Proceedings of the National Academy of Sciences, USA*, **83**, 1140–4.

Friston, K. J., Frith, C. D., Liddle, P. F. & Frackowiak, R. S. J. (1991). Comparing functional (PET) images: the assessment of significant change. *Journal of Cerebral Blood Flow and Metabolism*, **11**, 690–9.

Gibb, W. R. G., Luthert, P. J. & Marsden, C. D. (1989). Corticobasal degeneration. *Brain* **112**, 1171–92.

Gilman, S., Markel, D. S., Koeppe, R. A., Junck, L., Kluin, K. J., Gegarski, S. S. & Hichwa, R. D. (1988). Cerebellar and brainstem hypometabolism in olivopontocerebellar atrophy detected with positron emission tomography. *Annals of Neurology*, **23**, 223–30.

Goffinet, A. M., De Volder, A. G., Gillain, C., Rectem, D., Bol, A., Michel, C., Cogneau, M., Labar, D. & Laterre, C. (1989). Positron tomography demonstrates frontal lobe hypometabolism in progressive supranuclear palsy. *Annals of Neurology*, **25**, 131–9.

Hägglund, J., Aquilonius, S.-M., Eckernas, S.-A., Hartvig, P., Lundquist, H., Gullberg, P. & Långström, B. (1987). Dopamine receptor properties in Parkinson's disease and Huntington's chorea evaluated by positron emission tomography using 11C-N-methylspiperone. *Acta Neurologica Scandinavica*, **75**, 87–94.

Hardie, R. J., Pullon, H. W. H., Harding, A. E., Owen, J. S., Pires, M., Daniels, G. L., Imai, Y., Misra, V. P., King, R. H. M., Jacobs, J. M., Tippett, P., Duchen, L. W., Thomas, P. K. & Marsden, C. D. (1991). Neuroacanthocytosis: a clinical, haematological, and pathological study of 19 cases. *Brain*, **114**, 13–50.

Hawkins, R. A., Mazziotta, J. C. M. & Phelps, M. E. (1987). Wilson's disease studied with FDG and positron emission tomography. *Neurology*, **37**, 1707–11.

Herskovitz, E. & Blackwood, W. (1969). Essential (familial, hereditary) tremor: a case report. *Journal of Neurology, Neurosurgery and Psychiatry*, **32**, 509–11.

Hornykiewicz, O. (1981). Brain neurotransmitter changes in Parkinson's disease. In *Movement Disorders*, ed. C.D. Marsden & S. Fahn, pp. 41–58. London: Butterworth Scientific.

Hornykiewicz, O., Kish, S. J., Becker, L. E., Farley, I. & Shannak, K. (1986). Brain neurotransmitters in dystonia musculorum deformans. *New England Journal of Medicine*, **315**, 347–53.

Hosokawa, S., Ichiya, Y., Kuwabara, Y., Ayabe, Z., Mitsuo, K., Goto, I. & Kato, M. (1987). Positron emission tomography in cases of chorea with differing underlying diseases. *Journal of Neurology, Neurosurgery and Psychiatry*, **50**, 1284–7.

Kadekaro, M., Vance, W. H., Terrell, M. L., Gary, H., Eisenberg, H. M. & Sokoloff, L. (1987). Effects of antidromic stimulation of the ventral root on glucose utilization in the ventral horn of the spinal cord in the rat. *Proceedings of the National Academy of Sciences, USA*, **84**, 5492–5.

Kuhl, D. E., Phelps, M. E., Markham, C. H., Metter, E. J., Riege, W. H. & Winter, J. (1982). Cerebral metabolism and atrophy in Huntington's disease determined by 18FDG and computed tomographic scan. *Annals of Neurology*, **12**, 425–34.

Kuwert, T., Lange, H. W., Langen, K.-J., Herzog, H., Aulich, A. & Feinendegen, L. E. (1990). Cortical and subcortical glucose consumption measured by PET in patients with Huntington's disease. *Brain*, **113**, 1405–23.

Lamarre, Y. (1984). Animal models of physiological, essential and parkinsonian-like tremors. In *Movement Disorders: Tremor*, ed. L.J. Findley & R. Capildeo, pp. 183–94. London: Macmillan.

Functional imaging in motor disorders 167

Lang, A. E., Garnett, E. S., Firnau, G., Nahmias, C. & Tallala, A. (1988). Positron tomography in dystonia. In *Dystonia 2. Advances in Neurology*, vol. 50, ed. S. Fahn, C. D. Marsden & D. B. Calne, pp. 249–53. New York: Raven Press.

Laplane, D., Levasseur, M., Pillon, B., Dubois, B., Baulac, M., Mazoyer, B., Tran Dinh, S., Sette, G., Danze, F. & Baron J. C. (1989). Obsessive-compulsive and other behavioural changes with bilateral basal ganglia lesions. *Brain*, **112**, 699–725.

Leenders, K. L., Frackowiak, R. S. J. & Lees, A. J. (1988a) Steele-Richardson-Olszewski syndrome. *Brain*, **111**, 615–30.

Leenders, K. L., Frackowiak, R. S. J., Quinn, N. & Marsden, C. D. (1986*a*). Brain energy metabolism and dopaminergic function in Huntington's disease measured *in vivo* using positron emission tomography. *Movement Disorders*, **1**, 69–77.

Leenders, K. L., Palmer, A. J., Quinn, N., Clark, J. C., Firnau, G., Garnett, E. S., Nahmias, C., Jones, T. & Marsden, C. D. (1986*b*). Brain dopamine metabolism in patients with Parkison's disease measured with positron emission tomography. *Journal of Neurology, Neurosurgery and Psychiatry*, **49**, 853–60.

Leenders, K. L., Quinn, N., Frackowiak, R. S. J. & Marsden, C. D. (1988*b*). Brain dopaminergic system studied in patients with dystonia using positron emission tomography. In *Dystonia 2. Advances in Neurology*, vol. 50, ed. S. Fahn, C.D. Marsden & D.B. Calne, pp. 243–53. New York: Raven Press.

Lees, A. J. (1987). The Steele-Richardson-Olszewski syndrome (progressive supranuclear palsy). In *Movement Disorders 2*, ed. C. D. Marsden & S. Fahn, pp. 272–87. London: Butterworths.

Marsden, C. D. (1986). Movement disorders and the basal ganglia. *Trends in Neurosciences*, **9**, 512–15.

Marsden, C. D. (1987). Summary of part II: rigidity/dystonia. In *Motor Disturbances 1*, ed. R. Benecke, B. Conrad & C. D. Marsden, pp. 145–52. London: Academic Press.

Marsden, C. D., Obeso, J. A., Zarranz, J. J. & Lang, A. E. (1985). The anatomical basis of hemidystonia. *Brain*, **108**, 463–83.

Martin, W. R. W. (1989). Imaging techniques in Parkinson's disease. *Movement Disorders*, **4**, (Suppl. 1), S63–9.

Martin, W. R. W., Palmer, M. R., Patlak, C. S. & Calne, D. B. (1989). Nigrostriatal function in humans studied with positron emission tomography. *Annals of Neurology*, **26**, 535–42.

Mazziotta, J. C., Phelps, M. E., Pahl, J. J., Huang, S.-C., Baxter, L. R., Riege, W. H., Hoffman, J. M., Kuhl, D. E., Lanto, A. B., Wapenski, J. A. & Markham, C. H. (1987). Reduced cerebral glucose metabolism in asymptomatic subjects at risk for Huntington's disease. *New England Journal of Medicine*, **316**, 357–62.

Nakashima, K., Rothwell, J. C., Day, B. L., Thompson, P. D., Shannon, K. & Marsden, C. D. (1989). Reciprocal inhibition between forearm muscles in patients with writer's cramp and other occupational cramps, symptomatic hemidystonia and hemiparesis due to stroke. *Brain*, **112**, 681–97.

Neary, D., Snowdon, J. S., Shields, R. A., Burjan, A. W. I., Northen, B., MacDermott, N., Prescott, M. C. & Testa, H. J. (1987). Single photon emission tomography using 99mTc-HM-PAO in the investigation of dementia. *Journal of Neurology, Neurosurgery and Psychiatry*, **50**, 1101–9.

Nygaard, T. G., Marsden, C. D. & Fahn, S. (1991). Dopa responsive dystonia: long term treatment reponse and prognosis. *Neurology*, **41**, 174–81.

Oppenheimer, D. R. (1984). Diseases of the basal ganglia, cerebellum and motor neurons. In *Greenfield's Neuropathology*, 4th edn, ed. J. H. Adams, J. A. N. Corsellis & L. W. Duchen, pp. 699–747. London: Edward Arnold.

Parker, F., Tzourio, N., Blond, S., Petit, H., Mazoyer, P. B. & Syrota, A. (1991). Cerebellar and cortical involvement in Parkinson's tremor arrest during thalamic stimulation of the Vim nucleus: a blood flow PET study. *Neurology*, **41** (Suppl. 1), 360.

Passingham, R. E., Playford, E. D., Nutt, J., Frackowiak, R. S. J. & Brooks, D. J. (1991). Medial frontal and striatal activation are impaired in Parkinson's disease. *Neurology*, **41** (Suppl. 1), 359.

Penney, Y. & Young, A. B. (1986). Striatal inhomogeneities and basal ganglia function. *Movement Disorders*, **1**, 3–15.

Perlmutter, J. S. & Raichle, M. E. (1984). Pure hemidystonia with basal ganglion abnormalities on positron emission tomography. *Annals of Neurology*, **15**, 228–33.

Perlmutter, J. S. & Raichle, M. E. (1985). Regional blood flow in hemiparkinsonism. *Neurology,* **35**, 1127–34.

Quinn, N. (1989). Multiple system atrophy – the nature of the beast. *Journal of Neurology, Neurosurgery and Psychiatry* (Suppl.), 78–89.

Raichle, M. E., Martin, W. R. W., Herscovitch, P., Mintun, M. A. & Markham, J. (1983). Blood flow measured with intravenous H215O. II. Implementation and validation. *Journal of Nuclear Medicine,* **24**, 790–8.

Roy, C. S. & Sherrington, C. S. (1890). On the regulation of the blood-supply of the brain. *Journal of Physiology (London)*, **11**, 85–108.

Sawle, G. V., Brooks, D. J., Marsden, C. D. & Frackowiak, R. S. J. (1991*a*). Corticobasal degeneration. *Brain*, **114**, 541–56.

Sawle, G. V., Colebatch, J. G., Shah, A., Brooks, D. J., Marsden, C. D. & Frackowiak, R. S. J. (1991*b*). Striatal function in normal aging: implications for Parkinson's disease. *Annals of Neurology*, **28**, 799–804.

Sawle, G. V., Leenders, K. L., Brooks, D. J., Harwood, G., Lees, A. J., Frackowiak, R. S. J. & Marsden, C. D. (1991c). Dopa-responsive dystonia: 18F-dopa positron emission tomography. *Annals of Neurology,* **30**, 24–30.

Shoulson, I. & The Parkinson Study Group (1989). Effect of Deprenyl on the progression of disability in early Parkinson's disease. *New England Journal of Medicine,* **321**, 1364–71.

Snow, B. J., Bhatt, M., Martin, W. R. W., Li, D. & Calne, D. B. (1991). The nigrostriatal pathway in Wilson's disease studied with positron emission tomography. *Journal of Neurology, Neurosurgery and Psychiatry,* **54**, 12–17.

Stoessl, A. J., Martin, W. R. W., Clark. C., Adam, M. J., Ammann, W., Beckman, J. H., Bergstrom, M., Harrop, R., Rogers, J. G., Ruth, T. J., Sayre, C. I., Pate, B. D. & Calne, D. B. (1986). PET studies of cerebral glucose metabolism in idiopathic torticollis. *Neurology,* **36**, 653–7.

Suchowersky, O., Hayden, M. R., Martin, W. R. W., Stoessl, A. J., Hildebrand, A. M. & Pate B. D. (1986). Cerebral metabolism of glucose in benign hereditary chorea. *Movement Disorders,* **1**, 33–44.

Talairach, J. & Tournoux, P. (1988). *Co-Planar Stereotaxic Atlas of the Human Brain.* New York: Thieme Medical Publishers.

Temple, L. W. & Perlmutter, J. S. (1990). Abnormal vibration-induced cerebral blood flow responses in idiopathic dystonia. *Brain,* **113**, 691–707.

Trugman, J. M. & Wooten, G. F. Functional consequences of dopamine receptor stimulation. *Current Opinion in Neurology and Neurosurgery*, **3**, 548–51.

Wolfson, L. I., Leenders, K. L., Brown, L. L. & Jones, T. (1985). Alterations in regional cerebral blood flow and oxygen metabolism in Parkinson's disease. *Neurology,* **35**, 1399–405.

Wooten, G. F. & Collins, R. C. (1981). Metabolic effects of unilateral lesions of the substantia nigra. *Journal of Neuroscience*, **1**, 285–91.

11

Tonic stretch relfex in normal subjects and in cerebral palsy

P. D. NEILSON

*Cerebral Palsy Research Unit, Institute of Neurological Sciences, The Prince Henry Hospital
and
School of Electrical Engineering, University of New South Wales, Sydney NSW, Australia*

Introduction

The term 'cerebral palsy' implies injury to parts of the brain concerned with control of movement. It does not imply epilepsy, autism, blindness, deafness, mental retardation or other aspects of the more general brain damage syndrome. Various mixtures of these, however, occur in combination with cerebral palsy. The injury occurs around the time of birth, sometimes in the first few days after birth in premature babies. However, little is known about the nature of the injury or even what parts of the brain are involved. Recent advances in the neuropathology of cerebral palsy based on brain scans (see Volpe, 1987) have implicated periventricular leukomalacia (PVL), or cystic lesions in the deep periventricular white matter, in the genesis of cerebral palsy.

The consequences of cerebral palsy imply a lifetime of physical disability. A person with cerebral palsy can usually appreciate precisely the movement he/she wishes to make but the central nervous system is unable to translate this into appropriately coordinated motor commands. Inappropriate motor commands are generated, producing bizarre postures and movements. As well as the negative symptoms of cerebral palsy, there are many positive symptoms including unintelligible dysarthric speech, drooling, spasm, spasticity, involuntary movements, tremors, abnormal postural responses, muscle contractures and joint deformities. The current state of play is such that the positive symptoms of cerebral palsy are more amenable to therapy than are the negative ones. Understanding the mechanisms of spasm and spasticity and the role of stretch reflexes in motor behaviour has not been without much debate. This chapter discusses this controversy and puts forward a view of the role of tonic stretch reflex (TSR) mechanisms in controlling muscle stiffness in both normal and cerebral palsied subjects. The importance of distinguishing between phasic and tonic reflex mechanisms was emphasized in early papers on the reflex effects of vibration (Lance, 1965; DeGail, Lance & Neilson, 1966). The demonstration that vibration can suppress tendon jerks at the same time as it evokes a slowly developing tonic contraction (tonic vibration reflex or TVR) indicated that phasic and tonic reflex mechanisms might contribute differently to abnormalities of muscle tone. The discussion covers the pathophysiology of spasm and spasticity in cerebral palsy and the use of biofeedback training for reduction of spasm and spasticity.

Central control of muscle stiffness

The normal human nervous system can tune the force-displacement characteristics of muscles to suit different conditions such as running on a concrete surface or running on a sandy beach. It can defend the inherently unstable upright posture against the influences of gravity by increasing the stiffness of joints. In tasks requiring accurate control of position, the nervous system can increase the stiffness of muscles to minimize the influence of external disturbances. The same muscles can be used for tasks, in which control of position is secondary to control of pressure such as, for example, holding a paper cup. For such tasks the nervous system reduces the stiffness of muscles and generates forces with relatively compliant muscles. The ability of the nervous system to tune the 'suspension system' characteristics of muscles (i.e. their stiffness and viscosity) to suit the prevailing conditions is of fundamental importance. What is the physiological mechanism used to tune stiffness and viscosity of muscles?

A likely mechanism for tuning the stiffness of a muscle involves descending modulation of the gain of the stretch reflex. If the central part of the stretch reflex circuit acts like a simple gain constant, the reflex force will be proportional to the applied stretch and consequently, the stretch reflex will cause the muscle to display a linear force displacement relationship analogous to that of a spring. The slope of the force displacement graph represents the stiffness of the contracting muscle and the component of the stiffness contributed by reflex action is proportional to the gain of the reflex.

This is an attractive hypothesis, but does it actually happen? The question is far from new. Liddell and Sherrington (1924) hypothesized that both simple and complex reflex pathways form integral components in voluntary motor control and emphasized the possibility of reflex adaptation via descending supraspinal and suprabulbar influences.

Evidence against modulation of reflex gain

It has been argued that the stretch reflex has a gain too small for it to function effectively as a feedback control system (Rack, 1981), and that it is subject to transmission time delays of a magnitude which threaten its stability and preclude its use during fast movement (see Matthews, 1959*a,b*; Houk, Rymer & Crago, 1977; Jack & Roberts, 1978; Rymer & Hasan, 1980; Hoffer & Andreassen, 1981*a,b*; Stein & Lee, 1981). Based on evidence from the soleus muscle in the decerebrate cat where near-maximal stiffness has been observed after dorsal roots have been sectioned it has been claimed (Houk & Rymer, 1981) that the stiffness of a muscle servo (that is, a muscle and its autogenetic reflex circuitry), particularly at moderate to high contraction levels, is determined predominantly by the mechanical properties of the contracting muscle fibres themselves rather than by reflex action. Control of stiffness in a single muscle is seen to be achieved by recruitment of muscle fibres, while for a group of muscles operating across a joint, the net stiffness of the joint is seen as determined by both recruitment and cocontraction.

No evidence has been obtained from studies in the decerebrate cat that stiffness of muscles can be varied by descending modulation of reflex pathways (for review see Houk & Rymer, 1981). Studies of the static force length characteristics of isolated muscles in decerebrate cats with and without stretch reflex circuitry intact have indicated that the stretch reflex serves to linearize the length tension curves of muscles and to hold the stiffness constant. Electrical stimulation of the decerebrate vestibular nucleus, medial medullary reticular formation and pyramidal tract produces changes in stretch reflex threshold with no evidence of change in the slope of the force–length graph (Feldman & Orlovsky, 1972). Similarly, activation of synergistic, antagonistic and crossed extensor reflex pathways as well as procaine block of gamma efferent nerve fibres and spontaneous variations in rigidity, produce variation in the threshold of the stretch reflex with little or no change in the slope of the force–length graph (Matthews, 1959*a,b*). From such studies in decerebrate cats, it has been concluded that the force-length graph of an isolated muscle servo is invariant except for translations of the curve to the left or right along the length axis corresponding to changes in reflex threshold.

These findings could be attributed to anomalous behaviour of reflex circuits in decerebrate cats were it not for the fact that similar results have been observed in human subjects. Feldman (1966*a,b*) measured the static torque-angle curves at the elbow joint in human subjects (instructed not to intervene consciously) by graphing the equilibrium elbow-angles associated with various torques applied about the elbow. He found that if the muscle servos in human subjects were assumed to behave in the same way as observed in decerebrate cats, one could account for the measured torque-angle characteristics. The curve is linearized and the slope of the graph (net stiffness of the joint) could be varied by changing the level of cocontraction of agonist and antagonist muscles. He concluded that, just as for the decerebrate cat, the force length graphs of muscle servos in human subjects are invariant except for translations to the left or right along the length axis caused by descending supraspinal influences.

Evidence for modulation of reflex pathways

Using continuous, low frequency stretching of muscles in human subjects, a number of authors have shown that the gain and timing of TSRs can be adaptively tuned by central influences to suit the task in progress (Inbar, 1972; Inbar & Yafe, 1976; Dufresne, Soechting & Terzuolo, 1978; Neilson & Lance, 1978; Muller, Abbs & Kennedy, 1981; Soechting, Dufresne & Lacquaniti, 1981; Terzuolo, Soechting & Dufresne, 1981; Cannon & Zahalak, 1981; Chan & Kearney, 1982; Lacquaniti & Soechting, 1983). These experiments have lead to a very different view of the role of stretch reflexes in motor behaviour from that described above. The data reveal a feedback control system with a high gain and a 90–180 degree phase lead of the electromyogram (EMG) response ahead of muscle stretch. Central influences can modulate the gain and phase of reflex transmission independently of the operating force and length of the muscle and this, in turn, changes the effective mechanical stiffness and viscosity of the muscles to suit the task in which the subject is engaged. The next section assesses the extent to which these differences in results and interpretations can be attributed to methodological factors.

Methodological considerations

Understanding of stretch reflex behaviour has been derived from a range of experimental approaches. Different species, different types of animal preparations and different levels of cooperation on the part of the subjects have been involved; the subjects (human and nonhuman) have been variously anaesthetised or relaxed, making slow movements or rapid movements, controlling force or controlling position. Various types of perturbations have been employed to evoke stretch reflex responses, ranging through electrical stimuli, tendon taps, steps of torque, ramp stretches, sinusoidal displacements, irregular displacements, suddenly changing loads, slowly changing loads and vibration. Much of the divergence of opinion concerning stretch reflex behaviour can probably be resolved if differences in methodological factors are taken into account.

Human versus animal studies

When a muscle is contracted voluntarily, descending signals activate not only α and γ motoneurones but also many interneurons involved in reflex transmission (Baldissera, Hultborn & Illert, 1981). Thus the stretch reflex responses evoked by perturbations applied to an actively contracting muscle may be mediated by multiple, parallel, multisynaptic pathways through various levels of the nervous system, including spinal cord, brainstem, cerebellum, basal ganglia and transcortical loops. Although interneurons are activated during experiments with anaesthetised or decerebrate animals, it cannot be assumed that the descending activation is similar to that in intact human subjects.

Linear versus triggered reflex responses

It is important to distinguish between reflex mechanisms such as the TSR in which the response can be described as a linear transformation of the stimulus waveform, and reflex mechanisms such as a sneeze which generate triggered responses not linearly related to the stimulus. The relationship between the stimulus waveform and the response waveform in the former case can be described mathematically. For a triggered reflex, the response waveform may have little or no relationship to the stimulus waveform. For example, although the intensity of a sneeze may have a graded relationship with the intensity of the stimulus, the coordinated muscle contractions involved appear to have been preprogrammed and, except for being triggered by the stimulus, need have no further mathematical relationship with it. A triggered response may be gated by afferent activity but it is not the afferent activity which determines the pattern of the response. Since linear reflex mechanisms and triggered reflex mechanisms behave differently from each other, conclusions drawn from investigations of one will differ from those drawn from investigations of the other.

Tonic versus phasic stretch reflexes

The terms 'tonic stretch reflex' and 'phasic stretch reflex' distinguish between the respective reflex responses elicited by slow stretch perturbations and those elicited by rapid stretch perturbations. The full importance of this distinction cannot be overemphasised since, as discussed below, rapid perturbations and slow perturbations may activate different reflex pathways. When a slowly changing length perturbation elicits a TSR, the length and rate of change of length are encoded into a temporospatial neural code by muscle spindle and other afferents. The modulated neural signals are then transmitted through TSR pathways before being demodulated by the muscle back into EMG and tension. Thus the TSR behaves like a modulated feedback control system. Modulated systems have a finite modulation bandwidth. If the high frequency content of the applied length signal exceeds the modulation bandwidth, the encoding system saturates and the modulated signal becomes distorted.

Measurements of the input and output sides of the TSR (i) in individual group Ia and group II afferents, (ii) in the discharges of individual α motoneurones and (iii) in myotatic EMG responses have indicated that the bandwidth for linear modulation of muscle displacement signals does not exceed about 12 Hz (Matthews & Stein, 1969; Berthoz & Metral, 1970; Eysel & Grüsser, 1970; Poppele & Bowman, 1970; Rosenthal et al., 1970; Matthews, 1972; Hasan & Houk, 1975a,b; Newsom-Davis, 1975; Neilson & Lance, 1978). Responses at all three levels follow sinusoidal stretching with low harmonic distortion if there is a sustained background activity, the amplitude of stretch is small and the stretching frequency does not exceed 6–8 Hz. The linear range is best expressed as a percentage of the mean discharge level or contraction level. If, for example, the mean contraction level of the biceps brachii muscle is increased by cocontraction of elbow flexor and extensor muscles, the bandwidth of linear modulation increases. With a sustained background contraction 10–20% of maximum, the EMG linearly follows sinusoidal displacement perturbations of the elbow angle (0–20 degrees peak-to-peak displacement) up to frequencies as high as 8–10 Hz (Neilson & Lance, 1978) before breaking into a nonlinear response consisting of a burst of EMG phase locked to the stretching cycle.

It might seem that a modulation bandwidth of only 0–12 Hz could limit the functional value of linear TSR responses, but this is not the case. There would be little advantage in including a wide bandwidth modulation system in TSR pathways when high frequency movement is limited to less than 10 Hz by the low-pass filter characteristics of muscles and mechanical loads (Partridge, 1965, 1966). Even the fastest movements are highly attenuated at frequencies above 6.0 Hz and the minimal duration of a ballistic movement is limited to about 100 ms by the durations of EMG bursts (e.g. Freund & Büdingen, 1978). Thus although muscle length can be forced to change at frequencies greater than 12 Hz (during a tendon tap for example), a modulation bandwidth of 0–12 Hz is more than adequate for the TSR to provide continuous linear feedback control during even the fastest natural movements.

The tendon jerk illustrates the behaviour of the stretch reflex when a displacement containing frequencies in excess of the modulation bandwidth of the TSR is applied to a muscle. Spindle endings no longer encode the length change into a linear frequency code, but produce a synchronous burst of afferent activity which travels around a segmental reflex arc producing an almost synchronous discharge of α motoneurones and a brief contraction of the muscle. Although the amplitude of the response may have a graded relationship with the intensity of the tendon tap, there is no linear mathematical relationship between the waveform of the displacement perturbation and the waveform of the contraction. Indeed, a 'twitch' contraction can be elicited equally well by

Tonic stretch reflex 175

an electrical pulse or a sudden onset of vibration (Cheung *et al.*, 1983) as by a tendon tap. The tendon jerk should be regarded therefore as a triggered reflex response rather than as a linear transformation of the muscle stretch waveform. A similar argument applies to long-latency reflex responses. Chan & Kearney (1982) demonstrated that long-latency reflexes to rapid displacements of the human ankle are independent of the displacement parameters and that long-latency responses to ramp displacements are almost identical to those evoked by pulse displacements. They did not reflect a linear transformation of muscle stretch but more closely resembled triggered responses. Thus, tendon jerks and long-latency reflex responses evoked by sudden displacements behave differently from TSR responses and are therefore likely to lead to different conclusions about the role of reflexes in control of movement. Indeed, the synchronous burst of afferent activity produced by rapid stretch may activate different reflex pathways from those excited by continuous, asynchronous afferent input associated with slow stretching.

Stimulus–response latency measurement technique

Most investigations of segmental and long-latency stretch reflex mechanisms have applied a force or displacement to a muscle and the latency of the EMG or tension response is measured. If the latency is less than a voluntary reaction time the response is regarded as reflex since the system has had insufficient time to generate a voluntary response. The technique requires an abrupt stimulus able to elicit a discrete EMG or tension response so that the stimulus-to-response latency can be measured. Electrical pulses, tendon taps, steps of torque and ramp stretches have all been employed. Given that these stimulus waveforms contain high frequencies in excess of 12 Hz and are sufficiently rapid to excite phasic (tendon-jerk like) responses as well as long latency responses, it should not be surprising that the stretch reflex behaviour based on experiments employing stimulus-response latency measurements differs from that based on continuous stretching within the linear bandwidth of TSRs.

Properties of TSRs during voluntary activity

Production and analysis of the resting and action TSR

A continuously changing torque is applied about the elbow while the subject attempts to hold the arm in a fixed position. The resulting torque, elbow-angle and EMG signals from biceps brachii, brachialis, brachioradialis and triceps muscles are recorded. The amplitude of the elbow-angle is 5–20 degrees

Elbow angle

Biceps surface EMG

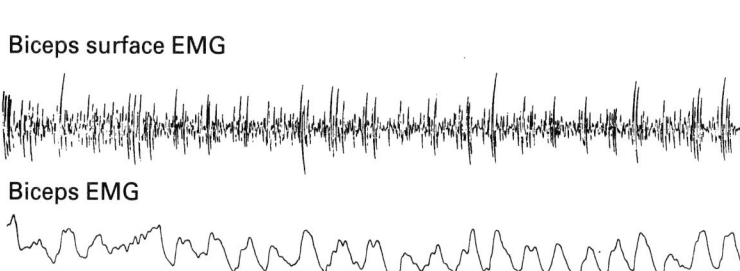

Biceps EMG

Fig. 11.1. Section of polygraph tracing showing action TSR response from biceps brachii muscle in a normal subject during sinusoidal stretching at 4.0 Hz. The amplitude of the elbow angle perturbation was 10 degrees peak to peak and the average contraction level of the biceps brachii was 30% of maximum.

(peak-to-peak) and the frequency of stretching is 2–8 Hz. Dynamic relationships between applied torque and angular displacement, and between angular displacement and integrated EMG (IEMG) are computed using cross correlation, spectral analysis and least squares techniques (Bendat & Piersol, 1968; Box & Jenkins, 1976; Neilson, 1972b). These estimate the mechanical stiffness and viscosity with which the elbow joint resists displacement and the transmission characteristics of the TSR for each.

When normal subjects relax, the elbow-angle can be displaced continuously (5–20 degrees at 0–8 Hz) without evoking EMG responses. However, when normal subjects sustain a voluntary cocontraction of elbow flexor and extensor muscles, small displacement perturbations (10 degrees peak-to-peak) produce a 20–60% modulation of the IEMG in each of the elbow muscles. This modulation of EMG is an 'action TSR' response since it is only evoked in normal subjects by slow stretching during voluntary activity. Typical EMG and IEMG action TSR responses are shown in Fig. 11.1. Descending supraspinal influences not only activate α and γ motoneurones to cause the sustained contraction of the biceps muscle but also bring action TSR pathways into play. During sinusoidal stretching the power spectrum of the IEMG signal shows a large peak at the stretching frequency with no peaks at harmonics or subharmonics of the stretching frequency. Although the IEMG contains fluctuations unrelated to elbow angle changes (IEMG provides only a noisy measure of muscle

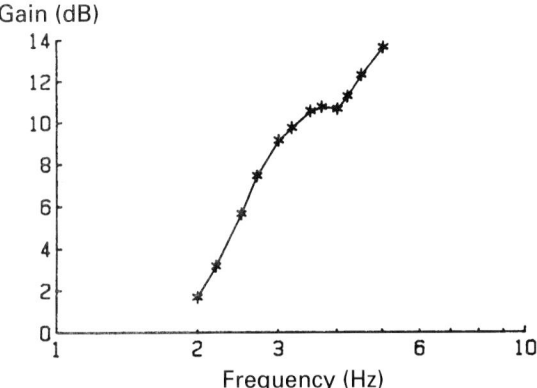

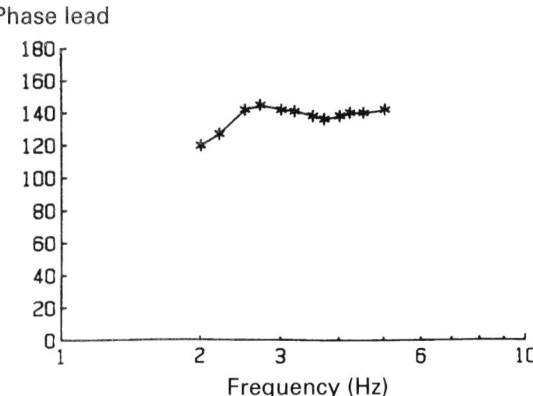

Fig. 11.2. Average gain and phase frequency response curves describing the relationship between sinusoidal elbow angle perturbations and IEMG responses from biceps brachii muscle. Gain and phase curves have been corrected for the attenuation and phase distortion introduced by the EMG filter and the gain is plotted in arbitrary dB (logarithmic) units. Gain increases with increasing frequency of stretch and EMG has a phase lead of 90–180 degrees ahead of muscle stretch.

excitation), the action TSR response follows the sinusoidal elbow angle change throughout the entire sinusoidal cycle with negligible distortion.

Gain and phase of action TSR

Typical gain and phase describing the dynamic relationship between elbow angle changes and the resulting EMG in biceps brachii muscle for normal subjects are presented in Fig. 11.2. Sustained background contraction level as well as amplitude and frequency of sinusoidal elbow angle displacements have been carefully controlled and correction has been made for the attenuation and phase distortion introduced by the EMG filter. Gain of the action TSR increases with increasing frequency of stretch and the EMG has a 90–180 degree phase lead ahead of muscle stretch. This is an important finding because transmission time delays around the reflex loop introduce phase lag into the relationship between applied stretch and IEMG response. Measurement of a 90–180 degree phase lead reveals that the action TSR introduces sufficient phase advance to more than compensate for phase lag due to transmission delay.

Action TSR

It could be argued that the IEMG responses described above are not action TSR responses but rather voluntary responses generated by the subject consciously tracking either the force or displacement perturbation in an effort to hold the arm in a fixed position. This is countered by data from kinesthetic, auditory and visual tracking studies (Ellson & Gray, 1948; Poulton, 1974; Neilson, 1972 a,c, 1974a,b; Neilson & Neilson, 1980; Neilson, Neilson & O'Dwyer, 1984; Neilson, O'Dwyer & Nash, 1990) demonstrating that subjects are unable to generate voluntary tracking responses coherent or correlated with target signals changing at frequencies greater than 2.0 Hz. Although normal subjects are able to contract muscles on and off voluntarily at frequencies as high as 4–6 Hz, the responses lose synchronization or correlation with the target signal at frequencies above 2.0 Hz. Action TSR responses, on the other hand, remain highly correlated (coherence square of 0.8–0.95) even at stretching frequencies as high as 8–10 Hz. It can be concluded that action TSR responses cannot be voluntary responses and so must be involuntarily or reflexly generated.

Gain and phase of the action TSR depend on the prior instructions to the subject. For example, if the subject is instructed to hold the force exerted against the arm-frame constant rather than the position of the arm, the gain of the action TSR is decreased and the phase lead of EMG ahead of muscle stretch is reduced, particularly at high stretching frequencies (Cannon & Zahalak, 1981). Thus descending influences can adaptively tune the transmission characteristics of action TSR pathways to suit the motor task.

If the amplitude of a sinusoidal elbow perturbation is changed, the amplitude of the IEMG response also changes, but not in proportion to the amplitude of the angle change. A decrease in the peak-to-peak amplitude of elbow displacement from 10.0 degrees to 1.6 degrees leads to a 4–6 fold increase in the

Tonic stretch reflex

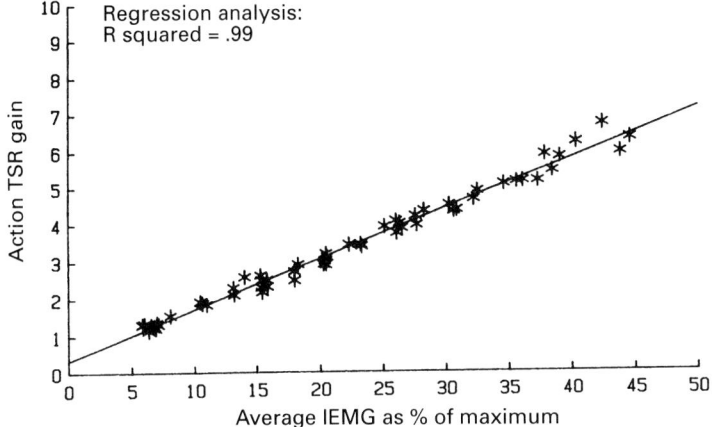

Fig. 11.3. Straight line relationship between the gain of the action TSR of biceps brachii muscle in a normal subject and the average contraction level of the muscle. Each point was computed from a 1-minute test run with frequency of stretching constant at 4.0 Hz and the amplitude of elbow angle perturbation constant at 10 degrees peak to peak. Gain is expressed in arbitrary units.

gain of the action TSR. The smaller the amplitude of stretching, the more sensitive the action TSR. This has been demonstrated in both normal subjects and in cerebral palsy (Neilson & McCaughey, 1981). The reduced gain of the action TSR to large amplitude stretching cannot be attributed to saturation because the IEMG response remains sinusoidal with no evidence of harmonic distortion. This implies the existence of a nonlinear automatic gain control mechanism able to reset the gain of the action TSR in response to a change in the amplitude of the perturbation. A similar resetting of gain in response to changes in amplitude of stretch has been observed in the responses of primary spindle endings in decerebrate cats (Matthews & Stein, 1969).

When a normal subject is instructed to resist external disturbances and to hold the arm fixed, the gain of the action TSR increases linearly with the level of sustained contraction in the muscle (Neilson & McCaughey, 1981). This is demonstrated in Fig. 11.3 which shows the straight-line relationship between gain of the action TSR in a normal subject and the average level of IEMG

Central control of the gain of the action TSR

By using biofeedback of average contraction level and action TSR gain it is possible to train both normal and cerebral palsied subjects consciously to alter the gain of the action TSR without changing either the average length or the

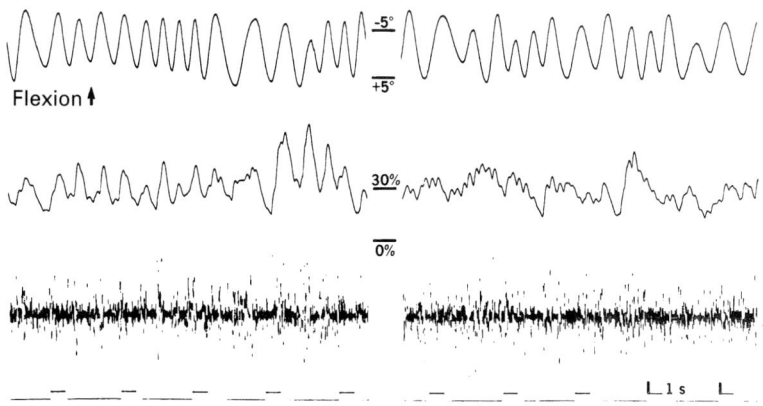

Fig. 11.4. Polygraph tracings of elbow angle, IEMG and EMG signals from triceps brachii muscle in a normal subject (*a*) with the gain of the action TSR high and (*b*) with the gain of the action TSR low. The average contraction level of the muscle was constant at 30% of maximum, the average elbow angle was constant at 90 degrees and the applied elbow angle perturbation was irregular with a bandwidth of 5 Hz and a standard deviation of 4 degrees.

average contraction level of the muscle (Fig. 11.4). Within less than one hour, normal subjects learned to produce a fourfold change in action TSR gain while holding the contraction level constant at 30–50% of maximum (Neilson & Lance, 1978). Cerebral palsied subjects required many months of training but eventually produced a two to threefold change in action TSR gain while holding the contraction level constant at 10% of maximum (Neilson & McCaughey, 1982).

The IEMG responses correlated with changes in elbow angle are two to fivefold bigger in Fig. 11.4(*a*) than in Fig. 11.4(*b*) when the subject attempted to increase the gain of the TSR. As the contraction strength was constant, 30% of maximum, this implies the existence of descending influences within the CNS which can modulate the gain of action TSR pathways independently of the drive to α motoneurones.

Action TSR and Muscle Stiffness

The mechanical stiffness and viscosity with which the elbow joint resists displacement by an externally applied force are determined predominantly by the gain and phase characteristics of the action TSRs (Fig. 11.5). In Fig. 11.5(*a*), the subject consciously increased the gain of action TSRs while in Fig. 11.5(*b*),

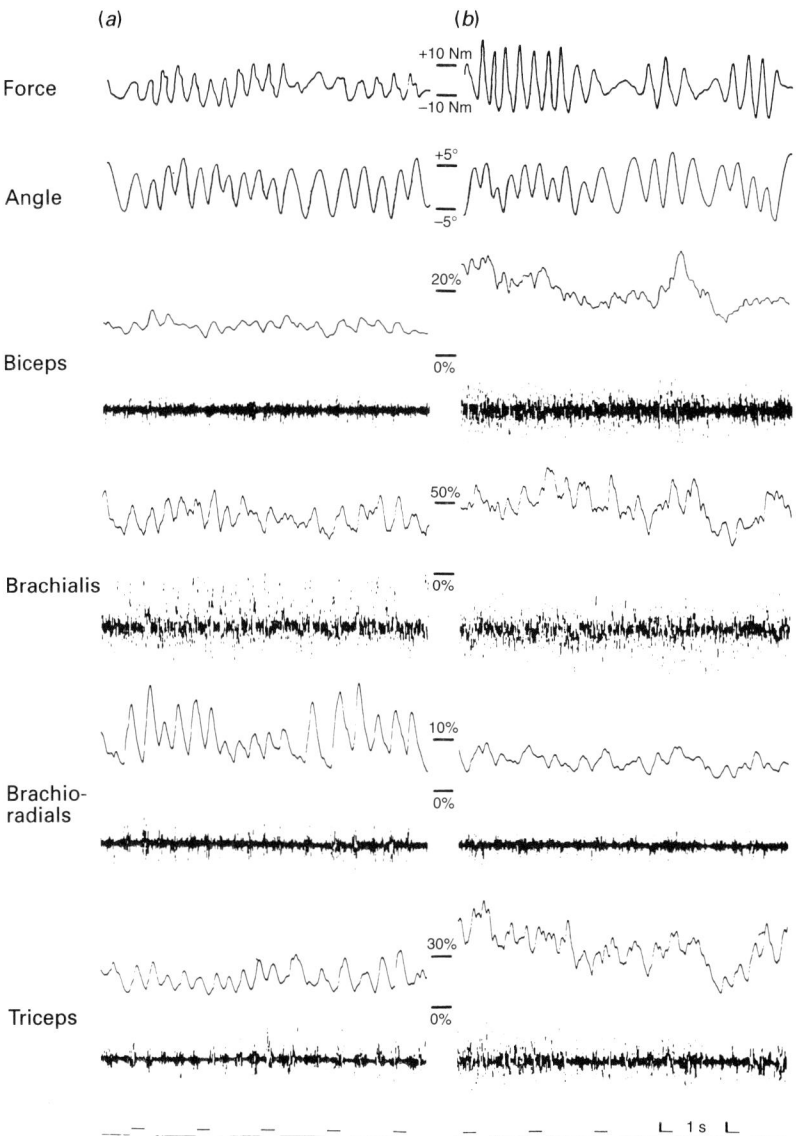

Fig. 11.5. Polygraph tracings of applied torque, elbow angle changes, EMG and IEMG signals from biceps brachii, brachialis, brachioradialis and triceps muscles in a normal subject (*a*) with action TSR gains high and (*b*) action TSR gains low. Despite higher levels of cocontraction in (*b*), stiffness of the elbow is 4.3 fold greater in (*a*).

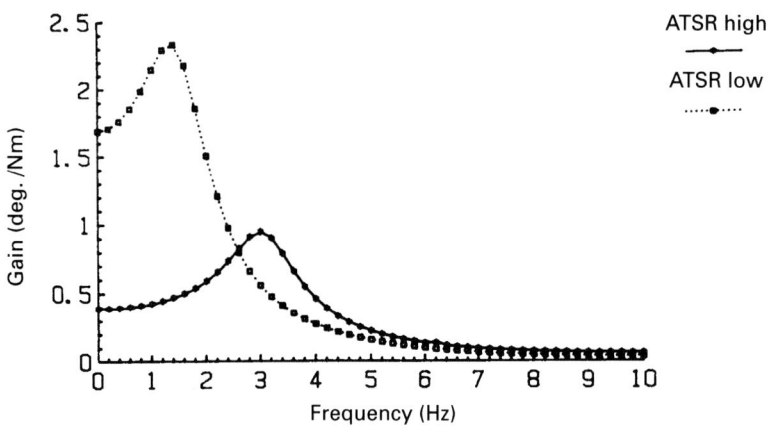

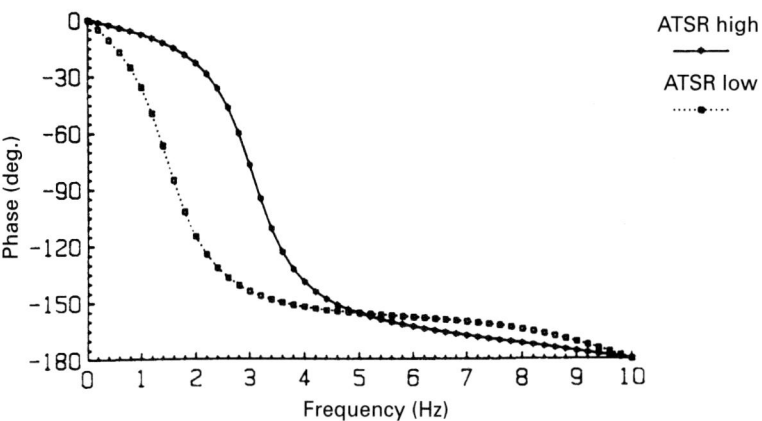

Fig. 11.6. Gain and phase frequency response curves computed from the second order difference equations describing the dynamic torque-angle relationships for the data in Fig. 11.5. Solid lines were computed for data in Fig. 11.5(a) with action TSR gains high and the dotted lines were computed for data in Fig. 11.5(b) with action TSR gains low. Gain is expressed in degrees of elbow angle displacement per Newton meter of applied torque and phase is expressed in degrees. The change in gain and phase is consistent with a 4.3-fold reduction in elbow stiffness between run (a) and run (b).

the subject consciously decreased the gain of action TSR but increased cocontraction of muscles about the elbow. If joint stiffness is determined predominantly by cocontraction, the stiffness should be greater in Fig. 11.5(b) than in Fig. 11.5(a). On the other hand, if stiffness is determined predominantly by the gains of the action TSRs then the stiffness should be greater in Fig. 11.5(*a*) than in Fig. 11.5(*b*).

Fig. 11.6 shows the computed gain and phase frequency response characteristics describing the best fit dynamic relationship between applied torque and elbow angle displacement for the data in Fig. 11.5. The measured torque-angle relationships correspond to a second order difference equation equivalent to a rotational mass-spring system. The analysis computed the coefficients of the continuous-time equivalent differential equation; that is, the rotational inertia, viscosity and stiffness of the elbow. The gain at 0 Hz corresponds to the rotational compliance or inverse rotational stiffness of the elbow. The rotational stiffness was 4.3 fold greater in Fig. 11.5(*a*) than Fig. 11.5(*b*). The rotational viscosity was 1.9 fold greater in the same direction. This shows that for small displacements of the elbow about the 90 degree position, the rotational stiffness contributed by action TSRs is at least 4.3 times greater than the stiffness contributed by the muscle fibres themselves. It shows that central influences can change the stiffness and viscosity of the joint independently of the contraction levels and lengths of the muscles.

Reduction of spasticity in cerebral palsy

Resting TSR in cerebral palsy

For subjects with cerebral palsy, EMG recordings frequently reveal sustained involuntary activity which cannot be easily relaxed. Even when the EMG can be reduced to zero, passive sinusoidal displacement at the elbow of either spastic or athetoid subjects usually evokes a burst of EMG during the early part of the stretching cycle (110–180 degree phase lead ahead of muscle stretch) with negligible activity during the remainder of the cycle. Since, as described above, similar stretching evokes no EMG responses in the relaxed muscles of normal subjects, the resting TSR in spastic and athetoid cerebral palsy is pathologically increased. Magnitudes of the abnormal EMG bursts depend on the resting length of the muscle. The EMG burst is smaller when the arm is oscillated about an extended position and bigger when oscillated about a flexed position. Abnormal resistance to passive movement of the 'relaxed' limb can be attributed to abnormal resting TSR responses.

Action TSR in cerebral palsy

When subjects with cerebral palsy generate a sustained contraction in the muscle being stretched, either voluntarily or by spasm, the TSR pathways are functionally different from those of the resting TSR in the same subject (Neilson,

1972a,b; Andrews, Neilson & Knowles, 1973; Neilson & Andrews, 1973; Neilson & Lance, 1978; Neilson & McCaughey, 1982). Amplitude of the action TSR response is insensitive to changes in resting length of the muscle. Angular displacements produce larger EMG responses than when the subject is at rest. The EMG follows the entire sinusoidal stretching cycle with negligible harmonic distortion and the phase lead is 0–90 degrees ahead of muscle stretch compared with 110–180 degrees for the resting TSR (that is, peak EMG for the action TSR occurs late in the stretching cycle, whereas the peak of the small amplitude EMG burst of the resting TSR occurs early in the stretching cycle).

The standard method of assessing muscle tone at the bedside involves passively displacing the relaxed limb through a range of movement at the joint and estimating whether or not resistance is increased. Subjects are instructed to relax because, if muscles are activated by either voluntary or involuntary influences, it is difficult for the clinician to separate the resistance caused by the sustained contraction of the muscles from that due to abnormal reflex behaviour. The finding that TSR pathways are functionally reorganised in cerebral palsy in association with descending activation of the muscle raises questions about the relevance of the assessment of muscle tone when the subject is at rest. For example, in a drug trial to determine the influence of phenoxybenzamine on athetosis (Andrews, Neilson & Knowles, 1973), it was found that the drug had no influence on action TSR responses or on patterns of involuntary athetoid movement even though it produced a reduction of resting TSRs and of abnormal muscle tone as assessed clinically.

Reduction of spasticity to control muscle contracture

Spasticity and spasm in cerebral palsy resist stretch and tend to maintain muscles in shortened positions. Spasticity and spasm of powerful antigravity muscles pull the body into stereotyped postures. In the case of the ankle joint, for example, involuntary contractions of the calf muscles prevent the heel from touching the ground and result in toe-walking. A person severely disabled with cerebral palsy might spend many hours each day with arms flexed and legs extended and adducted. Such frequent misuse of the muscles interferes with normal muscle growth (for review see O'Dwyer, Neilson & Nash, 1989) and produces exactly the conditions which in experimental animals result in loss of sarcomeres, muscle fibre shortening and reduced extensibility. In other words, spasm and spasticity contribute to the development of muscle contractures in cerebral palsy.

Plaster casting is used in the treatment of cerebral palsy to promote muscle growth and to prevent or correct muscle contracture. Surgical interventions, such as recession of the origin of the gastrocnemius-soleus muscle or lengthening or even division of the Achilles tendon, are sometimes employed. An increased range of joint movement may be a result of lengthening of either muscle or tendon, or both. However, only muscle lengthening increases the range over which the muscle can actively generate tension. Recurrence of symptoms following plaster casting or surgical intervention may be a problem since the involuntary contractions of spasm and spasticity usually persist after treatment. The best treatment, as suggested by Tabary *et al.* (1981), would be suppression of the abnormal muscle contractions responsible for muscle shortening. This would prevent not only the development of contracture but also the disruption of limb function caused by abnormal muscle contractions. For these reasons, the goal of a number of our studies has been to determine whether or not it is possible to teach children and adults with cerebral palsy consciously to suppress spasm and reduce spasticity.

Four adults with cerebral palsy were trained in three one-hour sessions each week for 18 months (Neilson & McCaughey, 1982). Each subject was severely disabled with a mixture of spasticity and athetosis, two were predominantly athetoid and two predominantly spastic. The study showed that adults with cerebral palsy can be taught to suppress spasm and reduce spasticity at the elbow while at rest. They also learned to produce a two- to threefold reduction in the sensitivity of the action TSR while sustaining a voluntary contraction of 10% of maximum. This suggests that individuals with cerebral palsy can learn to inhibit pathways involved in TSR transmission while maintaining an excitatory drive to α motoneurones and thus are capable of reducing spasticity during voluntary activity. Moreover, the study showed that, after training, they were able to reduce spasticity at the elbow without biofeedback display and could transfer this ability from the trained arm to the untrained arm. The skill was still present 8–12 months later. Despite the ability to reduce spasm and spasticity at the elbow, however, there was no improvement in ability to control purposive movement about the elbow as assessed by a visual pursuit tracking task.

The aim of a subsequent study (Nash, Neilson & O'Dwyer, 1989) was to determine whether or not the biofeedback training could be adapted for children (4–8 years) with cerebral palsy at risk of developing contractures of the calf muscles. To hold the attention and cooperation of the three children in this study, a biofeedback video game was developed. For example, the size of a smile on an animated face displayed on a computer screen responded in real-time to the sensitivity of the TSR computed from the ankle angle and calf muscle EMG data. As the level of spasticity in the muscle decreased, the size of the smile increased. In addition, a threshold could be adjusted by the experimenter.

When the level of spasticity decreased below threshold, the child was automatically rewarded with recorded music, stories or video cartoons. This motivated the children to play the game and hence, to practise reducing spasm and spasticity. The children were given 4 weeks pretraining assessment, 10 weeks training using biofeedback and 4 weeks posttraining assessment with no biofeedback. Both the variability and average value of spasticity were reduced during posttraining assessment relative to pretraining measures. The average size of the reflex EMG peaks before training was about 40% of maximum contraction and this was reduced to about 5% after training. Decreased EMG activity evoked by stretch was observed in the untrained leg. Apparently the children learned the subjective sensations associated with muscle relaxation sufficiently to transfer control of spasm and spasticity to the other leg without biofeedback.

These results are very encouraging. It is possible to teach adults and children with cerebral palsy consciously to suppress spasm and reduce spasticity not only when they are at rest, but also during voluntary activity. Whether this will help to prevent muscle contractures and joint deformities has yet to be determined. Techniques must be developed to ensure that children with cerebral palsy practise the ability to suppress spasm and reduce spasticity for most of the day rather than just during the laboratory training session. We have observed children reduce spasticity at the ankle and lower their heel to the floor during the training session, but walk about on their toes later in the day. Since it is the average pattern of activity of the muscles over prolonged periods of time that controls muscle growth, it is the average pattern of spasm and spasticity over each day which must be reduced to prevent the development of muscle contractures.

Summary and conclusions

This chapter presents a picture of the role of tonic stretch reflexes in control of movement. It is based on findings from experiments involving both normal and cerebral palsied subjects. When a normal subject is relaxed, the limbs can be moved passively without evoking tonic stretch reflex responses (TSR) in the EMG. However, when the subject voluntarily contracts muscles, TSR pathways are brought into play and EMG responses to applied muscle stretch appear. The amplitude and timing of the EMG responses depend on the task in which the subject is engaged. When the subject attempts to hold the limb in a fixed position, EMG responses have large amplitudes and a phase advance of 90–180 degrees ahead of muscle stretch. Such a large phase advance assures stability of the reflex loop. The gain of the TSR increases with the level of background activity. High gains are obtained when the subject cocontracts muscles about the joint or pulls against an external load. The TSR includes a nonlinear gain-control mechanism which automatically increases the sensitivity of the reflex as the amplitude of the displacement decreases. When the subject is engaged in a task requiring control of force rather than limb position, either the gain of the TSR is reduced, the phase lead of EMG ahead of muscle stretch is reduced, or both. Such changes in TSR result in a four- to fivefold reduction in rotational stiffness and a one- to twofold reduction in rotational viscosity of the elbow joint. In other words, central control of TSR transmission can tune the effective shock-absorber characteristics (stiffness and viscosity) of the joints. In cerebral palsy, descending control of motor activity is disrupted, including control of TSR transmission. Movements applied to relaxed muscles produce abnormal bursts of EMG. and are termed 'resting TSR' responses to distinguish them from 'action TSR' responses during voluntary or involuntary activity. Both resting TSRs and action TSRs are abnormal in cerebral palsy. By using biofeedback training, it is possible to teach adults and children with cerebral palsy consciously to reduce the level of involuntary contraction and the sensitivity of both resting TSRs and action TSRs. This corresponds to a reduction in spasm and spasticity. Such training may prove useful in preventing the development of muscle contractures and joint deformities. However, it does not improve functional control of movement about the joint as assessed by visual pursuit tracking tasks.

Acknowledgements

I wish to thank the Spastic Centre of New South Wales and the National Health and Medical Research Council of Australia for financial support over many years. I thank also Professor J.W. Lance, for encouraging my interest in movement disorders.

References

Andrews, C. J., Neilson, P. D. & Knowles, L. (1973). Electromyographic study of the rigidospasticity of athetosis. *Journal of Neurology, Neurosurgery, and Psychiatry*, **36**, 94–103.

Baldissera, F., Hultborn, H. & Illert, M. (1981). Integration in spinal neuronal systems. In: *Handbook of Physiology. Section I: The Nervous System, Volume II, Part 1*. pp. 509–95. Bethesda, Md: American Physiological Society.

Bendat, J. S. & Piersol, A. G. (1968). *Measurement and Analysis of Random Data*. New York: John Wiley & Sons.

Berthoz, A. & Metral, S. (1970). Behaviour of a muscle group subjected to a sinusoidal and trapezoidal variation of force. *Journal of Applied Physiology*, **29**, 378–84.

Box, G. E. P. & Jenkins, G. M. (1976). *Time Series Analysis: Forecasting and Control*. San Francisco: Holden-Day.

Cannon, S. C. & Zahalak, G. I. (1981). Reflex feedback in small perturbations of a limb. *ASME 1981 Biomechanics Symposium*, AMD-43, 117–20.

Chan, C. W. Y. & Kearney, R. E. (1982). Is the functional stretch response servo controlled or preprogrammed? *Electroencephalography and Clinical Neurophysiology*, **53**, 310–24.

Cheung, Y., Cozens, A., Hammond, G. R., Hewit, J. R., Miller, S. & Veitch, M. E. (1983). Amplitude-modulated pseudo-random excitation of stretch reflexes in arm and trunk muscles in normal subjects and stroke patients. *Journal of Physiology (London)*, **355**, 102P.

De Gail, P., Lance, J. W. & Neilson, P. D. (1966). Differential effects on tonic and phasic reflex mechanisms produced by vibration of muscles in man. *Journal of Neurology, Neurosurgery, and Psychiatry*, **29**, 1–11.

Dufresne, J. R., Soechting, J. F. & Terzuolo, C. A. (1978). Electromyographic response to pseudo-random torque disturbances of human forearm position. *Neuroscience*, **3**, 1213–26.

Ellson, D. G. & Gray, F. E. (1948). Frequency response of human operators following a sine wave input. *USAF Air Material Command, Memorandum Report MCREXD-6942N*. Wright-Patterson Air Force Base, Dayton, Ohio.

Eysel, U. Th. & Grüsser, O.-J. (1970). The impulse pattern of muscle spindle afferents. A statistical analysis of the response to static and sinusoidal stimulation. *Pfluegers Archives*, **313**, 1–26.

Feldman, A. G. (1966a). Functional tuning of the nervous system with control of movement or maintenance of a steady posture. 2. Controllable parameters of the muscle. *Biophysics*, **11**, 565–78.

Feldman, A. G. (1966b). Functional tuning of the nervous system with control of movement or maintenance of a steady posture. 3. Mechanographic analysis of the execution by man of the simplest motor task. *Biophysics*, **11**, 766–75.

Feldman, A. G. & Orlovsky, G. N. (1972). The influence of different descending systems on the tonic stretch reflex in the cat. *Experimental Neurology*, **37**, 481–94.

Freund, H.-J. & Büdingen, H. J. (1978). The relationship between the speed and amplitude of the fastest voluntary contractions of human arm muscles. *Experimental Brain Research*, **31**, 1–12.

Hasan, Z. & Houk, J. C. (1975a). Analysis of response properties of deafferented mammalian spindle receptors based on frequency of response. *Journal of Neurophysiology*, **38**, 663–89.

Hasan, Z. & Houk, J. C. (1975b). Transition in sensitivity of spindle receptors that occurs when the muscle is stretched more than a fraction of a millimetre. *Journal of Neurophysiology*, **38**, 673–89.

Hoffer, J. A. & Andreassen, S. (1981a). Regulation of soleus muscle stiffness in premammillary cats: intrinsic and reflex components. *Journal of Neurophysiology*, **45**, 267–85.

Hoffer, J. A. & Andreassen, S. (1981b). Limitations in the Servo Regulation Soleus Muscle Stiffness in Premammillary Cats. In *Muscle Receptors and Movement*, ed. A. Taylor & A. Prochazka, pp. 311–24. London: Macmillan.

Houk, J. C. & Rymer, W. Z. (1981). Neural control of muscle length and tension. In: *Handbook of Physiology. Section I: The Nervous System. Volume II. Motor Control, Part 1*. pp. 257–323. Bethesda, Md: American Physiological Society.

Houk, J. C., Rymer, W. Z. & Crago, P. E. (1977). Complex velocity dependence of the electromyographic component of the stretch reflex. *Proceedings of the 27th International Congress of Physiological Sciences*, Paris, vol. 13. p. 334.

Inbar, G. F. (1972). Muscle spindles in muscle control. III. Analysis of adaptive system model. *Kybernetik*, **11**, 130–41.

Inbar, G. F. & Yafe, A. (1976). Parameter and signal adaptation in the stretch reflex loop. *Progress in Brain Research*, **44**, 317–37.

Jack, J. J. B. & Roberts, R. C. (1978). The role of muscle spindle afferents in stretch and vibration reflexes of the soleus muscle of the decerebrate cat. *Brain Research*, **146**, 366–72.

Lacquaniti, F. & Soechting, J. F. (1983). Changes in mechanical impedance and gain of the myotatic response during transitions between two motor tasks. In *Neural Coding of Motor Performance*, eds. J. Massion, J. Palliad, W. Schultz & M. Wiesendanger, pp. 135–9. Heidelberg: Springer-Verlag.

Lance, J.W. (1965). The mechanism of reflex irradiation. *Proceedings of the Australian Association of Neurologists*, **3**, 77–81.

Liddell, E. G. T. & Sherrington, C. (1924). Reflexes in response to stretch (myotatic reflexes). *Proceedings of the Royal Society*, **B96**, 212–42.

Matthews, P. B. C. (1959*a*). The dependence of tension upon extension in the stretch reflex of the soleus muscle of the decerebrate cat. *Journal of Physiology (London)*, **147**, 521–46.

Matthews, P. B. C. (1959*b*). A study of certain factors influencing the stretch reflex of the decerebrate cat. *Journal of Physiology (London)*, **147**, 547–64.

Matthews, P. B. C. (1972). *Mammalian Muscle Receptors and their Central Actions*. London: Arnold.

Matthews, P. B. C. & Stein, R. B. (1969). The sensitivity of muscle spindle afferents to small sinusoidal changes of length. *Journal of Physiology (London)*, **200**, 723–43.

Muller, E. M., Abbs, J. H. & Kennedy, J. G. (1981). Some systems Physiology Considerations for Vocal Control. In *Proceedings of the Conference on Vocal Fold Physiology*, eds. M. Hirano & K. Stevens, Tokyo: University of Tokyo Press.

Nash, J., Neilson, P. D. & O'Dwyer, N. J. (1989). Reducing spasticity to control muscle contracture of children with cerebral palsy. *Developmental Medicine and Child Neurology*, **31**, 471–80.

Neilson, P. D. (1972*a*). Speed of response or bandwidth of voluntary system controlling elbow position in intact man. *Medical and Biological Engineering*, **10**, 450–9.

Neilson, P. D. (1972*b*). Frequency-response characteristics of the tonic stretch reflexes of biceps brachii muscle in intact man. *Medical and Biological Engineering*, 10, 460–72.

Neilson, P. D. (1972*c*). voluntary and reflex control of the biceps brachii muscle in spastic-athetotic patients. *Journal of Neurology, Neurosurgery, and Psychiatry*, **35**, 853–60.

Neilson, P. D. (1974*a*). Voluntary control of arm movement in athetotic patients. *Journal of Neurology, Neurosurgery, and Psychiatry*, **37**, 162–70.

Neilson, P. D. (1974*b*). Measurement of involuntary arm movement in athetotic patients. *Journal of Neurology, Neurosurgery and Psychiatry*, **37**, 171–7.

Neilson, P. D. & Andrews, C. J. (1973). Comparison of the tonic stretch reflex in athetotic patients during rest and voluntary activity. *Journal of Neurology, Neurosurgery, and Psychiatry*, **36**, 547–54.

Neilson, P. D. & Lance, J. W. (1978). Reflex Transmission Characteristics during Voluntary Activity in Normal Man and in Patients with Movement Disorders. In *Cerebral Motor Control in Man: Long Loop Mechanisms. Progress in Clinical Neurophysiology*, vol. 4. ed. J. E. Desmedt, pp. 263–299. Basel: Karger.

Neilson, P. D. & McCaughey, J. (1981). Effect of contraction level and magnitude of stretch on tonic stretch reflex transmission characteristics. *Journal of Neurology, Neurosurgery, and Psychiatry*, **44**, 1007–12.

Neilson, P. D. & McCaughey, J. (1982). Self-regulation of spasm and spasticity in cerebral palsy. *Journal of Neurology, Neurosurgery, and Psychiatry*, **45**, 320–330.

Neilson, P. D. & Neilson, M. D. (1980). Influence of control-display compatibility on tracking performance. *Quarterly Journal of Experimental Psychology*, **32**, 125–35.

Neilson, P. D., Neilson, M. D. & O'Dwyer, N. J. (1985). Acquisition of motor skill in tracking tasks: learning internal models. In *Motor Memory and Control*, eds. D. G. Russell & B. Abernathy, pp. 25–36. Dunedin: Human Performance Associates.

Neilson, P. D., O'Dwyer, N. J. & Nash, J. (1990). Control of isometric muscle activity in cerebral palsy. *Developmental Medicine and Child Neurology*, **32**, 778–88.

Newsom-Davis, J. (1975). The response to stretch of human intercostal muscle spindles studied in vitro. *Journal of Physiology (London)*, **204**, 561–79.

O'Dwyer, N. J., Neilson, P. D. & Nash, J. (1989). Mechanisms of muscle growth related to muscle contracture in cerebral palsy. *Developmental Medicine and Child Neurology*, **31**, 543–7.

Partridge, L. D. (1965). Modification of neural output signals by muscles: a frequency response study. *Journal of Applied Physiology*, **20**, 150–6.

Partridge, L. D. (1966). Signal-handling characteristics of load-moving skeletal muscle. *American Journal of Physiology*, **210**, 1178–91.

Poppele, R. E. & Bowman, R. J. (1970). Quantitative description of linear behaviour of mammalian muscle spindles. *Journal of Neurophysiology*, **33**, 59–72.

Poulton, E. C. (1974). *Tracking Skill and Manual Control*. New York: Academic Press.

Rack, P. M. (1981). Limitations of somatosensory feedback in control of posture and movement. In *Handbook of Physiology. Section I: The Nervous System. Volume II. Motor Control, Part 1.* pp. 229–56. Bethesda, Ma: American Physiological Society.

Rosenthal, N. P., McKean, T. A., Roberts, W. J. & Terzuolo, C. A. (1970). Frequency analysis of stretch reflex and its main subsystems in triceps surae muscle of the cat. *Journal of Neurophysiology*, **28**, 713–49.

Rymer, W. Z. & Hasan, Z. (1980). Absence of force-feedback in soleus muscle of the decerebrate cat. *Brain Research*, **184**, 203–9.

Soechting, J. F., Dufresne, J. R. & Lacquaniti, F. (1981). Time-varying properties of myotatic response in man during some simple motor tasks. *Journal of Neurophysiology*, **46**, 1226–43.

Stein, R. B. & Lee, R. G. (1981). Tremor and clonus. In: *Handbook of Physiology. Section I: The Nervous System. Volume II. Motor Control, Part 1.* pp. 325–44. Bethesda, Ma: American Physiological Society.

Tabary, J. C., Tardieu, C., Tardieu, G. & Tabary, C. (1981). Experimental rapid sarcomere loss with concomitant hypoextensibility. *Muscle and Nerve*, **4**, 198–203.

Terzuolo, C. A., Soechting, J. F. & Dufresne, J. R. (1981). Operational characteristics of reflex responses to changes in muscle length during different motor tasks and their functional utility. In *Brain Mechanisms and Perceptual Awareness*, eds. O. Pompeiano & C. Ajmone Marsan. pp. 183–209. New York: Raven Press.

Volpe, J. J. (1987). *Neurology of the Newborn*. Philadelphia: W.B. Saunders Company.

12
Measurement of physiological and essential tremor

J. D. GILLIES

Department of Neurology,
Institute of Neurological Sciences,
The Prince Henry and Prince of Wales Hospitals,
Sydney NSW, Australia

Tremor during muscle activity was recognized by Galen (AD 187), who noted that 'if someone lifts a weight too heavy with his hands then there is tremor in his hands'. He described the augmentation of tremor by fatigue and fear and clearly recognized both 'physiological' tremor and its superficial similarity to 'essential' tremor. In spite of numerous observations and dissertations, the possible relationship of physiological tremor to essential tremor and the underlying tremor mechanisms have been the subject of debate. This has been complicated by the difficulty in distinguishing essential and physiological tremor (Elble, 1986). Clinical classifications of tremor group exaggerated physiological tremor with subtypes of essential tremor (Marsden, Obeso & Rothwell, 1983), however, there are no clear objective parameters for distinguishing the tremor types, assuming of course that they do not form a continuum of motor disorders.

Most recent investigations of tremor have concentrated on accelerometer recordings of tremor oscillations in the outstretched finger, wrist or arm. This method of recording tremor has the advantage of simplicity, but the disadvantages of being influenced by the inertia of the limb (Elble & Randall, 1978), and not recording the tremor in circumstances when it creates the most functional problems for the subject, that is during voluntary muscle activity of the whole limb.

Measurement of tremor while the subject is exerting force such as that during everyday activities, rather than when maintaining a posture, should give a better indication of the functional incapacity of the subject during normal purposive motor activity. Further, the tremor may be recorded at varying levels of muscle activity, rather than during the minimal contractions necessary to maintain a posture. Such recording requires the detection of a small variation in a large force. In the data presented here, tremor has been recorded from the fingers during a static visual force tracking procedure involving

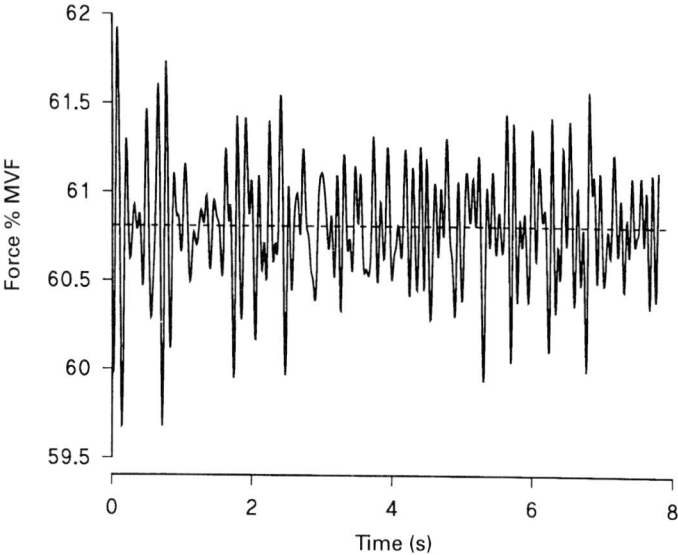

Fig. 12.1. Recording of force in the outstretched arm. There is an irregular tremor at 6–8 Hz and a mean peak-to-peak amplitude of about 1% MVF.

the whole arm. The subject pressed with the index and middle fingers on a low compliance strain gauge (0.2 mm/N). The fingers and wrist were aligned with the pronated forearm, the elbow flexed to 90 degrees, and the upper arm held 45 degrees forward and 45 degrees abducted. Frequencies above 1 Hz were separated from the force signal by a filter, amplified, and then recorded digitally with the force signal. To avoid the effects of fatigue, the subject maintained a predetermined level of force for only 8 s with at least 30 s rest between contractions. At high force levels, 2 min rest was allowed between each recording.

Variations of force during isometric contractions

At all levels of force, small variations in force can be detected in the strain gauge output (Fig. 12.1). It is clear on visual inspection that the tremor is not uniform, being irregular in both frequency and amplitude. Counting the number of peaks each second, using arbitrary criteria to determine how small a deviation is regarded as a peak, gives an approximate frequency range of 8.5–10 Hz. The amplitude of the tremor was dependent on the force; when the applied force was increased, the amplitude of tremor increased. To measure the proportion of the force involved in the tremor, it was assumed that the force

Measurement of physiological and essential tremor 193

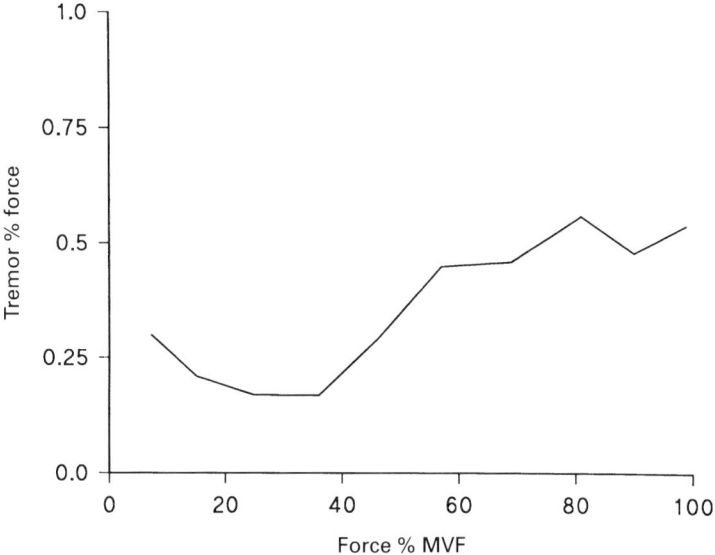

Fig. 12.2. Physiological tremor expressed as average deviation % force (as % MVF) in one subject.

had two components: an intended steady force upon which was imposed a varying tremor. The mean force was expressed as a percentage of maximal voluntary force (MVF). The deviation of the force from the mean force was measured every 1/128 s and the average deviation expressed as a percentage of the mean force. Using this representation, the proportion of the applied force which is involved in tremor can be determined at any force level. The results from one subject are shown in Fig. 12.2. As the force increased the deviation of the tremor increased, but when expressed as a percentage of applied force it remained constant at force levels below about 60% MVF. At greater force levels there was an increment in the component of force involved in tremor to about twice the previous level. This increased level then remained constant up to MVF.

In all subjects who had no history of significant tremor, a similar pattern was found. The average tremor deviation expressed as a percentage of force from 15 subjects (age range: 21–46 years) is plotted in Fig. 12.3. Of note is the relatively small proportion of the force (<0.5%) that is involved in tremor at the more usual levels of muscle activity, and the increase in tremor above 60% MVF. Both sympathomimetic amines and fatigue were shown to increase the tremor recorded in this manner, but it was not affected by an oral dose of 10 g alcohol.

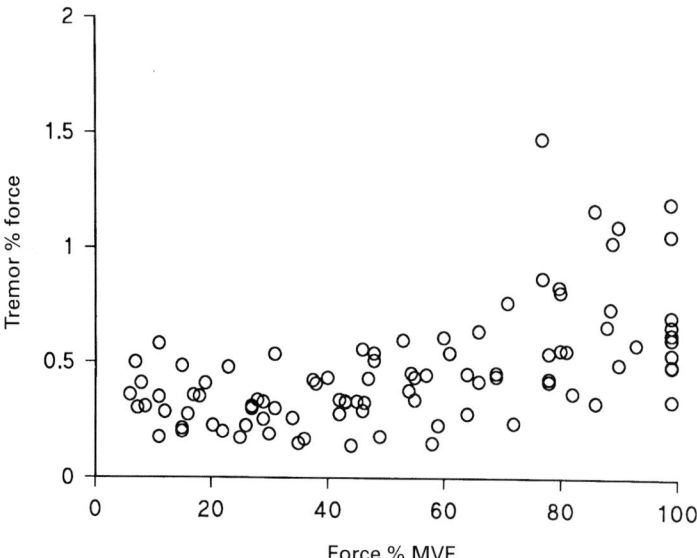

Fig. 12.3. Physiological tremor expressed as average deviation % force (as % MVF) in 15 normal subjects. At force levels less than 60% MVF, the tremor mean is 0.24% rising to 0.62% at higher force levels.

Frequency power spectra

To analyse the frequency of the tremor, the averaged power spectrum was calculated with a resolution of 0.25 Hz from 4 s overlapping segments of the tremor signal. While there were obvious peaks in each spectrum, no consistent tremor frequency could be identified in any normal subject at any level of force (Fig. 12.4). The only consistent observation was that frequency components above 10 Hz were better represented at higher force levels. When recordings were repeated at one force level, the position of the peaks in the spectra was inconsistent, i.e. there was not a reproducible preferred frequency for the tremor. When these multiple spectra were averaged by normalising the power at each frequency as a percentage of the total power, and then computing the mean and confidence limits at 0.5 Hz intervals, no preferred frequency was found (Fig. 12.5, lower graph). In most subjects there were usually broad peaks in the averaged spectra at about 2 Hz, and 5–10 Hz, and these peaks differed significantly from the intervening and higher frequencies. In a few subjects, the intervening decrease in power was not apparent. The lower frequency peak was well below the tremor frequency estimated from any unprocessed tremor recording, and probably represented the voluntary activity of the subject attempting to maintain a constant force output. The broad peak above 5 Hz was no more than the summation of the apparently random multiple frequency peaks observable in individual spectra.

Measurement of physiological and essential tremor 195

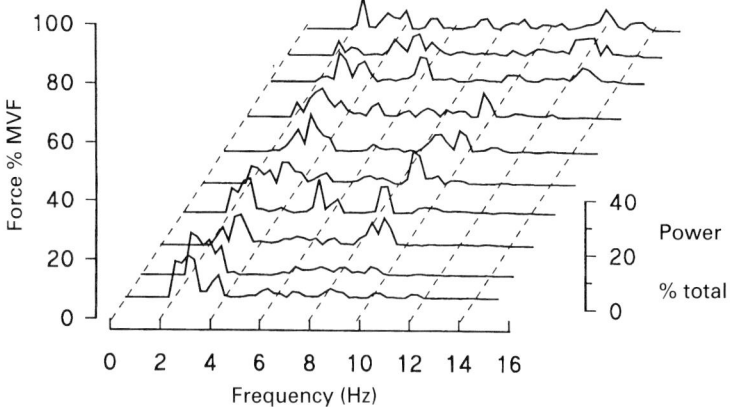

Fig. 12.4. The power spectrum of physiological tremor recorded under isometric conditions, with a resolution of 0.25 Hz and normalized to 100% power (scale right), is plotted at various force levels (scale left). There is an irregular spectrum with numerous inconsistent peaks. Note the tendency for higher frequencies to be better represented at higher force levels.

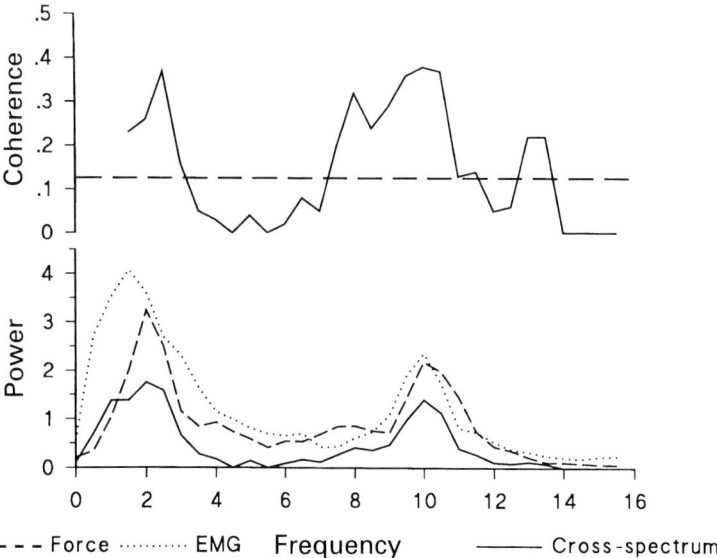

Fig. 12.5. Physiological tremor. The average spectrum of force from the outstretched hand, rectified EMG from the forearm muscles and their averaged cross spectra are plotted in the lower graph. There are peaks at 2 Hz and from 7 to 11 Hz. The coherence function, with 95% confidence limit show as a broken line is plotted in the upper graph. There is significant coherence in the 1–3 Hz range, probable related to voluntary tracking movements, and broad range of coherence from 7 to 10 Hz, establishing a relationship between the force variations and EMG.

Origin of the tremor

There are several possible physiological and artefactual sources for these variations in muscle force during contraction. Cardioballistic oscillations may account for physiological tremor in a limb at rest (Brumlick, 1962; Van Buskirk & Fink, 1962; Van Buskirk, Wolbarsht & Stecher, 1966; Yap & Boshes, 1967), however they make only a small contribution to physiological tremor in the outstretched arm (Carrie & Bickford, 1969; Marsden et al., 1969; Elble & Randall, 1978), and such cardiac derived oscillations would not increase with increasing muscle contraction. Mechanical oscillation in the limb should also be considered (Robson, 1959; Stiles & Randall, 1967; Stiles, 1976, 1980; Elble, 1986; Homberg et al., 1987). However, mechanical oscillations should be reproducible at the one level of contraction, and change their frequency in a predictable manner with limb stiffness (Robson, 1959), inertia and increasing muscle force (Stiles & Randall, 1967). Such oscillations are therefore inconsistent with the broad, variable range of frequencies found with isometric recording.

Electromyographic activity

To determine whether these multiple frequency peaks in the tremor spectra represented force output from muscle contraction, rather than noise arising from mechanical vibrations in the arm and strain gauge, or electrical noise in the recording system, the surface electromyogram (EMG) of the forearm muscles was recorded along with the tremor. The EMG signal was full-wave rectified and filtered in the same way as the tremor signal. The cross-spectrum and coherence function (Bendat & Piersol, 1986) of the EMG and tremor signals was then computed (Fig. 12.5). In the lower graph, averaged tremor and EMG spectra and the tremor-EMG crossed spectrum are shown. The EMG and tremor spectrum and crossed spectra are similar and all show the typical broad peaks seen with averaged spectra, suggesting that there is a relationship between the main frequencies represented in the EMG and the tremor recordings. This is evaluated further by the coherence function (upper graph, Fig. 12.6) in which the coherence of the EMG and tremor spectra exceed the 95% confidence limits in the 1–3 and 6–10 Hz bands, consistent with an association between EMG and tremor within those frequency bands. Since there is a clear statistical association between the force of the tremor and the EMG, the variations in force represent variation in motor unit activity. The possibility that the tremor originates from the unfused tetanus of motor units just recruited (Allum, Dietz & Freund, 1978; Freund, 1983), with or without some transient

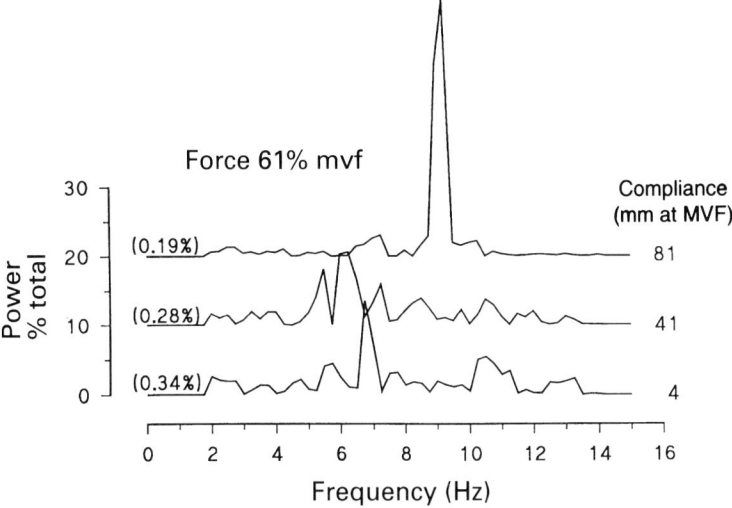

Fig. 12.6. The spectra of the force variations at 61% MVF recorded at three compliance levels. At low compliances (given as 4 and 41 mm at MVF) there are irregular spectra with multiple peaks. At higher compliance (81 mm at MVF), one single spectral peak dominates a 8.8 Hz. The amplitude of the tremor (indicated on the left as % average deviation/mean force) is lower when there is one frequency band, suggesting that the increased compliance has permitted synchronization of the multiple peaks seen in the other spectra.

entrainment from sensory feedback (Allum & Hulliger, 1982), is unlikely because the tremor amplitude relative to force increases at higher frequencies when there are few, if any motor units recruited and those recently recruited are firing at frequencies above that of the tremor. More likely the tremor reflects noise in the signal driving the motoneurone pools.

Is this physiological tremor?

At this point one should consider whether the signal recorded in normal subjects under nearly isometric conditions should be called a tremor. If a tremor is required to be a repetitive variation in muscle force with a definite frequency and approximately sinusoidal waveform (Elble, 1986), then the variations in force recorded isometrically do not meet these criteria because of their broad and variable frequency characteristics and should not be regarded as tremor. They are more in the nature of band-limited noise, the amplitude of which is a constant proportion of the force generated by the muscles with the frequency restricted to 5–10 Hz, with the upper frequency and amplitude increasing slightly with increasing muscle activity.

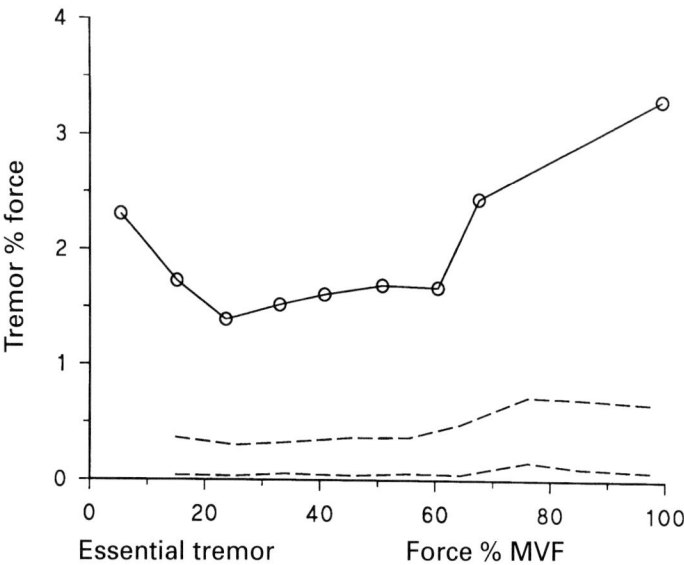

Fig. 12.7. Essential tremor. The tremor amplitude, expressed as average deviation (%). Mean force is plotted at various force levels. The broken lines indicate ± SEM for physiological tremor recorded under the same conditions. The essential tremor has a much greater amplitude than physiological tremor.

Effect of compliance on tremor spectrum

In contrast to the noisy signal described above, the usual finding with spectral analysis of accelerometer recordings of physiological tremor is a distinct, reproducible frequency peak. The essential difference between isometric force transducer and accelerometer recordings is that in the latter, the limb is maintaining a posture against gravity and is thus free to move. The effect of movement on isometric force variations can be assessed by introducing compliance into the connection between the strain gauge and the limb with the interposition of springs. At low levels of force, no difference was found in the tremor spectra. As the force increased, the power in the tremor spectra became concentrated in a narrow band, initially at high compliance levels (Fig. 12.6). At force levels close to maximal, the tremor became sinusoidal with most power concentrated in one band. The appearance of the power spectra was then similar to that recorded with an accelerometer. During these recordings, the total amount of tremor (expressed as a percentage of force) did not increase, suggesting that the effect of introducing compliance was to synchronise the previously broad-band noise to one frequency, rather than to introduce additional variations in motor activity.

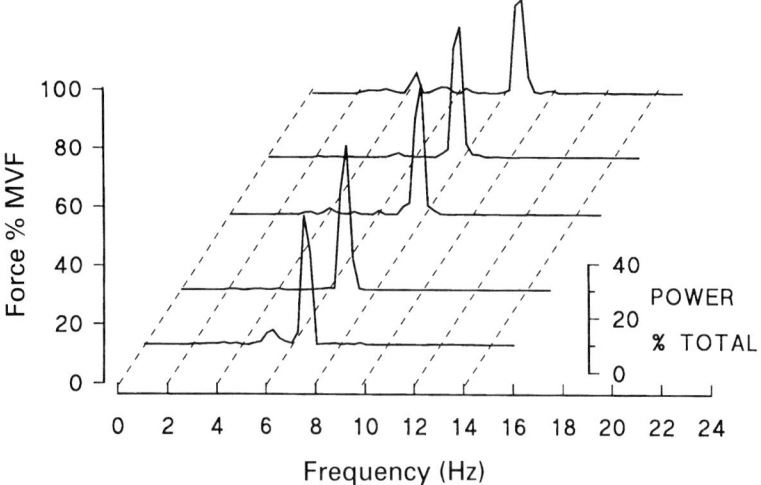

Fig. 12.8. Essential tremor. Power spectra of tremor at various force levels. The tremor power is concentrated in a single band, event at low force levels. There is an increase in tremor frequency from 6.25 to 8.5 Hz as the force is increased above 15% MVF.

Essential tremor

When isometric recordings were made from subjects with known essential tremor, the relationship of tremor to mean force and the spectra differed from that of physiological tremor. The amplitude of the tremor was greater, usually 4–10% of the mean force (Fig. 12.7). Further, at force levels above 60% MVF, the proportion of tremor to force usually did not increase. When the subject was treated with either propanolol (40–120 mg/day) or primadone (62.5–250 mg/day) or given 10 gm of alcohol, the amplitude of the tremor decreased. In all subjects with established essential tremor, the tremor was present at one frequency (Fig. 12.8). With higher force levels there was a small increase in the frequency of the tremor, usually 1–1.5 Hz comparing the force levels of 20 and 100% MVF.

Since under isometric conditions, essential tremor is present as a single frequency peak, and is a constant proportion of the mean force over a large range of force, peripheral input is unlikely to play a major part in its generation. Indeed, the presence of essential tremor during isometric contraction would suggest that there is a central oscillator whose output affects a constant proportion of the central drive to the motoneurone pools. There may, however, be some influence of peripheral input on the central oscillator and this might account for the increasing frequency of the tremor with increasing force output.

Conclusion

During isometric recording of force in the outstretched arm, small variations in force can be recorded. The amplitude of these variations is a constant proportion of the applied force. Spectral analysis of these variations suggests that they have the characteristics of band-limited white noise. These force variations are due to variations in motor unit activity. When the strict isometric conditions are relaxed by the introduction of compliance, this noise is synchronized into a unimodal tremor frequency, suggesting that peripheral input related to movement is responsible. Certainly peripheral input is necessary for rhythmic oscillation in enhanced physiological tremor since it is absent in severe deafferentation with peripheral neuropathy (Sanes, 1985) and in tabes dorsalis (Halliday & Redfearn, 1958).

Essential tremor is characterized by a stable frequency band during isometric muscle activity, and thus, at least in subjects with well-developed tremor, can easily be distinguished from physiological tremor. Movement of the limb is not necessary for the tremor to form one frequency band. In contrast to physiological tremor, essential tremor is not dependent on peripheral input, the frequency being unaffected by slowed nerve conduction in Charcot Marie Tooth neuropathy (Shahani, Young & Adams, 1973) and not correlated with reflex latencies (Elble, Higgins & Moody, 1987). These findings would support a dominant role for a central oscillator in essential tremor.

References

Allum, J. H. J., Dietz, V. & Freund, H.-J. (1978). Neuronal mechanisms underlying physiological tremor. *Journal of Neurophysiology*, **40**, 557–71.

Allum, J. H. J. & Hulliger, M. (1982). Presumed reflex responses of human first dorsal interosseus muscle to naturally occurring twitch contractions of physiological tremor. *Neuroscience Letters*, **28**, 309–14.

Bendat, S. & Piersol, A. G. (1986). *Random Data: Analysis and Measurement Procedures*, pp. 361–424 New York: Wiley & Sons.

Brumlick, J. (1962). On the nature of normal tremor. *Neurology*, **12**, 159–79.

Carrie, J. R. G. & Bickford, R. G. (1969). Cardiovascular factors in limb tremor. *Neurology*, 19, 116–27.

Elble, R. J. (1986). Physiologic and essential tremor. *Neurology*, **36**, 225–231.

Elble, R. J., Higgins, C. & Moody, C. J. (1987). Stretch reflex oscillations and essential tremor. *Journal of Neurology, Neurosurgery, and Psychiatry*, **50**, 691–8.

Elble, R. J. & Randall, J. E. (1978). Mechanistic components of normal hand tremor. *Electroencephalography and Clinical Neurophysiology*, **44**, 72–82.

Freund, H.-J. (1983). Motor unit and muscle activity in voluntary muscle control. *Physiological Reviews*, **63**, 387–436.

Galen, C. (187). De locus affectis. Kuhn, C. G. ed. Clavidius Galenus Opera Omnia In *Medicorum graecorum opera quae exstant*, vol. 4, 1964. Olms, Hildesheim, vol. 4 pp. 586–9.

Halliday, A. M. & Redfearn, J. W. T. (1958). Finger tremor in tabetic patients and its bearing on the mechanism producing the rhythm of physiological tremor. *Journal of Neurology, Neurosurgery, and Psychiatry*, **21**, 101–8.

Homberg, V., Hefter, H., Reiners, K. & Freund, H.-J. (1987). Differential effects of changes in mechanical limb properties on physiological and pathological tremor. *Journal of Neurology, Neurosurgery, and Psychiatry*, **50**, 568–79.

Marsden, C. D., Meadows, J. C., Lange, G. W. & Watson, R. S. (1969). The role of the ballistocardiac impulse in the genesis of physiological tremor. *Brain*, **92**, 647–62.

Marsden, C. D., Obeso, J. A. & Rothwell, J. C. (1983). Benign essential tremor is not a single entity. In *Current Concepts of Parkinson's Disease and Related Disorders*, ed. M. D. Yahr, pp. 31–46. Amsterdam: Excerpta Medica.

Robson, J. G. (1959). The effect of loading upon the frequency of muscle tremor. *Journal of Physiology (London)*, **149**, 29–30P.

Sanes, J. N., (1985). Absence of enhanced physiological tremor in patients without muscle or cutaneous afferents. *Journal of Neurology, Neurosurgery, and Psychiatry*, **48**, 645–9.

Shahani, B. T., Young, R. R. & Adams, R. D. (1973). The tremor in Roussy–Levy syndrome. *Neurology*, **23**, 425–6.

Stiles, R. N. (1976). Frequency and displacement amplitude relations for normal hand tremor. *Journal of Applied Physiology*, **40**, 44–54.

Stiles, R. N. (1980). Mechanical and neural feedback factors in postural hand tremor of normal subjects. *Journal of Neurophysiology*, **44**, 40–59.

Stiles, R. N. & Randall, J. E. (1967). Mechanical factors in human tremor frequency. *Journal of Applied Physiology*, **23**, 324–30.

Van Buskirk, C. & Fink, R. A. (1962). Physiological tremor: an experimental study. *Neurology*, **12**, 361–70.

Van Buskirk, C., Wolbarsht, M. L. & Stecher, K. (1966). The nonnervous causes of normal physiological tremor. *Neurology*, **16**, 217–20.

Yap, C. B. & Boshes, B. (1967). The frequency and pattern of normal tremor. *Electroencephalography and Clinical Neurophysiology*, **22**, 197–203.

Part III
Autonomic function and vascular disorders

13
Disorders of the autonomic nervous system
J. G. McLEOD

*Department of Medicine,
University of Sydney,
Sydney NSW, Australia*

The autonomic nervous system consists of sympathetic and parasympathetic divisions which are controlled centrally by the hypothalamus and its connections with the cerebral cortex and limbic system. Small-diameter myelinated preganglionic fibres arise from the cells in the intermediolateral columns of the spinal cord which extend from T1 to L2. Postganglionic unmyelinated fibres arise from sympathetic ganglia and join the nerve trunks in which they travel to blood vessels in the skin and muscle and to sweat glands. Parasympathetic fibres leave the brainstem in the 3rd, 7th, 9th and 10th cranial nerves and from the sacral cord in the 2nd, 3rd and 4th sacral nerves. Afferent fibres travelling with parasympathetic nerves are important in the autonomic reflex pathways for control of heart rate and blood pressure. Most of the myelinated fibres and all the unmyelinated fibres in these reflex pathways are of small diameter.

Diseases of the autonomic nervous system are manifested clinically by impaired control of blood pressure, heart rate, sweating and temperature regulation, and by impaired function of the bladder, sexual organs, gastrointestinal tract, pupils and lacrimal glands. They may be classified into those that affect primarily the central nervous system and those that affect primarily the peripheral autonomic nervous system (Table 13.1) (McLeod & Tuck 1987*a*; McLeod, 1992*a*).

Diseases affecting primarily the central nervous system
The most important of these conditions are those that are primary degenerative disorders; they may be classified as pure autonomic failure (PAF, idiopathic orthostatic hypotension), autonomic failure with multiple system atrophy (AF-MSA, Shy–Drager syndrome) and autonomic failure with Parkinson's disease. Autonomic failure may also be secondary to Parkinson's disease, spinal cord lesions, Wernicke's encephalopathy and a number of other miscellaneous diseases affecting the central nervous system.

Table 13.1. *Classification of autonomic disorders*

I. *Diseases affecting central nervous system*
    A. Primary autonomic failure
        1. Pure autonomic failure (PAF) (idiopathic orthostatic hypotension, IOH)
        2. Autonomic failure with Parkinson's disease (AF-PD)
        3. Autonomic failure with multiple-system atrophy (Shy–Drager syndrome) (AF-MSA)
    B. Parkinson's disease
    C. Spinal cord lesions
    D. Wernicke's encephalopathy
    E. Miscellaneous diseases
        1. Cerebrovascular disease
        2. Brainstem tumors
        3. Multiple sclerosis
        4. Adie's syndrome
        5. Tabes dorsalis

II. *Diseases affecting the peripheral autonomic nervous system*
    A. Disorders with no associated peripheral neuropathy
        1. Acute and subacute autonomic neuropathy
            a. Pandysautonomia
            b. Cholinergic dysautonomia
        2. Botulism
    B. Disorders associated with peripheral neuropathy
        1. Autonomic dysfunction clinically important
            a. Diabetes
            b. Amyloidosis
            c. Acute inflammatory neuropathy
            d. Acute intermittent porphyria
            e. Familial dysautonomia (Riley-Day syndrome; HSAN III)
            f. Chronic sensory and autonomic neuropathy
        2. Autonomic dysfunction usually clinically unimportant
            a. Alcohol-induced neuropathy
            b. Toxic neuropathies (vincristine sulphate, acrylamide, heavy metals, perhexiline maleate, organic solvents)
            c. HMSN I, II, and V
            d. Malignancy
            e. Vitamin B_{12} deficiency
            f. Rheumatoid arthritis
            g. Chronic renal failure
            h. Systemic lupus erythematosus
            i. Mixed connective tissue disease
            j. Fabry's disease
            k. Chronic inflammatory neuropathy
            l. HIV infections

Source: Modified from McLeod and Tuck, 1987*a*.

Pure autonomic failure (PAF)

Patients usually present with symptoms of orthostatic hypotension which include dizziness, light-headedness, blurred vision, and faintness or syncope on standing or after a period of exercise. These symptoms may be worse after meals (postprandial hypotension), the more pronounced blood pressure fall in these circumstances probably resulting from vasodilatation brought about by the abnormal release of gastrointestinal and pancreatic hormones (Mathias *et al.*, 1989). Clinical examination and investigations may reveal postural hypotension, impairment of sweating and disturbances of heart rate and blood pressure control. PAF is more common in males and the age of onset is usually 40–60 years (Bannister & Oppenheimer, 1972). It has a better prognosis than other forms of primary autonomic failure. There is increasing evidence that, in most cases, there is degeneration of the peripheral autonomic nervous system (Petito & Black, 1978; Goldstein *et al.*, 1989), although in some cases degeneration of cells in the intermediolateral columns in the spinal cord has been demonstrated (Low, Thomas & Dyck, 1978).

Autonomic failure with Parkinson's disease (AF-PD)

Features of Parkinson's disease are associated with severe orthostatic hypotension (Bannister & Oppenheimer, 1972). There is diffuse infiltration of Lewy bodies with hyaline structures in the sympathetic ganglia, substantia nigra and locus coeruleus, as well as degeneration of the intermediolateral columns, dorsal vagal nuclei, substantia nigra and locus coeruleus (Vanderhaegen, Perier & Sternon, 1970; Bannister & Oppenheimer, 1972).

Autonomic failure associated with multiple system atrophy (AF-MSA, Shy-Dräger syndrome)

These patients develop severe orthostatic hypotension, genito-urinary disturbances, impaired sweating and heart rate control as well as evidence of a multisystem degeneration. Pathologically, there are features of olivo-ponto-cerebellar degeneration, striatonigral and pyramidal tract degeneration as well as loss of cells in the intermediolateral columns and dorsal vagal nuclei. There is variable degeneration of anterior horn cells, dorsal columns and spinocerebellar pathways (Oppenheimer, 1988). AF-MSA has a worse prognosis than the other primary degenerative causes of autonomic failure.

Other diseases affecting the central nervous system

In Parkinson's disease there is frequently a mild degree of postural hypotension but there are not the severe manifestations of autonomic failure seen in AF-PD. Spinal cord injuries can cause severe postural hypotension if the lesion is above the level of the splanchnic outflow at T6. Postural hypotension is seen in Wernicke's encephalopathy and less commonly in cerebrovascular disease, brainstem tumours, Adie's syndrome and tabes dorsalis. Abnormalities of sweating, blood pressure and heart rate control have been described in multiple sclerosis (Anema *et al.*, 1991; Yokota *et al.*, 1991).

Diseases affecting the peripheral nervous system

Diseases that affect the peripheral somatic nervous system may also damage the peripheral autonomic system causing autonomic dysfunction. Since the nerve fibres in both sympathetic and parasympathetic divisions that participate in cardiovascular control are mainly small myelinated (2–6 μm) and unmyelinated fibres, conditions that affect small fibres are more likely to cause autonomic dysfunction than those that affect large fibres predominantly. However, some degree of autonomic dysfunction occurs in most peripheral neuropathies; from a practical viewpoint they can be classified into those in which autonomic neuropathy is frequently of clinical importance and those in which it is usually of minor importance (Table 13.1). There are some diseases that affect the peripheral autonomic nervous system in which there is no associated peripheral neuropathy, e.g. acute and subacute autonomic neuropathy and botulism.

Autonomic disorders with no associated peripheral neuropathy

Acute and subacute autonomic neuropathy

Two types of autonomic neuropathy with no associated peripheral neuropathy are recognized: pure pandysautonomia and pure cholinergic dysautonomia. A pure pandysautonomia was first reported by Young *et al.* (1969) and many cases have since been described. It affects both sexes and all ages and there are no known familial or hereditary factors. The onset may be rapid over a period of days or more gradual over weeks or months. Symptoms consist of postural hypotension, abdominal pain and constipation, loss of sweating, urinary retention and incontinence, impotence and dry mouth and eyes. On examination the pupils may be fixed or poorly reactive, the skin is dry, lacrimation is impaired, and the abdomen and bladder may be distended. Tests of autonomic function

Disorders of the autonomic nervous system 209

reveal impairment of both sympathetic and parasympathetic systems. Recovery occurs over a variable period ranging from months to years.

Pure cholinergic dysautonomia presents with similar symptoms although, since the sympathetic system is spared, there is no postural hypotension. The illness tends to be more chronic and recovery less complete than is usual with pure pandysautonomia. The cause of these conditions remains uncertain but they are possibly a form of acute idiopathic polyneuritis restricted to autonomic nerves.

Botulism

Botulism is characterized by muscle paralysis, acute autonomic dysfunction, and gastrointestinal symptoms caused by the absorption from the alimentary tract of toxins produced by strains of Clostridium botulinum. The toxin impairs the release of acetylcholine from nerve terminals. Acute autonomic dysfunction may be present without the associated muscular weakness.

Peripheral neuropathies in which autonomic dysfunction is frequently clinically important

Diabetes

Autonomic neuropathy is common in diabetes, clinical features including impaired sweating, postural hypotension, impaired control of heart rate, diarrhea, impaired oesophogeal and gastric motility, impotence and sphincter and pupillary disturbances. Pathological changes in the vagus, carotid sinus and sympathetic nerves have been described (Guo, McLeod & Baverstock, 1987; Low *et al.*, 1975*a*; Tamura, Baverstock & McLeod, 1988). Damage to the splanchnic outflow may be an important factor in causing postural hypotension (Low *et al.*, 1975*a*) and it is usually a later manifestation of diabetic autonomic neuropathy than cardiac denervation and other effects of vagal involvement such as impaired heart rate response to standing and loss of sinus arrhythmia (Wheeler & Watkins, 1973; Ewing, Campbell & Clarke, 1980).

Amyloidosis

Autonomic dysfunction is most commonly seen in primary amyloidosis and the Portuguese type of familial amyloid polyneuropathy (FAP Type I). The autonomic disturbances may be disabling, frequently accompanying and often preceding the motor and sensory manifestations of peripheral neuropathy. Impotence, postural hypotension and abnormalities with sweating are common. Autonomic dysfunction is attributable to predominant loss of unmyelinated and small myelinated fibres in the peripheral autonomic nerves and reduction in the number of cells in the intermediolateral columns. Widespread deposition of amyloid in the autonomic nerves and ganglia has been frequently noted (Low *et al.*, 1981; Hersch & McLeod, 1987).

Guillain–Barré syndrome

Disturbances of heart rate and blood pressure control and abnormalities of sweating result from demyelination of fibres in the vagus and glossopharyngeal nerves and the preganglionic sympathetic efferent fibres, respectively (Asbury *et al.*, 1969; Matsuyama & Haymaker, 1967; Tuck & McLeod, 1981). Orthostatic hypotension may lead to syncope and occasionally to irreversible brain damage in the paralysed patient. Sudden death due to cardiac arrhythmias or asystole may be a consequence of damage to the autonomic nervous system.

Porphyria

Postural hypotension may complicate both acute intermittent and variegate porphyria, although hypertension is more common and may precede the clinical manifestations of peripheral neuropathy. Persistent tachycardia may be an early feature of an attack and may also precede the onset of neuropathy. Autonomic function studies have demonstrated abnormalities of sympathetic and parasympathetic function (Yeung Laiwah *et al.*, 1985).

Hereditary sensory and autonomic neuropathy

Postural hypotension and other features of autonomic dysfunction may be present in the hereditary sensory and autonomic neuropathies, particularly familial dysautonomia (Riley–Day syndrome, HSAN Type III) and the Swanson type (HSAN Type IV). The Riley–Day syndrome has an autosomal recessive mode of inheritance. The main clinical features are diminished lacrimation, hyperhydrosis, transient blotching of the skin, lability of blood pressure, postural hypotension and poor temperature control (Riley, 1957). Preganglionic neurons in the intermediolateral columns are reduced in number as are small myelinated fibres in the ventral roots. The sural nerves show a loss of unmyelinated and small myelinated fibres (Dyck *et al.*, 1978).

Peripheral neuropathies in which autonomic dysfunction is usually of minor clinical importance

Hereditary neuropathies

Charcot–Marie–Tooth disease. The pupillary reflexes may be abnormal and there is impairment of sweating distally. A recent study found no abnormality of cardiovascular reflexes in patients with HMSN types I and II when compared with control subjects (Ingall & McLeod, 1991*a*).

Friedreich's ataxia. Autonomic function studies, including sweat tests, are normal with no evidence of postural hypotension or impairment of baroreflex function (Ingall & McLeod, 1991*b*). The normal autonomic function is consistent with the relative preservation of small myelinated and unmyelinated fibres.

Other neuropathies

Chronic inflammatory demyelinating polyradiculoneuropathy (CIDP). In contrast to the common presence of autonomic disturbances in the Guillain–Barré syndrome, only minor abnormalities of autonomic function have been noted in CIDP (Ingall, McLeod & Tamura, 1990). It is presumed that the more severe disturbances in Guillain–Barré syndrome are related to acute conduction block and possibly more extensive involvement of unmyelinated fibres in the acute stages.

Metabolic disorders. Autonomic dysfunction, including postural hypotension, may be seen in patients on chronic hemodialysis (Nies, Robertson & Stone, 1979). Autonomic dysfunction, predominantly affecting the parasympathetic system is present in about 40% to 50% of patients with both alcoholic and non-alcoholic chronic liver disease (Thuluvath & Triger, 1989). Orthostatic hypotension may be the initial manifestation of pernicious anemia (McCombe & McLeod, 1984).

Postural hypotension is common in Wernicke's encephalopathy, and is probably caused by impaired sympathetic outflow at central or peripheral levels. Clinical manifestations of autonomic dysfunction are unusual in uncomplicated alcoholic peripheral neuropathy (Low *et al.*, 1975*b*; Duncan *et al.*, 1980) although postural hypotension may be present in patients who are severely affected (Novak & Victor, 1974). These findings indicate that peripheral sympathetic vasomotor control is relatively well preserved in alcoholics until the peripheral neuropathy reaches an advanced stage, even though abnormal sweat tests indicate that there is early involvement of sympathetic efferent pathways

(Low et al., 1975b). By contrast vagal damage may be demonstrated by autonomic function tests before postural hypotension occurs (Duncan et al., 1980). Morphometric studies have demonstrated loss of fibres in the most distal parts of the vagus nerve consistent with a dying back neuropathy (Guo et al., 1987) and some loss of fibres in the carotid sinus nerve (Tamura et al., 1988) but no significant changes in the splanchnic nerve (Low et al., 1975b). The absence of postural hypotension and the relatively normal baroreflex function are consistent with the lack of pathology in the splanchnic nerves since postural hypotension is more likely to occur if the splanchnic outflow is involved.

Malignancy. Impairment of sweating occurs in association with the peripheral neuropathy of remote malignancies. Pandysautonomia has been reported as a paraneoplastic manifestation of small cell carcinoma of the bronchus (Chiappa & Young, 1973) and other manifestations of autonomic dysfunction have been reported in carcinoma and lymphoma. Local hyperhydrosis and piloerection may result from direct irritation of nerves or roots (Walsh, Low & Allsop, 1976).

Toxic neuropathies. Postural hypotension has been reported as a neurotoxic side-effect of vincristine, and constipation, abdominal pain, paralytic ileus, urinary retention, and other bladder disturbances are well-recognized complications (Hancock & Naysmith, 1975; McLeod & Penny, 1969). Autonomic dysfunction has also been reported in heavy metal poisoning due to thallium (Bank et al., 1972), arsenic (Goldstein, McCall & Dynck 1975; Le Quesne & McLeod, 1977) and mercury (Kark, 1979). Autonomic function has been shown to be disturbed in some workers who have experienced prolonged chronic occupational exposure to a variety of organic solvents containing mainly aliphatic, aromatic and other hydrocarbons, alcohols, ketones, esters and ethers as well as to carbon disulphide and toluene (Matikainen & Juntunen, 1985). Industrial exposure to acrylamide may result in excessive sweating (Auld & Bedwell, 1967). Perhexiline maleate may cause postural hypotension (Fraser, Campbell & Miller, 1977).

Connective tissue diseases. Impairment of sweating on the extremities is relatively common in rheumatoid arthritis and is probably related to postganglionic sympathetic efferent fibre damage. In addition, there may be vagal nerve involvement causing impaired heart rate control (Bennett & Scott, 1965; Edmonds et al., 1979). Autonomic neuropathy has been reported as a complication of systemic lupus erythematosus (Gledhill & Dessein, 1988; McCombe et al., 1987) and mixed connective tissue diseases (Gudesblatt et al., 1985).

Infections. Impaired sweating, postural hypotension, cardiac denervation and other features of autonomic dysfunction have been reported in *leprosy* (Radhakrishnan *et al.*, 1978; Gadoth *et al.*, 1979; Khattri *et al.*, 1979).

Autonomic dysfunction affecting both sympathetic and parasympathetic divisions may be associated with human immunodeficiency virus (HIV) infections. Abnormalities of autonomic function tests may be demonstrated in asymptomatic patients or there may be specific symptoms of postural hypotension, disorders of sweating, diarrhoea, impotence and disturbed bladder function. Autonomic dysfunction is more frequent and of greater severity in patients with AIDS but may be present in the early stage of HIV infection and appears to progress during the illness (Freeman *et al.*, 1990). The mechanism of disordered autonomic function is not clear; it is not necessarily related to the presence of peripheral neuropathy (Cohen & Laudenslager, 1989).

Chagas' disease is common in South America. It is caused by the protozoal parasite, *Trypanasoma cruzi* and, in its chronic form, it produces cardiac failure, arrhythmias, disturbances of gastrointestinal motility including megaesophagus and megacolon, and other autonomic disturbances including postural hypotension. Tests of autonomic function are abnormal (Iosa *et al.*, 1990). There are pathological changes in the myocardium and conducting system, autonomic ganglia, postganglionic nerves and myenteric plexus. Multifocal inflammatory lesions associated with demyelination are present in the peripheral nervous system in humans and in the murine experimental model (Said *et al.*, 1985).

Investigation of autonomic dysfunction

There are many tests of autonomic dysfunction, both invasive and noninvasive (McLeod & Tuck, 1987*b*; McLeod, 1992*a*). A comprehensive range of noninvasive tests that may be used for evaluating sympathetic and parasympathetic function is listed in Table 13.2. They include tests of cardiovascular function, sweat tests and tests of vasomotor function.

Cardiovascular tests

Heart rate variation

In normal resting subjects the heart rate is determined mainly by background vagal activity and the laboratory tests of heart rate variation are therefore mainly tests of parasympathetic function.

Table 13.2. *Noninvasive Electrophysiological Tests of Autonomic Function*

Tests	Normal Response	Part of Reflex Arc Tested
1. *Cardiovascular function*		
Heart-rate variation with respiration	Maximum–minimum heart rate ≥15 beats/min E:I ratio ≥1.2	Parasympathetic afferent and efferent pathways
Valsalva ratio	≥1.4	Parasympathetic afferent and efferent, sympathetic efferent
Heart-rate response to standing	Increase 11-29 beats/min 30:15 ratio min ≥1.04	Parasympathetic afferent and efferent, sympathetic efferent
Heart-rate response to tilting	5–30 beats/min	Parasympathetic afferent and efferent, sympathetic efferent
Blood pressure response to tilting	≤20/10 mm Hg	Parasympathetic afferent and efferent sympathetic efferent
Isometric exercise	≥15 mm Hg	Sympathetic efferent
Valsalva manoeuvre (full haemodynamic response)	Phase I – Rise in BP Phase II – Gradual reduction of BP to plateau; tachycardia Phase III – Fall in BP Phase IV – Overshoot of BP; bradycardia	Parasympathetic afferent and efferent; sympathetic efferent
2. *Sweat tests*		
Sympathetic skin response	Response present	Sympathetic efferent Somatic afferent
3. *Vasomotor tests*		
Laser Doppler velocimetry skin blood flow measurement with inspiratory gasp, Valsalva manoeuvre, cold pressor test	Blood flow reduction ≥10%	Sympathetic efferent

Source: Modified from McLeod, 1992a.

Heart rate variation with breathing (sinus arrhythmia). Heart rate increases during inspiration because of decreased cardiac vagal activity. In practice, the variation in R–R interval on the electrocardiogram is measured during deep breathing and either the difference between the longest and shortest R–R interval during one minute of deep breathing is calculated (Wheeler & Watkins,

Disorders of the autonomic nervous system

1973) or the ratio of the mean of the maximum R–R intervals during deep expiration to the mean of the minimum R–R intervals during deep inspiration (E:I ratio) (Sundqvist, Almer & Lilja, 1979). The normal values are related to age.

Valsalva ratio. The change in heart rate that occurs in response to a brief period of forced expiration against a closed glottis or mouthpiece (Valsalva's manoeuvre) is a useful screening test for abnormalities of autonomic control of the cardiovascular system. It is calculated by dividing the longest R–R interval after the manoeuvre by the shortest R–R interval during the manoeuvre (Levin, 1966). It is a measure of both parasympathetic and sympathetic function and is age-dependent.

Heart rate response to change in posture.
Immediate increase in heart rate with standing. When changing from the supine to the standing position the heart rate increases usually by 10–20 beats a minute; the changes are related to age.

30:15 ratio. On standing, the heart rate increases until it reaches a maximum at about the 15th beat after which it slows to a relatively stable level at about the 30th beat. The ratio of the R-R intervals, corresponding to the 30th and 15th heart beats (the 30:15 ratio) is a measure of parasympathetic function (Ewing *et al.*, 1978). The ratio decreases with age.

Blood pressure variation
Change in posture. The change in blood pressure on assuming the upright posture after rest in the supine position for at least 10 minutes is measured. A fall in blood pressure of greater than 20/10 mm Hg is generally regarded as abnormal (Ingall, McLeod & O'Brien, 1990). The blood pressure can be measured with a standard sphygmomanometer or by continuous recording from digital arteries using a servo plethysmomanometer (Finapres), a noninvasive device that faithfully records transmural arterial pressure changes (Molhoek *et al.*, 1984). The change in blood pressure is not significally related to age (Ingall *et al.*, 1990).

Isometric exercise. Sustained handgrip against resistance causes an increase in heart rate and in arterial blood pressure. The normal response to maintaining 30% of the maximal handgrip pressure for five minutes is an increase in diastolic pressure of at least 15 mm Hg (Ewing *et al.*, 1974). The response is mainly a test of sympathetic efferent function and is not dependent on age (Ingall *et al.*, 1990).

Hemodynamic changes during the Valsalva manoeuvre. The full range of the hemodynamic response during and after the Valsalva manoeuvre can be measured noninvasively with a servo plethysmomanometer.

Sweat tests

There are several tests of sudomotor function including the thermoregulatory sweat tests (TST), the quantitative sudomotor axon reflex test (Q-SART), the sympathetic skin response (SSR) and the sweat imprint method (Low & Fealey, 1986). The TST is the classic sweat test. Sweating is detected on the body by chemicals that change colour when moist following the elevation of the body temperature by 1 °C by application of radiant heat to the trunk. It is a sensitive index of disorders of preganglionic and postganglionic sympathetic function and provides information about the distribution of the neuropathy (Fealey, Low & Thomas, 1989). The Q-SART is a sensitive test of postganglionic sympathetic function and measures the sweat response to acetylcholine iontophoresis (Low *et al.*, 1983).

The SSR is mediated by unmyelinated sympathetic fibres. It is measured by placing surface electrodes on the hands and feet and measuring the electrical potential that follows an electrical stimulus to the contralateral wrist or ankle (Shahani *et al.*, 1984). An absent SSR indicates impaired sympathetic efferent function.

Tests of peripheral vasomotor function

Reflex changes in skin blood flow in response to stimuli such as local cooling (cold pressor tests), sudden inspiration (inspiratory gasp), radiant heating of the trunk or immersion of the hand in hot water can be assessed with laser Doppler velocimetry (Low *et al.*, 1983). Muscle blood flow can be measured using venous occlusion or strain-gauge plethysmography in the forearm.

Emotional stress (mental arithmetic, a sudden loud noise, painful or emotional stimuli) causes a transient increase in the sympathetic vasomotor nerve activity that can be measured indirectly by its effect on reducing blood flow in the skin or extremities (Delius & Kellerova, 1971) or by increasing the arterial blood pressure (Ludbrook, Vincent & Walsh, 1975). These tests, which do not involve a reflex arc, evaluate sympathetic efferent activity.

Other tests of autonomic function

Other tests of autonomic function include measurement of plasma noradrenaline levels and their response to change in posture; tests of pupillary and blad-

Disorders of the autonomic nervous system 217

der function; the effect of infusion of pressor drugs on heart rate (baroreflex sensitivity); and intraneural recording of postganglionic sympathetic activity.

Screening tests of autonomic function

A useful battery of tests to screen non-invasively for autonomic dysfunction consists of measurement of: heart rate response to breathing, heart rate response to standing or tilting, and to Valsalva manoeuvre; blood pressure response to standing or tilting, sustained handgrip; and the sympathetic skin response (see Table 13.2).

Treatment of autonomic failure

The treatment of autonomic failure has been discussed in a number of reviews (see McLeod & Tuck, 1987*b*). If there is an underlying treatable cause of the autonomic dysfunction (e.g. acute inflammatory neuropathy, diabetes or toxins), the appropriate therapy should be administered, but in most cases treatment is directed towards optimal management of symptoms, the most important of which is orthostatic hypotension.

Orthostatic hypotension

The treatment of orthostatic hypotension is summarized in Table 13.3 and consists of general measures and drug therapy (McLeod, 1992*b*). 9-α-fluorohydrocortisone (fludrocortisone acetate) is the most reliable and effective agent for treatment. It acts mainly by increasing the blood volume but it also has some additional effect on noradrenaline release from sympathetic endings, sensitizing vascular receptors to pressor amines, and increasing vascular fluid content and vasoreactivity. Doses should be commenced at 0.1 mg/day and increased every three or four days by 0.1 mg/day until the blood pressure is controlled. The maximum dose is usually about 1 mg/day.

Sympathomimetic drugs that may also be employed include ephedrine (25 mg three or four times daily); dihydroergotamine, an α-agonist that produces its main vasoconstrictor effect on the venous capacitance vessels (40–60 mg/day orally); clonidine; and tyramine. Other drugs that have been employed include prostaglandin synthesis inhibitors, β-adrenergic blocking agents, dopaminergic antagonists, D,L-threo-3,4-dihydroxyphenylserine, caffeine, somatostatin (particularly to prevent postprandial hypotension) and desmopressin.

Table 13.3. *Treatment of Orthostatic Hypotension*

GENERAL MEASURES

General Advice
    Rise slowly, avoid straining
    Avoid extremes of temperature
    Wear light clothes
    Take regular moderate exercise
    Treat infections

Diet
    Small, frequent, low carbohydrate meals
    Low alcohol
    Increase daily salt intake (150 mEq or 3.5 g/day)
    Drink ample fluids
    Avoid diuretics and drugs causing hypotension

Mechanical measures
    Raise head of bed 15 cm
    Elastic garments

Atrial pacemaker

DRUG THERAPY

Mineralocorticoids
    9-α-fluorohydrocortisone (fludrocortisone)

Sympathominetic drugs
    Ephedrine
    Dihydroergotamine
    Tyramine and MAO inhibitors
    Midodrine
    Clonidine
    DOPA

β-adrenergic blocking agents
    Propranolol
    Pindolol

Dopaminergic antagonists
    Metoclopramide
    Domperidone

Prostaglandin inhibitors
    Indomethacin
    Ibuprofen
    Flurbiprofen

Other drugs
    Caffeine

Source: From McLeod, 1992*b*.

References

Anema, J. R., Heijenbrok, M. W., Faes, T. J. C., Heimans, J. J., Lanting, P. & Polman, C. H. (1991). Cardiovascular autonomic function in multiple sclerosis. *Journal of the Neurological Sciences*, **104**, 129–34.

Asbury, A. K., Arnason, B. G., Astrom, K. E. & Adams, R. D. (1969). The inflammatory lesion in idiopathic polyneuritis. Its role in pathogenesis. *Medicine*, **48**, 173–215.

Auld, R. B. & Bedwell, S. F. (1967). Peripheral neuropathy with sympathetic activity from industrial contact with acrylamide. *Canadian Medical Association Journal*, **96**, 652–4.

Bank, W. J., Pleasure, D. E., Suzuki, K., Nigro, M. & Katz, R. (1972). Thallium poisoning. *Archives of Neurology*, **26**, 456–64.

Bannister, R. & Oppenheimer, D. R. (1972). Degenerative diseases of the autonomic nervous system associated with autonomic failure. *Brain*, **95**, 457–74.

Bennett, P. H. & Scott, J. T. (1965). Autonomic neuropathy in rheumatoid arthritis. *Annals of Rheumatic Diseases*, **24**, 161–8.

Chiappa, K. A. & Young, R. R. (1973). A case of paracarcinomatous pandysautonomia. *Neurology*, **23**, 423.

Cohen, J. A. & Laudenslager, M. (1989). Autonomic nervous system involvement in patients with human immunodeficiency virus infection. *Neurology*, **39**, 1111–12.

Delius, W. & Kellerova, E. (1971). Reaction of arterial and venous vessels in the human forearm and hand to a deep breath or mental strain. *Clinical Science,* **40**, 271–82.

Duncan, G., Johnson, R. H., Lambie, D. G. & Whiteside, E. A. (1980). Evidence of vagal neuropathy in chronic alcoholics. *Lancet*, **2**, 1053–6.

Dyck, P. J., Kawamura, Y., Low, P. A. & Shimono, M. (1978). The number and sizes of reconstructed peripheral, autonomic, sensory and motor neurons in a case of dysautonomia. *Journal of Neuropathology and Experimental Neurology*, **37**, 741–55.

Edmonds, M. E., Jones, T. C., Saunders, W. A. & Sturrock, R. D. (1979). Autonomic neuropathy in rheumatoid arthritis. *British Medical Journal*, **2**, 173–5.

Ewing, D. J., Campbell, I. W., Kerr, F., Wildesmith, J. A. W. & Clarke, E. B. F. (1974). Cardiovascular responses to sustained handgrip in normal subjects and in patients with diabetes mellitus: a test of autonomic function. *Clinical Science and Molecular Medicine*, **46**, 295–306.

Ewing, D. J., Campbell, I. W. & Murray, A., Neilson, J. M. M. & Clarke, B. F. (1978). Immediate heart-rate response to standing: simple test for autonomic neuropathy in diabetes. *British Medical Journal*, **1**, 145–7.

Ewing, D. J., Campbell, I. W. & Clarke, B. F. (1980). Assessment of cardiovascular effects in diabetic autonomic neuropathy and prognostic implications. *Annals of Internal Medicine*, **92**, 308–11.

Fealey, R. D., Low, P. A. & Thomas, J. E. (1989). Thermoregulatory sweating abnormalities in diabetes mellitus. *Mayo Clinic Proceedings*, **64**, 617–28.

Fraser, D. M., Campbell, I. W. & Miller, H. C. (1977). Peripheral and autonomic neuropathy after treatment with perhexiline maleate. *British Medical Journal*, **3**, 675–6.

Freeman, R., Roberts, M. S., Friedman, L. S. & Broadbridge, C. (1990). Autonomic function and human immunodeficiency virus infection. *Neurology*, **40**, 575–80.

Gadoth, N., Bechar, M., Kushnir, M., Davidovitz, S. & Sandbank, U. (1979). Somatosensory and autonomic neuropathy as the only manifestation of long standing leprosy. *Journal of the Neurological Sciences*, **43**, 471–7.

Gledhill, R. F. & Dessein, P. H. M. C. (1988). Autonomic neuropathy in systemic lupus erythematosus. *Journal of Neurology, Neurosurgery, and Psychiatry*, **51**, 1238–40.

Goldstein, D. S., Polinsky, R. J., Garty, M., Robertson, D., Brown, R. T., Biaggioni, I., Stull, R. & Kopin, I. J. (1989). Patterns of plasma levels of catechols in neurogenic orthostatic hypotension. *Annals of Neurology*, **26**, 558–63.

Goldstein, N. P., McCall, J. T. & Dyck, P. J. (1975). Metal neuropathy. In *Peripheral Neuropathy*, ed. P. J. Dyck, P. K. Thomas & E. H. Lambert, pp. 1227–62. Saunders: Philadelphia.

Gudesblatt, M., Goodman, A. D., Rubenstein, A. E., Bender, A. N. & Choi, H.-S. H. (1985). Autonomic neuropathy associated with autoimmune disease. *Neurology*, **35**, 261–4.

Guo, Y.-P., McLeod, J. G. & Baverstock, J. (1987). Pathological changes in the vagus nerve in diabetics and chronic alcoholics. *Journal of Neurology, Neurosurgery, and Psychiatry*, **50**, 1449–53.

Hancock, B. W. & Naysmith, A. (1975). Vincristine-induced autonomic neuropathy. *British Medical Journal*, **3**, 207.

Hersch, M. I. & McLeod, J. G. (1987). Peripheral neuropathy associated with amyloidosis. In *Handbook of Neurology. Neuropathies*, vol. 1, ed. P. J. Vinken, G. W. Bruyn & H. L. Klawans, pp. 413–28. Amsterdam: Elsevier.

Ingall, T. J. McLeod, J. G. & Tamura, N. (1990). Autonomic function and unmyelinated fibres in chronic inflammatory demyelinating polyradiculoneuropathy. *Muscle and Nerve*, **13**, 70–6.

Ingall, T. J., McLeod, J. G. & O'Brien, P. C. (1990). The effect of ageing on autonomic nervous system function. *Australian & New Zealand Journal of Medicine*, **20**, 570–7.

Ingall, T. J. & McLeod, J. G. (1991*a*). Autonomic function in hereditary motor and sensory neuropathy (Charcot-Marie-Tooth disease). *Muscle and Nerve*, **14**, 1080–3.

Ingall, T. J. & McLeod, J. G. (1991*b*). Autonomic function in Friedreich's ataxia. *Journal of Neurology, Neurosurgery, and Psychiatry*, **54**, 162–4.

Iosa, D., Dequattro, V., Lee, D., Elkayam, U., Caeiro, T. & Palmero, H. (1990). Pathogenesis of cardiac neuro-myopathy in Chagas' disease and the role of the autonomic nervous system. *Journal of the Autonomic Nervous System*, **30**, 583–8.

Kark, R. A. P. (1979). Clinical and neurochemical aspects of inorganic mercury intoxication. In *Handbook of Clinical Neurology, Intoxication of the nervous system*, Part I, vol. 36, ed. P. J. Vinken & G. W. Bruyn, pp. 147–97. North Holland: Amerstam.

Khattri, H. N., Radhakrishnan, K., Kaur, S., Kumar, B. & Wahi, P. L. (1979). Cardiac dysautonomia in leprosy. *International Journal of Leprosy*, **46**, 172–4.

Le Quesne, P. M. & McLeod, J. G. (1977). Peripheral neuropathy following a single exposure to arsenic. *Journal of the Neurological Sciences*, **32**, 437–51.

Levin, A. B. (1966). A simple test of cardiac function based upon the heart-rate changes during the Valsalva maneuver. *American Journal of Cardiology*, **18**, 90–9.

Low, P. A., Caskey, P. E, Tuck, R. R., Fealey, R. D. & Dyck, P. J. (1983). Quantitative sudomotor axon reflex test in normal and neuropathic subjects. *Annals of Neurology*, **14**, 573–80.

Low, P. A., Dyck, P. J., Okazaka, I. H., Kyle, R. & Fealey, R. D. (1981). The splanchnic autonomic outflow in amyloid neuropathy and Tangier disease. *Neurology*, **31**, 461–3.

Low, P. A. & Fealey, R. D. (1986). Sudomotor neuropathy. In *Diabetic Neuropathy*, ed. P. J. Dyck, P. K. Thomas, A. K. Asbury, A. I. Winegrad & D. Porte, pp. 140–5. Philadelphia: W.B. Saunders.

Low, P. A., Neumann, C., Dyck, P. J. & Fealey, R. D. (1983). Evaluation of skin vasomotor reflexes by using laser Doppler velocimetry. *Mayo Clinic Proceedings*, **58**, 583–92.

Low, P. A., Thomas, J. E. & Dyck, P. J. (1978). The splanchnic autonomic outflow in Shy-Drager syndrome and idiopathic orthostatic hypotension. *Annals of Neurology*, **4**, 511–14.

Low, P. A., Walsh, J. C., Huang C. Y. & McLeod, J. G. (1975a). The sympathetic nervous system in diabetic neuropathy – a clinical and pathological study. *Brain*, **98**, 341–56.

Low, P. A., Walsh, J. C., Huang C. Y. & McLeod, J. G. (1975b). The sympathetic nervous system in alcoholic neuropathy. A clinical and pathological study. *Brain*, **98**, 357–64.

Ludbrook, J., Vincent, A. & Walsh, J. A. (1975). Effects of mental arithmetic on arterial blood pressure and hand blood flow. *Clinical and Experimental Pharmacology*, Suppl 2, 67–70.

Mathias, C. J., da Costa, D. F., Fosbraey, P., Bannister, R., Wood, S. M., Bloom, S. R. & Christensen, N. J. (1989). Cardiovascular, biochemical and hormonal changes during food-induced hypotension in chronic autonomic failure. *Journal of the Neurological Sciences*, **94**, 255–69.

Matikainen, E. & Juntunen, J. (1985). Autonomic nervous system dysfunction in workers exposed to organic solvents. *Journal of Neurology, Neurosurgery, and Psychiatry*, **32**, 297–304.

Matsuyama, H. & Haymaker, W. (1967). Distribution of lesions in the Guillain-Barré syndrome with emphasis on involvement of the sympathetic system. *Acta Neuropathologica (Berlin)*, **8**, 203–41.

McCombe, P. A. & McLeod, J. G. (1984). The peripheral neuropathy of vitamin B12 deficiency. *Journal of the Neurological Sciences*, **66**, 117–26.

McCombe, P. A., McLeod, J. G., Pollard, J. D., Guo, Y.-P. & Ingall, T. J. (1987). Peripheral sensimotor and autonomic neuropathy associated with systemic lupus erythematosus. *Brain*, **110**, 533–49.

McLeod, J. G. (1992a). Evaluation of the autonomic nervous system. In *Electrodiagnosis in Clinical Neurology*, 3rd edn, ed. M. J. Aminoff, New York: Churchill-Livingstone (in press)

McLeod, J. G. (1992b). Autonomic, sexual and sphincter disorders - Syncope. In: *Drug Therapy in Neurology*, ed. M. J. Eadie, pp. 227–46, Edinburgh: Churchill Livingstone.

McLeod, J. G. & Penny, R. (1969). Vincristine neuropathy: an electrophysiological and histological study. *Journal of Neurology, Neurosurgery, and Psychiatry*, **32**, 297–304.

McLeod, J. G. & Tuck, R. R. (1987a). Disorders of the autonomic nervous system: Part I: Pathophysiology and clinical features. *Annals of Neurology*, **21**, 419–30.

McLeod, J. G. & Tuck, R. R. (1987b). Disorders of the autonomic nervous system: Part II. investigation and treatment. *Annals of Neurology*, **21**, 519–29.

Molhoek, G. P., Wesseling, K. H., Settels, J. J. M., Weedah, W. H., de Wit, B. & Arntzenious, A. C. (1984). Evaluation of the Penàz servo-plethysmo-manometer for the continuous, non-invasive measurement of finger blood pressure. *Basic Research in Cardiology*, **79**, 597–609.

Nies, A. S., Robertson, D. & Stone, W. J. (1979). Hemodialysis hypotension is not the result of uremic peripheral autonomic neuropathy. *Journal of Laboratory and Clinical Medicine*, **94**, 395–402.

Novak, D. J. & Victor, M. (1974). The vagus and sympathetic nerves in alcoholic neuropathy. *Archives of Neurology*, **30**, 273–84.

Oppenheimer, D. (1988). Neuropathology of autonomic failure. In *Autonomic Failure*, 2nd edn, ed. R. Bannister, pp. 451–63. Oxford: Oxford University Press.

Petito, C. K. & Black, I. B. (1978). Ultrastructure and biochemistry of sympathetic ganglia in idiopathic orthostatic hyptension. *Annals of Neurology*, **4**, 6–17.

Radhakrishnan, K., Shenoy, K. T., Kumar, B., Kaur, S. & Khattri, H. N. (1978). Orthostatic hypotension in lepromatous leprosy. *Neurology (India)*, **26**, 25–7.

Riley, C. M. (1957). Familial dysautonomia. *Advances in paediatrics*, 9, 157–90.

Said, G., Joskowicz, M., Barreira, A. A. & Eisea, H. (1985). Neuropathy associated with experimental Chagas' disease. *Annals of Neurology*, **18**, 676–83.

Shahani, B. Halperin, J. J., Bolu, P. J. & Cohen, J. (1984). Sympathetic skin response - A method of assessing unmyelinated axon dysfunction in peripheral neuropathies. *Journal of Neurology, Neurosurgery, and Psychiatry*, **47**, 536–42.

Sundqvist, G., Almer, L.-O. & Lilja, B. (1979). Respiratory influence on heart rate in diabetes mellitus. *British Medical Journal*, **1**, 924–5.

Tamura, N., Baverstock, J. & McLeod, J. G. (1988). A morphometric study of the carotid sinus nerve in patients with diabetes mellitus and chronic alcoholism. *Journal of the Autonomic Nervous System*, **23**, 9–15.

Thuluvath, P. J. & Triger, D. R. (1989). Autonomic neuropathy in chronic liver disease. *Quarterly Journal of Medicine*, **72**, 737–47.

Tuck, R. R. & McLeod, J. G. (1981). Autonomic dysfunction in Guillain-Barré syndrome. *Journal of Neurology, Neurosurgery, and Psychiatry*, **44**, 983–90.

Vanderhaegen, J. J., Perier, O. & Sternon, J. E. (1970). Pathological findings in idiopathic orthostatic hypotension. Its relationship with Parkinson's disease. *Archives of Neurology*, **22**, 207–14.

Walsh, J. C., Low, P. A. & Allsop, J. L. (1976). Localized sympathetic overactivity: an uncommon complication of lung cancer. *Journal of Neurology, Neurosurgery, and Psychiatry*, **39**, 93–5.

Wheeler, T. & Watkins, P. J. (1973). Cardiac denervation in diabetes. *British Medical Journal*, 4, 584–6.

Yeung Laiwah, A. C., Macphee, G. J. A., Boyle, P., Moore, M. R. & Goldberg, A. (1985). Autonomic neuropathy in acute intermittent porphyria. *Journal of Neurology, Neurosurgery, and Psychiatry*, **48**, 1025–30.

Yokota, T., Matsunaga, T., Okiyama, R., Hirose, K., Tanabe, H., Furukawa, T. & Tsukagoshi, H. (1991). Sympathetic skin response in patients with multiple sclerosis compared with patients with spinal cord transection and normal controls. *Brain*, **114**, 1381–94.

Young, R. R., Asbury, A. K., Adams, R. D. & Corbett, J. L. (1969). Pure pandysautonomia with recovery. *Transactions of the American Neurological Association*, **94**, 355–7.

14

Autonomic innervation of the face

P. D. DRUMMOND

Department of Psychology
Murdoch University
Perth WA, Australia

Introduction

There is considerable confusion regarding the anatomy of individual autonomic pathways to the head and upper part of the body and the nature and mechanisms of disturbances resulting from damage to these pathways.

This chapter reviews both the anatomy and physiology of sympathetic and parasympathetic pathways to the head and related structures and describes the various clinical syndromes resulting from disturbance of their function.

Sympathetic pathways of the face

The sympathetic supply of the pupils, eyelids, blood vessels, sweat glands, salivary glands, and lacrimal glands descends from the hypothalamus in the anterolateral columns of the spinal cord to synapse in the intermediolateral section of the upper-thoracic spinal cord (ciliospinal centre of Budge) (Fig. 14.1) (Nathan & Smith, 1986). Most ocular fibres leave the spinal cord through the first thoracic root, whereas vasomotor and sudomotor fibres leave through the second and third thoracic roots (Goetz, 1948). This dissociation accounts for normal facial sweating and flushing in patients with ocular sympathetic disturbances after injury to the first thoracic root (Morris, Lee & Lim, 1984; Drummond & Lance, 1987). Conversely, surgical resection of sympathetic fibres between T2 and T4 results in hemifacial loss of sweating and flushing without affecting ocular sympathetic outflow (Drummond & Lance, 1987; Drummond & Lance, 1992). A similar pattern of disturbance can develop spontaneously (harlequin syndrome), possibly because of a vascular occlusion in the region of the second or third thoracic roots (Lance *et al.*, 1988).

Preganglionic fibres travel up the sympathetic chain to synapse in the superior cervical ganglion. Postganglionic ocular fibres and vasomotor and sudomotor fibres to the lower part of the forehead then follow the internal carotid

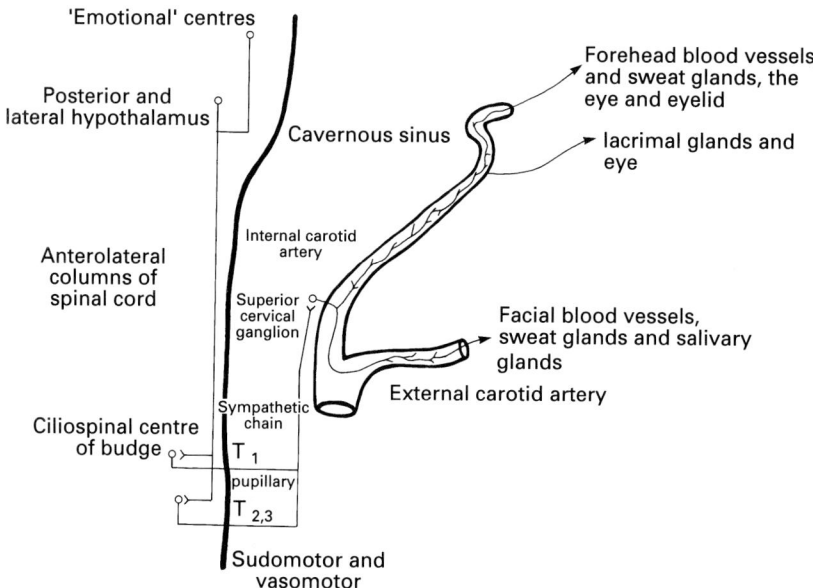

Fig. 14.1. Cervical sympathetic pathway to the face. The first neurone descends from the hypothalamus to the upper thoracic region of the spinal cord. Preganglionic neurones project to the superior cervical ganglion in the neck. Postganglionic neurones follow the internal or external carotid arteries before joining peripheral branches of the trigeminal nerve, with which they are distributed to the skin.

artery through the carotid canal to the cavernous sinus. Some sympathetic ocular fibres leave the internal carotid plexus and enter the middle ear to form a tympanic plexus with the tympanic branch of the glossopharyngeal nerve before rejoining the internal carotid nerve (Walsh & Hoyt, 1969). Sympathetic, parasympathetic and sensory fibres make their way separately to a retro-orbital plexus in the cavernous sinus which supplies various orbital structures, including the lacrimal gland (Ruskell, 1970; Ruskell & Simons, 1987); sympathetic fibres then follow branches of the ophthalmic nerve to the periphery. A bundle of sympathetic fibres leaves the internal carotid nerve in the deep petrosal nerve; these fibres unite with parasympathetic fibres from the greater superficial petrosal nerve to form the nerve of the pterygoid canal (Vidian nerve). Injury to fibres following the internal carotid artery produces the ocular signs of Horner's syndrome (miosis and ptosis) and loss of flushing and sweating on the symptomatic side of the forehead, extending sometimes to the hairline (Drummond & Lance, 1987; Morris et al., 1984).

Sympathetic fibres to other parts of the face follow the external carotid artery before joining branches of the maxillary and mandibular nerves (List &

Peet, 1938). Injury to these fibres produces local loss of sweating and flushing in the cheek and chin.

Sympathetic responses in the face

During the startle reaction, ocular sympathetic outflow increases pupil diameter and retracts the eyelids. This response is diminished in patients with a peripheral sympathetic lesion but not, curiously, in patients with central Horner's syndrome (Reeves & Posner, 1969). Thus, pupillary responses to light and to startling stimuli appear to be mediated by different central pathways. Arieff and Pyzik (1953) reported that pupillary dilatation could be induced by a spinal cord reflex. In patients with cervical spinal cord injuries, electrical stimulation of the chest below the level of the lesion caused pupillary dilatation, whereas painful stimulation of the face did not.

Sympathetic outflow dilates the pupils in darkness; a sluggish response is characteristic of Horner's syndrome (Pilley & Thompson, 1975). In normal illumination, a sympathetically denervated pupil is usually slightly smaller than its counterpart. The eyelid on the symptomatic side may also droop, but the pupillary disturbance is often more noticeable in conditions such as syringomyelia and syringobulbia (Nathan & Smith, 1986) and cluster headache (Drummond, 1988; Riley & Moyer, 1971). The fibres to the pupils may follow a slightly different course than those to the eyelids, and may be more susceptible to injury.

Injury near the cervical sympathetic pathway occasionally produces signs of sympathetic overactivity in the face. For example, Ottomo and Heimburger (1980) reported that facial sweating and pupillary dilatation alternated between the left and right sides in a patient 8 years after a cervical spinal cord injury. Signs subsided after adhesions between the spinal cord and the dura mater were freed, suggesting that sympathetic activity had been induced by traction on the spinal cord during shifts in posture. Shift of cystic fluid within the spinal cord may provoke a similar disturbance (Freeman & Russell, 1955). Teeple *et al.* (1981) observed sympathetic overactivity (lid retraction, pupillary dilatation and vascular disturbances) in a woman 2 days after she was hit on the neck by the boom of a sailboat. Pain in the left arm and autonomic signs in the face resolved after a series of stellate ganglion blocks. A similar autonomic disturbance (termed the Pourfour Du Petit syndrome by Teeple *et al.*, 1981) was observed recently after a parotidectomy (Byrne & Clough, 1990). These signs were attributed to injury to the sympathetic plexus around the internal carotid artery during the operation.

Blood flow through the ears, lips, eyes and nose is regulated by sympathetic vasoconstriction (Blair, Glover & Roddie, 1961; Fox, Goldsmith & Kidd, 1962; Drummond & Finch, 1989). Active sympathetic vasodilatation increases blood flow during prolonged body heating in other parts of the face (Drummond & Lance, 1987; Drummond & Finch, 1989). The sympathetic control of blood flow to the face was studied recently in 10 patients undergoing diagnostic blockade of the stellate ganglion (Drummond & Finch, 1989). Release of vasoconstrictor tone after stellate ganglion blockade increased ipsilateral cheek and orbital temperature, and increased vascular pulsations on the forehead and cheek. Furthermore, sympathetic blockade also prevented a far larger increase in vascular pulsations during body heating, indicating that an active vasodilator mechanism had been interrupted. Active sympathetic vasodilatation is triggered by increases in central body temperature, whereas warmth receptors in the skin cause the release of vasoconstrictor tone (Rowell, 1977).

Whether thermoregulatory flushing and sweating are induced by the same mechanism is uncertain. Sweating but not flushing is blocked by atropine, indicating that spillover of acetylcholine from sweat glands to nearby blood vessels is not the mechanism for flushing. In the nasal mucosa and salivary glands, interaction between vasoactive intestinal polypeptide (VIP) and acetylcholine, released simultaneously from parasympathetic nerves, increases blood flow (Lundberg, 1981; Lundberg et al., 1981). Since VIP is present in sudomotor fibres (Landis & Fredieu, 1986; Vaalasti, Tainio & Rechardt, 1985), release of VIP during sweating could contribute to flushing.

Facial sweating and flushing form part of the autonomic response to psychological stimuli. In 23 patients with cervical sympathetic deficit, sweating and blushing while singing children's songs were diminished on the symptomatic side of the forehead in patients with a pre- or postganglionic sympathetic lesion (Drummond & Lance, 1987). Nordin (1990) reported recently that forehead blood flow and electrodermal activity increased in parallel with sympathetic neural outflow during electrical stimulation of the wrist (an 'arousal' stimulus) and mental arithmetic. Hence, active sympathetic vasodilatation apparently increased facial blood flow during psychological stimulation. In our study, sweating and blushing were symmetrical in patients with a central sympathetic lesion, even though heat-induced sweating and flushing were diminished on the symptomatic side of the face (Drummond & Lance, 1987). Thus psychological and thermoregulatory responses appear to be mediated by separate central sympathetic pathways.

The sympathetic innervation of the salivary glands has secretory and excretory roles (Emmelin, 1981). In the presence of background parasympathetic activity, sympathetic stimulation induces secretion via beta-adrenergic recep-

Autonomic innervation of the face 227

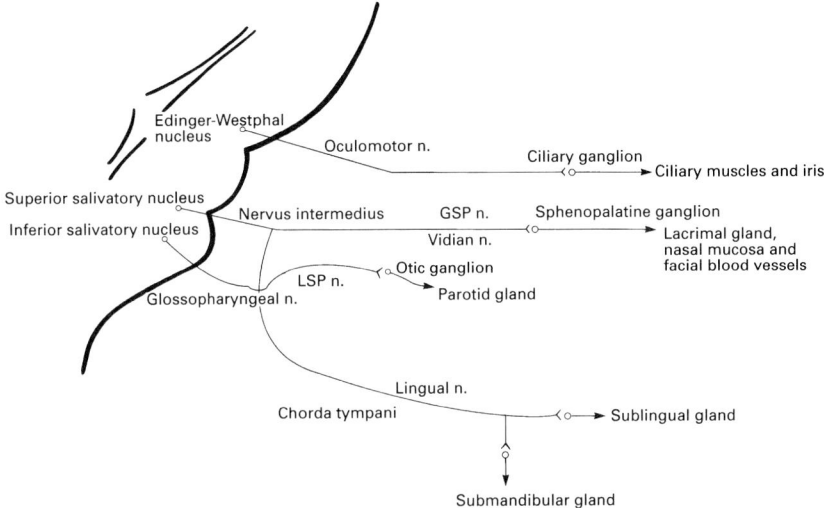

Fig. 14.2. Parasympathetic pathways to the face. Parasympathetic fibres leave the brainstem with the 3rd, 7th (in the nervus intermedius) and 9th cranial nerves, and synapse in ganglia close to their target organs.

tors. Sympathetic stimulation also provokes a brief, rapid expulsion of saliva into the mouth, due to contraction of myoepithelial cells surrounding the secretory cells. This excretory response is mediated by alpha-adrenergic receptors. Sympathetic fibres terminating on blood vessels surrounding salivary glands regulate vasoconstriction, and could be responsible for "speaker's dry mouth". Sympathetic fibres supplying the lacrimal glands may have similar secretory and vascular roles (Walsh & Hoyt, 1969).

Parasympathetic pathways to the face

Parasympathetic axons projecting from the Edinger–Westphal nucleus in the rostral midbrain leave the brainstem with the oculomotor nerve and synapse in the ciliary ganglion (Fig. 14.2). Postganglionic fibres travel with the short ciliary nerves to the iris and ciliary body. The ciliary body receives most of the parasympathetic fibres distributed by the ciliary ganglion (Walsh & Hoyt, 1969).

Another set of parasympathetic fibres originates in the superior salivatory nucleus (Fig. 14.2). These emerge from the brainstem in the nervus intermedius and travel with the facial nerve to the geniculate ganglion. Here, some fibres branch off without synapsing in the greater superficial petrosal (GSP)

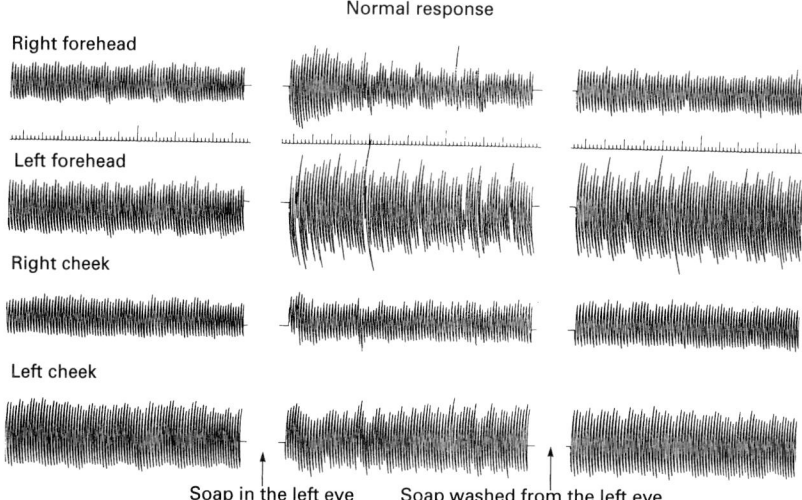

Fig. 14.3. The effect of irritating the left eye with soapy water on the facial microcirculation. Vascular pulsations were detected with photoplethysmographs, placed on the right and left sides of the forehead 1 cm above the eyebrows and 4 cm from the midline, and on the cheeks 3 cm below the eyes and 5 cm from the midline.

nerve, and unite with sympathetic fibres in the deep petrosal nerve to form the Vidian nerve. Parasympathetic neurones synapse in the sphenopalatine ganglion, which distributes postganglionic fibres to the lacrimal gland (Walsh & Hoyt, 1969), nasal mucosa (Lundberg et al., 1981), and facial skin (Lambert et al., 1984).

Parasympathetic neurones projecting to the submandibular and sublingual ganglia branch from the facial nerve in the chorda tympani. Postganglionic fibres innervate the submandibular and sublingual salivary glands (Walsh & Hoyt, 1969).

Parasympathetic neurones in the inferior salivatory nucleus arrive at the otic ganglion via the glossopharyngeal and lesser superficial petrosal nerves (Fig. 14.2). Postganglionic fibres from the otic ganglion reach the parotid gland via the auriculotemporal nerve (Walsh & Hoyt, 1969).

Parasympathetic responses in the face

The role of the parasympathetic nervous system in control of pupillary diameter and secretion of tears and saliva is well known (Walsh & Hoyt, 1969). Parasympathetic vasodilator fibres also innervate exocrine glands in the face,

Autonomic innervation of the face

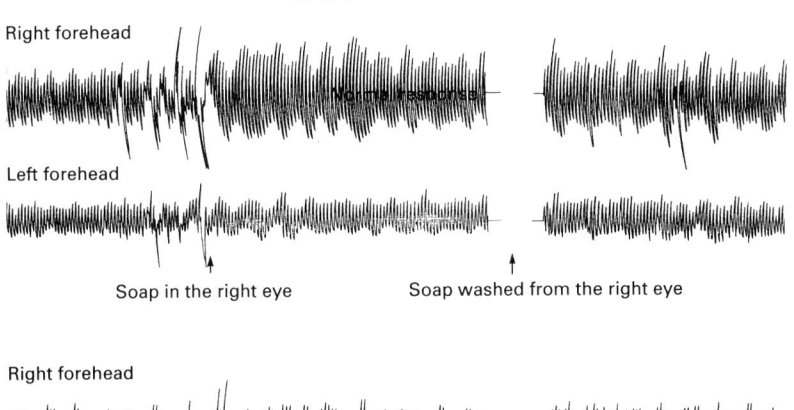

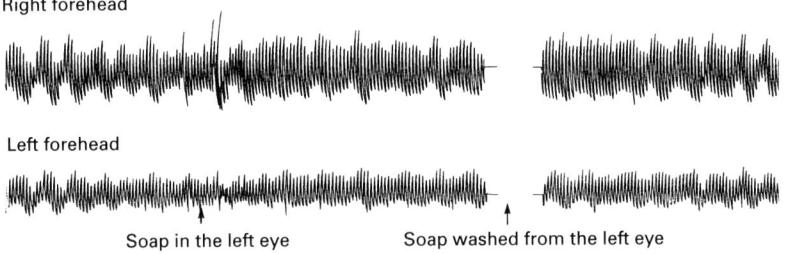

Fig. 14.4. Effect of ocular irritation on the forehead microcirculation in a patient with left-sided facial palsy and a dry left eye.

including the salivary glands, nasal mucosa, and lacrimal glands (Lundberg, 1981). Release of VIP from these fibres causes atropine-resistant vasodilatation, and potentiates secretion induced by acetylcholine. Thus, the increase in blood flow required by actively secreting exocrine glands appears to be regulated by VIP.

A similar mechanism increases blood flow in large cranial arteries and facial skin (Lambert *et al.*, 1984; Goadsby & Macdonald, 1985). The trigeminal nerve forms the afferent limb of the cutaneous vasodilator parasympathetic reflex, thus suggesting that the reflex could be induced by facial pain. This hypothesis was investigated recently in 15 patients with a unilateral facial nerve lesion compromising parasympathetic outflow (Drummond, 1992). In normal subjects, irritating the eye with soapy water provoked an increase in forehead blood flow which persisted for several minutes on the stimulated side after the soap had been washed from the eye (Fig. 14.3). This response was absent on the lesioned side in patients with facial palsy (Fig. 14.4), indicating that the response in normally-innervated skin was mediated by a trigeminal-parasympathetic reflex.

Vasodilatation to ocular irritation was greater in the forehead than in the cheeks (Fig. 14.3). In cats, stimulating the facial nucleus and the spinal tract of the trigeminal nucleus increases facial temperature limited to the cutaneous distribution of one or two divisions of the trigeminal nerve (Gonzalez, Onofrio & Kerr, 1975). Thus separate trigeminal-parasympathetic vasodilator reflexes could serve different facial regions.

The functional role of cutaneous trigeminal-parasympathetic vasodilator reflexes is uncertain. Local blood flow must increase to maintain tear flow during ocular irritation, and to maintain salivation in response to oral stimulants. The local increase in blood flow could spill over to surrounding skin, or might even have a protective function in the skin itself. Injury to the skin induces an inflammatory reaction with release of neuropeptides from sensory nerve endings (Holtzer, 1988; Chahl, 1988). This type of response occurred during intense stimulation of the trigeminal nerve (thermocoagulation of the trigeminal ganglion) (Drummond, Gonski & Lance, 1983), and was mediated by release of calcitonin gene-related peptide from trigeminal nerve terminals (Goadsby, Edvinsson & Ekman, 1988). When the face is injured, trigeminal-parasympathetic vasodilator reflexes could complement the inflammatory reaction provoked by antidromic release of vasoactive peptides.

Curiously, ocular irritation normally induced slightly greater increases in sweating (measured by changes in electrodermal activity) on the ipsilateral side of the forehead than on the contralateral side (Drummond, 1992). Furthermore, electrodermal responses were diminished on the lesioned side of the forehead in patients with facial palsy. VIP interacts with acetylcholine to modulate sweating (Lundberg, 1981), and induces atropine-dependent sweating when injected into rats' paws (Stevens & Landis, 1987). Thus pain-induced release of VIP from parasympathetic vasodilator terminals in the skin might potentiate local sweating to ocular irritation.

Effects of injury to autonomic nerves

Compensatory mechanisms quickly come into play after injury to autonomic nerves. Nearby autonomic fibres sprout to form collateral connections, regeneration proceeds rapidly, and denervated effector tissues and nerves develop supersensitivity to autonomic neurotransmitters. Unfortunately, sprouting and regeneration of autonomic fibres often proceeds so haphazardly that inappropriate connections develop. Furthermore, denervated effector tissues often become supersensitive to a wide range of compounds (Fleming & Westfall, 1988), further increasing the likelihood of inappropriate responses.

Denervation supersensitivity

After injury to autonomic fibres, the sensitivity of the target organ to autonomic neurotransmitters increases; the nature of the supersensitivity is influenced by the site of the lesion (Trendelenberg, 1966). After injury to postganglionic sympathetic fibres, a decrease in metabolism or reuptake of the neurotransmitter into the damaged neurone causes an accumulation of the neurotransmitter in the synaptic cleft and an increase in the response of effector tissues (denervation supersensitivity). Thus, dysfunction of metabolic or reuptake mechanisms will produce a specific supersensitivity to the target neurotransmitter. In contrast, preganglionic or central denervation of the target tissue induces a gradual, often nonspecific increase in sensitivity which is thought to be due to changes in receptor and postreceptor activity (adaptive supersensitivity) (Fleming & Westfall, 1988). For example, a moderate, nonspecific increase in sensitivity of smooth muscle membranes develops because changes in sodium–potassium exchange induce partial depolarization (Fleming & Westfall, 1988).

Ramsay (1986) studied denervation supersensitivity of the pupils in relation to the site of the lesion in patients with Horner's syndrome; a reduced response to hydroxyamphetamine eyedrops, a noradrenaline releaser, was taken as evidence of a postganlionic lesion. Pupillary dilatation to 1% phenylephrine eyedrops (an alpha-1 adrenergic agonist) was found to be greater in patients with a postganglionic lesion than in patients with a preganglionic or central sympathetic lesion (Ramsay, 1986). On the other hand, the degree of dilatation to 1% adrenaline eyedrops was not helpful in distinguishing between patients with a pre- or postganglionic lesion in a larger series (Maloney, Younge & Moyer, 1980). Furthermore, pupillary dilatation to 1% phenylephrine eyedrops was found to be greater on the symptomatic side in five patients with a central Horner's syndrome (Salvesen *et al.*, 1987), indicating that central denervation is a sufficient stimulus for adaptive supersensitivity of the sympathetic system in the iris. A convincing supersensitivity reaction was also demonstrated in two patients with a postganglionic sympathetic lesion (Salvesen, de Sousa & Sjaastad, 1989) but not, surprisingly, in three patients with a preganglionic lesion (Salvesen *et al.*, 1987). The reason for this discrepancy is not clear.

After a parasympathetic lesion, the pupil becomes unusually sensitive to cholinergic agents such as methacholine (Ponsford, Bannister & Paul, 1982) and pilocarpine (Ramsay, 1986). The presence of cholinergic supersensitivity is often regarded as a sign of a postganglionic parasympathetic lesion, but this is now controversial. For example, pupillary constriction to methacholine and pilocarpine eyedrops was found to be greater than normal in patients with a preganglionic lesion of the oculomotor nerve (Ponsford et al., 1982; Jacobson, 1990). Supersensitivity to cholinergic agents in preganglionic third nerve palsies could develop in response to reduced turnover of acetylcholine in iris muscles (Jacobson, 1990). Dissociation between the light and near vision reflexes occasionally develops in patients with a preganglionic oculomotor nerve lesion, suggesting that trans-synaptic degeneration and aberrant regeneration of pre- or postganglionic fibres could also contribute to iris supersensitivity (Jacobson, 1990). Transsynaptic degeneration might even progress centrally to affect contralateral pathways (Ponsford et al., 1982). The mere fact that large pupils constrict more than small pupils to pilocarpine eyedrops (Jacobson, 1990; Drummond, 1991) could also mimic an apparent cholinergic supersensitivity in patients with a tonically dilated pupil.

Denervation supersensitivity does not usually develop in sweat glands in the limbs after injury to postganglionic sympathetic fibres (Low et al., 1983). In contrast, facial sweat glands usually become more sensitive to cholinergic agents after injury to postganglionic sympathetic fibres, presumably because of partial innervation or re-innervation by parasympathetic fibres (List & Peet, 1938; Gardner & McCubbin, 1966; Glaister et al., 1958). Sweating after a subcutaneous injection of the cholinergic agent pilocarpine (0.1 mg/kg body weight) was investigated in four patients with a central Horner's syndrome, in three patients with a preganglionic lesion resulting from a brachial plexus injury (Salvesen et al., 1987), and in two patients with a probable postganglionic lesion (Salvesen et al., 1989). The sweating response to pilocarpine was greater on the symptomatic side in two patients with central Horner's syndrome and in one patient with a postganglionic lesion. Although the number of patients studied was small, these findings suggest that facial sweat glands develop supersensitivity even after a central sympathetic lesion.

Studies many years ago in laboratory animals established that sympathetically denervated cranial blood vessels develop supersensitivity to sympathetic agents (Grant, 1935). However, there has been little investigation of supersensitivity in the human facial circulation. Triple innervation of some cranial blood vessels (by sympathetic, parasympathetic and trigeminal-sensory fibres) complicates the study of this field. Nevertheless, further studies are required because of the relevance of findings for the mechanism of headache and facial pain.

Pathological gustatory facial flushing and sweating

Pathological responses to gustatory stimulation develop in over 70% of patients after bilateral preganglionic cervical sympathectomy (Kurchin et al., 1977). Months or years after the operation, spicy food or other strong tastes begin to provoke facial sweating, flushing and paresthesiae in sympathetically denervated skin and, less commonly, piloerection and pain. Pathological responses can be prevented by stellate ganglion blockade (Ashby, 1960; Haxton, 1948). Thus inappropriate connections apparently develop between preganglionic salivatory fibres and sudomotor, vasomotor and pilomotor cells in the superior cervical ganglion.

Pathological gustatory responses also develop after injury to postganglionic sympathetic fibres in the face. Injury to the auriculotemporal nerve (Gardner & McCubbin, 1966), submental nerve (Uprus, Gaylor & Carmichael, 1934; Young, 1956), and buccal nerve (Drummond, Boyce & Lance 1987), which each contain sympathetic and parasympathetic elements, can result in pathological gustatory responses. The response is thought to be mediated by cross-stimulation or sprouting of parasympathetic fibres into vacant sympathetic pathways, because it can be prevented by blocking the appropriate parasympathetic pathway with local anesthetic (Gardner & McCubbin, 1966; Uprus *et al.*, 1934; Young, 1956).

Sensory fibres can also grow into denervated sweat glands. In rats subjected to a chemical sympathectomy shortly after birth, fibres containing substance P-like immunoreactivity grew into the sympathetically denervated iris and sweat glands (Yodlowski, Fredieu & Landis, 1984; Kessler, Bell & Black, 1983). Kessler *et al.* (1983) suggested that sensory and sympathetic fibres compete for target nerve growth factor, resulting in reciprocal regulation of innervation. Cross-innervation of sympathetically denervated sweat glands by a mixture of sensory and parasympathetic fibres could account for paresthesiae and pain during gustatory reactions.

Crocodile tears

Excessive lacrimation during eating (crocodile tears) sometimes develops after injury to nerves which contain parasympathetic salivatory and lacrimal fibres, such as the nervus intermedius. Boyer and Gardner (1949) described two patients who developed crocodile tears after section of the GSP nerve. The syndrome was attributed to cross-regeneration or cross-stimulation between salivatory fibres in the lesser superficial petrosal nerve and lacrimal fibres in the GSP nerve, because the crocodile tears were prevented by dividing the glossopharyngeal nerve.

Pathological lacrimal sweating

After injury to cervical sympathetic fibres, facial sweating to ocular irritation may increase greatly. This was first described in a patient with a 3-year history of continuous mild pain over the right eye and temple (Van Weerden *et al.*,

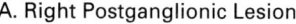
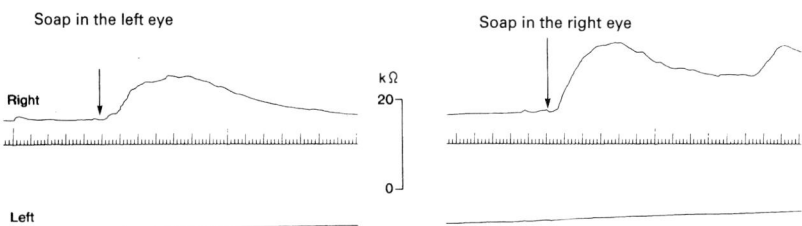
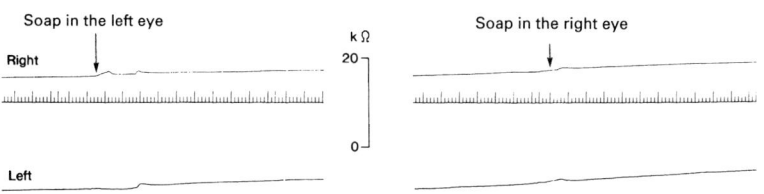

Fig. 14.5. Forehead sweating in response to ocular irritation. A. Pathological lacrimal sweating in a patient with postganglionic Horner's syndrome induced by an aneurysm of the right internal carotid artery. Even contralateral stimulation provokes sweating on the sympathetically denervated side of the forehead (upper left figure). (*b*) Normal (slight) lacrimal sweating in a patient with preganglionic right-sided Horner's syndrome.

1979). Abnormal sweating on the right side of the forehead developed 6 months after the onset of headaches. The presence of ocular sympathetic deficit, suggested by right-sided miosis and ptosis, was confirmed by the lack of response of the right pupil to 4% cocaine eyedrops. Cocaine blocks the reuptake of noradrenaline into sympathetic terminals; after sympathetic denervation, no transmitter is released and thus no response occurs. Irritating the eye during Schirmer's test provoked profuse sweating in the right frontal region, whereas lacrimation was symmetrical. Thus, faulty connections apparently had developed between lacrimal fibres and sympathetically denervated sweat glands in the forehead.

We recently investigated pathological lacrimal sweating in 24 patients with a lesion in the cervical sympathetic pathway, and in another 11 patients with cluster headache (Drummond & Lance, 1992). Seven of the cluster headache patients showed signs of a postganglionic sympathetic lesion (loss of sweating and flushing on the symptomatic side of the forehead, and a diminished pupillary response to tyramine eyedrops, a noradrenaline releaser). The soapy eyedrop generally induced a slight increase in electrodermal activ-

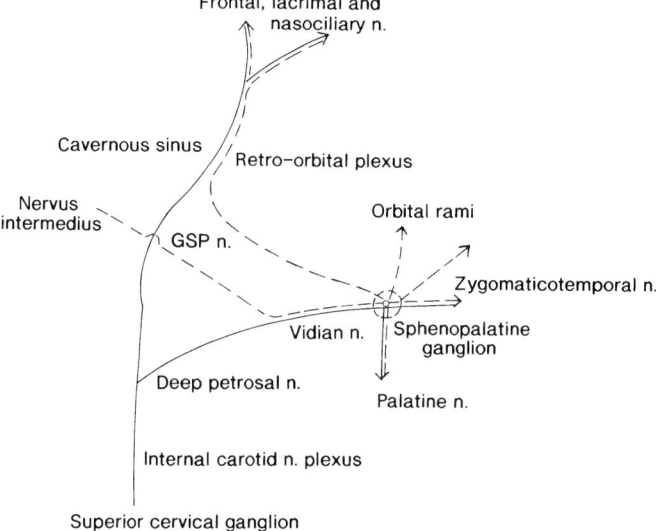

Fig. 14.6. Possible sites of convergence of sympathetic (solid lines) and parasympathetic fibres (dashed lines) mediating lacrimation, as described by Walsh and Hoyt (1969) and Ruskell (1970). Some fibres may travel to the lacrimal gland via the zygomaticotemporal nerve, whereas other fibres reach the gland via the retro-orbital plexus.

ity on the normally innervated side of the forehead. In most patients with signs of a postganglionic sympathetic lesion, the soapy eyedrop provoked a far greater increase in electrodermal activity on the sympathetically denervated side of the forehead (Fig. 14.5). Hence, lacrimal sweating was exaggerated in a region of diminished thermoregulatory sweating. These findings suggest that parasympathetic lacrimal fibres had grown into sympathetically denervated sweat glands, so that stimuli which normally induced lacrimation (e.g. soapy eyedrops) also induced sweating on the symptomatic side of the forehead. Pathological lacrimal sweating was detected in patients with a postganglionic sympathetic lesion caused by an aneurysm of the internal carotid artery (three patients), migraine (one patient), Ross's syndrome (one patient), and cluster headache (seven patients). The close proximity of sympathetic and lacrimal fibres in the cavernous plexus, Vidian nerve and lacrimal nerve provides opportunities for lacrimal fibres to sprout into vacant sympathetic pathways (Fig. 14.6). This mechanism could account for paradoxical sweating on the symptomatic side of the forehead during attacks of cluster headache, in spite of the presence of an ocular Horner's syndrome (Sjaastad et al., 1981).

In most patients with a central or preganglionic sympathetic lesion, lacrimal sweating was similar on both sides of the forehead, although in one case with a preganglionic sympathetic lesion, lacrimal sweating was greater on the symptomatic side. In this patient preganglionic lacrimal fibres could have developed faulty connections with denervated sudomotor neurons in the superior cervical ganglion.

The vascular response to ocular irritation was greater on the sympathetically denervated side of the forehead than on the other side in patients with a lesion anywhere in the cervical sympathetic pathway. In contrast to these findings in patients with a long-standing lesion, the vascular response in 14 patients shortly after stellate ganglion blockade was not increased on the blocked side of the forehead (Drummond, 1993). Thus, the increase in response in patients with a long-standing lesion might be due to adaptive supersensitivity of sympathetically denervated smooth muscle. Denervated smooth muscle develops increased sensitivity to a wide variety of agents (Fleming & Westfall, 1988). Nonspecific supersensitivity to VIP released by trigeminal-parasympathetic vasodilator reflexes could increase the response to ocular irritation in patients with a long-standing cervical sympathetic lesion (Drummond, 1992; Goadsby & Macdonald, 1985).

Pathological lacrimal sweating was observed in one patient with facial palsy (Drummond, 1992). In her case, the previously dry eye now watered more than the normally innervated eye, particularly when she was eating (crocodile tears). Thus lacrimal fibres apparently had regenerated, and salivary fibres may also have grown into the lacrimal gland. Unlike other patients with facial palsy, irritating the symptomatic eye with soapy water provoked symmetrical increases in pulse amplitude on each side of the forehead, and larger increases in electrodermal activity on the symptomatic side. These findings suggest that regenerating lacrimal fibres had made faulty connections with sweat glands. Unlike the situation in Horner's syndrome, the sympathetic supply to the sweat glands was intact, because facial sweating during body heating was symmetrical. Thus, regenerating lacrimal fibres apparently made inappropriate connections with normally functioning sweat glands.

Tonic pupils

Tonic pupils react poorly to light, and constrict slowly but extensively to near vision (Loewenfeld & Thompson, 1967). The condition develops after damage to pre- or postganglionic ocular parasympathetic fibres, and can be explained by cross-innervation of the iris by fibres originally innervating the ciliary muscle.

In the Holmes–Adie syndrome, the tonic pupil is usually unilateral but over the course of time the other pupil is often affected as well. Muscle stretch reflexes are disturbed in approximately 50% of cases (Loewenfeld & Thompson, 1967), indicating that pathology is not limited to parasympathetic ocular pathways. Neuropathological studies confirmed that the ciliary ganglia were depleted of nerve cells in patients with Holmes–Adie syndrome, and that dorsal root ganglia and dorsal columns in the lumbosacral region showed signs of ongoing neural degeneration (Ulrich, 1980; Selhorst, Madge & Ghatak, 1984).

In rare cases, tonic pupils are associated with a progressive disturbance in sympathetic activity (Ross's syndrome) (Ross, 1958). Cholinergic agents fail to induce local sweating in the affected limb (e.g. Spector & Bachman, 1984), consistent with a lesion of postganglionic sympathetic fibres (Low et al., 1983). We recently investigated the source of cervical sympathetic deficit in a woman with bilateral tonic pupils, loss of muscle stretch reflexes in the limbs, and hemifacial loss of sweating and flushing (Drummond & Edis, 1990). The pupil on the side that did not flush or sweat dilated less to 2% tyramine and 5% cocaine eyedrops than the other pupil, suggesting that the patient had a postganglionic ocular sympathetic lesion. Hemifacial loss of sweating and flushing during body heating and embarrassment indicated that the sympathetic lesion was probably in or close to the superior cervical ganglion. The autonomic and motor disturbances in Ross's syndrome could be caused by a degenerative process which attacks autonomic and motor ganglia. This obviously does not rule out independent sources of autonomic disturbance in individual cases. For example, Trend, Smith and Wiles (1986) reported that a patient with Horner's syndrome from an apical carcinoma also had a tonic pupil, presumably due to a preexisting parasympathetic lesion in the ciliary ganglion.

Acute idiopathic polyneuritis (Guillain–Barré syndrome) occasionally involves ophthalmoplegia as well as loss of muscle stretch reflexes and ataxia (Fisher's syndrome) (Fisher, 1956). A tonic pupil in such cases indicates that the polyneuritis affects parasympathetic outflow in the oculomotor nerve (Keane, 1977; Okajima et al., 1977). The polyneuritis may also compromise sympathetic activity. For example, a patient described by Okajima et al. (1977) showed a weak pupillary response to 5% cocaine eyedrops, suggesting that ocular sympathetic activity was also impaired. Young et al. (1975) presented a case with virtually complete (but reversible) paralysis of sympathetic and parasympathetic activity, with preservation of normal sensory, motor and cognitive functions. Young et al. concluded that the patient had an acute polyneuritis, restricted to the autonomic nervous system.

We recently studied four patients who complained of hemifacial loss of sweating and flushing, but who had no sign of ocular sympathetic deficit (harlequin syndrome) (Lance et al., 1988). In fact, the pupil on the nonflushing side of the face was consistently larger than the other pupil in two of these patients, thus resembling a tonically dilated pupil. However, the light and near reflexes were intact. Curiously, in these two patients and in another two patients studied more recently, the symptomatic and asymptomatic pupil both constricted extensively to weak (0.0625%) pilocarpine eyedrops (unpublished observations). These findings imply a subclinical ocular parasympathetic disturbance; in this respect, the mixture of sympathetic and parasympathetic deficit in harlequin syndrome resembles that in Ross's and Fisher's syndromes.

Hedges and Gerner (1975) suggested that autonomic disturbances in Holmes–Adie syndrome, and greater dysfunction in conditions such as the Shy–Dräger syndrome and acute pan-dysautonomia, could represent a spectrum of disorders. The severity of these conditions corresponds well with the extent of degeneration in central and peripheral autonomic pathways (McLeod & Tuck, 1987), but the underlying cause of autonomic degeneration has not been identified. This seems to be a promising area for future research.

Summary

Ocular sympathetic outflow dilates the pupils in darkness and in response to psychological stimuli. Sympathetic vasomotor activity mediates vasoconstriction in the ears, lips, eyes and nose, and cutaneous vasodilatation in other parts of the face. Vasomotor and sudomotor fibres to the forehead travel with ocular sympathetic fibres along the internal carotid artery. The rest of the face receives fibres distributed from the external carotid artery. Outflow through these pathways mediates thermoregulatory reflexes, and influences facial blood flow and sweating during emotional reactions.

Separate parasympathetic pathways supply the pupils, salivary glands, lacrimal glands and facial skin. Interaction between sympathetic and parasympathetic outflow to these organs regulates specialized reflexes, and adjusts local blood flow to meet on-going requirements. Trigeminal-parasympathetic reflexes increase cutaneous blood flow in a segmental distribution, possibly complementing inflammatory responses induced by antidromic trigeminal discharge.

After injury to autonomic nerves, inappropriate connections sometimes develop between target organs and regenerating fibres. Thus, sprouting of parasympathetic fibres into sympathetically denervated pathways can cause gustatory sweating and flushing, and pathological lacrimal sweating. Cross-innervation or cross-stimulation of parasympathetic fibres induces crocodile tears and tonic pupils. These autonomic disturbances are aggravated by denervation supersensitivity of target organs.

Acknowledgements

My work in this field was initiated by support and encouragement from Professor J.W. Lance. Our studies received support from the National Health and Medical Research Council of Australia, the J.A. Perini Family Trust, the Australian Brain Foundation, the Australian Research Council, and Murdoch University.

References

Arieff, A. J. & Pyzik, S. W. (1953). The ciliospinal reflex in injuries of the cervical spinal cord in man. *Archives of Neurology and Psychiatry*, 70, 621-9.

Ashby, W. B. (1960). Gustatory sweating and pilomotor changes. *British Journal of Surgery*, **47**, 406-10.

Blair, D. A., Glover, W. E. & Roddie, I. C. (1961). Cutaneous vasomotor nerves to the head and trunk. *Journal of Applied Physiology*, **16**, 119-22.

Boyer, F. C. & Gardner, W. J. (1949). Paroxysmal lacrimation (syndrome of crocodile tears) and its surgical treatment. *Archives of Neurology and Psychiatry*, **61**, 56-64.

Byrne, P. & Clough, C. (1990). A case of Poufour Du Petit syndrome following parotidectomy. *Journal of Neurology, Neurosurgery and Psychiatry*, **53**, 1014.

Chahl, L. A. (1988). Antidromic vasodilatation and neurogenic inflammation. *Pharmacology and Therapeutics*, **37**, 275-300.

Drummond, P. D. (1988). Autonomic disturbances in cluster headache. *Brain*, **111**, 1199-209.

Drummond, P. D. (1991). The effect of light intensity and dose of dilute pilocarpine eyedrops on pupillary constriction in healthy subjects. *American Journal of Ophthalmology*, **112**, 195-9.

Drummond, P. D. (1992). The mechanism of facial sweating and cutaneous vascular responses to painful stimulation of the eye. *Brain*, **115**, 1417–28.

Drummond, P. D. (1993). The effect of sympathetic blockade on facial sweating and cutaneous vascular responses to painful stimulation of the eye. *Brain*, **116**, 233–41.

Drummond, P. D., Boyce, G. M. & Lance, J. W. (1987). Postherpetic gustatory flushing and sweating. *Annals of Neurology*, **21**, 559-63.

Drummond, P. D. & Edis, R. H. (1990). Loss of facial sweating and flushing in Holmes-Adie syndrome. *Neurology*, **40**, 847-9.

Drummond, P. D. & Finch, P. M. (1989). Reflex control of facial flushing during body heating in man. *Brain*, **112**, 1351-8.

Drummond, P. D., Gonski, A. & Lance, J. W. (1983). Facial flushing after thermocoagulation of the Gasserian ganglion. *Journal of Neurology, Neurosurgery and Psychiatry*, **46**, 611-6.

Drummond, P. D. & Lance, J. W. (1987). Facial sweating and flushing mediated by the sympathetic nervous system. *Brain*, **110**, 793-803.

Drummond, P. D. & Lance, J. W. (1992). Pathological seating and flushing accompanying the trigeminal lacrimal reflex in patients with cluster headache and in patients with a confirmed site of cervical sympathetic deficit. *Brain*, **115**, 1429–45.

Ekbom, K. & Greitz, T. (1971). Carotid angiography in cluster headache. *Acta Radiologica Diagnosis*, **10**, 177-86.

Emmelin, N. (1981). Nervous control of salivation. *The Alabama Journal of Medical Sciences*, **18**, 294-9.

Fisher, M. (1956). An unusual variant of acute idiopathic polyneuritis (syndrome of ophthalmoplegia, ataxia and areflexia). *New England Journal of Medicine*, **255**, 57-65.

Fleming, W. & Westfall, D. P. (1988). Adaptive supersensitivity. In *Catecholamines. I. Handbook of Experimental Pharmacology*, vol. 90, ed. U. Trendelenberg & N. Weiner, pp. 509-59. Berlin: Springer-Verlag.

Freeman, L. W. & Russell, J. R. (1955). The significance of temporary and alternating ptosis, miosis and anhidrosis. *Journal of Neurosurgery*, **12**, 584-90.

Fox, R. H., Goldsmith, R. & Kidd, D. J. (1962). Cutaneous vasomotor control in the human head, neck and upper chest. *Journal of Physiology (London)*, **161**, 298-312.

Gardner, W. J. & McCubbin, J. W. (1966). Auriculotemporal syndrome: gustatory sweating due to misdirection of regenerated nerve fibres. *Journal of the American Medical Association*, **160**, 272-7.

Glaister, D. H., Hearnshaw, J. R., Heffron, P. F., Peck, A. W. & Patey, D. H. (1958). The mechanism of post-parotidectomy gustatory sweating (the auriculotemporal syndrome). *British Medical Journal*, **2**, 942-6.

Goadsby, P. J., Edvinsson, L. & Ekman, R. (1988). Release of vasoactive peptides in the extracerebral circulation of humans and the cat during activation of the trigeminovascular system. *Annals of Neurology*, **23**, 193-6.

Goadsby, P. J. & Macdonald, G. J. (1985). Extracranial vasodilatation mediated by vasoactive intestinal polypeptide (VIP). *Brain Research*, **329**, 285-8.

Goetz, R. H. (1948). The surgical physiology of the sympathetic nervous system with special reference to cardiovascular disorders. *International Abstracts of Surgery*, **87**, 417-39.

Gonzalez, G., Onofrio, B. M. & Kerr, F. W. L. (1975). Vasodilator system for the face. *Journal of Neurosurgery*, **42**, 696-703.

Grant, R. T. (1935). Further observations on the vessels and nerves of the rabbit's ear, with special reference to the effects of denervation. *Clinical Science*, **2**, 1-33.

Haxton, H. A. (1948). Gustatory sweating. *Brain*, **71**, 16-25.

Hedges, T. R. & Gerner, E. W. (1975). Ross' syndrome (tonic pupil plus). *British Journal of Ophthalmology*, **59**, 387-91.

Holzer, P. (1988). Local effector functions of capsaicin-sensitive sensory nerve endings: involvement of tachykinins, calcitonin gene-related peptide and other neuropeptides. *Neuroscience*, **24**, 739-68.

Jacobson, D. M. (1990). Pupillary responses to dilute pilocarpine in preganglionic 3rd nerve disorders. *Neurology*, **40**, 804-8.

Keane, J. R. (1977). Tonic pupils with acute ophthalmoplegic polyneuritis. *Annals of Neurology*, **2**, 393-6.

Kessler, J. A., Bell, W. O. & Black, I. B. (1983). Interactions between the sympathetic and sensory innervation of the iris. *Journal of Neuroscience*, **3**, 1301-7.

Kurchin, A., Adar, R., Zweig, A. & Mozes, M. (1977). Gustatory phenomena after upper dorsal sympathectomy. *Archives of Neurology*, **34**, 619-23.

Lambert, G. A., Bogduk, N., Goadsby, P. J., Duckworth, J. W. & Lance, J. W. (1984). Decreased carotid arterial resistance in cats in response to trigeminal stimulation. *Journal of Neurosurgery*, **61**, 307-15.

Lance, J. W., Drummond, P. D., Gandevia, S. C. & Morris, J. G. L. (1988). Harlequin syndrome: the sudden onset of unilateral flushing and sweating. *Journal of Neurology, Neurosurgery and Psychiatry*, **51**, 635-42.

Landis, S. C. & Fredieu, J. R. (1986). Coexistence of calcitonin gene-related peptide and vasoactive intestinal polypeptide in cholinergic sympathetic innervation of rat sweat glands. *Brain Research*, **377**, 177-81.

List, C. F. & Peet, M. M. (1938). Sweat secretion in man. IV. Sweat secretion of the face and its disturbances. *Archives of Neurology and Psychiatry*, **40**, 443-70.
Loewenfeld, I. E. & Thompson, H. S. (1967). The tonic pupil: a re-evaluation. *American Journal of Ophthalmology*, **63**, 46-87.
Low, P. A., Caskey, P. E., Tuck, R. R., Fealey, R. D. & Dyck, P. J. (1983). Quantitative sudomotor axon reflex test in normal and neuropathic subjects. *Annals of Neurology*, **14**, 573-80.
Lundberg, J. M. (1981). Evidence for coexistence of vasoactive intestinal polypeptide (VIP) and acetylcholine in neurons of cat exocrine glands: morphological, biochemical and functional studies. *Acta Physiologica Scandinavica*, Supplement **496**, 1-57.
Lundberg, J. M., Anggard, A., Emson, P., Fahrenkrug, J. & Hokfelt, T. (1981). Vasoactive intestinal polypeptide and cholinergic mechanisms in cat nasal mucosa: studies on choline acetyltransferase and release of vasoactive intestinal polypeptide. *Proceedings of the National Academy of Sciences, USA*, **78**, 5255-9.
Maloney, W. F., Younge, B. R. & Moyer, N. J. (1980). Evaluation of the causes and accuracy of pharmacologic localization in Horner's syndrome. *American Journal of Ophthalmology*, **90**, 394-402.
McLeod, J. G. & Tuck, R. R. (1987). Disorders of the autonomic nervous system: part I. Pathophysiology and clinical features. *Annals of Neurology*, **21**, 419-30.
Morris, J. G. L., Lee, J. & Lim, C. L. (1984). Facial sweating in Horner's syndrome. *Brain*, **107**, 751-8.
Nathan, P. W. & Smith, M. C. (1986). The location of descending fibres to sympathetic neurons supplying the eye and sudomotor neurons supplying the head and neck. *Journal of Neurology, Neurosurgery and Psychiatry*, **49**, 187-94.
Nordin, M. (1990). Sympathetic discharges in the human supraorbital nerve and their relation to sudo- and vasomotor responses. *Journal of Physiology (London)*, **423**, 241-55.
Okajima, T., Imamura, S., Kawasaki, S., Ideta, T. & Tokuomi, H. (1977). Fisher's syndrome: a pharmacological study of the pupils. *Annals of Neurology*, **2**, 63-5.
Ottomo, M. & Heimburger, R. F. (1980). Alternating Horner's syndrome and hyperhidrosis due to dural adhesions following cervical spinal cord injury. *Journal of Neurosurgery*, **53**, 97-100.
Pilley, S. F. J. & Thompson, H. S. (1975). Pupillary 'dilatation lag' in Horner's syndrome. *British Journal of Ophthalmology*, **59**, 731-5.
Ponsford, J. R., Bannister, R. & Paul, E. A. (1982). Methacholine pupillary responses in third nerve palsy and Adie's syndrome. *Brain*, **105**, 583-97.
Ramsay, D. A. (1986). Dilute solutions of phenylephrine and pilocarpine in the diagnosis of disordered autonomic innervation of the iris. *Journal of the Neurological Sciences*, **73**, 125-34.
Reeves, A. G. & Posner, J. B. (1969). The ciliospinal response in man. *Neurology*, **19**, 1145-52.
Riley, F. C. & Moyer, N. J. (1971). Oculosympathetic paresis associated with cluster headaches. *American Journal of Ophthalmology*, **72**, 763-8.
Ross, A. T. (1958). Progressive selective sudomotor denervation: a case with coexisting Adie's syndrome. *Neurology*, **8**, 809-17.
Rowell, L. B. (1977). Reflex control of the cutaneous vasculature. *Journal of Investigative Dermatology*, **69**, 154-66.
Ruskell, G. L. (1970). The orbital branches of the pterygopalatine ganglion and their relationship with internal carotid nerve branches in primates. *Journal of Anatomy*, **106**, 323-39.

Ruskell, G. L. & Simons, T. (1987). Trigeminal nerve pathways to the cerebral arteries in monkeys. *Journal of Anatomy*, **155**, 23-37.

Salvesen, R., de Souza, C. D. & Sjaastad, O. (1989). Horner's syndrome: sweat gland and pupillary responsiveness in two cases with a probable 3rd neurone dysfunction. *Cephalalgia*, **9**, 63-70.

Salvesen, R., Fredriksen, T. A., Bogucki, A. & Sjaastad, O. (1987). Sweat gland and pupillary responsiveness in Horner's syndrome. *Cephalalgia*, **7**, 135-46.

Selhorst, J. B., Madge, G. & Ghatak, N. R. (1984). The neuropathology of the Holmes-Adie syndrome. *Annals of Neurology*, **16**, 138.

Sjaastad, O., Saunte, C., Russell, D., Hestnes, A. & Marvik, R. (1981). Cluster headache. The sweating pattern during spontaneous attacks. *Cephalalgia*, **1**, 233-44.

Spector, R. H. & Bachman, D. L. (1984). Bilateral Adie's tonic pupil with anhidrosis and hyperthermia. *Archives of Neurology*, **41**, 342-3.

Stevens, L. M. & Landis, S. C. (1987). Development and properties of the secretory response in rat sweat glands: relationship to the induction of cholinergic function in sweat gland innervation. *Developmental Biology*, **123**, 179-90.

Teeple, E., Ferrer, E. B., Ghia, J. N. & Pallares, V. (1981). Poufour Du Petit syndrome – hypersympathetic dysfunctional state following a direct non-penetrating injury to the cervical sympathetic chain and brachial plexus. *Anesthesiology*, **55**, 591-2.

Trend, P. St. J., Smith, S. E. & Wiles, C. M. (1986). A tonic pupil with Horner's syndrome. *Journal of Neurology, Neurosurgery and Psychiatry*, **49**, 841.

Trendelenberg, U. (1966). I. Mechanisms of supersensitivity and subsensitivity to sympathomimetic amines. *Pharmacological Reviews*, **18**, 629-40.

Ulrich, J. (1980). Morphological basis of Adie's syndrome. *European Neurology*, **19**, 390-5.

Uprus, V., Gaylor, J. B. & Carmichael, E. A. (1934). Localized abnormal flushing and sweating on eating. *Brain*, **57**, 443-53.

Vaalasti, A., Tainio, H. & Rechardt, L. (1985). Vasoactive intestinal polypeptide (VIP) -like immunoreactivity in the nerves of human axillary sweat glands. *Journal of Investigative Dermatology*, **85**, 246-8.

Van Weerden, T. W., Houtman, W. A., Schweitzer, N. M. J. & Minderhoud, J. M. (1979). Lacrimal sweating in a patient with Raeder's syndrome. *Clinical Neurology and Neurosurgery*, **81**, 119-21.

Walsh, F. B. & Hoyt, W. F. (1969). *Clinical Neuro-Ophthalmology*, 3rd edn, vol 1. Baltimore: Williams and Wilkins.

Yodlowski, M. L., Fredieu, J. R. & Landis, S. C. (1984). Neonatal 6-hydroxydopamine treatment eliminates cholinergic sympathetic innervation and induces sensory sprouting in rat sweat glands. *Journal of Neuroscience*, **4**, 1535-48.

Young, A. G. (1956). Unilateral sweating of the submental region after eating (chorda tympani syndrome). *British Medical Journal*, **2**, 976-9.

Young, R. R., Asbury, A. K., Corbett, J. L. & Adams, R. D. (1975). Pure pan-dysautonomia with recovery. *Brain*, **98**, 613-36.

15

Stroke and antiphospholipid antibodies

K. M. A. WELCH AND S. R. LEVINE

Center for Stroke Research,
Department of Neurology,
Henry Ford Hospital and Health Sciences Center
Detroit, Michigan, USA

Introduction

In recent years, antiphospholipid antibodies have been implicated in ischemic stroke, transient ischemic attacks (TIA), visual disturbances (including ischemic optic neuropathy and amaurosis fugax), migraine-like syndromes, and ischaemic encephalopathy (Harris *et al.*, 1984; Levine *et al.*, 1987; Levine & Welch, 1987*a,b*; Alarcon-Segovia, 1988; Case records of the Massachusetts General Hospital, 1988; Levine *et al.*, 1988; Briley, Coull, & Goodnight, 1989; Digre *et al.*, 1989). The lupus anticoagulant (LA) and anticardiolipin antibodies (aCL) are the two best characterized antiphospholipid antibodies (aPL). They are acquired or familial serum immunoglobulins that recognize and bind negatively charged phospholipid moieties (Thiagarajan, Shapiro & DeMarco, 1980; Freyssinet *et al.*, 1986; Pengo *et al.*, 1987). LA and aCL are related though not necessarily identical (Exner, Sahman & Trudinger, 1988; Triplett *et al.*, 1988). They have an overlap of structure but with distinct epitopes. Different protein fractions responsible for LA and aCL activities have recently been isolated (Exner *et al.*, 1988). The specific epitopes with which aPL react have not been fully defined, but may be the phosphodiester-linked phosphate groups. Phospholipids are a major constituent of cellular membranes and many coagulation factors. Specifically, phosphatidylcholine, phosphatidic acid, phosphatidylinositol, phosphatidylethanolamine, phosphatidylserine, and cardiolipin can be bound by the antibody. This is important in cerebrovascular disease because aPL may bind phospholipids in vascular endothelium and platelet membranes.

Currently, the most readily available methods to detect aPL include the Venereal Disease Research Laboratory (VDRL) serologic test for syphilis (cardiolipin is in the assay), phospholipid dependent-coagulation studies such as the activated partial thromboplastin time (PTT) and prothrombin time (PT), and the aCL test using either the enzyme-linked immunosorbent assay

(ELISA) or a radioimmunoassay (RIA). Recent data suggest that VDRL results are only positive in approximately 25% of patients with aPL (Case records of the Massachusetts General Hospital, 1988). The VDRL test is thus too insensitive to be used as a screening study. The yield for detecting LA may be increased by more sensitive coagulation assays (employing less phospholipid) such as the Russell viper venom time (RVVT), the kaolin clotting time (KCT), and the dilute activated PTT (aPTT).

Prevalence and stroke risk

Antiphospholipid antibodies may be present in 2–5% of healthy people under 60 years of age (Love & Santoro, 1990) and 12–50% in those over 60 (Manoussakis et al, 1987; Fields et al., 1989). These antibodies have been described in syphilis (Levy et al., 1990a), HIV (Brey, Arroyo & Boswell, 1991) and other viral infections (Vaarala et al., 1986), Lyme disease (Mackworth-Young et al., 1988), malignancies, inflammatory and bacterial disorders and in association with drugs such as phenothiazines, penicillin and phenytoin (Lillicrap et al., 1990). The clinical syndrome associated with aPL is rarely seen in these patients. On the other hand antibodies directed against phospholipids are commonly described in patients with autoimmune diseases such as SLE and other collagen vascular diseases in which aPL-associated clinical symptomatology does occur. In studies evaluating unselected patients with SLE, aPL prevalence estimates range from 7–58% (Love & Santoro, 1990). The presence of aPL in these patients is associated with approximately a five-fold greater frequency of thrombotic events than in patients without them (Love & Santoro, 1990).

Patients with aPL and thrombosis, thrombocytopenia or fetal loss, but who do not have another definable autoimmune disease are said to have the 'primary' aPL syndrome (Mackworth-Young, Loizou & Walport, 1989). Other patients falling into this category include those with a variety of cardiac abnormalities (most commonly left-sided valvular heart lesions), and those with a number of neurological syndromes (Levine & Welch, 1987b). Stroke is an important feature of the primary aPL syndrome. The prevalence of aPL in unselected series of stroke patients ranges from 6 to 46% (Brey et al., 1990; Trimble et al., 1990).

Recently, consecutive patients with their first ischemic stroke and age- and sex-matched hospitalized controls were prospectively screened for aCL at 12 centres in the United States (Anonymous, 1992). Two-hundred and forty-eight cases were included in the analysis, 119 with prior strokes and 257 controls. The aCL was present in 24 stroke cases and 11 controls giving an odds ratio of

2.4 (95% confidence intervals 1.1–5.0, $p = 0.016$). After adjusting for other stroke risk factors in this population, the adjusted odds ratio of aCL for ischemic stroke remained at 2.4 ($p = 0.024$), higher than diabetes mellitus, for example, which in this population gave an odds ratio of 1:58 ($p = 0.045$). Thus the presence of aCL is now emerging as a risk factor for stroke which is equal to if not greater than diabetes mellitus.

To assess recent stroke and thromboembolic events in an aPL population, aPL-positive stroke or TIA patients ($n = 74$) were followed for 72.1 patient-years of prospective follow-up (Levine et al., 1992). Forty percent of patients had a prior systemic thromboembolic event. The combined rate for stroke plus systemic thromboembolic events equaled 0.21 events/patient–year. Stroke, systemic thromboembolic events, TIA and death had a combined rate of 0.45 events/patientyear. Thirty-six percent of patients had a subsequent stroke or TIA during a mean follow-up of 14.5 months.

Putative mechanisms of thrombosis

The pathogenetic significance of aPL is still uncertain. Antibody characteristics which are strongly associated with clinical symptoms seem to be persistently high titre, IgG isotype and cross-reactivity with multiple negatively charged phospholipids (and occasionally DNA) (Harris et al., 1988). Patients with aPL-associated stroke generally have a false-positive VDRL, low-positive ANA, thrombocytopenia, low fourth component of the complement cascade (C-4) and an elevated sedimentation rate.

The antigen responsible for the autoimmune aPL response is uncertain. The phosphodiester linkage uncommon to phospholipids and DNA has been suggested (Lafer et al., 1981) explaining the cross-reactivity often seen. The chain length of phospholipid fatty acids may also be important (Levy et al., 1990b). Membrane phospholipids can exist in an immunogenic, hexagonal form during the course of remodelling. Phospholipids so exposed may well be the self-antigen against which pathogenetic aPL are directed (Levy et al., 1990a).

Beta-2-lipoprotein I or apolipoprotein H appears to enhance aPL binding in the ELISA assay system in a way similar to the potentiating effects described in coagulation assays (Galli et al., 1990; Matsuura et al., 1990; McNeil et al., 1990). Apolipoprotein H may regulate the coagulation system as an inhibitor of the intrinsic pathway and by altering prothrombinase activity in platelets (Nimpf et al., 1986). Inhibition of apolipoprotein H by aPL could lead to hypercoagulability and the clinical features of the aPL syndrome.

Several other aPL effects could lead to thrombosis. Serum containing aPL decreases the production of prostacyclin (Carreras & Vermylen, 1982; Schorer,

Wickham & Watson, 1989). aPL may interfere with natural anticoagulants such as proteins C and S and antithrombin III (Nelson & Goodnight, 1988). If platelet activation coexists, aPL could contribute to platelet-initiated thrombosis. Complement activation occurs in stroke patients with aPL, but not in age-matched stroke patients without them (Davis & Brey, 1991). Complement-mediated endothelial cell damage could contribute to thrombosis.

Clinical features of stroke assosciated with aPL

To illustrate the clinical association of cerebral ischemia and the aPL, we present the findings in 48 patients who attended the Department of Neurology at Henry Ford Hospital over a consecutive period of 48 months from July 1985 to April 1989 (Levine *et al.*, 1990). The patients presented with symptoms of brain and ocular ischaemia and were followed prospectively.

Patients with aPL suffered either from TIAs or cerebral infarction. Importantly, they were generally younger (mean age, 43 years) than patients with atherothrombotic stroke. Recurrent stereotypic spells included visual, motor and somatosensory symptoms, with or without headache. Rostral brainstem events associated with encephalopathy and minimal CSF pleocytosis seen in two of our patients has not, to our knowledge, been described previously with aPL although Briley *et al.* (1989) have described a case of acute ischemic encephalopathy.

Most of our patients suffered recurrent strokes prior to the detection of aPL. Nearly half also had a subsequent transient or permanent thrombotic event after the study event in just over 1 year of follow-up. The high number of recurrent thrombotic events indicate that aPL is a marker for such recurrence. Recurrent events were significantly more common among patients who smoked cigarettes, were hyperlipidaemic or had a positive ANA. Approximately half our patients had either a false-positive VDRL or a low platelet count as a clue to aPL. Importantly, these laboratory studies are generally obtained on routine screening of patients with stroke and TIA. Furthermore, obtaining a history of spontaneous abortion was another important clue in several women (Hughes, 1986).

Although many of our patients had other stroke risk factors coexisting with aPL, approximately 20% had no other known cause of their syndrome, which thus may be more readily called the 'primary antiphospholipid syndrome' (Asherson *et al.*, 1989), lending support to an immunologic mechanism for aPL-associated stroke.

The clinical and radiologic features of the strokes were similar to those of embolic events: medium- to large-vessel arteriopathy, or *in situ* thrombosis was usually noted. Lacunar infarcts were distinctly unusual, although seen. Vascular pathology supported a nonvasculitic thromboembolic process.

Cardiac disease, primarily left-sided cardiac valvular disease, was relatively common, seen in over 1/3 of the 48 patients. Verrucous (Libman-Sacks) endocarditis with cardiac valvular abnormalities has been linked previously to the presence of aPL (Asherson & Lubbe, 1988; Ford, Foro & Lillicrap, 1988).

Patients with migraine headaches, with or without aura, as well as those with aura without headache, have had documented aPL (Levine *et al.*, 1987*a*, *b*; Hogan *et al.*, 1988; Briley *et al.*, 1989; Shuaib *et al.*, 1989). Some of these patients had suffered cerebral infarcts (Hogan *et al.*, 1988; Shuaib *et al.*, 1989). Our data further support this association, although the mechanisms are uncertain (Levine *et al.*, 1987*a*, *b*) and coincidental illness cannot be excluded. It is difficult to differentiate transient cerebral ischemia or TIA with headache from a primarily migrainous event with associated aura or prolonged aura. Only detailed studies of prospective prevalence will ascertain a link between migraine and aPL.

Treatment

In the absence of a clear cardiac source of embolism in patients with aPL-associated ischemic cerebrovascular disease, we recommend an extensive hematological evaluation to ascertain a specific coagulation or platelet deficit or both that may indicate a specific therapy. Warfarin should be considered in the presence of cardiac disease. Currently, however, there is no good evidence to support one type of therapy (i.e. aspirin, warfarin, corticosteroids) over another, and individual patient management remains empirical. In the absence of a documented source of cardiac embolic or large-vessel occlusive disease, we generally initiate treatment with aspirin and aggressive modification of any other risk factors for stroke, especially cigarette smoking.

Conclusion

We suggest that patients of any age with cerebrovascular or ocular syndromes who do not have an obvious etiology should be evaluated for aPL. Those aged 50 years or younger should even be evaluated if concomitant stroke risk factors are present or any of the following: SLE, prolonged aPTT or PT without explanation, miscarriages, thrombocytopenia, systemic thrombotic events or prior stroke of unknown cause, positive ANA, left-sided cardiac valvular abnormality, family history of premature stroke and atypical recurrent visual disturbances. We suggest both an aPTT and quantitative aCL by ELISA. Based on history and laboratory data, if an LA is suspected even with a normal aPTT, a more sensitive test should be performed: kaolin clotting time, Russell viper venom time or dilute aPTT. In summary, aPL can now be regarded as a distinct stroke risk factor but the mechanisms of the associated stroke are unknown. The choice of treatment remains uncertain.

References

Alarcon-Segovia, D. (1988). Pathogenetic potential of antiphospholipid antibodies. *Journal of Rheumatology*, **15**, 390–3.

Anonymous (1992). Antiphospholipid antibodies in Stroke Study Group: the association of anticardiolipin antibodies with first ischemic stroke: a multicenter case-control study. *Stroke*, **23**, 28.

Asherson, R. A., Kamashta, M. A., Gil, A., Vazquez, J.-J., Chan, O., Bagular, E. & Hughes, G. R. V. (1989). Cerebrovascular disease and antiphospholipid antibodies in systemic lupus erythematosus, lupus-like disease, and the primary antiphospholipid syndrome. *American Journal of Medicine*, **86**, 391–9.

Asherson, R. A. & Lubbe, W. F. (1988). Cerebral and valve lesions in SLE: association with antiphospholipid antibodies (Editorial). *Journal of Rheumatology*, **15**, 539–43.

Brey, R. L., Arroyo, R. & Boswell, R. N. (1991). Cerebrospinal fluid anticardiolipin antibodies in patients with HIV-I infection. *Journal of Acquired Immune Deficiency Syndromes*, **4**, 435–41.

Brey, R. L., Hart, R. G., Sherman, D. G. & Tegeler, C. H. (1990). Antiphospholipid antibodies and cerebral ischemia in young people. *Neurology*, **40**, 1190–6.

Briley, D. P., Coull, B. M. & Goodnight, S. H. (1989). Neurological disease associated with antiphospholipid antibodies. *Annals of Neurology*, **25**, 221–7.

Carreras, L. O. & Vermylen, J. G. (1982). 'Lupus' anticoagulant and thrombosis: possible role of inhibition of prostacyclin formation. *Thrombosis and Haemostasis*, **48**, 38–40.

Case records of the Massachusetts General Hospital (1988). Weekly clinicopathological exercises. Case 37-1988, eds. R. E. Scully, E. J. Mark, W. F. McNeely & B. U. McNeely. *New England Journal of Medicine*, **319**, 699–712.

Davis, W. D. & Brey, R. L. (1992). Complement activation in stroke associated with antiphospholipid antibodies (abstract). *Neurology*, in press.

Digre, K. B., Duncan, F. J., Branch, D. W., Jacobson, D. M., Varner, M. W. & Baringer, J. R. (1989). Amaurosis fugax associated with antiphospholipid antibodies. *Annals of Neurology*, **25**, 228–32.

Exner, T., Sahman, N. & Trudinger, B. (1988). Separation of anticardiolipin antibodies from lupus anticoagulant on a phospholipid-coated polystyrene column. *Biochemical and Biophysical Research Communications*, **155**, 1001–7.

Fields, R. A., Toubbeh, H., Searles, R. P. & Bankhurst, A. D. (1989). The prevalence of anticardiolipin antibodies in a healthy elderly population and its association with antinuclear antibodies. *Journal of Rheumatology*, **16**, 623–5.

Ford, P. M., Foro, S. E. & Lillicrap, D. P. (1988). Association of lupus anticoagulant with severe valvular heart disease in systemic lupus erythematosus. *Journal of Rheumatology*, **15**, 597–600.

Freyssinet, J. M., Wiesel, M. L., Gauchy, J., Boneu, B. & Cazenave, J. P. (1986). An IgM lupus anticoagulant that neutralizes the enhancing effect of phospholipid on purified endothelial thrombumodulin activity. A mechanism for thrombosis. *Thrombosis Haemostasis*, **55**, 309–13.

Galli, M., Comfurius, P., Maassen, C., Hemker, H. C., Be Baets, M. H., Van Breda-Vriesman, P. J. C., Barbui, T., Swaal, R. F. A., Bevers, E. M. (1990). Anticardiolipin antibodies directed not to cardiolipin but to a plasma protein cofactor. *Lancet*, **35**, 1544–7.

Harris, E. N., Gharavi, A. E., Asherson, R. A., Boey, M. L. & Hughes, G. R. V. (1984). Cerebral infarction is systemic lupus: association with anticardiolipin antibodies. *Clinical Experimental Rheumatology*, **2**, 47–51.

Harris, E. N., Gharavi, A. E. Wasley, G. D. & Hughes, G.R. (1988). The use of an enzyme-linked immunosorent assay and inhibition studies to distinguish between antibodies to cardiolipin from patients with syphilis or autoimmune disorders. *Journal of Infectious Diseases*, **157**, 23–31.

Hogan, M. J., Brunet, D. G., Ford, P. M. & Lillicrap, D. (1988). Lupus anticoagulant, antiphospholipid antibodies and migraine. *Canadian Journal of Neuroscience*, **15**, 420–5.

Hughes, G. N. (1986). Thrombosis, recurrent fetal loss and thrombocytopenia. Predictive value of the anticardiolipin antibody test. *Archives of Internal Medicine*, **146**, 2153–6.

Lafer, E. M., Rauch, J., Andrzejewski, C., Mudd, D., Furie, B., Schwartz, R. S. & Stollar, B. D. (1981). Polyspecific monoclonal lupus autoantibodies reactive with both polynucleotides and phospholipids. *Journal of Experimental Medicine*, **153**, 897–909.

Levine, S. R., Brey, R. L., Joseph, C. L. M. & Havstad, S. (1992). The risk of current thromboembolic events in patients with focal cerebral ischemia and antiphospholipid antibodies. *Stroke*, **23** (Suppl. 1), 129–32.

Levine, S. R., Crofts, J. W., Lesser, G. R., Floberg, J & Welch, K. M. A. (1988). Visual symptoms associated with the presence of a lupus anticoagulant. *Ophthalmology*, **95**, 686–92.

Levine, S. R., Deegan, M. J., Futrell, N. & Welch, K. M. A. (1990). Cerebrovascular and neurologic disease associated with antiphospholipid antibodies. 48 cases. *Neurology*, **40**, 1181–9.

Levine, S. R., Joseph, R., D'Andrea, G & Welch, K. M. A. (1987). Migraine and the lupus anticoagulant. Report of cases and review of the literature. *Cephalalgia*, **7**, 93–9.

Levine, S. R. & Welch, K. M. A. (1987*a*). Cerebrovascular ischemia associated with lupus anticoagulant. *Stroke*, **18**, 257–63.

Levine, S. R. & Welch, K. M. A. (1987*b*). The spectrum of neurological disease associated with antiphospholipid antibodies: Lupus anticoagulants and anticardiolipin antibodies. *Archives of Neurology*, **543**, 876–83.

Levy, R. A., Gharavi, A. E., Sammaritano, L. R., Habina, L., Qamar, T & Lockshin, M. D. (1990*a*). Characteristics of IgG antiphospholipid antibidles in patients with systemic lupus erythematosus and syphilis. *Journal of Rheumatology*, **17**, 1036–41.

Levy, R. A., Gharavi, A. E., Sammaritano, L. R., Rabina, L., Lockshin, M. D. (1990*b*). Fatty acid chain is a critical epitope for antiphospholipid antibody. *Journal of Clinical Immunology*, **10**, 141–5.

Lillicrap, D. P., Pinto, M., Benford, K., Ford, P. M. & Ford, S. (1990). Heterogeneity of laboratory test results for antiphospholipid antibodies in patients treated with chlorpromaine and other phenothiazines. *American Journal of Clinical Pathology*, **93**, 771–5.

Love, P. E. & Santoro, S. A. (1990). Antiphospholipid antibodies: anticardiolipin and the lupus anticoagulant in systemic lupus erythematosus (SLE) and in non-SLE disorders. *Annals of Internal Medicine*, **112**, 682–98.

Mackworth-Young, C. G., Harris, E. N., Steere, A. C., Rizvi, R., Malawista, S. E., Hughes, G. R. & Gharavi, A. E. (1988). Anticardiolipin antibodies in Lyme disease. *Arthritis and Rheumatism*, **31**, 1052–6.

Mackworth-Young, C. G., Loizou, S. & Walport, M. J. (1989). Primary antiphospholipid syndrome: features of patients with raised anticardiolipin antibodies and no other disorder. *Annals of Rheumatic Diseases*, **48**, 362–7.

Manoussakis, M. N., Tzioufas, A. G., Silis, M. P., Pange, P. J. E., Goudevenous, J. & Moutsopoulos, H. M. (1987). High prevalence of anticardiolipin and other autoantibodies in a healthy elderly population. *Clinical and Experimental Immunology*, **69**, 557–65.

Matsuura, E., Igarashi, Y., Fujimoto, M., Ichikawa, K. & Koike, T. (1990). Anticardiolipin co-factor(s) and differential diagnosis of autoimmune disease (letter). *Lancet*, **336**, 177–8.

McNeil, H. P., Simpson, R. J., Chesterman, C. V. & Krillis, S. A. (1990). Antiphospholipid antibodies are directed against a complex antigen that includes a lipid-binding inhibitor of coagulation: Beta-2-glycoprotein I (apoprotein H). *Proceedings of the National Academy of Science*, **87**, 4120–4.

Nelson, D. & Goodnight, S. H. (1988). Pathology of antiphospholipid antibodies and thrombosis. In *Thrombosis and Haemastasis Check Sample*, vol. 10, no TH88–6, Chicago: American Society of Clinical Pathologists.

Nimpf, J., Bevers, E. M., Bomans, P. H. H., Till, U., Wurm, H., Kostner, G. M. & Zwaal, R. F. A. (1986). Prothrombinase activity of human platelets is inhibited by beta-w-Glycoprotein I. *Biochimica et Biophysica Acta*, **884**, 142–9.

Pengo, V., Thiagarajan, P., Shapiro, S. S. & Heine, M. J. (1987). Immunological specificity and mechanisms of action of IgG lupus anticoagulants. *Blood*, **70**, 69–76.

Rauch, J. & Janoff, A. S. (1990). Phospholipid in the hexagonal II phase is immunogenic: evidence for immunorecognition of nonbilayer lipid phases in-vivo. *Proceedings of the National Academy of Science*, **87**, 4112–4.

Schorer, A. E., Wickham, N. R. & Watson, K. V. (1989). Lupus anticardiolipin induces a selective defect in thrombin-mediated endothelial prostacyclin release and platelet aggregation. *British Journal of Haematology*, **71**, 399–407.

Shuaib, A., Barklay, L., Lee, M. A. & Suchowersky, O. (1989). Migraine and antiphospholipid antibodies. *Headache*, **29**, 43–5.

Thiagarajan, P., Shapiro, S. S. & DeMarco, L. (1980). Monoclonal immunoglobulin MA coagulation inhibitor with phospholipid specificity. *Journal of Clinical Investigations*, **66**, 397–405.

Trimble, M., Ell, D. A., Brien, W. *et al.* (1990). The antiphospholipid syndrome: prevalence among patients with stroke and transient ischemic attacks. *American Journal of Medicine*, **40**, 1190–6.

Triplett, D. A., Brandt, J. T., Musgrave, K. A. & Orr, C. A. (1988). the relationship between lupus anticoagulant and antibodies to phospholipid. *Journal of the American Medical Association*, **259**, 550–4.

Vaarala, O., Palosuo, T., Kleemola, M. & Aho, K. (1986). Anticardiolipin response in acute infections. *Clinical Immunology and Immunopathology*, **41**, 491–8.

Part IV

Migraine and other headaches

16

Experimental headaches. A useful tool in the investigation of migraine mechanisms

J. OLESEN AND H. IVERSEN

Department of Neurology,
University of Copenhagen,
Gentofte Hospital
DK-2900 Hellerup, Denmark

Introduction

The immense value of animal experimental models for diabetes, epilepsy, Parkinson's disease and many other diseases needs no emphasis. For vascular headache there are, unfortunately, no animal models which have a documented relation to the clinical entities of migraine and cluster headache. There are, however, ways in which a vascular headache can be reliably induced in human beings. Histamine and nitroglycerine may do so, and this has been known for more than 50 years. The first systematic attempt to use such models in a pharmacological analysis of the origin of head pain was published as late as 1980 (Krabbe & Olesen, 1980).

Spontaneous attacks of migraine or cluster headache are difficult to study because they occur unpredictably, often at odd hours. When patients are admitted to a hospital, the attack frequency decreases dramatically and, if they get attacks at home, they are reluctant to go to hospital because physical activity aggravates their pain. They are also often reluctant to postpone necessary treatment because of severe pain. The mechanisms of the migraine attack furthermore, are, complex and it is often unclear if measured abnormalities are cause or effect of the attack. An experimental headache paradigm is not beset with the above mentioned problems. It allows induction of headache via a single chemical substance, it can be scheduled under controlled conditions and studied in many different ways so that its site of origin (intracranial or extracranial; arteriolar, arterial or venous) and pharmacological mechanisms may be determined.

An experimental model is only useful if relevance to the clinical condition can be shown. Better clinical definition of migraine has been achieved through the work of the International Headache Society (Headache Classification Committee, 1988). Changes in cerebral and extracranial blood flow and responses of cerebral and extracranial arteries during migraine attacks have

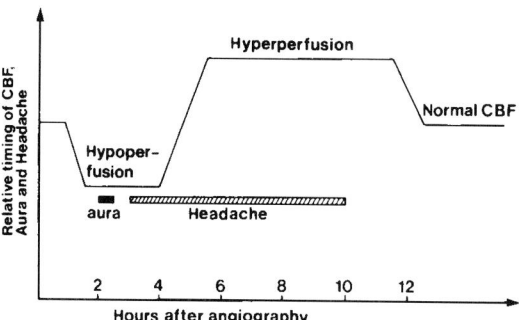

Fig. 16.1. Schematic drawing illustrating the temporal relation between angiography (time 1 hour), hypoperfusion, aura, headache, hyperperfusion, disappearance of headache, and disappearance of hyperperfusion. The time axis is chosen to illustrate what is typical. It does not depict actually recorded times in a single patient or in a group of patients, because the picture only becomes clear by synthesizing information from many patients. It is the sequence of events rather than the exact timing that is important. The angulation of the flow curve is to show that we do not know details about how fast flow changes. The real course of regional cerebral blood flow (CBF) is of course smooth. Olesen *et al.* (1990) reproduced with permission.

also been worked out (Jensen & Olesen, 1985; Iversen *et al.*, 1990; Olesen *et al.*, 1990; Friberg *et al.*, 1991). Distinct clinical characteristics and vascular changes during spontaneous attacks of migraine and of cluster headache can therefore now be compared with changes during experimental headaches to show the validity of the latter.

Recent advances in understanding vascular mechanisms of migraine

During migraine with aura (previously termed 'classical migraine') rCBF is decreased in the posterior part of one cerebral hemisphere. This area gradually enlarges, it is associated with decreased PCO_2 reactivity, abolished functional activation and normal autoregulation, and it is followed by delayed and long-lasting hyperemia (Olesen, Larsen & Lauritzen, 1981; Lauritzen *et al.*, 1983; Lauritzen & Olesen, 1984; Andersen *et al.*, 1988; Olesen *et al.*, 1990). The observed changes suggest cortical spreading depression of Leao (CSD) as the underlying mechanism (Olesen, 1991). A reasonable correlation between the site of origin of aura symptoms, rCBF abnormalities and headache has been established (Olesen *et al.*, 1990). This research shows that Wolff's vasospastic theory cannot be maintained for the following reasons: (i) spasm of major cerebral arteries is very rare, (ii) headache begins while rCBF is reduced, (iii) headache remains unchanged or improves late in the attack when hyperperfu-

Experimental headaches 255

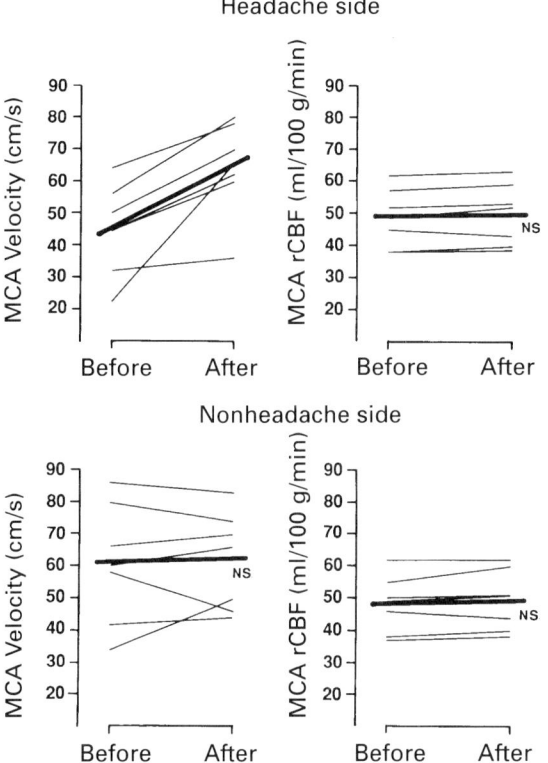

Fig. 16.2. Individual velocity in the middle cerebral artery and corresponding rCBF in the territory of the middle cerebral artery values during a migraine attack before treatment and after recovery from headache. Thick lines are mean values. Friberg *et al.* (1991) reproduced with permission.

sion occurs, and (iv) hyperperfusion may persist after the headache has disappeared (Fig. 16.1) (Olesen *et al.*, 1990). Dilatation of the middle cerebral and superficial temporal arteries occurs on the painful side as compared to the nonpainful side during migraine attacks (Fig. 16.2) (Iversen *et al.*, 1990; Friberg *et al.*, 1991). Aura and associated rCBF changes may occur without headache (Headache Classification Committee, 1988) and convexity pial arteries are relatively insensitive to pain (Ray & Wolff, 1940). Pain in migraine with aura is therefore unlikely to be caused by cortical vascular changes, but it is more likely to be related to the pathophysiological mechanisms which dilate the large basal cerebral arteries. In migraine without aura, rCBF changes are absent (Olesen *et al.*, 1981), but arterial dilatation occurs as in migraine with aura (Iversen *et al.*, 1990; Friberg *et al.*, 1991). Migraine without aura also has the same pain characteristics as migraine with aura, and it responds to the same drugs. Pain mechanisms in both forms of migraine are thus related to arterial dilatation. The role of the aura and associated rCBF changes is probably just to trigger changes in the cerebral and, possibly, meningeal arteries.

Animal experiments have shown that the trigeminal nerve supplies intra- and extracranial blood vessels (Mayberg, Zervas & Moskowitz, 1984). Cerebrovascular and extracranial nerve content of peptide transmitters and peptidergic as well as mono-aminergic receptors have been described in detail (for review see Olesen & Edvinsson, 1988). Neurogenic inflammation may be an important source of nociception (Markowitz, Saito & Moskowitz, 1987; Moskowitz, 1991). Stimulation of the trigeminal ganglion (which causes neurogenic inflammation) is associated with increased CGRP concentration in the cat superior sagittal sinus and in the human external jugular vein (Goadsby, Edvinsson & Ekman, 1988). Both this and the inflammation itself may be blocked pharmacologically using drugs active in the treatment of migraine attacks (Moskowitz, 1991). During migraine attacks CGRP increases in the external jugular blood on the headache side (Goadsby, Edvinsson & Ekman, 1990).

Histamine-induced headache

More than ten years ago we tested an experimental model of headache using continuous intravenous infusion of histamine, 0.16, 0.33 and 0.66 μg/kg/min (Krabbe & Olesen, 1980). The experimental plan was always the same. Subjects rested supine on a comfortable bed for 10 minutes or more after placement of an intravenous line. Blood pressure was measured repeatedly with a cuff and heart rate by palpation. Any signs of discomfort plus side-effects were noted. The headache response was rated with regard to duration, character, location and intensity on a scale from 0 to 3. Infusion was then given starting with the lowest dose of 0.16 μg/kg/min for ten minutes, then increasing to 0.33 μg/kg/min for ten minutes and finally 0.66 μg/kg/min for 10 minutes. We studied 48 subjects, 25 with common migraine, 10 with muscle-contraction headache and 13 headache-free controls. To 18 patients, who developed headache during the histamine infusion, mepyramine maleate (0.5 mg/kg) was administered intravenously during the last 2 minutes of the histamine infusion. This part of the study was unblinded. In a single blind, cross-over study, cimetidine was subsequently administered to 10 patients suffering from common migraine who had developed severe, throbbing headache during a former histamine infusion. An initial dose of 3.3 mg/kg was injected intravenously, followed by continuous infusion of 1.66 mg/kg for 45 minutes. Correspondingly, the placebo test was performed with an initial dose of 5 ml sodium chloride (9 mg/ml) followed by continuous infusion 0.5 ml/min for 45 minutes. After the

Table 16.1. *Headache characteristics, blood pressure and pulse rate during i.v. histamine*

Diagnosis	Dose (μg/kg/min)	Headache (number of patients)					BP	Pulse rate
		None	pressing	Pulsating				
				+	++	+++		
Control subjects N = 13	Baseline	13	0	0	0	0	130/80	66
	0.16	12	1	0	0	0	125/78	67
	0.33	10	3	0	0	0	125/75	75[b]
	0.66	9	4	0	0	0	120[a]/67[b]	87[b]
Muscle contraction headache N = 10	Baseline	4	6	0	0	0	137/84	79
	0.16	2	5	3	0	0	142/86	82
	0.33	1	4	5	0	0	133/78	86[a]
	0.66	1	4	4	1	0	133/73	91[b]
Common migraine N = 25	Baseline	16	9	0	0	0	131/81	76
	0.16	6	10	9	0	0	127[b]/79	86[a]
	0.33	2	2	11	8	2	122[b]/74[b]	89[b]
	0.66	1	0	2	9	13	120[b]/70[b]	97[b]

Notes: [a] $P < 0.05$.
[b] $P < 0.01$.

45 minutes the infusion of histamine chloride (0.66 μg/kg/min) was given for 10 minutes along with infusion of cimetidine or placebo.

The main results of the study are shown in Table 16.1. There was clearly a relationship between the dose of histamine and the incidence and intensity of headache. There was also a highly significant difference between the headache response of migraine sufferers and normal controls. Thus medium or severe pulsating headache developed in 22 of 25 migraineurs on the highest dose, but only in one of ten patients with tension-type headache, and in none of the 13 normal individuals. The lowest dose resulted in mild headache in 19 migraineurs, whereas none of the normal individuals developed headache on this dose. The infusion of mepyramine during the histamine infusion diminished or abolished the headache in 15 of 18 migraine patients within 1–2 minutes despite continuous infusion of histamine. However, four patients had a short-lasting initial worsening of the headache before it was relieved and in three patients the headache persisted unchanged. Pretreatment with cimetidine decreased the headache slightly, but significantly compared to placebo pretreatment ($p < 0.05$, Wilcoxon test).

This early study of experimental vascular headache was not up to present day methodological standards. Nevertheless, it resulted in several interesting findings: histamine infusion and headache were temporally closely related, mild headaches were pressing while greater headache severity was associated with pulsating pain, and control subjects were less sensitive to histamine than migraine sufferers. Subjects with tension-type headache responded in an intermediate way. It was also established that a steady-state headache could be maintained using a substance present in the normal human body and that analyzing receptor mechanisms in vivo was possible.

Experimental nitroglycerine-induced headache

Five years ago we decided to revisit the area of experimental vascular headaches. Nitroglycerin-induced headache (NTGH) was selected for the following reasons: (i) previous publications had shown NTGH to occur commonly, (ii) there was an indication that migraine sufferers were more sensitive than nonheadache sufferers, (iii) nitroglycerine (NTG) was then considered to have a purely vascular site of action, (iv) NTGH is an important side-effect of some drugs used used for the treatment of cardiovascular disease. NTG is the probable cause of hot-dog induced headache when eating cured meat and, (v) NTG may induce attacks of cluster headache and possibly also migraine-like attacks in migraineurs.

The first step was to validate a good experimental model (continuous intravenous infusion was selected to assure 100% bioavailability). The advantage of nitroglycerine is its short half-life in plasma; this means that steady-state is rapidly achieved by intravenous infusion, and the short duration of its effects after discontinuation of infusion. This allows repeated experiments before and after intervention. It is especially important that, if a subject develops intolerable headache, it will disappear within a few minutes after discontinuation of the infusion. Another important aspect is the possibility of maintaining a steady-state headache which is necessary in the further study of rCBF and arterial responses.

In ten human volunteers, not subject to migraine or frequent tension-type headache, NTG was infused intravenously for periods of 10 minutes separated by periods of saline infusion (wash-out) using a volume-directed pump. Doses of 0.25, 0.50, 1.0 and 2.0 µg/kg/min were given (Iversen, Olesen & Tfelt-Hansen, 1989c). A placebo period was randomly inserted in a double-blind fashion. Every 2 minutes the subjects scored headache on a scale from 0 to 10; 1 representing a very mild headache (including a feeling of pressing or throbbing), 5 a moderate headache, and 10 the worst possible headache. Subjects were asked about pain location, quality and aggravation by coughing, and to report any side-effects. The doctor scored the headache while a technician,

Experimental headaches 259

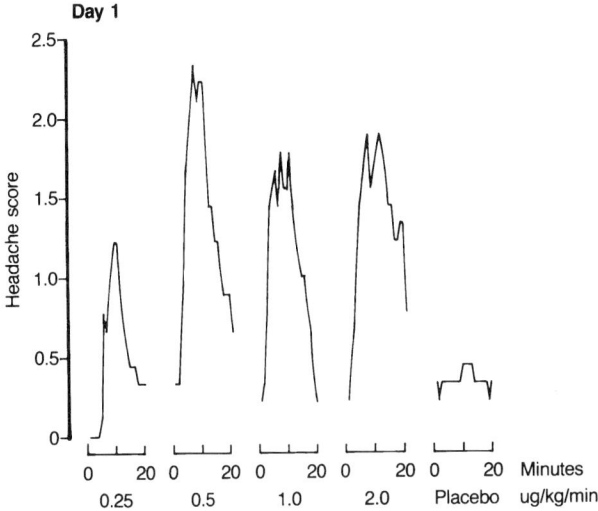

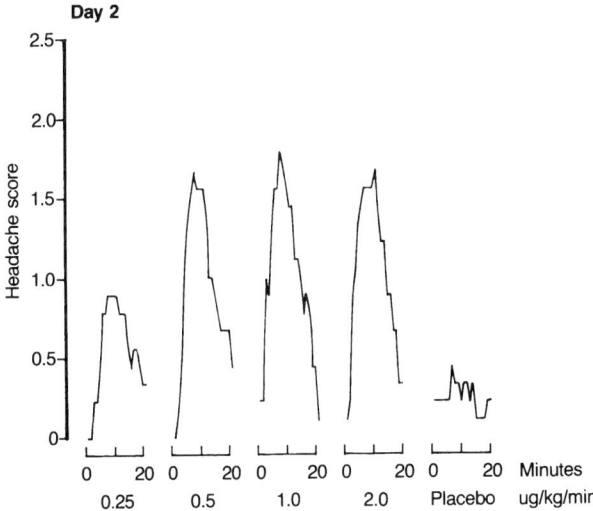

Fig. 16.3. Average headache scores, during the nitroglycerin infusions and the first 10 minutes of wash-out, day 1 and day 2 ($n=9$). The nitroglycerin infusions were discontinued at 10 minutes. Iversen *et al.* (1989c) reproduced with permission.

behind a screen, to achieve double-blinding, controlled the infusion pump and measured heart rate and blood pressure. After 1–8 weeks a retest was performed. On both occasions nine out of ten volunteers developed NTGH. The lowest dose, 0.25 µg/kg/min, caused headache in all nine responders on day 1 and in eight on day 2. Fig. 16.3(*a*), (*b*) show the average headache scores in the

NTG infusion periods and in the first 10 minutes of the wash-out on day 1 and day 2, respectively. Headache scores increased rapidly after NTG and decreased rapidly after its discontinuation. Maximal headache score during infusion was 5 out of 10. In more than half of the subjects NTGH disappeared completely within ten minutes after infusion. The dose of 0.25 µg/kg/min, NTG caused significantly less headache than the higher doses, but there was no significant difference in headache score between 0.5, 1.0 and 2.0 µg/kg/min. Thus, a ceiling effect occurs at 0.5 µg/kg/min. The headache was located bifrontally in five, biparietally in one and diffusely in two subjects on both days. In only one subject did the localization differ on the two days. Headache quality was pulsating in seven, pressing in two and did not vary between days. Headache was aggravated by coughing in five on day 1 and seven on day 2. Comparing maximal headache scores in the ten subjects during each of the five infusion periods on day 1 and day 2 (a total of 50 comparisons), 50% did not differ, 40% differed by one, 8% by two, and 2% differed by three in headache score. Three subjects had a family history of migraine without aura. They all developed headache 3–4 hours after NTG (delayed NTGH), two on both days and one only on one day. Our model thus had a rather low test/retest variability for both headache intensity and other headache characteristics. A feeling of warmth and blushing in the cheeks was most frequently reported at the highest doses and lasted only a few minutes. At doses 0.25 and 0.5 µg/kg/min such complaints were relatively infrequent. Nasal congestion and pain behind the eyes were quite frequent complaints using doses of 0.5 and upwards but not at the 0.25 µg/kg/min dose. No accompanying flushing (such as seen with histamine) was observed. Investigator blindness could thus be preserved.

Infusion periods of 15 minutes are necessary to achieve complete steady state and to leave time for examinations. Even longer infusions may be necessary for special purposes. Tolerance to NTG may, however, develop rapidly (Zimrin et al., 1985). In six healthy volunteers (four female, two male) we therefore evaluated the effect of 0.5 µg/kg/min of NTG or saline infused for 7 hours in a double blind fashion (Iversen et al., 1989b). The diameter of the radial artery was measured using high-frequency ultrasound. During NTG infusion the diameter increased rapidly and stabilized after 20 minutes. Between the period 30 and 420 minutes the median dilatation was 40% of baseline and did not vary significantly. All subjects were headache free before the study. Within 5 minutes of NTG infusion all subjects developed headache 5–30 minutes. After discontinuation of NTG, headache began to diminish after a median of 6 min (range 4–10 minutes). In summary, this study showed that headache and dilatation of the radial artery are stable for a prolonged period of time.

Experimental headaches 261

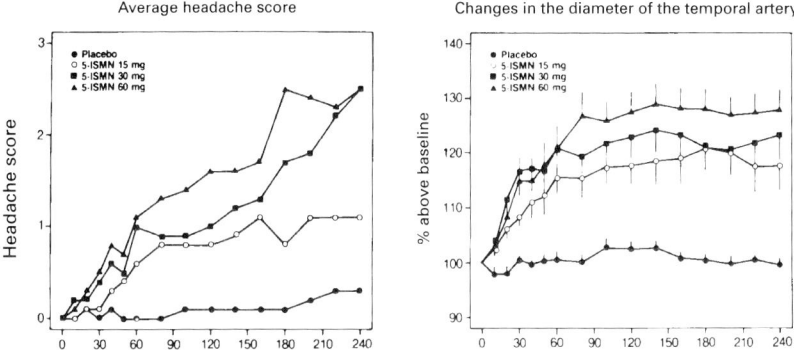

Fig. 16.4. Headache and diameter of the temporal artery during three doses of 5-isosorbide mononitrate and placebo (n=10). Iversen *et al.* (1992) reproduced with permission.

NTG is converted to nitric oxide in the body and the effects of the former are believed to be due to this metabolite. The synthetic compound, 5-isosorbide-mononitrate is used in treating angina patients. It gives rise to nitric oxide formation but its other metabolites are different from those of NTG. A dose response relationship between headache and arterial diameter was shown (Fig. 16.4) (Iversen *et al.*, 1992). This indicated that headache is indeed caused by the action of nitric oxide and not by other metabolites, that tolerance does not develop over the course of hours, and that dose relations between headache and arterial diameters can be shown in small numbers of patients.

Extracranial arterial responses are relatively easy to measure and interesting in relation to NTGH, but intracranial arterial responses during NTGH are probably more directly relevant to migraine (Iversen *et al.*, 1990; Friberg *et al.*, 1991). In a small methodological study, we measured rCBF by 133-xenon inhalation SPECT and, for the first time, introduced the simultaneous measurement of regional cerebral blood volume (rCBV) using Tc-labelled erythrocytes and a new, dynamic, brain-dedicated SPECT system (Tomomatic 232) (Iversen, Holm & Friberg, 1989a; Holm *et al.*, 1990). This equipment has dual-energy window facilities allowing simultaneous measurements of two different isotopes. In addition, transcranial doppler (TCD) was applied. The use of simultaneous TCD and rCBF allows an estimate of the diameter of the intracranial arteries, as previously discussed (Friberg *et al.*, 1991). Subjects were studied before and during NTG infusion 0.5 µg/kg/min. During infusion, headache rapidly increased to a plateau level and blood flow velocity decreased and rCBV increased while rCBF remained constant. After infusion rCBV normalized but blood velocity in the MCA remained low

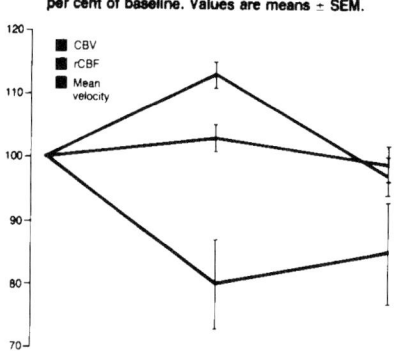

Fig. 16.5. Changes in cerebral blood volume and cerebral blood flow in the perfusion territory of the middle cerebral artery and velocity in the middle cerebral artery before, during and after nitroglycerin infusion.

(Fig. 16.5). This study was too small to allow correlation of the various experimental parameters and headache in a statistically acceptable way, but it demonstrated the usefulness of these measurement parameters and the ability to measure rCBV, rCBF and MCA velocity during NTG infusion.

A summary of our results with the intravenous nitroglycerin infusion model is given in Table 16.2.

Conclusions

We have established experimental models of headache using continuous intravenous infusion of histamine and nitroglycerine in normal human volunteers. These headaches are of low to moderate intensity and seldom cause real suffering in normal subjects. Headaches are generally much worse in migraineurs. With both models, steady-state is rapidly achieved and maintained for the duration of the infusion, after which it abates in a few minutes. In the histamine model, we have demonstrated that the H1 receptor is of prime importance for the development of headache, while the H2 receptor plays a minor role. In normal volunteers, there is a ceiling effect for nitroglycerine induced headache (NTGH) at a dose of 0.5 µg/kg/min. Characteristics of NTGH resemble migraine as defined by the International Headache Society, but NTGH is milder and virtually always bilateral. We have introduced measurement of the diameter of the superficial temporal and the radial arteries which closely parallel headache both at its onset and its termination. The combined measurements of rCBF using ^{133}Xe inhalation SPECT, cerebral blood volume (rCBV) by

Table 16.2. *Characteristics of Nitroglycerin induced headache (NTGH; 0.5 mg/kg/min) and Associated Vascular Responses*

Immediate NTGH	Immediate vascular responses
Onset: Immediate	Dilatation radial/temporal art.: 35–45%
Time to peak: 5–15 min	MCA velocity: decreased: 20%
Termination: <–20 min	rCBV: increased: 13%
Steady headache possible 3 h	rCBF: unchanged
Severity: mild to moderate	Onset: immediate
Location: bifrontotemporal	Time to peak: 5-15 min
Character: steady or pulsating,	Termination: 10-? min
Aggravated by physical activity	Steady responses possible ≈7 h
Nausea and vomiting: rare	
Photo- and phonophobia: rare	
Reproducibility: acceptable	
Delayed NTGH:	*Delayed vascular responses:*
Only in migraineurs?	Unknown
Occurs one to several hours after NTG	
Probably lasts for hours	
Character: unknown	

SPECT and transcranial doppler (TCD) measurement of arterial velocity, give an estimate of the diameter of the large cerebral arteries. In accordance with the findings of others (Dahl *et al.*, 1990), we have shown that NTG dilates the large arteries without appreciable effect on rCBF. However, rCBV is increased during NTGH. With the establishment of the nitroglycerine infusion model, and the measurement of vascular diameters, blood flow and blood volume, we now have the opportunity to determine the site of nociception in NTGH, its pharmacological mechanisms, and its relationship to migraine and other headaches.

References

Andersen, A.R., Friberg, L., Olsen, T.S. & Olesen, J. (1988). SPECT demonstration of delayed hyperemia following hypoperfusion in classic migraine. *Archives of Neurology*, **45**, 154–9.

Dahl, A., Russell, D., Nyberg-Hansen, R. & Rootwelt, K. (1990). Cluster headache: transcranial doppler ultrasound and rCBF studies. *Cephalalgia*, **10**, 87–94.

Friberg, L., Olesen, J., Iversen, H,K. & Sperling, B. (1991). Migraine pain associated with middle cerebral artery dilatation: reversal by sumatriptan. *Lancet*, **338**, 13–7

Goadsby, P.J., Edvinsson, L. & Ekman, R. (1988). Release of vasoactive peptides in the extracerebral circulation of humans and the cat during activation of the trigeminovascular system. *Annals of Neurology*, **23**, 193–6.

Goadsby, P.J., Edvinsson, L. & Ekman, R. (1990). Vasoactive peptide release in the extracerebral circulation of humans during migraine headache. *Annals of Neurology*, **28**, 183–7.

Headache Classification Committee of the International Headache Society (1988). Classification and diagnostic criteria for headache disorders, cranial neuralgias and facial pain. *Cephalalgia*, **8** (Suppl. 7), 1–96.

Holm, S., Friberg, L., Iversen, H.K. & Lassen, N.A. (1990). Dual energy brain SPECT - methods and applications. In *Nuclear Medicine. Quantitative Analysis in Imaging and Function*, eds. H.A.E. Schmidt & J. Chambion, pp. 11–3. Stuttgart – New York: Schattauer.

Iversen, H.K., Holm, S. & Friberg, L. (1989*a*). Intracranial hemodynamics during intravenous nitroglycerine infusion. *Cephalalgia*, **9** (Suppl. 10), 84–5.

Iversen, H.K., Nielsen, T.H., Garre, K., Tfelt-Hansen, P. & Olesen, J. (1992). Dose-dependent headache response and dilatation of extremity and extracranial arteries after three doses of 5-isosorbide mononitrate. *European Journal of Clinical Pharmacology*, **42**, 31–5.

Iversen, K.H., Nielsen, T.H., Olesen, J. & Tfelt-Hansen, P. (1990). Arterial responses during migraine headache. *Lancet*, **336**, 837–9.

Iversen, H.K., Nielsen, T.H., Tfelt-Hansen, P. & Olesen, J. (1989*b*). Headache and changes in the diameter of the radial artery during 7 hours intravenous nitroglycerine infusion. *Journal of Clinical Pharmacology*, **44**, 47–50.

Iversen, H.K., Olesen, J. & Tfelt-Hansen, P. (1989*c*). Intravenous nitroglycerin as an experimental model of vascular headache. Basic characteristics. *Pain*, **38**, 17–24.

Jensen, K. & Olesen, J. (1985). Temporal muscle blood flow in common migraine. *Acta Neurologica Scandinavica*, **72**, 561–70.

Krabbe, A.Æ. & Olesen, J. (1980). Headache provocation by continuous intravenous infusion of histamine. Clinical results and receptor mechanisms. *Pain*, **8**, 253–9.

Lauritzen, M. & Olesen, J. (1984). Regional cerebral blood flow during migraine attacks by Xenon– 133 inhalation and emission tomography. *Brain*, **107**, 447–61.

Lauritzen, M., Skyhøj Olsen, T., Lassen, N.A. & Paulson, O.B. (1983). The changes of regional cerebral blood flow during the course of classical migraine attacks. *Annals of Neurology*, **13**, 633–41.

Markowitz, S., Saito, K. & Moskowitz, M.A. (1987). Neurogenically mediated leakage of plasma protein occurs from blood vessels in dura mater but not brain. *Journal of Neuroscience*, **7**, 4129–36.

Mayberg, M.R., Zervas, N.T. & Moskowitz, M.A. (1984). Trigeminal projections to supratentorial pial and dural blood vessels in cats demonstrated by horseradish peroxidase histochemistry. *Journal of Comparative Neurology*, **223**, 46–56.

Moskowitz, M.A. (1991). Receptors on sensory fibers provide a locus for antimigraine drug action. In *Migraine and Other Headaches: The Vascular Mechanisms*, ed. J. Olesen, pp. 153–60. New York: Raven Press.

Olesen, J. (1991). Cerebral and extracranial circulatory disturbances in migraine: Pathophysiological implications. *Cerebrovascular and Brain Metabolism Reviews*, **1**, 1–28.

Olesen, J. & Edvinsson, L. eds. (1988). *Basic Mechanisms of Headache*. Amsterdam: Elsevier.

Olesen, J., Friberg, L., Olsen, T.S., Iversen, H.K., Lassen, N.A., Andersen, A.R. & Karle, A. (1990). Timing and topography of cerebral blood flow, aura and headache during migraine attacks. *Annals of Neurology*, **28**, 791–8.

Olesen, J., Larsen, B. & Lauritzen, M. (1981). Focal hyperemia followed by spreading oligemia and impaired activation of rCBF in classic migraine. *Annals of Neurology*, **9**, 344–52.

Olesen, J., Tfelt-Hansen, P., Henriksen, L. & Larsen, B. (1981). The common migraine attack may not be initiated by cerebral ischemia. *Lancet,* **ii**, 438–40.

Ray, B.S. & Wolff, H.G. (1940). Experimental studies on headache: pain sensitive structures of the head and their significance in headache. *Archives of Surgery,* **41**, 813–56.

Zimrin, D., Reichek, N., Bogin, K., Cameron, S., Douglas, P. & Fung, H.L. (1985). Antianginal effects of i. v. nitroglycerin. *Circulation,* **72** (Suppl. 3), 111–460.

17
Unilateral headache
M. ANTHONY

Institute of Neurological Sciences,
The Prince Henry and Prince of Wales Hospitals and
School of Medicine,
University of New South Wales,
Sydney, Australia

Unilateral headache, whether recurrent or persistent, poses diagnostic and therapeutic problems for the treating physician. For instance, the patient with constant and severe unilateral headache, not responding to conventional antimigraine treatment, is a problem in diagnosis. This has to be resolved quickly lest the patient become depressed, is referred to pain clinics for behavioural modification or ends up under a psychiatric regime, none of which specifically addresses the pain. On the other hand, the patient with cluster headache may be a therapeutic problem: he needs prompt and effective treatment to abort one of the most severe pains the human being is required to bear. Thus, it is necessary to know what types of headache are predominantly unilateral, to understand their pathophysiology and to know how to treat them. Before dealing with the types of unilateral headache, it is important to understand the pathophysiology of headache in general (see also Goadsby, this volume)

Pathophysiology of headache

Headache is a composite of changes due to stimulation of neural and vascular structures related to the head and neck, and mediated through various neurotransmitters.

Neural mechanisms

The main afferent pathways for headache are the trigeminal nerve, particularly the ophthalmic division, and the second and third cervical roots, which innervate the posterior fossa and the upper cervical region. The spinal tract and nucleus of the trigeminal nerve descend to the second cervical segment of the spinal cord (Taren & Kahn, 1962) where pain afferents from the second cervical root and trigeminal system converge. This pain centre in the upper segments of the spinal cord facilitates the referral of pain from the frontotemporal area to the neck, and vice versa, and is commonly referred to as the 'cervicotrigeminal relay' (Lance *et al.*, 1983). This relay is modulated by the endogenous pain control system, consisting of the median raphe nuclei and the locus coeruleus. The predominant neurotransmitter is 5-hydroxytryptamine (5-HT), except in the locus coeruleus which releases noradrenaline (Buda *et al.*, 1975).

Vascular changes

First-order neurones carrying pain impulses from the superior sagittal sinus, middle meningeal artery, middle cerebral artery, and superficial temporal artery synapse at the trigeminal nucleus caudalis and the dorsolateral area of the upper cervical cord (Goadsby, Zagami & Lambert, 1989). In migraine and cluster headache, there is cranial dilatation involving the extra- and intracranial circulations. Afferent impulses from these vessels are conveyed by the trigeminal nerve and also independently by the component nuclei of the endogenous pain control system (Lance *et al.*, 1983). These impulses converge at the second and third cervical segments of the spinal cord. Second-order neurones from this area project to the ventrobasal area of the contralateral thalamus (Zagami, Lambert & Lance, 1989). Synapses in these pain pathways are modulated by 5-HT and noradrenergic pathways (Lance *et al.*, 1983).

Biochemical changes

The most significant biochemical changes in migraine are a fall in plasma 5-HT and its increased urinary excretion during an attack (Anthony, Hinterberger & Lance, 1969). In tension-type headache, platelet 5-HT is significantly lower than in migrainous subjects or normal controls (Anthony & Lance, 1989). It is assumed that similar changes in 5-HT levels occur in cerebral neurones, particularly those of the endogenous pain control system, reducing its efficiency in suppressing pain.

In cluster headache, blood levels of histamine rise significantly during each attack and this probably contributes to the sudden and significant vasodilatation observed during an attack. Suppression of this rise with steroids prevents recurrent episodes of headache (Anthony & Daher, 1990).

Varieties of unilateral headache

The varieties of unilateral headache are few and include: migraine, 'occipital neuralgia', cervicogenic headache, cluster headache, chronic paroxysmal hemicrania, and hemicrania continua.

Migraine

The term developed from the word "hemicrania", meaning literally one half of the head, and the component of pain or headache is implied. The most recent definition of common migraine (migraine without aura) by the Classification Committee of the International Headache Society (1988) is:

...'Idiopathic, recurring headache disorder, manifesting in attacks lasting 4–72 hours. Typical characteristics of headache are unilateral location, pulsating quality, moderate or severe intensity, aggravation by routine physical activity, and association with nausea, photo- and phonophobia.'

Classical migraine (migraine with aura) is defined as: "Idiopathic recurring disorder manifesting with attacks of neurological symptoms, unequivocally localizable to the cerebral cortex or brain stem, usually gradually developing over 5–20 min, and usually lasting less than 60 min. Headaches, nausea and/or photophobia usually follow neurological aura symptoms directly or after a free interval of less than an hour. The headache usually lasts 4–72 hours, but may be completely absent."

The unilateral nature of migraine

The pain of migraine is not always confined to one-half of the head. It is unilateral in about two-thirds of patients (Lance, 1982; Selby, 1983), alternating from side to side in the majority. A significant minority of patients experience hemicranial headaches always on the same side of the head.

In classical migraine, the headache tends to be more frequently unilateral, usually opposite to the side of neurological symptoms and more commonly on the same side of the head than in common migraine (Sjaastad & Saunte, 1983). Cerebral blood flow differs during attacks of headache in the two varieties of migraine. In classical migraine, regional cerebral blood flow (rCBF) is decreased during the aura and the headache phase, the reduction beginning in the occipital area and proceeding in an occipitoparietal direction independently of vascular territories supplied by the large cerebral arteries. Further, the spreading oligaemia was calculated to travel across the visual cortex at the rate of 2–3 mm/min, that is at the same rate as the spreading depression of Leão, and this phenomenon was similar in experimental animals and humans (Lauritzen *et al.*, 1982; Olesen, 1985). This physiological process develops in the cerebral cortex in a contiguous fashion. The reduction of rCBF could be due to either reduced cortical metabolism or primary vasoconstriction. If the latter, the vasoconstriction would need to be localised to the cortical arterioles rather than the major arteries, because the oligemia is not confined to the territories supplied by such vessels. In common migraine there were no changes in rCBF during headache when compared to the headache-free period in the 12 patients studied by Olesen (l985).

No satisfactory explanation exists as to why the majority of migrainous attacks are unilateral. Paired and unpaired nuclei exist in the brainstem and upper cervical cord, all part of the endogenous pain control system which modulates pain from the craniocervical area. How neural discharges from such structures end up as a unilateral headache is not understood. The increase of tyrosine hydroxylase in one locus coeruleus in the rat following destruction of the contralateral locus (Buda *et al.*, 1975), may be no more than a compensatory mechanism of the brainstem noradrenergic system and cannot be taken as proof of reciprocal innervation between the two sides of the brainstem. Rather, the hemicranial location of migraine attacks may represent unilateral irritation of upper cervical sensory roots and activation of the corresponding cervical cord segments, the nucleus caudalis and dorsolateral area on one side, due to prolonged maintenance of abnormal or unusual neck postures during sleep. This could well lead to activation of the cervicotrigeminal relay and a unilateral headache. This assumption is based on the clinical observation that patients with 'occipital neuralgia' experience hemicranial headaches, always on the same side with signs of irritation of the occipital nerve (tenderness and sensory changes in its area of distribution) (Anthony, 1989). This assumption leads to the next type of unilateral headache.

Occipital neuralgia *(Chronic unilateral headache with occipital nerve irritation)*

In only a minority of patients does the character of migrainous headaches remain unchanged throughout a patient's life. Commonly, the headaches change in frequency, duration and severity and sometimes in the quality of the pain. The character of the attacks may so change as to have little resemblance to those of earlier years. There is, however, a particular group of patients in whom, insidiously over several months, the pain becomes dull, aching and nonpulsatile in quality, is more severe than previously, and spreads to the occipital area. The new attack may be situated in the occipital, frontotemporal and ocular regions, even spreading into the same side of the neck, and may have few or no accompaniments. Attacks are usually more frequent than in the past, often almost daily, and always occur on the same side of the head. The only migrainous feature that remains is the hemicranial nature of the headache. Preventive antimigrainous therapy is of little value and neither oral analgesics

Table 17.1. *Clinical features of 500 cases of idiopathic headache*

Headache type	Total number	Female/ Male	Unilateral headache	Bilateral headache	Mean	Headache frequency/ month
Migraine	285	216/69	157	128	33.0	4.9
Migraine + GON irritation	98	73/25	98	-	43.6	10.8
Occipital neuralgia	86	65/21	86	-	48.2	18.9
Chronic tension headache	31	16/15	-	31	35.9	24.6
Total	500					
Cluster headache	28	4/24	28	-	44.1	49.0

nor ergotamine-containing preparations are as effective as previously. Such headaches may also begin spontaneously, without a previous history of migraine.

The question that arises is: does this change in the character of the headache represent the natural course in the evolution of migraine, or does it represent the operation of a trigger factor, which so far had played a minor role in initiating headache attacks? If that is so, what is the nature of such a trigger factor?

It is quite possible that the new factor operating in such cases is irritation of neural structures in the upper part of the cervical spine, the role of which has been underestimated in the causation of headache (see Bogduk, this volume). It has long been accepted that cervical spondylosis can cause pain referred to the orbit and frontal regions, and experimental stimulation of various structures of the neck in man can produce pain in the anterior parts of the head (Cyriax, 1938; Feinstein *et al.*, 1954). This projection of pain has been explained on the basis of an overlap of central connections between the spinal nucleus of the trigeminal nerve and the upper cervical nerves, in the 'cervicotrigeminal relay'.

The main sensory nerve in the upper neck and posterior scalp area is the greater occipital nerve (GON), and a study was designed to assess the incidence of headaches triggered or caused by irritation of that structure (Anthony, 1989). It comprised 500 new patients with 'idiopathic headache' classified as migraine, migraine with GON irritation (MON), occipital neuralgia (ON), and tension-type headache. The clinical profile of the patients is given in Table 17.1. The diagnostic criteria for 'migraine with GON irritation' (MON) and 'occipital neuralgia' (ON), are listed in Table 17.2. If the patients with tension-type headache are excluded, 469 remained with a diagnosis of migraine, and of

Table 17.2. *Diagnostic Criteria*

Migraine with GON irritation (MON)
1. History of established migraine.
2. Recent increase in frequency/severity of headache with occipital radiation/origin of pain
3. Headache always or almost always on the same side of the head
4. Tenderness/reduced pain threshold over GON on the affected side
5. Absence of sensory changes in the area of distribution of the GON on that side

Occipital neuralgia (ON)
1. Unilateral occipital headache, continuous or paroxysmal (neuralgic), always on the same side
2. Circumscribed tenderness over the GON as it crosses the superior nuchal line
3. Hypo- or hyper-algesia or dyesthesiae in the area of distribution of the GON
4. Relief of acute attacks by infiltration of the GON with local anesthetic

Table 17.3. *Idiopathic headache – evidence of GON dysfunction in patients with MON and ON*

	Migraine + GON irritation	'Occipital neuralgia'	Total
Number	98	86	184
Previous history of migraine	All	44 (51.2%)	
Sensory changes	None	All	
– Hypoalgesia	-	33	
– Hyperalgesia	-	21	
– Dysesthesiae	-	32	
Pain threshold			
– mean (Kg)			
HA side	2.3	1.8	
HAF side	4.1	3.9	
Statistical significance	$p = 0.001$	$p = 0.001$	

Notes: HA = headache HAF = headache-free

these 86 (17.2%), were cases of ON. They all suffered from frequent (mostly daily) unilateral headaches and all showed evidence of irritation of the GON.

In all patients with MON and ON, the exit of the GON onto the scalp was located with a nerve stimulator as it crossed the superior nuchal line, at about the midpoint between the occipital protuberance and the mastoid tip, and the pain threshold of the nerve at that point was measured with a spring-loaded pressure algometer. The GON on each side was examined twice within 5 min and both readings were used for statistical comparison. Results are shown in Table 17.3.

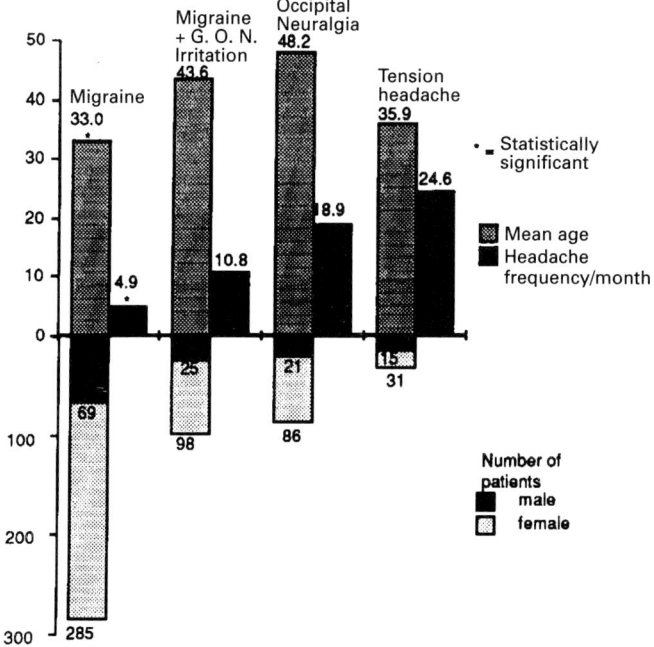

Fig. 17.1. A summary of data from 500 patients with 'idiopathic headache' (see text).

Appreciation of pain in the distribution of the GON on the scalp (medial two-thirds of each occipital area) was assessed by the patient in response to pin prick and compared to the anterior part of the vertex on each side. Responses were classified as no sensory change, hypoalgesia, hyperalgesia, and dysesthesiae (if the pin prick caused a sensation of crawling ants, unpleasant pins and needles, or sharp, shooting jabs of pain).

Apart from the group with tension-type headache, in which the sexes are equally represented, females predominate in the other three groups. Patients with pure migraine were younger, with a lower frequency of headaches per month than those with MON and ON (Fig. 17.1). This suggests that with advancing years, cervical osteoarthritis particularly of the apophysial and atlanto-axial joints, disc degeneration and muscle spasm are common at middle age (Edmeads, 1988), the latter contributed to by similar changes at lower levels of the cervical spine (Sluijter & Koetsveldt-Baart, 1980). Since these joints fall within the receptive field of the cervicotrigeminal relay, pain can also be referred to the trigeminal territory producing occipito-frontal or hemicranial headache of the type described earlier.

Table 17.4. *'Occipital neuralgia' Comparison of results of local block and occipital neurectomy*

	Number of patients	Female/ male	Sensory changes GON/LON	Headache frequency	Response Number	Range	Mean
GON/LON block (Local injection Depomedrol – 160 mg)	180	114/66	ALL	3/week- daily	169	10–77 days	23.5 days
Occipital neurectomy (GON/LON)	60	28/15	ALL	daily	42	1–28 months	8.5 months

Previous studies have demonstrated that attacks of unilateral headache can be arrested by injection of local anaesthetic into the region of the GON, and recurrent attacks can be prevented by injection of methylprednisolone acetate (Depomedrol) or surgical division of the nerve (Anthony, 1987). This was attributed to blocking the GON by the local anesthetic, demyelination of the nerve caused by Depomedrol (Selby, 1983), or to interruption of the nerve trunk. The assumed mechanism in each case was a reduction of input into the cervicotrigeminal relay. Table 17.4 summarizes the results of both chemical (injection of 120–160 mg Depomedrol into the ipsilateral GON) (personal observations), and surgical occipital neurectomies (Anthony & Blum, 1991). Chemical neurectomy was performed in patients with MON or ON, whilst surgical neurectomy only in patients with intractible ON. Injection of Depomedrol had a success rate of 84%, producing a mean period of headache-freedom of 23.5 days. An advantage of the procedure is that it can be repeated as required. Occipital neurectomy had a success rate of 70%. However, the mean period of relief of 8.5 months suggests that the majority of patients experienced benefit for more than a few months. Recurrences of headache were attributed to re-growth of the proximal cut end of the nerve and eventual reconstitution of the continuity of the nerve trunk, as shown by the disappearance of the occipital numbness, hypoalgesia or hyperalgesia.

Thus the question arises as to whether one can continue to refer to all cases of persistently unilateral headache with otherwise migrainous features as migraine, or whether tenderness, reduced pain threshold and sensory changes in the distribution of the ipsilateral GON should suggest a diagnosis of occipital neuralgia. An accurate diagnostic decision is of great relevance, given that such cases do not usually respond to antimigrainous therapy but respond to measures that reduce irritation of the GON, e.g. nonsteroidal antiinflammatory drugs (NSAIDs), systemic corticosteroids, local corticosteroid injections, physiotherapy to the neck with graduated traction (as suggested by Graft-Radford, Reeves & Jaeger, 1987), or even occipital neurectomy. It is almost certain that equivocal results in clinical trials on the effectiveness of antimigrainous drugs in the past were due to contamination of the trial population with patients suffering from irritation of the GON and not from typical migraine headaches.

Cervicogenic Headaches

This type of headache was initially described by Sjaastad *et al.* (1983), and the diagnostic criteria were redefined again recently (Sjaastad, Fredriksen & Pfaffenrath, 1990). The headache profile includes:

1. Unilateral headache, without sideshift
2. Infrequent but long-lasting attacks
3. Nonclustering of pain episodes
4. Provocation of attacks by:
    a) neck movements or sustained abnormal neck postures, and
    b) external pressure over the upper neck or occipital region ipsilaterally.
5. Vague neck, shoulder and arm pain on the side of headache
6. Anesthetic block of the greater occipital nerve or the C2 root abolishes the pain
7. Previous history of head or neck trauma

The number of cases so far reported has been rather small, compared to the huge number of patients with headaches allegedly arising from the neck, i.e. 'occipital neuralgia', common migraine, posttraumatic headaches or cervical migraine (Bartschi-Rochaix, 1968). Whether such headaches constitute a separate entity, or are part of the spectrum described here as 'occipital neuralgia' (chronic unilateral headache with evidence of irritation of the GON), will be decided by further clinical investigation.

Cluster Headache

The etiology and pathogenesis of cluster headaches (CH) or 'periodic migrainous neuralgia' continue to elude us. The condition is often misdiagnosed, leav-

Table 17.5. *Clinical Profile of an Attack of Cluster Headache*

Frequency	Majority: 1–3 daily. Minority: up to 8 daily
Duration	Majority: 15 min–2 hours. Minority: 6–9 hours
Site	- Always unilateral on same side. Rarely may change sides from bout to bout, and exceptionally during the same bout
	- Eyes, forehead, temple, cheek, upper gum, nostril, neck, ear - in that order of frequency
Quality	- Severe and constant. Rarely pulsating Boring, burning, piercing, screwing, tearing
Onset	- Majority nocturnal (62%) - evening and during sleep
	- Circadian regularity in 87% (particular time of day or night)
Precipitating factors (during bout only)	
	- Alcohol - most common
	- Glyceryl trinitrate, 1 mg sublingually, will always precipitate typical attack, usually within 30–90 min. For best result give at least 4 hours after previous attack.
	- Other vasodilating drugs for the treatment of ischemic heart and peripheral vascular disease
	- Small doses of histamine
Relieving factors	- Pressure on superficial temporal artery and on eye
	- Application of heat or cold to scalp
Associated factors (most common)	
	Lacrimation, mostly unilateral 87%; conjunctival injection 45%; partial Horner's syndrome 32%; rhinorrhea/nasal stuffiness 70%; nausea 43%; vomiting (rarely regularly) 25%; flushing of face 20%; tender temporal artery 17%; and tender scalp 17%

Source: From Anthony 1984; Kudrow 1990.

ing the patient to suffer daily attacks of severe headaches for months or years. The vast majority of patients are male. The clinical profiles of an attack of cluster headache, and a bout of cluster headache are shown in Tables 17.5 and 17.6.

Types of cluster headache

Episodic cluster headache

This is the commonest variety. During a bout or cluster, headaches occur with almost monotonous regularity, usually 1–3 times per day. Bouts tend to recur once or twice annually, each lasting 4–8 weeks (range 2 weeks to 6 months). Between bouts the patient is completely symptom free.

Table 17.6. *Clinical Profile of a Bout of Cluster Headache*

Annual frequency	Majority:	1–2
	Minority:	4 to one every 2 years
Duration	Majority:	2–12 weeks
	Minority:	10 days–6 months
Precipitating factors	Cessation of heavy drinking	
	Stress	
	Change in work environment	
	Change in shift work	
	Change in sleep routine	
	Long vacation	
	Prolonged rage, anger, frustration	
Seasonal periodicity	Not proven	
Type	Episodic:	90%
	Chronic:	10%
Headache-free periods	6 weeks–2 years	

Chronic cluster headache

This term refers to cases in which headaches recur daily or almost daily, without remissions, for more than 12 months. About 20% of patients suffer from this variety of the disorder (Kudrow, 1990). Some cases are chronic from the beginning (about 10%), whilst others evolve to that state having previously suffered periodic attacks.

Chronic paroxysmal hemicrania

The condition is rare. It resembles chronic CH, in that there are no remissions, though it is seen predominantly in women. It was originally described by Sjaastad & Dale (1976) and by 1987 about 80 cases had been identified worldwide (Sjaastad, 1987). In the most recent study of CPH, mean frequency of attacks was 13.6 per 24 hours, with a range of 4–38. Mean duration of attacks was 13.5 min range 3–46 min (Russell, 1984). Its interest lies in the fact that it responds promptly and completely to indomethacin therapy, to which cluster headaches do not.

Atypical variants of cluster headache

Cluster-tic

This is an infrequent association of typical trigeminal neuralgia occurring on the same side as CH attacks. Whilst the attack of CH can start before, during or after attacks of trigeminal neuralgia, both conditions appear to retain their own characteristics. The trigeminal territory involved is mostly the second division, attacks are precipitated by the usual triggers (Watson & Evans, 1985) and the condition is very resistant to treatment (Medina, 1990).

Cluster-migraine

Two types of this condition have been distinguished clinically. Cluster-migraine, in which patients experience typical cluster attacks during migraine, and migraine-cluster, in which attacks of migraine occur in a cluster pattern. Seven such cases have been reported by Medina and Diamond (1977), linking the two disorders.

Cluster-vertigo

Attacks of vertigo associated with attacks of CH have been reported by Gilbert (1965), but in 900 cases of CH patients some had vertigo due to labyrinthitis or Ménière's syndrome, none had vertigo temporally related to attacks or bouts of CH (Kudrow, 1987).

Pathophysiology

Vascular mechanisms

Dilatation of the internal carotid artery is thought to be the main mechanism of pain, because the pain is felt mostly in the orbital and frontotemporal areas and the lumen of the artery is narrowed at angiography during an attack of headache (Ekbom & Greitz, 1970), a change considered to be due to oedema of the arterial wall. Further, the ophthalmic artery (a branch of the internal carotid) dilates, as reflected by the increased intraocular pressure (Horven, Nornes & Sjaastad, 1972). Increased flow velocity in the middle cerebral artery and reduced reactivity to CO_2 of the anterior cerebral artery on the headache side, confirm the disturbed control of the intracranial circulation (Gawel et al., 1990).

Extracranial blood flow increased by 250% during cluster headache (Sakai & Meyer, 1978). Conjunctival injection and nasal blockage also suggest capillary and arteriolar dilatation on the headache side during attacks. Segments of superficial temporal artery removed from patients with chronic CH dilated more in response to histamine than those from patients with episodic CH or from non-headache subjects (Hardebo, Krabbe & Gjerris, 1980).

Finally, the importance of vascular dilatation in CH is demonstrated by the ease with which vasodilator agents can induce headache attacks (see Table 17.4).

Neurogenic mechanisms

Activation of the cervico-trigeminal relay probably occurs during attacks of CH, because blocking the GON with local anaesthetic will arrest an attack of headache within a few minutes (Anthony, 1987). However, what actually triggers the attack is of greater importance. It has been suggested that, in its complete form, an attack of CH is characterized by pain in the first and second divisions of the trigeminal nerve, sympathetic dysfunction (partial Horner's Syndrome), sympathetic activation (sweating of face and forehead), and parasympathetic activation (lacrimation, rhinorrhea and nasal congestion). The pathophysiological focus for the production of this constellation of symptoms and signs could be the superior pericarotid cavernous sinus plexus, where nerve fibres from all the above neural structures converge and join together (Moskowitz, 1988). The above hypothesis draws support from the clinical observation that aneurysms of the proximal intradural carotid artery segment may produce cluster-like symptoms (Greve & Mai, 1988). In addition, Gallium SPECT scanning (single photon emission computerized tomography) has shown in the region of the cavernous sinus, a lesion which fades as the patient moves out of the cluster bout, and which was interpreted as being a localized area of sterile inflammation. This may explain the prompt response of CH to steroid therapy (Gawel *et al.*, 1990). The question posed by the above authors is whether CH is a periodic inflammatory disorder similar to the Tolosa–Hunt syndrome, affecting the nerves of the cavernous sinus rather than those of the superior orbital fissures. These hypotheses appear complementary and could be close to the truth about the mechanism of CH.

Biochemical mechanisms

Elevation of whole blood histamine during attacks of CH occurs in about 90% of patients (Anthony, Lance & Lord, 1978). The fact that histamine administration during the bout can induce an attack of CH (Horton, 1961) and, in larger amounts, cause generalized headaches when given intraveniously to non-headache volunteers (Pickering, 1933) suggests that histamine release plays a significant role in the causation of the individual attacks, the most likely mechanism of action being intracranial vasodilatation. Steroids are highly effective in arresting recurrent attacks of CH (Kudrow, 1990). They do so by suppressing the rise of blood histamine during attacks (Anthony & Daher, 1990), possibly by inhibiting histidine decarboxylase, which converts histidine to histamine, and by preventing histamine release from mast cells and basophils. This is suggested by mast cell degranulation demonstrated by Dimitriadou *et al.* (1990). Further, histamine causes greater dilatation of the superficial temporal arteries in patients with chronic CH, than in normal controls (Hardebo, Krabbe & Gjerris, 1980).

Hemicrania continua (HC)

This very rare type of headache was first described by Sjaastad & Spierings (1984) and reviewed again by Bordini *et al.* (1991). It is a constant unilateral headache, of moderate severity, with few associated symptoms, no side-shifts and complete response to indomethacin. Until the end of 1990, only 18 cases had been recognised world wide, 15 of whom were female. The site of pain is mostly in the frontotemporal, ocular and retro-auricular areas and the age of onset is almost equally distributed among the third, fourth and fifth decades. Relevant investigations are negative and other forms of treatment, usually antimigrainous drugs, are ineffective.

Some questions remain unanswered in the above description of HC. Could not the condition be another form of presentation of 'cervical headache', as described under the heading of 'occipital neuralgia', in which headaches can be continuous and persistently unilateral without side-shifts? Further, what is the meaning of the response of the headache to indomethacin? Which property of the drug is likely to be responsible for its beneficial effect and why is it not effective in other varieties of unilateral headache, such as cluster and cervicogenic headaches, though it is effective in CPH? These questions remain unanswered and raise doubts about whether this type of headache represents a distinct clinical entity.

Post-traumatic headaches

In some patients who have sustained trauma to the neck in the form of whiplash injury, recurrent or continuous unilateral headaches develop without side-shifts, and these headaches are indistinguishable from occipital neuralgia. Whereas flexion-extension are the commonest movements occurring during whiplash, hyperextension is the most extensive and possibly the most damaging, responsible for a variety of cervical injuries, e.g. strain or ruptures of the paraspinal muscles or ligaments, avulsion of cartilagenous plates or intervertebral discs, and injury to zygo-apophysial joints. Underlying cervical osteoarthritis would aggravate the situation (Appenzeller, 1987). Such headaches are associated with tenderness and sensory changes in the distribution of the ipsilateral GON and are relieved by the administration of systemic steroids or local injection into the ipsilateral GON, and for longer periods by occipital neurectomy (personal observations).

In the majority of cases, the severity of the injury bears no direct relationship to the severity of the headache. It may well be that such patients are cases of 'traumatic' occipital neuralgia and that the mechanisms of the two conditions are similar.

Conclusions

Apart from migraine, persistently unilateral headaches are not common, but frequently present a dilemma in both diagnosis and treatment. It is important that they are diagnosed correctly and treated effectively. This refers particularly to cluster headache and chronic paroxysmal hemicrania.

This chapter describes a new subclass of migraine-like headaches, which are persistently unilateral and are thought to be due to irritation of neural structures in the upper neck and/or of the occipital nerve(s). For convenience, they have been called 'occipital neuralgia', until their final mechanism is established. Their clinical relevance is that they do not respond to measures and treatments commonly used in migraine, but do respond to systemic steroids, NSAIDs, large doses of ergotamine, neck physiotherapy, local steroid injections into the region of the ipsilateral GON and even occipital neurectomy.

Such headaches comprise about one-third of cases previously diagnosed as migraine. Whether the pain in such cases is arising from dysfunction of the cervical zygoapophysial joints as suggested by Bogduk (this volume) or is due to direct damage to the occipital nerve(s) and its branches, remains to be resolved.

Acknowledgements

This work was supported by the Arthur Yenebis Foundation and The Prince Henry Hospital, Centenary Research Fund. The secretarial assistance of Mrs. Ann Traynor is gratefully appreciated.

References

Anthony, M. (1984). Cluster headache: understanding the problem. *Patient Management*, **8**, 49–60.

Anthony, M. (1987). The role of the occipital nerve in unilateral headaches. In *Advances in Headache Research*, ed. F. Clifford Rose, pp. 257–62. London: John Libbey.

Anthony, M. (1989). Unilateral migraine or occipital neuralgia? In *Advances in Headache Research*, ed. F. Clifford Rose, pp. 30–44. London: Smith-Gordon.

Anthony, M. & Blum, P.W. (1991). Occipital neurectomy for occipital neuralgia. *Australian and New Zealand of Medicine*, **21** (Suppl. 9), 602.

Anthony, M. & Daher, N. B. (1990). Mechanism of action of steroids in cluster headache. *Proceedings, 8th Migraine Trust International Symposium*, London, p. 18.

Anthony, M., Hinterberger, H. & Lance, J. W. (1969). The possible relationship of serotonin to the migraine syndrome. In *Research and Clinical Studies in Headache*, vol. 2, ed. A. P. Friedman, pp. 29–59. Basel: Karger.

Anthony, M. & Lance, J. W. (1989). Plasma serotonin in patients with chronic tension headache. *Journal of Neurology, Neurosurgery, and Psychiatry*, **52**, 182–4.

Anthony, M., Lance, J. W. & Lord, G. D. R. (1978). Migrainous Neuralgia - Blood histamine levels and clinical response to Hl and H2 receptor blockade. In *Current Concepts in Migraine Research*, ed. R. Greene, pp. 149–51. New York: Raven Press.

Appenzeller, O. (1987). Post-traumatic headaches. In *Wolffe's Headache and Other Head Pain*, ed. D. J. Dalessio, pp. 289–303. New York: Oxford University Press.

Bartschi-Rochaiz, N. (1968). Headache of cervical origin. In *Handbook of Clinical Neurology, vol. 5, Headache*, eds. P. J. Vinken & G. W. Bruyn, pp. 192–203. Amsterdam: North-Holland.

Bordini, C., Antonaci, F., Stovner, L. J., Schrader, H. & Sjaastad, O. (1991). 'Hemicrania continua'. A clinical review. *Headache*, **31**, 20–6.

Buda, M., Roussel, B., Renaud, B. & Pujol, J. F. (1975). Increase in tyrosine hydroxylase activity in the locus coeruleus of the rat brain after contralateral lesioning. *Brain Research*, **93**, 564–9.

Classification Committee of the International Headache Society (1988). Classification and diagnostic criteria for headache disorders, cranial neuralgias and facial pain. *Cephalalgia*, **8** (Suppl. 7), 9–96.

Cyriax, J. (1938). Rheumatic headache. *British Medical Journal*, **ii**, 1367–8.

Dimitriadou, V., Henry, P., Brochet, B., Matteran, P. & Aubrineau, P. (1990). Cluster headache: ultra-structural evidence for mast cell degranulation and interaction with nerve fibres in the human temporal artery. *Cephalalgia*, **10**, 221–8.

Edmeads, J. (1988). The cervical spine and headache. *Neurology*, **38**, 1874–78.

Ekbom, K. & Greitz, T. (1970). Carotid angiography in cluster headache. *Acta Radiologica*, **10**, 177–86.

Feinstein, B., Langton, J. N. J. K., Jameson, R. M. & Schiller, F. (1954). Experiments of pain referral from deep tissues. *Journal of Bone and Joint Surgery*, **36A**, 981–97.

Gawel, M. J., Krajewski, A., Luo, Y. M. & Ichise, M. (1990). The cluster diathesis. *Headache*, **30**, 652–5.

Gilbert, G. J. (1965). Ménière's syndrome and cluster headache. Recurrent paroxysmal vasodilatation. *Journal of the American Medical Association*, **191**, 691–4.

Goadsby, P. J., Zagami, A. S. & Lambert, G. A. (1989). Metabolic activity and blood flow in the caudal medulla and upper spinal cord of the cat during stimulation of the superior sagittal sinus. *Cephalalgia*, **9** (Suppl. 10), 14.

Graft-Radford, S. B., Reeves, J. L. & Jaeger, B. (1987). Management of chronic head and neck pain: effectiveness of altering factors perpetuating myofascial pain. *Headache*, **27**, 186–90.

Greve, E. & Mai, J. (1988). Cluster headache-like syndrome: a symptomatic feature? *Cephalalgia*, **8**, 79–82.

Hardebo, J. E., Krabbe, A. A. & Gjerris, F. (1980). Enhanced dilatory response to histamine in large extracranial vessels in chronic cluster headache. *Headache*, **20**, 316–20.

Horton, B. T. (1961). Histamine cephalgia (Horton's headache or syndrome). *Maryland State Medical Journal*, **10**, 178–203.

Horven, I., Nornes, H. & Sjaastad, O. (1972). Different corneal indentation pulse pattern in cluster headache and migraine. *Neurology*, **22**, 92–8.

Kudrow L (1987). Cluster headaches. In *Migraine: Clinical, Therapeutic, Conceptual and Research Aspects,* ed. J. N. Blau, pp. 113-33. London: Chapman & Hall.

Kudrow, L. (1990). Management of cluster headache: An American view. *Headache Quarterly*, **1**, 57–63.

Lance, J. W. (1982). *Mechanism and Management of Headache*. London: Butterworths.

Lance, J. W., Lambert, G. A., Goadsby, P. J. & Duckworth, J. W. (1983). Brainstem influences on the cephalic circulation: experimental data from the cat and monkey of relevance to the mechanism of migraine. *Headache*, **23**, 258–65.

Lauritzen, M., Gjedde, A., & Hanlen, A. J. (1982). Regional cerebral blood flow in classic migraine: a possible relationship to the spreading depression of Leão? In *Advances in Migraine Research and Therapy*, ed. F. Clifford Rose, pp. 117-20. New York: Raven Press.

Medina, J. L. (1990). Cluster headache disorders. The cluster headache variants. *Headache Quarterly*, **1**, 162–9.

Medina, J. L. & Diamond, S. (1977). The clinical link between migraine and cluster headache. *Archives of Neurology*, **34**, 470–2.

Moskowitz, M. A. (1988). Cluster headache: evidence for a pathophysiological focus in the superior pericarotid cavernous sinus plexus. *Headache*, **28**, 585–6.

Olesen, J. (1985). Migraine and regional blood flow. *Trends in Neurosciences*, **8**, 318–21.

Pickering, G. W. (1933). Observations on the mechanism of headache produced by histamine. *Clinical Science*, **1**, 77–101.

Russell, D. (1984). Chronic paroxysmal hemicrania: severity, duration and time of occurrence of attacks. *Cephalalgia*, **4**, 53–6.

Sakai, F. & Meyer, J. S. (1978). Regional cerebral haemodynamics during migraine and cluster headaches measured by the 133-Xe inhalation method. *Headache*, **18**, 122–32.

Selby, G. (1983). *Migraine and its Variants*, pp. 33–4. Sydney: Adis Press.

Sjaastad, O. (1987). Chronic paroxysmal hemicrania: clinical aspects and controversies. In Migraine: *Clinical, Therapeutic, Conceptual and Research Aspects*, ed. J.N. Blau, pp. 135–52. London: Chapman & Hall.

Sjaastad, O. & Dale, I. (1976). A new(?) clinical headache entity; 'chronic paroxysmal hemicrania'. *Acta Neurologica Scandinavica*, **54**, 140–59.

Sjaastad, O., Fredriksen, T. A. & Pfaffenrath, V. (1990). Cervicogenic headache: Diagnostic criteria. *Headache*, **30**, 725–6.

Sjaastad, O. & Saunte, C. (1983). Unilaterality of headache. Hauges' studies revisited. *Cephalalgia*, **3**, 201–5.

Sjaastad, O., Saunte, C., Hovdahl, H., Breivik, H., & Cronback, E. (1983). Cervicogenic headache: an hypothesis. *Cephalalgia*, **3**, 249–56.

Sjaastad, O. & Spierings, E. L. H. (1984). 'Hemicrania continua': another headache absolutely responsive to indomethacin. *Cephalalgia*, **4**, 65–70.

Sluijter, M. E. & Koetsveldt-Baart, C. C. (1980). Interruption of pain pathways in the treatment of cervical syndrome. *Anaesthesia*, **35**, 302–7.

Taren, J. A. & Kahn, E. A. (1962). Anatomic pathways related to pain in face and neck. *Journal of Neurosurgery*, **19**, 116–21.

Watson, P. & Evans, P. (1985). Cluster-tic syndrome. *Headache*, **25**, 123–6.

Zagami, A. S., Lambert, G. A. & Lance, J. W. (1989). Capsaicin applied to cranial vessels in the cat excites thalamic neurones. *Cephalalgia*, **9** (Suppl. 10), 296–7.

18
Pathways for headache

GEOFFREY A. LAMBERT

Institute of Neurological Sciences,
The Prince Henry and Prince of Wales Hospitals and
School of Medicine,
University of New South Wales,
Sydney, Australia

Pain sensitive structures important in headache

The principal systems involved in cranial sensory processing are the trigeminal nerve and upper cervical roots, although some sensory fibres occur in nearly all cranial nerves. There are few areas of the head, at least on the skin of the head, where overlap between the trigeminal and cervical innervations cannot be demonstrated. For example, Denny-Brown and Yanagisawa (1973) showed that subconvulsive doses of strychnine shrank the area of facial analgesia produced by sectioning the trigeminal root, an effect which they attributed to expansion of receptive fields of the occipital nerves and to contributions from the seventh and tenth cranial nerves. Likewise, the area of sensory loss following C2-4 rhizotomies constricted with strychnine.

Because of the poor somatotopic arrangements for registering deep head pain a definitive localization of what is hurting is hardly ever possible. The notion of just which intracranial structures are most important rests upon early anecdotal observations during neurosurgery and some recent studies.

Intracranial structures

Penfield (1935) and Ray and Wolff (1940) stimulated the dura and dural blood vessels of conscious patients, who reported that this procedure produced pain of an aching or throbbing nature. Wirth and van Buren (1971) analysed the pattern of convergence between cranial vessels and facial receptive fields. It seems that the overlap of both peripheral and dural innervation pathways, together with central convergence mechanisms, may be responsible for some of the referral patterns seen, such as headaches of cervical origin presenting as frontal pain. These studies showed that the the most pain-sensitive structures in the cranium were associated with the blood vessels of the dura. Vessels elsewhere were less sensitive, as was the dura remote from large vessels. Among the most sensitive vessels were the sagittal sinus, the middle meningeal artery, the proximal portion of the middle and anterior cerebral arteries and the vertebral artery. Pain could be produced from the large dural vessels through mechanical displacement or electrical stimulation and could be mimicked by arterial injections of vasodilator substances such as histamine. Although few, if any, of these studies were carried out in migraineurs, it is clear that the subjects experienced pain similar to that experienced by migraineurs.

Extracranial structures

Extracranial arteries and their branches also seem to play an important part in vascular headache (Ray & Wolff, 1940; Wolff, 1963). Wolff allowed his own superficial temporal artery to be exposed and manipulated, producing pain referred to the upper teeth, the temple and deep behind and in the eye. Injection of histamine into this artery produced a short-lasting deep throbbing pain in the same areas. Electrical or mechanical stimulation of most extracranial arteries produces headache-like pain, but similar treatment of extracranial veins does not (Ray & Wolff, 1940). Wolff claimed that compression of the external carotid artery relieved migraine pain in a large proportion of sufferers (Graham & Wolff, 1937). However, only about one-third of migraineurs appear to suffer pain which can be relieved this way (Drummond & Lance, 1983).

Peripheral pathways

The pain-sensitive structures described above have rich plexuses of nerves derived from many sources. McNaughton (1938) concluded that there was 'abundant anatomical evidence for a nerve supply to the dural, pial and intracerebral vessels'. It is not easy to distinguish the proportion of afferent and efferent fibres carried in these nerves. Gross anatomical studies suggest that virtually all cranial nerves send fibre bundles (albeit often small and inconsistent ones) to cerebral vessels. Trigeminal rhizotomy is generally successful in relieving head pain of 'peripheral' origin, but not always (White & Sweet, 1955). Ray and Wolff (1940) reported that pain from stimulation of the dura in the lateral wall of the posterior fossa disappeared after section of the ninth and tenth cranial nerves in one patient. Gardner, Sowell and Dutlinger (1947), sectioned the greater superficial petrosal nerve in migraine patients, partly on the basis that both sensory and vasodilator signals travelled in it and claimed some success for the operation. Earlier, superior cervical sympathectomy was touted as a treatment for migraine (Dandy, 1931), but the experience of others failed to support the early optimism. The rationale for this treatment rested on the assumption that sympathetic dennervation also represented section of afferent pathways (White & Sweet, 1955), but the association of sensory and sympathetic fibres in the head may be considerably more complicated than this.

Trigeminal innervation

The trigeminal and cervical nerves provide the greater bulk of the vascular sensory innervation. Those parts of the dura sensitive in normal patients were, in general, insensitive in patients with section of the trigeminal roots, particularly section of the first division (Penfield & McNaughton, 1940). This was followed by the use of trigeminal rhizotomy or of alcohol injections into the trigeminal ganglion as a treatment for migraine (White & Sweet, 1955).

Anatomical studies showed that, although all three divisions of the trigeminal contribute to the sensory innervation of the dura, the ophthalmic (first) division seems to be most important. It gives rise to the tentorial nerve which innervates the area in and around the superior sagittal sinus. The maxillary (second) division gives rise to the nervous spinosus, which sends a branch to accompany the middle meningeal artery, innervating it and the surrounding dura as far as the superior sagittal sinus. The mandibular (third) division of the trigeminal contributes little to the sensory innervation of the dura, mainly on the lateral aspects of the middle cranial (Steiger & Meakin, 1984) and posterior cranial fossa (Keller, Beduk & Saunders, 1985).

There are significant complexities of dural innervation. Branches of the internal carotid artery, for instance, are surrounded by nerve bundles containing populations of parasympathetic, sympathetic and sensory fibres (first and second division trigeminal), with mingling of the two latter populations having occurred separately in both the trigeminal ganglion and the cavernous sinus (Matthews & Robinson, 1980; Ruskell & Simons, 1987; Ruskell, 1988; Simons & Ruskell, 1988). This accounts for the complex structure of the tentorial nerve as revealed by microscopy, with the presence of myelinated and unmyelinated fibres in the nerve and in the walls of vessels (Andres *et al.*, 1987; Dahl & Nelson, 1964; Sato & Suzuki, 1975).

Occipital innervation

The first three cervical segments innervate both superficial and deep structures of the posterior third of the head, principally through the greater and lesser occipital nerves, and the greater auricular nerve. The dura of the posterior fossa receives its innervation mainly from the upper three cervical roots although many other possible sources have been suggested (see Kimmel, 1961). The earliest reliable reports of this came from Siwe (1931) and Feindel, Penfield and McNaughton (1960). Kimmel (1961) demonstrated that the nerves supplying the dura mater of the lateral and posterior walls of the posterior fossa were branches of the first and second cervical roots mixed with fibres from the superior cervical ganglia. These mixed nerves accompanied the vagal and hypoglossal nerves and entered the fossa through the jugular foramen and the hypoglossal canal.

Fibre composition

The dura mater of the cat is richly supplied by myelinated A-fibres and unmyelinated C-fibres (Andres *et al.*, 1987). In the supratentorial dura of the cat, there are about a thousand nerve fibres, with small unmyelinated fibres outnumbering the myelinated ones by about 3 to 1. The perineural sheath forms a tube-like net, especially dense along the course of the superior sagittal sinus. Some of the A-fibres had terminations resembling stretch receptors around the confluence of the venous sinuses, leading the authors to speculate that these terminations are responsible for the pain from stretching of the sinuses. The majority of A- and C-fibres form unencapsulated or 'free' terminals which could subserve nociception.

Modern techniques have added to knowledge of craniovascular innervation. Staining of the dura and dural arterial walls with specific immunochemicals or ligands has shown that the fibres seen after gross dissection or on microscopy contain a number of neurotransmitters, or are associated with specific binding sites. Among the neurochemicals and neurotransmitters with possible sensory significance are substance P (SP) (Hanko *et al.*, 1986; Matsuyama *et al.*, 1986; Liu-Chen *et al.*, 1984), calcitonin gene-related peptide (CGRP) (Hanko *et al.*, 1986; Nozaki, Kikuchi & Mizumo, 1989) and cholecystokinin-8 (CCK8) (Liu-Chen, Norregaard & Moskowitz, 1985).

Origin of fibres

Mayberg *et al.* (1981) demonstrated that application of horseradish peroxide (HRP) to the proximal part of the middle cerebral artery of the cat labelled cell bodies in the ophthalmic division of the trigeminal ganglion, exactly the same area as labelled by the application of HRP to the supraorbital nerve. Using similar techniques, Steiger and colleagues (1982) found that the anterior fossa (particularly its dorsal aspect) and the tentorium received projections mainly from the ophthalmic division. Labelled cells associated with the orbital roof were found mainly in the dorsal and intermediate layers of the second division of the ganglion, while the middle fossa was represented in the more dorsal layers of the third division. These neurones labelled by HRP had a mean cell diameter of 20 to 35 µm, somewhat smaller than unstained neurones in the ganglion (35–44 µm) and consistent with the idea that input from the dura was mainly nociceptive.

Mayberg, Zervas and Moskowitz (1984) examined not only the trigeminal ganglia, but most other ganglia that could conceivably send projections to the dura and its vessels: geniculate (VII), superior glossopharyngeal (IX), superior vagal (X), nodose (X) and superior cervical ganglia. They also applied HRP to a wider range of perivascular dura, including that of the middle cerebral artery, the superior sagittal sinus and the middle meningeal artery. Application of HRP to the two nonmidline sites produced generalized labelling in the ipsilateral superior cervical ganglion (SCG), confirming the earlier anatomical work on sympathetic innervation. Both SCGs were labelled after application of HRP to the sagittal sinus. Labelling of other ganglia was not commented upon extensively, but seems to have been minor or absent. The ophthalmic division of the trigeminal ganglion was heavily labelled after application to all three vascular sites, with minor labelling in the second and third divisions after HRP application to the middle meningeal artery only. In accord with the clinical observations that only the proximal portions of the cerebral arteries are pain sensitive, HRP applications only to those portions produced significant labelling in the ganglion. Steiger and Meakin (1984) reported on the distribution of sensory afferents of all three trigeminal divisions to the floor of the cranial fossa, showing that they displayed a similar somatotopic organisation to that seen in cutaneous receptive fields. Similar results were reported in 1985 by Tsai *et al.* (1985).

Keller, Beduk & Saunders (1985) showed that the basilar artery receives sensory nerves from the first division of the trigeminal and the superior cervical ganglia. A few fibres originating in the inferior ganglion of the vagus nerve were seen, but none arising from 7th or 9th nerve or from the dorsal root ganglia of the upper cervical cord. Most of the dura of the posterior fossa appeared to receive fibres from the C1-3 segments of the spinal cord and the superior cervical ganglion. The lateral convexity and the rostro-dorsal aspect of the dura also received fibres from the third and first trigeminal divisions.

O'Connor and van der Kooy (1986), in an elegant double labelling study, showed that trigeminal ganglion cells in the ophthalmic division send collaterals to various intracranial arteries, e.g. middle cerebral and middle meningeal arteries, and to sites on the dura not immediately overlying a vessel. Such ganglion cells do not themselves receive input from cutaneous receptive fields, but are closely associated with ophthalmic division cells that do. They conjectured

Pathways for headache 289

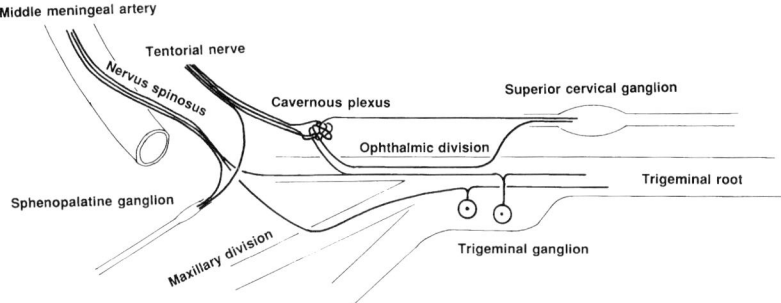

Fig. 18.1. Diagrammatic representation of the contributions to the nervus spinosus and tenorial nerve made by the sympathetic, parasympathetic and sensory systems. Fibres in the walls of the middle meningeal artery arise largely from the nervus spinosus, the sensory component of which arises mostly in the maxillary division of the trigeminal. The tentorial nerve, which supplies much of the dura and the superior sagittal sinus derives its trigeminal component from the ophthalmic division.

that the observed convergence and the close association with cutaneous input might partly underly the referral of pain seen with dural stimulation. Arbab, Wiklund and Svengaard (1986) used both anterograde and retrograde transport of HRP to demonstrate that the vertebral and basilar arteries receive sensory projections from the first and second cervical dorsal root ganglia. The latter also contributes some innervation to the intracranial part of the internal carotid artery. Both trigeminal and C2 dorsal root ganglion cells which receive input from the rat middle cerebral artery contain CGRP and a small proportion contain substance P (Edvinsson, Hara & Uddman, 1989).

There have been few anatomical or tracing studies of the afferent innervation of extracranial arteries or the paths and targets of these nerves (Borges & Moskowitz, 1983). We have examined this question using electrophysiological techniques. Electrical stimulation of the superficial temporal or supraorbital arteries activates neurones in the ipsilateral trigeminal ganglion (Lambert *et al*., 1979), with latencies consistent with A-delta and C-fibre conduction velocities.

A summary of the origin of fibres in the nervus spinosus and the tentorial nerve is shown in Fig. 18.1.

First-level processing

Where do the trigeminal and cervical nerves which convey pain from the dura terminate in the brain and spinal cord? Much data exist about the terminations of trigeminal pathways but, there is little anatomical evidence identifying central terminations of arterial and dural afferents. This is largely due to the technical difficulties of persuading tracers such as HRP to reveal themselves in these areas after the long transganglionic journey from the periphery.

Trigeminal and cervical dorsal horns

The first level of processing of input from the trigeminal system takes place in the trigeminal sensory nuclear complex, extending as far caudally as the C6 level of the spinal cord (Darian-Smith, 1973; Shigenaga *et al.* 1986). The complex has three distinct components: a principal sensory nucleus located in the pons, a mesencephalic nucleus and the nucleus of the descending spinal trigeminal tract, referred to below simply as the spinal trigeminal nucleus. The spinal trigeminal nucleus is further subdivided, in a rostrocaudal axis, into three subnuclei: oralis, interpolaris and caudalis (Olszewski, 1950). Electrophysiological and behavioural evidence indicates that trigeminal nociception is handled at all levels of the trigeminal complex (Broton & Rosenfeld, 1986; Hayashi, Sumino & Sessle, 1984; Wall & Taub, 1962), although the predominant role in facial nociception is ascribed to nucleus caudalis. The first level of processing for the cervical nerves takes place in the C1-C6 region of the cord.

Processing of cervical input takes place in the dorsal horn of the cervical cord (see Bonica, 1990). Nociceptive fibres terminate mainly in laminae I and V, the cells of which send projections to the contralateral spinal cord and other parts of the dorsal horn. Some dorsal root fibres from the cervical cord ascend to terminate in the trigeminal nucleus. As early as 1948 Escolar suggested that afferents from the C2 and C3 sensory roots overlapped with those of the trigeminal nerve, a suggestion later confirmed (Kerr, 1961; Kerr & Olafson, 1961). This overlap occurs at all cervical levels between C1 and at least C6 (Dejerine, 1914; Darian-Smith, Mutton & Procter, 1965; Kerr, 1972; Shigenaga *et al.*, 1986; Pfaller & Ardvinsson, 1988).

One nucleus which attracted our attention for its possible role in processing craniovascular information was the lateral cervical nucleus (LCN). This is a relay nucleus for the spino-cervico-thalamic tract and most studies have emphasized its role in relaying tactile information from the limbs and trunk (Willis, 1985). However, LCN neurones also respond to noxious stimuli (Kajander & Giesler, 1987) and receive cranio-facial input (Wall & Taub, 1962), including noxious facial input (Craig & Tapper, 1978). It is unclear whether such input reaches the LCN directly, or via a relay in higher structures such as the trigeminal nucleus, although the latter seems the more likely (Craig, 1978; Pfaller & Ardvinsson, 1988; Wall & Taub, 1962).

Pathways for headache 291

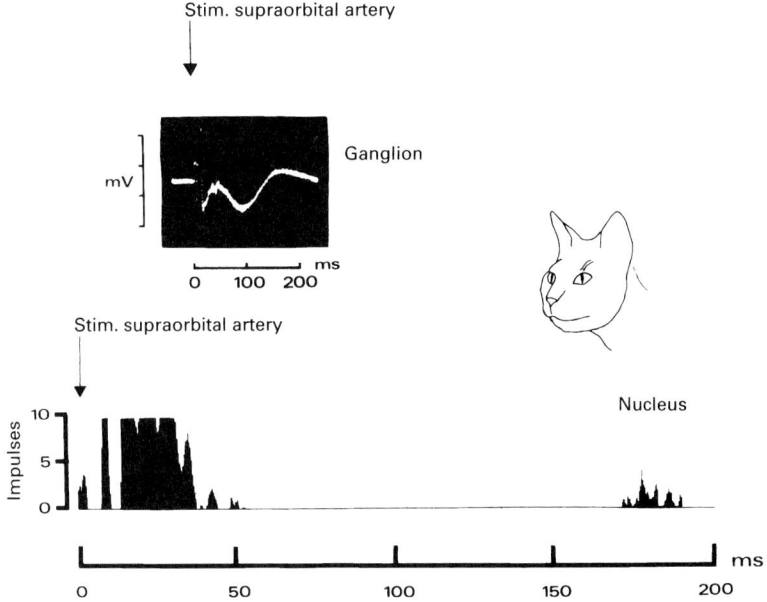

Fig. 18.2. Field potential recording made in the trigeminal ganglion and single unit recording made in the trigeminal nucleus caudalis showing both A-δ and C-fibre responses to electrical stimulation of the supraorbital artery. Based on data from Lambert *et al.* (1979).

Terminations of dural afferents

Arbab *et al.* (1988) reported that HRP placed on the proximal portion of the middle cerebral artery produced a low density of stained neuronal terminal profiles in the main trigeminal sensory nucleus and the pars oralis of the spinal trigeminal nucleus. Intense labelling was seen in the pars interpolaris and a small amount of labelling in the C2 dorsal horn, which they considered to be of trigeminal origin. In addition, labelling was seen in the nucleus tractus solitarius, dorsal raphe nucleus and other regions of the dorsal periaqueductal grey matter, and the dorsal motor nucleus of the vagus. Nearly all of the innervation (even that in the dorsal horn) was dependent upon the integrity of trigeminal pathways because this labelling failed to appear following a unilateral trigeminal rhizotomy prior to HRP application. An association of the innervation with peptide-containing sensory neurones was shown as labelling was much reduced in animals pretreated with the substance P depletor, capsaicin. Experiments on the basilar artery produced terminal labelling in the C2 dorsal horn of the spinal cord and in the cuneate nucleus, with no labelling at all in the trigeminal nuclei. Resection of the C2 and C3 dorsal roots prevented

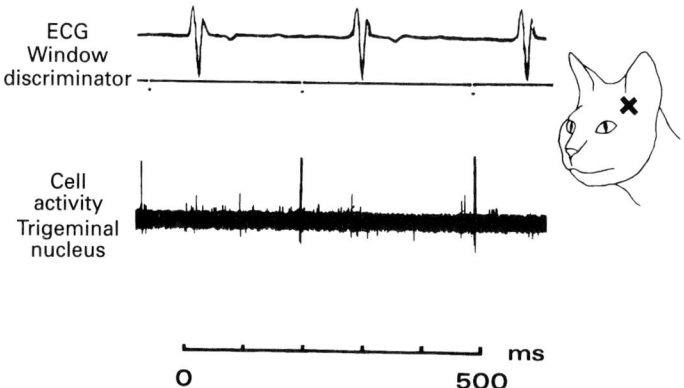

Fig. 18.3. Single-unit recording in the trigeminal nucleus caudalis showing a unit with discharges linked to the cardiac cycle. The unit had a receptive field in an area of the face supplied by the supraorbital nerve. Based on data from Lambert et al. (1979).

labelling in these areas, except for a small amount contralateral to the application site.

Much information on central terminations of dural afferents derived from electrophysiological experiments (Lambert et al., 1979). Stimulation of extracranial arteries such as the superficial temporal and supraorbital activates neurones in the nucleus caudalis. Cells in the nucleus respond to stimulation of the supraorbital artery at both A-delta and C-fibre latencies (Fig. 18.2). In addition, these cells discharge to an intracarotid injection of bradykinin. Some show rhythmic activity synchronized with the heart beat (Fig. 18.3). This synchronization is blocked reversibly by occlusion of the carotid artery (Fig. 18.4).

Neurones in the nucleus caudalis also respond to electrical stimulation of the middle cerebral artery, superior sagittal sinus and surrounding dura (Davis & Dostrovsky, 1986; Strassman et al., 1986). In cats, cells in the rostral nucleus caudalis or the caudal nucleus interpolaris respond to electrical stimulation of the dura overlying the middle meningeal artery, or of the artery itself (but not the dura remote from vessels) with discharges at a latency of about 10–15 ms. These cells receive convergent input from the face and can be activated by electrical stimulation of the infraorbital nerve but rarely respond to low-threshold mechanical stimulation.

The finding of Davis & Dostrovksy (1986) of a number of responsive neurones in the nucleus interpolaris and even in the nucleus oralis was surprising given the accepted idea that neurones responsible for nociceptor processing cluster in the caudalis. As many as 70% of interpolaris and oralis neurones

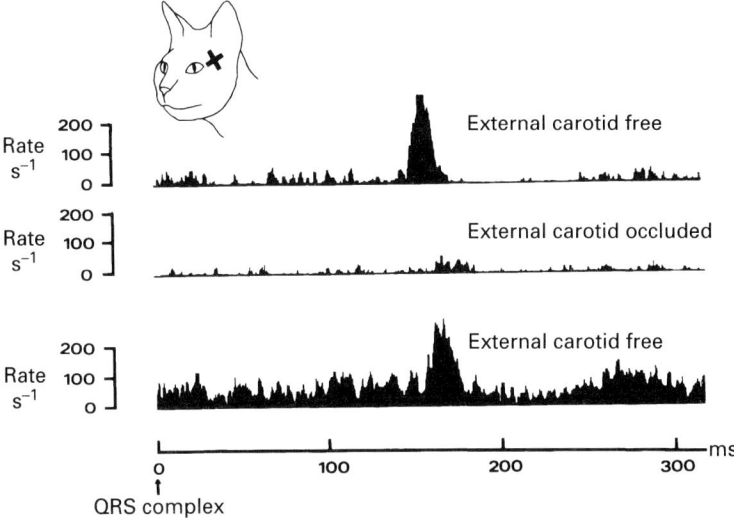

Fig. 18.4. Same unit as in Fig. 18.3, showing a reversible block of rhythmic discharges produced by occlusion of the common carotid artery. Based on data from Lambert *et al.* (1979).

received their input via the caudalis, since the responses of these cells was reversibly blocked by cooling it. The remainder of cells were unaffected, indicating that at least some oralis and interpolaris units receive a direct projection (see also Arbab *et al.*, 1988). It was interesting that Davis & Dostrovsky found that cells with convergent input from the dural vessels and facial receptive fields showed a differential response to cold block of the caudalis, with craniovascular input being affected, while the response to cutaneous facial inputs was spared. This is consistent with the results of human trigeminal tractotomy and with the results we have found in the thalamus.

Some cells in the trigeminal nucleus which respond to craniovascular stimulation have been studied microscopically (Potrebic *et al.*, 1989). They have extensive axon fields within nucleus interpolaris and in lamina IV and V of nucleus caudalis. Their cell bodies and dendrites are contacted by dendrites of various types. Cells have myelinated axons and collaterals that give rise to unmyelinated preterminal processes. The authors speculated that the connections revealed by their experiments might serve to sensitize other trigeminal brainstem neurones, including those not receiving a direct dural vascular input, thus explaining the cutaneous hyperalgesia seen in vascular headaches such as migraine and cluster headache (Lance 1982).

We found cells responsive to craniovascular electrical stimulation in the cervical cord close to and within the LCN of cats (Lambert, Zagami & Lance, 1986; Lambert et al., 1991). Two-thirds of these units responded to electrical stimulation of the superior sagittal sinus (SSS). Some also responded to stimulation of the middle meningeal artery (44% of units tested) or the superficial temporal artery (85%). Electrode tracks from which the highest proportion of units were obtained generally passed near or through the LCN, although few were found in the nucleus itself and only rarely in the adjacent dorsal horn. The LCN is a well-delineated nucleus, but electrophysiological experiments suggest that isolated units having receptive fields similar to those in the LCN can be found in surrounding white matter (Metherate et al., 1986). The LCN is a relay for forelimb sensation in cats and it is notable that an acupuncture point for treatment of headache is in the hand (the *ho-ku*) (Chapman & Gunn, 1990).

The discharge latency of these units was bimodal: 5–50 ms and 200–250 ms. We were unable to assign long-latency units to a sensory fibre category because we could not distinguish between a direct input to the cervical cord or a relayed input such as that described by Craig (1978) and Wall & Taub (1962). Units could be activated by mechanical stimulation of the superior sagittal sinus, reacting only to displacement or tugging of the superior sagittal sinus and not to touch alone, in accordance with the findings of Penfield (1935) and Ray & Wolff (1940) in humans.

Three-quarters of SSS-responsive units had cutaneous receptive fields on the face (25%), forelimb (5%), forepaw (43%) and hindpaw (2%). Most with facial receptive fields were wide dynamic-range units, while those with receptive fields on the limb or paw usually responded to low-threshold mechanoreceptor stimuli only. Referral of headache pain to the limbs in migraine and cluster headache has often been reported (Guiloff & Fruns, 1988), and may be a consequence of this convergence. These cells receive convergent input from ipsilateral tooth pulp (Olausson et al., 1990b). The nociceptive-specific nature of this input was confirmed as these cells responded to the purely nociceptive stimulus of cooling the teeth. This convergence may be important in the referral of pain to the teeth in cluster headache.

We also found cells in the cervical cord which responded to the application of bradykinin to intracranial vessels. Units responding to bradykinin were more often found close to the dorsal surface of the cord, the mean depth being 900 μm. Application of capsaicin, which releases neuropeptides such as substance P and produces pain in animals and man, increased the discharge rate of cells in the cervical cord (Lambert et al., 1991). Furthermore, electrical stimulation of the SSS increased glucose utilization in the spinal cord, maximal in the dorsolateral quadrant at the C2 level and in the trigeminal subnucleus caudalis (Goadsby, Zagami & Lambert, 1991).

Second-level processing

Spinal and trigeminal units send ascending projections to brainstem and midbrain structures, particularly the thalamus (Bonica, 1990), and it is not surprising that craniovascular-responsive neurones do so as well. For trigeminal nucleus caudalis cells relaying nociceptive information from the dura, the most important ascending tract is the neotrigeminothalamic tract, which projects to the ventroposteromedial (VPM) nucleus and to the posterior cell group (POm) of the thalamus.

Responses of thalamic neurones to dural stimulation

Davis & Dostrovksy (1988*a*) found responsive cells mostly in VPM, in the zona incerta (ZI), in POm and in the ventrolateral nucleus (VL) of cats. Some cells in these areas also responded to mechanical displacement of vessels or chemical stimulation. We found that, out of 56 thalamic cells which responded to electrical stimulation of the the cranial vasculature (middle meningeal artery and superior sagittal sinus), 26 were in the VPM and 6 in its ventral periphery (VPMvp) (Zagami & Lambert, 1990). All units in VPM had facial receptive fields, usually involving the first trigeminal division. Cells with nociceptive receptive fields, which responded to mechanical stimulation of the vessels or to application of bradykinin, were often found in the 'shell region' of VPM, an area considered by Yokota, Koyama & Matsumoto (1985) to be important in the reception of nociceptor input from facial skin. Such cells often had circumscribed receptive fields in the first division. Cells with low threshold mechanoreceptor input were, by contrast, in the body of VPM, did not respond to bradykinin or mechanical stimulation and often had receptive fields covering more than one trigeminal division. A relatively large proportion of VPM cells with input from the middle meningeal artery had receptive fields restricted to the second division of the trigeminal innervation. Other craniovascular nociceptive cells were found in the VPMvp, POm and ZI and in two areas not previously reported- the intralaminar complex (ILC) and the VPMvp.

Thalamic units received craniovascular input via both the trigeminal nucleus and the cervical cord. Units in the nucleus caudalis were activated antidromically by electrical stimulation of the ventrobasal complex (Strassman *et al*., 1986). Stimulation in the medial lemniscus and in the VPM also activated cervical cord neurones via an antidromic pathway (Lambert *et al*., 1986). Cooling the spinal cord in the C2-3 segment reversibly blocks the responses of neurones in the shell region of the VPM, leaving those in the body of the nucleus relatively unaffected (Angus-Leppan *et al*., 1989). The affected cells were more likely to have circumscribed nociceptive-specific or wide dynamic-range receptive fields than were those unaffected by cooling. Furthermore, even when the responses of cells with nociceptor input were blocked by cooling, their responses to stimulation of their cutaneous receptive fields were not. Thus, craniovascular nociceptive input is preferentially handled at the spinal level, while nonnociceptive and noncraniovascular input is more likely to be handled in the trigeminal nucleus, as would be expected.

Thalamic cells with dural afferents also receive a convergent input from the tooth pulp. Olausson *et al.* (1990*b*) demonstrated that cells in the VPM and ventrobasal complex displayed a high degree of convergence between facial receptive fields, craniovascular input and the contralateral tooth pulp. The proportion of cells showing this convergence is higher in the thalamic nuclei than in the spinal cord.

We have also undertaken complementary studies of the effect of sagittal sinus stimulation upon glucose utilization in the thalamus (Goadsby, Zagami & Lambert, 1991). In the VPM, glucose utilization triples during electrical stimulation of the SSS. Smaller changes are seen in POm, but not in the ventroposterolateral (VPL) nucleus, geniculate nucleus, or the overlying cortex.

Implications for headache pathogenesis

How can we use the knowledge we have gained about the pathways for headache to unravel the mysteries of the pathogenesis of headache? As Wolff (1963) emphasized, there is often little peripheral pathology to show for such excruciatingly painful conditions as migraine, cluster headache and trigeminal neuralgia. Does the pain of migraine and cluster headache arise at peripheral sites or in the central connections?

Extracranial and intracranial arteries may dilate during migraine and cerebral oedema has been reported in severe cases (Lance, 1982). It is not certain that these events cause the pain, are an epiphenomenon or even a consequence of the pain. In this latter regard, we have shown that electrical stimulation of the trigeminal ganglion produces vascular changes and changes in vasoactive peptide levels that could well account for many of the vascular phenomena seen in migraine (Drummond, Gonski & Lance, 1983; Lambert *et al.*, 1984; Zagami, Goadsby & Edvinsson, 1990).

Wolff's idea was that distension of extracranial and intracranial vessels, perhaps accompanied by the release of pain producing substances, activated the dural sensory nerves and all else followed from that. Modern versions of this idea hypothesise, for instance, that the primary defect is dilatation of arteriovenous anastomoses (Saxena, 1968). Mere dilatation of cranial vessels is not painful however; otherwise we would all get a migraine-like headache under a hot shower. On the other hand, powerful vasodilators such as nitrites and organic nitrates, even alcohol, can induce a headache (Olesen, this volume). Such pain may therefore require a complex interaction between vasodilatory and sensory mechanisms. Headaches occur with acute pressor responses and go away when the hypertension is controlled (albeit often with drugs that are antimigraine agents themselves), indicating that distension may be more important than dilatation.

Migraineurs are more disposed than nonmigraineurs to suffer headache following the use of vasodilator agents and many suffer 'ice-pick pains' and 'ice-cream headache' in the sensory fields in which they most often suffer also from migraine (Drummond & Lance, 1984). This suggests that some more central neural mechanism is at fault, perhaps a defect in descending pain control systems or some other increase in excitability. Sicuteri postulated that migraineurs suffer from a defect in the descending serotoninergic pain control system (see Lance, 1992). The results of stimulation of the periaqueductal grey and nucleus raphe magnus lend weight to this argument (Strassman *et al.*, 1986; Lambert *et al.*, 1986). Other brainstem nuclei such as the locus coeruleus might be important, as they not only influence nociceptive transmission in the trigeminal system, but also change the calibre of both extracranial and intracranial arteries (Lance *et al.*, 1983, 1993). Ascending cholinergic innervation to the thalamus, arising in the brainstem, exerts a facilitatory effect on thalamic responses to dural stimulation (Olausson, Angus-Leppan & Lambert, 1990*a*).

Moskowitz and colleagues (1986) have advanced the intriguing possibility that antidromic activation of the trigeminal nerve, with release of its vasodilator transmitters substance P and CGRP may help produce a reactive dilatation possibly as a compensatory mechanism in ischemia. Such an activation, were it to occur in migraine, would presumably also activate the central ends of trigeminal neurones, producing the pain. This might explain why the oligaemia seen with spreading depression, long held to be relevant to the prodromal stage of migraine (Lashley, 1941; Piper, Duckworth & Lambert, 1991) is followed by development of the pain of migraine.

References

Andres, K. H., von During, M., Muszynski, K. & Schmidt, R. F. (1987). Nerve fibres and their terminals in the dura mater encephali of the rat. *Anatomy and Embryology*, **175**, 289–301.

Angus-Leppan, H., Lambert, G. A., Boers, P., Lance, J. W. & Zagami, A. S. (1989). The cervical spinal cord is a relay centre for the central nervous system processing of input from the cranial vasculature. *Cephalalgia*, **9**, Suppl. 10, 137–8.

Arbab, M. A.-R., Delgado, T., Wiklund, L. & Svendgaard, N. A. (1988). Brain stem terminations of the trigeminal and upper spinal ganglia innervation of the cerebrovascular system: WGA-HRP transganglionic study. *Journal of Cerebral Blood Flow and Metabolism*, **8**, 54–63.

Arbab, M. A.-R., Wiklund, L. & Svengaard, N. A. (1986). Origin and distribution of cerebral vascular innervation from superior cervical, trigeminal and spinal ganglia investigated with retrograde and anterograde WGA-HRP tracing in the rat. *Neuroscience*, **3**, 695–708.

Bonica, J. J. (1990). Anatomic and physiologic basis of nociception and pain, In *The Management of Pain*, ed. J. J. Bonica, pp. 28–94. Philadelphia: Lee and Febiger.

Borges, L. F. & Moskowitz, M. A. (1983). Do intracranial and extracranial afferents represent divergent axon collaterals?. *Neuroscience Letters*, **35**, 265–70.

Broton, J. G. & Rosenfeld, J. P. (1986). Cutting rostral trigeminal nuclear complex projections preferentially affects perioral nociception in the rat. *Brain Research*, **397**, 1–8.

Chapman, C. R. & Gunn, C. C. (1990). Acupuncture, In *The Management of Pain*, ed. J.J. Bonica, pp. 1805–21. Philadelphia: Lee and Febiger.

Craig, A. D. Jr. (1978). Spinal and medullary input to the lateral cervical nucleus. *Journal of Comparative Neurology*, **181**, 729–44.

Craig, A. D. Jr & Tapper, D. N. (1978). Lateral cervical nucleus in the cat: functional organization and characteristics. *Journal of Neurophysiology*, **41**, 1511–34.

Dahl, E. & Nelson, E. (1964). Electron microscopic observations on human intracranial arteries. *Archives of Neurology*, **10**, 158–64.

Dandy, W. E. (1931). Treatment of hemicrania (migraine) by removal of the inferior cervical and first thoracic sympathetic ganglion. *Bulletin of the Johns Hopkins Hospital*, **48**, 357–61.

Darian-Smith, I. (1973). The trigeminal system. In *Handbook of Sensory Physiology*, ed. A. Iggo, pp. 271–314. Berlin: Springer-Verlag.

Darian-Smith, I., Mutton, P. & Proctor, R. (1965). Functional organization of tactile cutaneous afferents within the semilunar ganglion and trigeminal spinal tract of the cat. *Journal of Neurophysiology*, **28**, 682–94.

Davis, K. D. & Dostrovsky, J. O. (1986). Activation of trigeminal brainstem nociceptive neurons by dural artery stimulation. *Pain*, **25**, 395–401.

Davis, K. D. & Dostrovsky, J. O. (1988a). Properties of feline thalamic neurons activated by stimulation of the middle meningeal artery and sagittal sinus. *Brain Research*, **454**, 89–100.

Davis, K. D. & Dostrovsky, J. O. (1988b). Cerebrovascular application of bradykinin excites central sensory neurons. *Brain Research*, **446**, 401–6.

Dejerine, J. (1914). *Semiologie des Affections du Systeme nerveux*, Paris: Masson et Cie. **26**, 1212–24.

Denny-Brown, D. & Yanagisawa, N. (1973). The function of the descending root of the fifth nerve. *Brain*, **96**, 783–814.

Drummond, P. D., Gonski, A. & Lance, J. W. (1983). Facial flushing after thermocoagulation of the Gasserian ganglion. *Journal of Neurology, Neurosurgery, and Psychiatry*, **46**, 611–16.

Drummond, P. D. & Lance, J. W. (1983). Extracranial vascular changes and the source of pain in migraine headache. *Annals of Neurology*, **13**, 32–7.

Drummond, P. D. & Lance, J. W. (1984). Neurovascular disturbances in headache patients. *Clinical and Experimental Neurology*, **20**, 93–9.

Edvinsson, L., Hara, H. & Uddman, R. (1989). Retrograde tracing of nerve fibers to the rat middle cerebral artery with true blue: colocalization with different peptides. *Journal of Cerebral Blood Flow and Metabolism*, **9**, 212–8.

Escolar, J. (1948). The afferent connections of the 1st, 2nd, and 3rd cervical nerves in the cat. Journal of Comparative *Neurology*, **89**, 79–92.

Feindel, W., Penfield, W. & McNaughton, F. (1960). The tentorial nerves and localisation of intracranial pain in man. *Neurology*, **10**, 555–63.

Gardner, W. J., Sowell, A. & Dutlinger, R. (1947). Resection of the greater superficial petrosal nerve in the treatment of unilateral headache. *Journal of Neurosurgery*, **4**, 105–14.

Goadsby, P. J., Zagami, A. S. & Lambert, G. A. (1991). Neural processing of craniovascular pain: A synthesis of the central structures involved in migraine. *Headache*, **31**, 365–71.

Graham, J. R. & Wolff, H. G. (1937). Mechanism of migraine headache and action of ergotamine tartrate. *Proceedings of the Association for Research in Nervous and Mental Diseases*, **18**, 638–69.

Guiloff, R. J. & Fruns, M. (1988). Limb pain in migraine and cluster headache. *Journal of Neurology, Neurosurgery, and Psychiatry*, **51**, 1022–31.

Hanko, J., Hardebo, J. E., Kahrstrom, J., Owman, C. & Sundler, F. (1986). Existence and coexistence of calcitonin gene-related peptide (CGRP) and substance P in cerebrovascular nerves and trigeminal ganglion cells. *Acta Physiologia Scandanavica*, **127**, Suppl. 552, 29–32.

Hayashi, H., Sumino, R. & Sessle, B. J. (1984). Functional organisation of trigeminal subnucleus interpolaris: nociceptive and innocuous afferent inputs, projections to thalamus, cerebellum, spinal cord, and descending modulation from periaqueductal gray. *Journal of Neurophysiology*, **51**, 890–905.

Kajander, K. C. & Giesler, G. J. Jr. (1987). Responses of neurons in the lateral cervical nucleus of the cat to noxious cutaneous stimulation. *Journal of Neurophysiology*, **57**, 1686–704.

Keller, J. T., Beduk, A. & Saunders, M. C. (1985). Origin of fibres innervating the basilar artery of the cat. *Neuroscience Letters*, **58**, 263–8.

Keller, J. T., Marfurt, C. F., Dimlich, R. V. W. & Tierney, B. E. (1989). Sympathetic innervation of the supratentorial dura mater of the rat. *Journal of Comparative Neurology*, **290**, 310–21.

Keller, J. T., Saunders, M. C., Beduk, A. & Jollis, J. G. (1985). Innervation of the posterior fossa dura of the cat. *Brain Research Bulletin*, **14**, 97–102.

Kerr, F. W. L. (1961). Structural relation of the trigeminal spinal tract to upper cervical roots and the solitary nucleus of the cat. *Experimental Neurology*, **4**, 134–48.

Kerr, F. W. L. (1972). Central relationships of trigeminal and cervical primary afferents in the spinal cord and medulla. *Brain Research*, **43**, 561–72.

Kerr, F. W. & Olafson, R. (1961). Trigeminal and cervical volleys: convergence on single units in the spinal gray matter at C1 and C2. *Archives of Neurology*, **5**, 171–8.

Kimmel, D. L. (1961). Innervation of the spinal dura mater and the dura mater of the posterior cranial fossa. *Neurology*, **11**, 800–9.

Lambert, G. A., Bogduk, N., Duckworth, J. W. & Lance, J. W. (1979). Trigeminal correlates of cranio-vascular sensation. *Proceedings of the Australasian Physiology and Pharmacology Society*, **10**, 231P.

Lambert, G. A., Bogduk, N., Goadsby, P. J., Duckworth, J. W. & Lance, J. W. (1984). Decreased carotid arterial resistance in cats in response to trigeminal stimulation. *Journal of Neurosurgery*, **61**, 307–15.

Lambert, G. A., Zagami, A. & Lance, J. W. (1986). Physiology and pharmacology of cervical spinal cord elements activated by stimulation of the dura mater. *Proceedings of the Society for Neuroscience*, **12**, 230.

Lambert, G. A., Zagami, A. S., Bogduk, N. & Lance, J. W. (1991). Cervical spinal cord neurons receiving sensory input from the cranial vasculature. *Cephalalgia*, **11**, 75–85.

Lance, J. W. (1982). *Mechanism and Management of Headache*, 4th Edition. London: Butterworth Scientific.

Lance, J. W. (1992). The possible role of serotonin in the migraine syndrome. In *Advances in Pain Research and Therapy. Pain Versus Man,* vol. 20, ed. F. Sicuteri, pp. 225–31. New York: Raven Press.

Lance, J. W., Lambert, G. A., Goadsby, P. J. & Duckworth, J. W. (1983). Brainstem influences on the cephalic circulation: experimental data from cat and monkey of relevance to the mechanism of migraine. *Headache*, **23**, 258–65.

Lashley, K. S. (1941). Patterns of cerebral integration indicated by the scotomas of migraine. *Archives of Neurology and Psychiatry*, **46**, 331–9.

Liu-Chen, L-Y., Gillespie, S. A., Norregaard, T. V. & Moskowitz, M. A. (1984). Co-localisation of retrogradely transported wheat germ agglutinin and the putative neurotransmitter Substance P within trigeminal ganglion cells projecting to cat middle cerebral artery. *Journal of Comparative Neurology*, **225**, 187–92.

Liu-Chen, L-Y., Norregaard, T. V. & Moskowitz, M. A. (1985). Some cholecystokinin-8 immunoreactive fibers in large pial arteries originate from trigeminal ganglion. *Brain Research*, **359**, 166–76.

Matsuyama, T., Wanaka, A., Yoneda, S., Kimura, K., Shiosaka, S., Kamada, T., Emson, P. C. & Tohyama, M. (1986). Fine structure and interrelationship between peptidergic and catecholaminergic nerve fibers in the cerebral artery. *Acta Physiologia Scandinavica*, **127**, Suppl. 552, 17–20.

Matthews, B. & Robinson, P. P. (1980). The course of postganglionic sympathetic fibres distributed with the trigeminal nerve. *Journal of Physiology (London)*, **303**, 391–401.

Mayberg, M., Langer, R. S., Zervas, N. T. & Moskowitz, M. A. (1981). Perivascular meningeal projections from cat trigeminal ganglia: possible pathway for vascular headaches in man. *Science*, **213**, 228–30.

Mayberg, M. R., Zervas, N. T. & Moskowitz, M. A. (1984). Trigeminal projections to supratentorial pial and dural blood vessels in cats demonstrated by horseradish peroxidase histochemistry. *Journal of Comparative Neurology*, **223**, 46–56.

McNaughton, F. L. (1938). The innervation of the intracranial blood vessels and dural sinuses. *Association for Research in Nervous and Mental Diseases*, **18**, 178–200.

Metherate, R. S., Da Costa, D. C. N., Herron, P. & Dykes, R. W. (1986). A thalamic terminus of the lateral cervical nucleus: the lateral division of the posterior nuclear group. *Journal of Neurophysiology*, **56**, 1498–520.

Moskowitz, M. A., Henrikson, B. M. & Beyerl, B. D. (1986). Trigeminovascular connections and mechanisms of vascular headache. In *Handbook of Clinical Neurology*, ed. F. C. Rose, pp. 107–15. Amsterdam: Elsevier.

Nozaki, K., Kikuchi, H. & Mizumo, N. (1989). Changes of calcitonin gene-related peptide-like immunoreactivity in cerebrovascular nerve fibers in the dog after experimentally produced subarachnoid hemorrhage. *Neuroscience Letters*, **102**, 27–32.

O'Connor, T. P. & van der Kooy, D. (1986). Pattern of intracranial and extracranial projections of trigeminal ganglion cells. *Journal of Neuroscience* **6**, 2200–7.

Olausson, B., Angus-Leppan, H. & Lambert, G. A. (1990a). Cholinergic modulation of thalamic cells processing information from skin, teeth and cranial vessels in cat. *European Journal of Neuroscience*, (Suppl. 3), 290.

Olausson, B., Angus-Leppan, H., Lambert, G. A. & Boers, P. (1990b). Convergence of tooth pulp and craniovascular input to the spinal cord and thalamus. *Pain*, (Suppl. 5), S46.

Olszewski, J. (1950). On the anatomical and functional organization of the spinal trigeminal nucleus. *Journal of Comparative Neurology*, **92**, 401–13.

Penfield, W. (1935). A contribution to the mechanisms of intracranial pain. Research *Publications of the Association for Research into Nervous and Mental Diseases.*, **15**, 399–416.

Penfield, W. & McNaughton, F. L. (1940) Dural headache and the innervation of the dura mater. *Archives of Neurology and Psychiatry*, **44**, 43–75.

Pfaller, K. & Ardvinsson, J. (1988). Central distribution of trigeminal and upper cervical primary afferents in the rat studied by anterograde transport of horseradish peroxidase conjugated to wheat germ agglutinin. *Journal of Comparative Neurology,* **268**, 91–108.

Piper, R. D., Lambert, G. A. & Duckworth, J. W. (1991). Cortical blood flow changes during spreading depression in cats. *American Journal of Physiology*, **261**, H96–H102.

Potrebic, S., Strassman, H., Hartwig, E. A. & Maciewicz, R. (1989). Ultrastructure of intracellularly labeled trigeminal vascular convergence neurons. *Brain Research*, **507**, 317–20.

Ray, B. S. & Wolff, H. G. (1940). Experimental studies on headache. Pain sensitive structures of the head and their significance in headache. *Archives of Surgery*, **41**, 813–56.

Ruskell, G. L. (1988). The tentorial nerve in monkeys is a branch of the cavernous plexus. *Journal of Anatomy*, **157**, 67–77.

Ruskell, G. L. & Simons, T. (1987). Trigeminal pathways to the cerebral arteries in monkeys. *Journal of Anatomy*, **155**, 23–37.

Rydenhag, B., Shyu, B. C., Olausson, B. & Andersson, S. (1986). Influence of changes of tooth temperature on reflex and central activity evoked by stimulation of tooth pulp afferents. *Brain Research Bulletin*, **64**, 37–48.

Sato, S. & Suzuki, J. (1975). Anatomical mapping of the cerebral nervi vasorum in the human brain. *Journal of Neurosurgery*, **43**, 559–68.

Saxena, P. R. (1968). Arteriovenous shunting and migraine. *Research and Clinical Studies in Headache*, **6**, 89–102.

Shigenaga, Y., Chen, I. C., Suemune, S., Nishimori, T., Nasution, I. D., Yoshida, A., Sato, H., Okamoto, T., Sera, M. & Hosoi, M. (1986). Oral and facial representation within the medullary and upper cervical dorsal horns in the cat. *Journal of Comparative Neurology*, **243**, 388–408.

Simons, T. & Ruskell, G. L. (1988). Distribution and termination of trigeminal nerves to the cerebral arteries in monkeys. *Journal of Anatomy*, **159**, 57–71.

Siwe, S. A. (1931). The cervical part of the ganglionated cord, with special reference to its connections with spinal nerves and certain cerebral nerves. *American Journal of Anatomy*, **48**, 479–97.

Steiger, H. J. & Meakin, C. J. (1984). The meningeal representation in the trigeminal ganglion- an experimental study in the cat. *Headache*, **24**, 305–9.

Steiger, H. J., Tew, J. M. & Keller, J. T. (1982). The sensory representation of the dura mater in the trigeminal ganglion of the cat. *Neuroscience Letters*, **31**, 231–6.

Strassman, A., Mason, P., Moskowitz, M. & Maciewicz, R. (1986). Response of brainstem trigeminal neurons to electrical stimulation of the dura. *Brain Research*, **379**, 242–50.

Tsai, S-H., Lin, S-Z., S-D Wang, Liu, J-C. & Shih, C-J. (1985). Retrograde localization of the innervation of the middle cerebral artery with horseradish peroxidase in cats. *Neurosurgery*, **16**, 463–7.

Wall, P. D. & Taub, A. (1962). Four aspects of trigeminal nucleus and a paradox. *Journal of Neurophysiology*, **25**, 110–26.

White, J. C. & Sweet, W. H. (1955). Cephalic Pain-transmission pathways uncertain, In *Pain: Its Mechanisms and Neurosurgical Control*, pp. 494–532. Springfield: Charles C. Thomas.

Willis, W. D. Jr, (1985). Nociceptive transmission in the primate spinal cord. In *Development, Organization and Processing in Somatosensory Pathways*, ed. M. Rowe. & W. D. Willis Jr, pp. 333–45. New York: Alan R. Liss.

Wirth, F. P. Jr & Van Buren, J. M. (1971). Referral of pain from dural stimulation in man. *Journal of Neurosurgery*, **34**, 630–42.

Wolff, H. G. (1963). *Headache and Other Head Pain*. New York: Oxford University Press.

Yokota, Y., Koyama, N. & Matsumoto, N. (1985). Somatotopic distribution of trigeminal nociceptive neurons in ventrobasal complex of cat thalamus. *Journal of Neurophysiology*, **53**, 1387–1400.

Zagami, A. S., Goadsby, P. J. & Edvinsson, L. (1990). Stimulation of the superior sagittal sinus in the cat causes release of vasoactive peptides. *Neuropeptides*, **16**, 69–75.

Zagami, A. S. & Lambert, G. A. (1990). Stimulation of cranial vessels excites nociceptive neurones in several thalamic nuclei of the cat. *Experimental Brain Research*, **81**, 552–66.

19
Pathophysiology of migraine

PETER J. GOADSBY

Department of Neurology,
The Prince Henry Hospital,
Little Bay, Sydney NSW, Australia

Migraine pathophysiology has traditionally been viewed either from a clinical or a basic research perspective. In this chapter, laboratory data that characterize the cerebral circulation will be assessed with complementary results from clinical studies in an attempt to synthesize a coherent hypothesis for migraine. Of the main competing theories, the neural theory asserts that the fundamental defect is in the central nervous system (Lance *et al.*, 1983) or its connections with the vessels (Moskowitz *et al.*, 1979), and the vascular theory favours a disorder of the cranial vessels (Wolff, 1963).

Definition and classification

The conventions of the Headache Classification Committee of the International Headache Society are adopted throughout (Headache Classification Committee, 1988). Migraine is thus defined as headache lasting usually from 4 to 72 hours which is often unilateral, has a pulsating quality, is of moderate or severe intensity and can be aggravated by activity. It is episodic and may be associated with nausea, vomiting or photophobia. There should be no relevant neurological abnormality on either clinical examination or laboratory investigation. In practice, two varieties of migraine are commonly distinguished: those with an aura of neurological symptoms (previously called classical migraine), and a periodic headache without aura (common migraine). Wide acceptance of this classification has enabled clinical studies to be correlated from various countries with some precision.

Premonitory features

About a quarter of patients report symptoms of elation, irritability, depression, hunger, thirst or drowsiness during the 24 hours preceding headache. Most of these manifestations can be attributed to disturbances in the hypothalamus

(Kupfermann, 1985), and this has led to the suggestion of a central site for their evolution. In addition, the suprachiasmatic nucleus of the hypothalamus has been suggested to be one of two primary oscillators generating circadian rhythms (Moore-Ede, 1983) and thus could be implicated in the periodicity of migraine that is an important clinical feature. Physiological mechanisms in the brainstem suggests another site for the premonitory symptoms. Mood changes including elation and depression may be linked to the central noradrenergic system and thus implicate the locus coeruleus (Amaral & Sinnamon, 1977). Indeed, depression and suicide are increased in migraine with aura (Breslau, Davis & Andreski, 1991). Furthermore, the locus coeruleus has been implicated in directed attention and tuning the signal-to-noise ratio for afferent input (Aston-Jones et al., 1986). It is attractive to propose that such a tuning process is not operating correctly during migraine given that this structure can alter cerebral blood flow (Goadsby & Duckworth, 1989). Connections of the locus coeruleus with the hypothalamus are involved in the regulation of blood volume and thus interact with the thirst mechanisms (Foote, Bloom & Aston-Jones, 1983).

Prodrome of migraine with aura

Changes in cerebral blood flow associated with the aura

Numerous studies have confirmed that the aura phase of migraine is associated with a reduction in cerebral blood flow (Skinhøj & Paulson, 1969; O'Brien, 1971; Skinhøj, 1973; Simard & Paulson, 1973; Norris, Hachinski & Cooper, 1975; Mathew, Hrastnik & Meyer, 1976; Edmeads, 1977; Hachinski et al., 1977; Sakai & Meyer, 1978; Olesen, Larsen & Lauritzen, 1981a; Staehelin-Jensen et al., 1981; Lauritzen et al. 1983; Lauritzen & Olesen, 1984; Skyhoj-Olsen, Friberg & Lassen, 1987). Visual disturbances, paresthesiae or other focal neurological signs are associated with this reduction in cerebral blood flow (Olesen et al., 1981a). These neurological changes parallel what is seen when the brain is directly stimulated (Brindley & Lewin, 1968; Penfield & Perot, 1963). This change in flow moves across the cortex as a *spreading oligemia* at 2–3 mm/min (Lauritzen et al., 1983) corresponding to the rate that Lashley estimated from plotting the progression of his own visual aura (Lashley 1941). It also resembles the phenomenon of spreading cortical depression (Leao, 1944a,b). First, there is a focal reduction in flow that is usually posterior near Brodman area 7 and the superior part of area 19 although focal frontal oligemia without visual aura has been reported (Friberg et al., 1987). Secondly, this reduction enlarges and may involve the whole hemisphere. The progression of the oligemia across the cortex does not respect vascular territories and is thus unlikely to be primarily vasospastic. Furthermore, vasospasm is only rarely seen if patients have an angiogram during migraine (Symonds, 1952; Connor, 1962; Skinhøj, 1973; Lauritzen et al., 1983; Skyhoj-Olsen et al., 1987). Vasoconstriction, however, does occur as evidenced by retrograde flow from the carotid to vertebrobasilar circulation, as seen in some patients (Skinhøj, 1973; Norris, Hachinski & Cooper, 1975; Hachinski et al., 1977). Furthermore, the oligemia may be preceded by focal hyperemia (Olesen et al., 1981a; Friberg et al., 1987). Such a change is again exactly what would be expected if a phenomenon similar to cortical spreading depression were involved. Following the oligaemia, the cerebrovascular response to hypercapnia is blunted (Simard & Paulson, 1973; Sakai & Meyer, 1979; Harer & Von Kummer, 1991), while autoregulation remains intact (Lauritzen et al., 1983). Again, this pattern is repeated in spreading depression. Usually the flow change is accompanied by a contralateral aura and the unilateral headache is homolateral with respect to the oligaemia (Olesen et al., 1981a) but patients have reported unilateral headache with a homolateral aura suggesting a mismatch of the aura with the subsequent headache (Peatfield & Rose, 1991). Headache may begin while cortical blood flow is still reduced (Olesen et al., 1990) thus reducing the likelihood that the pain arises from a primary vascular abnormality.

Some methodological errors may account for the oligemia and affect whether the flow changes reach an ischemic level (Skyhoj-Olsen, 1990). Because of the errors involved, only metabolic studies will be able to determine if ischemia does occur. The best estimates of flow that take into account Compton scatter, an error generated by detectors inadvertently picking up a signal from a well perfused area when positioned over a poorly perfused area (Skyhoj-Olsen & Lassen, 1989), put the flow values during the aura at 20–25 ml/100 g/min (Skyhoj-Olsen et al., 1987). Electroencephalographic activity in humans undergoing carotid endarterectomy is reduced when flow drops below 23 ml/100 g/min (Trojaborg & Boysen, 1973), while at this level in the awake monkey a neurological deficit can be seen (Jones et al., 1981). Studies of oxygen utilization using positron emission tomography (PET) have shown that during the aura the flow reduction is balanced by an increase in oxygen extraction and thus oxygen metabolism is normal (Herold et al., 1985). This problem has been difficult to address because of the technological demands of PET but, for the moment, it can be said that there is no hard evidence for cerebral ischemia during the aura of migraine.

The neurophysiology of aura

What can be inferred from known cerebrovascular physiology concerning the aura? The cortical spreading depression of Leao was first reported in the exposed rabbit cortex as a negative shift in DC potential (Leao, 1944a,b). This shift corresponds to a redistribution of K^+, Na^+, Cl^+, Ca^{++} and H^+ ions (Kraig & Nicholson, 1978). The characteristic features of spreading depression include: propagation at 2–6 mm/min, limitation to one hemisphere, and a refractory period for further spreading depression of up to 3 min. Spreading depression occurs in a number of species including rat (Lauritzen *et al.*, 1982), cat (Marshall, 1959) and man (Sramka *et al.*, 1977), and the changes can even include subcortical structures. Initially, cerebral blood flow may double after spreading depression has been initiated (Hansen, Quistorff & Gjedde, 1980) and a prolonged moderate reduction in flow follows (Lauritzen *et al.*, 1982; Tomida *et al.*, 1989). This is associated with blunting of cerebrovascular responses to hypercapnia but normal autoregulation (Lauritzen, 1984; Hansen, Quistorff & Gjedde, 1980).

The reduction in cerebral blood flow seen after spreading depression may involve not only the cortex but also subcortical structures including the brainstem (Mraovitch, Calando & Seylaz, 1989). In a recent study, cerebral cortical perfusion was recorded in the anesthetized rat with laser Doppler flowmetry during induction of spreading depression. An electrode was placed in the centromedian parafasicular complex of the thalamus (CMPf), an area capable of altering cerebral blood flow independent of cerebral glucose utilization (Mraovitch *et al.*, 1986). Electrode placement produced a wave of hyperemia at a latency consistent with propagation at 3–5 mm/min. The increase was 566±124%, lasted only 2–3 mins and was followed by an oligaemic phase during which flow was reduced by 19±4%. Subsequently, the cortical vasodilator response to hypercapnia was markedly blunted, while electrical stimulation of the CMPf still powerfully increased flow (Goadsby, Seylaz & Mraovitch, 1991). These observations provide evidence for a functional neurogenic innervation of the cerebral cortex and indicate how abnormal cerebrovascular physiology becomes after electrode implantation. They reinforce the view that central mechanisms can still play a major role in the control of cerebrovascular tone even during the acute attack of migraine.

Experimentally induced changes in cerebral blood flow and the aura

Stimulation of a discrete nucleus in the brainstem, nucleus locus coeruleus (the main central noradrenergic nucleus) (Amaral & Sinnamon, 1977) reduces cerebral blood flow in a frequency-dependent manner (Goadsby, Lambert & Lance, 1982) through an α_2-adrenoceptor-linked mechanism (Goadsby, Lambert & Lance, 1985). This reduction is maximal in the occipital cortex (Goadsby & Duckworth, 1989). Importantly, while overall a 25% reduction in

Pathophysiology of migraine

cerebral blood flow is seen, extracerebral vasodilatation occurs in parallel (Goadsby, Lambert & Lance, 1983). From the clinical standpoint, aura can exist in isolation from the pain as 'migraine equivalent'; it is thus possible that the aura originates in the central nervous system with the vascular changes being a secondary feature.

Headache

Pain processing

The two features of migraine most feared by the patient are pain and nausea. There is no strong evidence as to whether nausea is mediated by the area postrema or by gut mechanisms. Mild attacks of migraine without aura may be treated successfully with an analgesic, such as aspirin, and an antiemetic, such as metoclopramide, which may act at either site since the area postrema is outside the blood–brain barrier and dopamine-2 receptor antagonists reduce nausea.

The pain of migraine may take many forms but it is usually identified by its severity, episodic nature and the pulsatile or throbbing character (Wolff, 1963). Migraine headache may be eased temporarily in one-third of patients by compression of the superficial temporal artery and in another third by compression of the common carotid artery (Drummond & Lance, 1983). Vascular pain implicates the sensory innervation of the vessels peripherally or centrally. The cranial vessels are innervated by sensory fibres containing substance P that arise from the trigeminal system (Saito & Moskowitz, 1989) and the upper cervical roots (Arbab *et al.*, 1988). The first (ophthalmic) division of the trigeminal nerve innervates predominantly the anterior cerebral circulation while the posterior circulation is innervated by the C2-3 cervical roots (Saito & Moskowitz, 1989).

Using the 2-deoxyglucose method in the cat, the second- and third-order neurons in the intracranial vascular pain pathway have been mapped. Electrical stimulation of the superior sagittal sinus, a pain-sensitive structure (Ray & Wolff, 1940) innervated by C fibres (Feindel, Penfield & McNaughton, 1960; Penfield & McNaughton, 1940) increases cerebral blood flow and glucose utilization in restricted and specific brain regions. Metabolic activation is seen in the second-order neurons in the trigeminal nucleus caudalis, the dorsal horn of the C2 region and in a curious group of cells in the dorsolateral white matter of the cervical cord again at the C2 level (Goadsby & Zagami, 1991). These cells then project via the quintothalamic tract to decussate in the pons or midbrain and synapse on third-order neurons at the level of the ventrobasal complex of

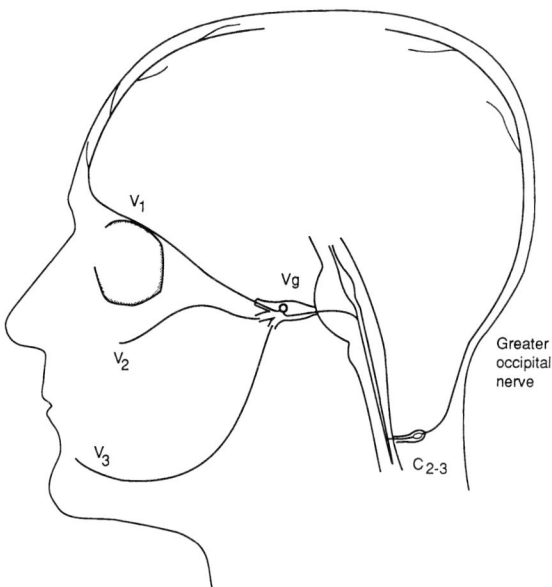

Fig. 19.1. Line drawing illustrating the interaction of the trigeminal system and the innervation of the posterior fossa and head by the high cervical nerves. The first (ophthalmic) division of the trigeminal nerve provides sensory innervation to pain-sensitive structures in the cranium, such as the dura mater and vessels, while a similar innervation of posterior fossa structures and the back of the head is provided by the second and third cervical nerves. Both these systems converge centrally at their first synapse in the central nervous system on second order neurones in the trigeminal nucleus caudalis and its functional extension in the C_2 spinal cord. This overlap on second-order neurones is the basis for fronto-occipital-cervical pain referral in headache. Modified from Lance (1982) with permission.

the thalamus in the ventroposteromedial nucleus, medial nucleus of the posterior complex and intralaminar complex of the thalamus (Goadsby & Zagami, 1990). These cells have the electrophysiological properties (latency and chemical nociceptive activation) of pain processing neurons in both the high cervical cord and the thalamus (Lambert *et al.*, 1991; Zagami & Lambert, 1990; see Lambert, this volume). In contrast, a non-nociceptive trigeminal input increases activity in trigeminal nucleus principalis, interpolaris and caudalis, ventroposteromedial thalamus and SI somatosensory cortex (Young *et al.*, 1990). It is likely that these lower brainstem and cord areas represent the anatomical site of overlap between the trigeminal and cervical systems and thus explain the fronto-occipital pain that so often characterizes headache. Whether the pain arises from the vessels and their sensory innervation or whether there is central instability remains undecided (Fig. 19.1).

Pathophysiology of migraine

Several studies have implicated a role for the trigeminal neurons in pain generation. While it is almost self-evident that the pain is mediated or at least transduced in the trigeminal system, does it play a more direct role? Stimulation of the trigeminal ganglion in the rat leads to plasma extravasation and mast cell degranulation in the dura mater (Markowitz, Saito & Moskowitz, 1987; Buzzi *et al.*, 1990), and this can be blocked by dihydroergotamine (Markowitz, Saito & Moskowitz, 1988) and sumatriptan (Buzzi & Moskowitz, 1990). These agents do not affect changes in vascular permeability or mast cell degranulation mediated by substance P or calcitonin gene-related peptide (CGRP) when applied directly to cells, and thus it has been suggested that their effect is prejunctional (Dimitriadou *et al.*, 1990). This system is relatively nonspecific as non-nociceptive, cutaneous afferents, are also stimulated. In some studies, the large stimulus currents could have induced activity in remote neuronal systems such as the facial nerve dilator system (Goadsby, 1991), and thus the results may not be exclusively interpreted as changes in the trigeminal system.

Clinical observations favour the concept of central instability. Sudden pain in the head after eating ice-cream or swallowing a cold drink is a common experience that is more frequently felt by migraineurs (Raskin & Knittle, 1976). In addition, one third of migrainous subjects locate the pain of ice-cream headache to the area habitually affected in their migraine headaches (Drummond & Lance, 1984). Ice-pick pains, sudden jabs of pain in the head (Raskin & Schwartz, 1980), are also commonly experienced by migrainous patients, 40% of whom report that these are felt on the same side as their headaches (Drummond & Lance, 1984). Thus there may be hyperexcitability of trigeminal pathways that could discharge spontaneously to initiate a migraine attack. The only operations which have produced lasting benefit in migraine are trigeminal rhizotomy and bulbar tractotomy (White & Sweet, 1955), while placement of electrodes into the region of the peri-aqueductal grey matter can induce migraine-like headaches (Raskin, Hosobuchi & Lamb, 1987)

Changes in cerebral blood flow

Interictal studies in man

Using both ^{133}Xe inhalation and HMPAO with SPECT, regional cerebral blood flow has been compared in migraine patients and age- and sex-matched controls. In 92 patients (60 with aura and 32 without aura), interictal flow asymmetries were found in about 40%. The largest asymmetries were seen in patients with migraine with aura (Friberg *et al.*, 1991). Further studies to characterize whether these abnormalities correlate with the flow changes seen in the attack and whether interictal physiology (such as hypercapnic vasodilatation) is normal are awaited.

Migraine with aura

The headache phase of migraine may be accompanied by hyperemia (Sakai & Meyer, 1978) although in some studies the headache was not sufficiently characterised to determine its type (Skinhøj, 1973). The pain of the headache may come during the oligemic phase; it is therefore unlikely that dilatation of the cerebral vessels is solely responsible for the pain (Olesen et al., 1981a; Lauritzen et al., 1983). Differences between studies in the flow changes during the headache phase may be due to assessment of the patients at different phases of their attacks. Patients studied serially may have oligemia, hyperaemia or no change during a headache, dependent upon when they are studied (Andersen et al., 1988).

Migraine without aura

One study suggested that in migraine without aura there is hyperemia in the headache phase (Sakai & Meyer, 1978) but this has not been observed by others (Olesen et al., 1981a; Olesen et al., 1982; Lauritzen & Olesen, 1984). Interestingly, hypercapnic vasodilatation may also be blunted in migraine without aura although reports to date have demonstrated changes that, while reduced, were symmetrical (Sakai & Meyer, 1978).

Clinical observations

There is a significant extracerebral vascular component to the headache in one third of patients (Drummond & Lance, 1984). The level of CGRP is elevated in the external jugular vein blood of migraineurs during headache (Goadsby, Edvinsson & Ekman, 1990), clearly demonstrating activation of trigeminovascular neurones during migraine with or without aura. It is uncertain whether the activity is peripherally generated, although such changes can be seen in both man and cat with direct stimulation of the trigeminal ganglion (Goadsby, Edvinsson & Ekman, 1988). The release of this peptide offers the prospect of a marker for migraine that can be determined from a venous blood sample.

Studies in the experimental animal

Experimental evidence suggests that the trigeminovascular system promotes vasodilatation. Nerves that innervate the cerebral vessels through the trigemi-

Pathophysiology of migraine 311

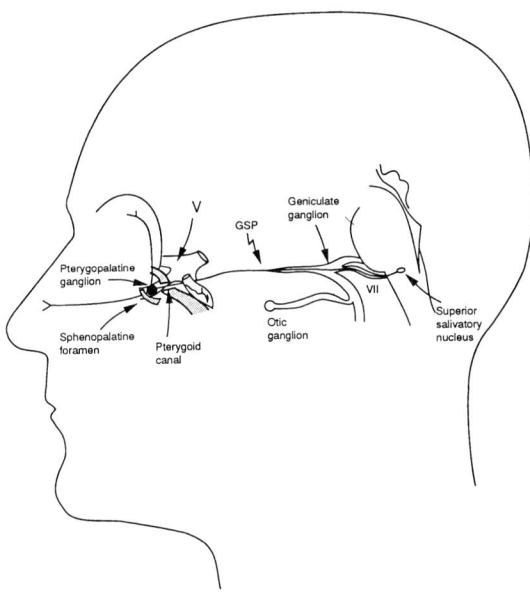

Fig. 19.2. Drawing of the origin and distribution of parasympathetic fibres that innervate the cranial circulation. The cells of origin are in the superior salivatory nucleus and pass out the facial (VIIth cranial) nerve through the geniculate ganglion without synapsing. They then distribute via the greater superficial petrosal nerve (GSP) to the pterygopalatine (sphenopalatine in lower mammals), otic and internal carotid miniganglia. The main innervation synapses in the pterygopalatine ganglion passing out the ethmoid foramen and loops back to travel along the internal carotid artery to the circle of Willis. The relationship of the distributing parasympathetic nerves to the second (maxillary) division of the trigeminal nerve (V) is shown as they pass adjacent to the sphenopalatine foramen.

novascular system contain almost exclusively vasodilator transmitters such as CGRP and substance P (SP) (Edvinsson *et al.*, 1988), release of which could *not* provoke spreading depression. Although these nerves and those of the parasympathetic innervation of the cranial vessels from the facial nerve may have a protective role in situations of perceived cerebrovascular threat (Fig. 19.2), lesions of the trigeminal ganglion do not affect resting cerebral blood flow or glucose utilization in the cat (Edvinsson *et al.*, 1986). They do, however, affect vasodilator protector mechanisms such as those seen during hyperemia following ischemia or epilepsy (Sakas *et al.*, 1989). Furthermore, in subarachnoid hemorrhage with threatened cerebrovascular compromise from vasospasm, venous CGRP levels are elevated in man (Edvinsson, Uddman & Juul, 1990; Juul *et al.*, 1990; Edvinsson *et al.*, 1990, 1991). Electrical stimulation of the trigeminal ganglion in man and the cat increases extracerebral blood

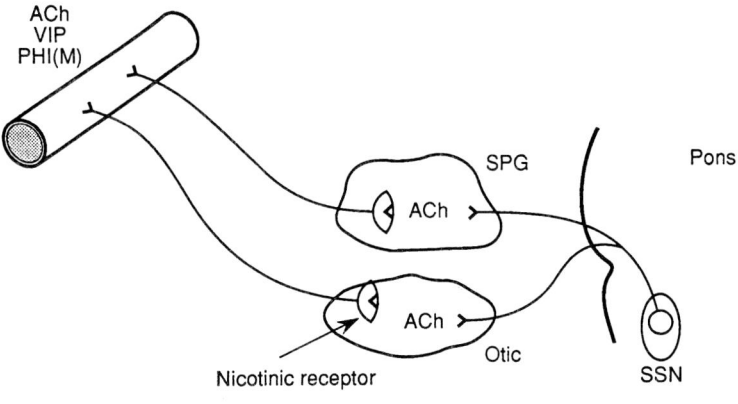

Fig. 19.3. The pharmacological characterisation of the parasympathetic innervation of the cranial circulation from the facial nerve. The first synapse in the sphenopalatine (SPG; pterygopalatine in man) and otic ganglia employs acetylcholine (ACh) to activate a nicotinic receptor. Each of acetylcholine, vasoactive intestinal polypeptide (VIP), peptide histidine isoleucine [methionine in man; PHI(M)] and pituitary adenylate cyclase activating peptide (PACAP) have been identified in nerves innervating the cranial vessels that arise from this system.

flow and local release of CGRP and SP (Goadsby et al., 1988). In the cat trigeminal ganglion, stimulation also increases cerebral blood flow by a pathway traversing the greater superficial petrosal branch of the facial nerve (Goadsby & Duckworth, 1987; Fig. 19.3), again releasing a vasodilator peptide, vasoactive intestinal polypeptide (VIP) (Goadsby & MacDonald, 1985). Interestingly, the VIP-ergic innervation of the cerebral vessels is predominantly anterior rather than posterior (Matsuyama et al., 1983) and this may contribute to the vulnerability of this region to spreading depression, and in part explain why the aura usually commences posteriorly. Stimulation of the more specifically vascular pain-sensitive superior sagittal sinus increases cerebral blood flow (Lambert et al., 1988) and jugular CGRP levels (Zagami, Goadsby & Edvinsson, 1990). Given that CGRP is elevated in the headache phase of migraine, the trigeminovascular system may be activated in a protective role in this condition.

The neuropharmocology of migraine

Treatment of migraine may be divided into two parts, prophylactic (interval) therapy and the management of the acute attack (see Clifford Rose & Anthony, this volume). Each is based on different neuropharmacological data concerning the condition and its underlying mechanisms (Goadsby & Lance, 1990). The vasoconstrictor action of ergotamine was once thought to explain the pathophysiology of migraine. The use of dihydroergotamine has made a vascular theory less tenable since dihydroergotamine is a poor vasoconstrictor compared with ergotamine (Lambert & Duckworth, 1986), affecting veins rather than arteries, but is equally effective as an antimigraine agent. There is circumstantial evidence that implicates serotonin (5-hydroxytryptamine, 5HT) in migraine: urinary excretion of 5-hydroxyindoleacetic acid, the main metabolite of serotonin is increased in migraine attacks (Sicuteri, Testi & Anselmi, 1961; Curran, Hinterberger & Lance, 1965); platelet 5HT drops during the onset of migraine (Curran *et al.*, 1965); intravenous injection of 5HT can abort either reserpine-induced or spontaneous headache (Kimball, Friedman & Vallejo, 1960; Anthony, Hinterberger & Lance, 1967); and blood 5HT levels drop at the onset of migraine and a low molecular weight serotonin-releasing factor is present during headache (Anthony, 1986).

Recent work (Hamblin *et al.*, 1987) has reconciled these two apparently disparate clinical observations by demonstrating that dihydroergotamine binds to α-adrenoceptors and serotonin receptors ($5HT_1$-like). The latter have recently been classified into four major types, $5HT_1$, $5HT_2$, $5HT_3$ and $5HT_4$ (Bradley *et al.*, 1986; Saxena, 1991). Parallel clinical experience with the newly synthesized $5HT_1$-like agonist sumatriptan (Doenicke, Brand & Perrin, 1988; Ferrari, 1991; Goadsby *et al.*, 1991) indicates that such agents provide a new acute-attack therapy (Clifford Rose & Anthony, this volume). It is likely that the serotonin-agonist action of a drug is an important determinant of its efficacy as an antimigraine compound.

The interval therapy of migraine is dominated by drugs that interact with monoaminergic receptors. Both methysergide and pizotifen are $5HT_2$ receptor antagonists and are effective, relatively safe, long-term prophylactic drugs. They may also modify the behaviour of central serotonergic neurons. Interestingly, methysergide is metabolized to methyl-ergometrine and thus its antimigraine action may be due to the activity of that metabolite on $5HT_{1D}$ receptors. The β-blockers lacking any agonist activity (propranolol, nadolol, atenolol, timolol and metoprolol) are effective prophylactic drugs and may interact with central noradrenergic pathways from the locus coeruleus that play a role in nociceptive control. Overall, the pharmacological profile of interval therapy drugs is different from that of drugs for attack therapy and suggests a complex interaction with monoaminergic systems that is likely to be of central origin since it is these, and not peripheral monoaminergic vascular fibres, that have an established role in the modulation of nociception.

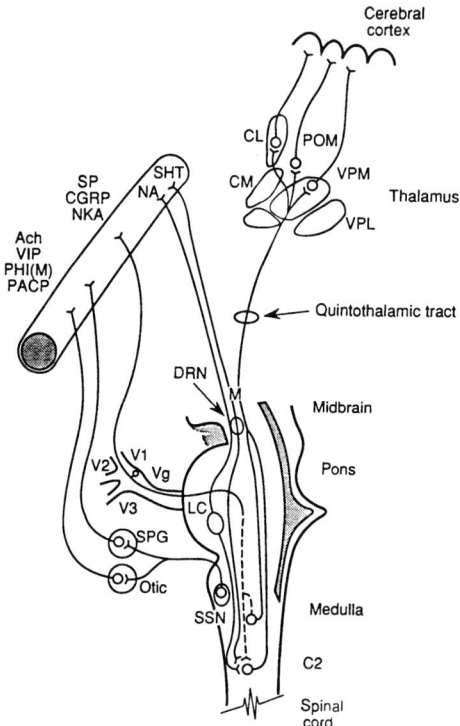

Fig. 19.4. A scheme for the pathophysiology of migraine. Input is by way of the trigeminal ganglion (Vg) which contains the bipolar cell bodies for pain-sensitive innervation of the dura mater and blood vessels. The peripheral transmitters in this system are substance P (SP), calcitonin gene-related peptide (CGRP) and neurokinin A (NKA). The second-order synapse is in the trigeminal nucleus caudalis in the medulla and on neurons at C2 level of the spinal cord (C2). The second-order neurones decussate and pass through the quintothalamic tract to the thalamus, specifically to the ventroposteromedial (VPM), centrolateral (CL) and medial nucleus of the posterior group nuclei (POM). A central reflex vasodilator response is also present and mediated largely by the facial nerve with its outflow through the sphenopalatine (SPG) and otic ganglia. The post-ganglionic nerves contain the dilator transmitters, VIP, ACh, PHI(M), and PACAP (see Fig. 19.3 for abbreviations). These systems are entrained or influenced by the central noradrenergic (locus coeruleus, LC) and serotoninergic (dorsal raphe nucleus, DRN) nuclei. They normally gate incoming pain signals but in migraine are likely to function abnormally.

For a neural theory to be plausible in migraine, drugs effective in treating migraine should pass through the blood-brain barrier. (Strictly speaking, this is not an absolute rule since it is not known whether the blood-brain barrier is normal during a migraine attack.) Given the abnormalities of brain blood flow

Pathophysiology of migraine

in migraine, dysfunction of the blood–brain barrier is also likely. Many drugs used in migraine prophylaxis (propranolol, pizotifen and methysergide) cross the blood–brain barrier and may cause central side-effects, such as drowsiness. What of acute-attack therapy? There are conflicting data for ergotamine derivatives: dihydroergotamine appears not to cross the blood–brain barrier (Kalberer, Schreier & Zehnder, 1971), but sensitive autoradiographic procedures suggest that it binds in the area postrema, nucleus of the tractus solitarius, and the dorsal raphe nucleus (Goadsby & Gundlach, 1991). Sumatriptan, however, accesses the cerebrospinal fluid poorly and does not seem to enter the central nervous system significantly (see Humphrey *et al.*, this volume). Hence, termination of an acute attack may depend upon its peripheral action on blood vessels or at a prejunctional serotonin receptor on the trigeminal nerves, while prophylactic therapy may act mainly in the central nervous system.

Hypothesis

A hypothesis for the pathophysiology underlying migraine is shown schematically in Fig. 19.4. Many of the premonitory symptoms of migraine suggest dysfunction of the hypothalamus before the headache. Activation of areas responsible for biological rhythms such as the hypothalamus or external triggers such as stress overload brainstem mechanisms which normally gate cranial nociception but which, due to some inherited defect, cannot deal with such afferent volleys. An increased discharge from the locus coeruleus, which would normally tune the signal-to-noise ratio for such inputs, would reduce cerebral blood flow and may thus initiate spreading depression and the prodromal phase of migraine with aura. Cerebral and extracerebral dilator nerves are excited as the firing increases in the locus coeruleus, or as a result of a protective reflex firing of the central serotonergic system or trigeminovascular system. The combined firing and degree of excitation of these systems would thus determine extracerebral vascular changes and changes in neuropeptide levels. The depletion of brainstem monoaminergic nociceptive control mechanisms with continued activity and the antidromic activation of the trigeminal system releasing mediators of inflammation are thus perceived by the patient through the trigeminal innervation of the vessels as a throbbing and severe pain. The distribution of the pain in a fronto-occipital radiation results from the overlap of both trigeminal and cervical pain afferents in the high cervical spinal cord.

Such a scheme does not exclude humoral factors. Indeed, there is experimental evidence that locus coeruleus stimulation can activate the adrenal gland

and thus release noradrenaline that could, in turn, induce platelet aggregation and serotonin release (Goadsby, 1985). In addition, trigeminal system activation may cause platelet aggregation locally in the cerebral circulation (Dimitriadou *et al.*, 1990).

Conclusion

The pathophysiology of migraine has been examined in the light of recent physiological, neuropharmacological and clinical observations. Migraine is presented as a dysnociceptive condition which may be triggered centrally, via the hypothalamus, or peripherally via the trigeminal system. Available information suggests that migraine is caused by the response to afferent stimuli or biological rhythms of regions of the central nervous system that are defective in the control of craniovascular nociception (the inherited trait). This hypothesis provides a basis for understanding the condition, a rational approach to therapy and direction for further basic research.

Acknowledgements

The work of the author has been supported by the National Health and Medical Research Council of Australia and by grants from Warren and Cheryl Anderson, The J.A. Perini Family Trust, the Basser Trust and the Australian Brain Foundation. PJG is a Wellcome Senior Research Fellow. The author thanks Sandoz Australia Ltd for access to the data concerning dihydroergotamine (Kalberer *et al.*, 1971).

References

Amaral, D. G. & Sinnamon, H. M. (1977). The locus coeruleus: neurobiology of a central noradrenergic nucleus. *Progress in Neurobiology*, **9**, 147–96.

Andersen, A. R., Friberg, L., Skyhoj-Olsen, T. & Olesen, J. (1988). SPECT demonstration of delayed hyperemia following hypoperfusion in classic migraine. *Archives of Neurology*, **45**, 154–9.

Anthony, M. (1986). The biochemistry of migraine. In *Handbook of Clinical Neurology*, vol. 48, ed. P.J. Vinken, G. W. Bruyn, H. L. Klawans & F. C. Rose, pp. 85–105. Amsterdam: Elsevier.

Anthony, M., Hinterberger, H. & Lance, J. W. (1967). Plasma serotonin in migraine and stress. *Archives of Neurology*, **16**, 544–52.

Arbab, M. A.-R., Delgado, T., Wiklund, L. & Svendgaard, N. A. (1988). Brain stem terminations of the trigeminal and upper spinal ganglia innervation of the cerebrovascular system: WGA-HRP transganglionic study. *Journal of Cerebral Blood Flow and Metabolism*, **8**, 54–63.

Aston-Jones, G., Ennis, M., Pieribone, V. A., Nickell, W. T. & Shipley, M. T. (1986). The brain nucleus locus coeruleus: restricted afferent control of a broad efferent network. *Science*, **234**, 734–7.

Bradley, P. B., Engel, G., Feniuk, W., Fozard, J. R., Humphrey, P. P. A., Middlemiss, D. N., Mylecharane, E. J., Richardson, B. P. & Saxena, P. R. (1986). Proposals for the classification and nomenclature of functional receptors for 5-hydroxytryptamine. *Neuropharmacology*, **25**, 563–76.

Breslau, N., Davis, G. C. & Andreski, P. (1991). Migraine, psychiatric disorders, and suicide attempts: an epidemiologic study of young adults. *Psychiatry Research*, **37**, 11–23.

Brindley, G. S. & Lewin, W. S. (1968). The sensations produced by electrical stimulation of the visual cortex. *Journal of Physiology (London)*, **196**, 479–93.

Buzzi, M. G. & Moskowitz, M. A. (1990). The antimigraine drug, sumatriptan (GR43175), selectively blocks neurogenic plasma extravasation from blood vessels in dura mater. *British Journal of Pharmacology*, **99**, 202–6.

Buzzi, M. G., Dimitriadou, V., Theoharides, T. C. & Moskowitz, M. A. (1990). Morphological effects of electrical trigeminal ganglion stimulation on intra- and extracranial vessels. *Proceedings of the Society for Neuroscience*, **16**, 160.

Connor, R. C. R. (1962). Complicated migraine. A study of permanent neurological and visual defects caused by migraine. *Lancet*, **ii** 1072–5.

Curran, D. A., Hinterberger, H. & Lance, J. W. (1965). Total plasma serotonin, 5–hydroxyindoleacetic acid and *p*-hydroxy-*m*-metnoxymandelic acid excretion in normal and migrainous subjects. *Brain*, **88**, 997–1010.

Dimitriadou, V., Buzzi, M. G., Lambracht-Hall, M., Moskowitz, M. A. & Theoharides, T. C. (1990). In vivo and in vitro ultrastructural evidence for stimulation of dural mast cells by neuropeptides. *Proceedings of the Society for Neuroscience*, **16**, 161.

Doenicke, A., Brand, J. & Perrin, V. L. (1988). Possible benefit of GR43175, a novel 5-HT1-like receptor agonist, for the acute treatment of severe migraine. *Lancet*, **i**, 1309–11.

Drummond, P. D. & Lance, J. W. (1983). Extracranial vascular changes and the source of pain in migraine headache. *Annals of Neurology*, **13**, 32–7.

Drummond, P. D. & Lance, J. W. (1984). Neurovascular disturbances in headache patients. *Clinical and Experimental Neurology*, **20**, 93–9.

Edmeads, J. (1977). Cerebral blood flow in migraine. *Headache*, **17**, 148–52

Edvinsson, L., Delgado-Zygmunt, T., Ekman, R., Jansen, I., Svendgaard, N. A. & Uddman, R. (1990). Involvement of perivascular sensory fibers in the pathophysiology of cerebral vasospasm following subarachnoid hemorrhage. *Journal of Cerebral Blood Flow and Metabolism*, **10**, 602–7.

Edvinsson, L., Ekman, R., Jansen, I., McCulloch, J., Mortensen, A. & Uddman, R. (1991). Reduced levels of calcitonin gene-related peptide-like immunoreactivity in human brain after subarachnoid haemorrhage. *Neuroscience Letters*, **121**, 151–4.

Edvinsson, L., MacKenzie, E. T., McCulloch, J. & Uddman, R. (1988). Nerve supply and receptor mechanisms in intra- and extracerebral blood vessels. In *Basic Mechanisms of Headache*, eds. J. Olesen & L. Edvinsson, pp. 127–44. Amsterdam: Elsevier.

Edvinsson, L., McCulloch, J., Kingman, T. A. & Uddman, R. (1986). On the functional role of the trigemino-cerebrovascular system in the regulation of cerebral circulation. In *Neural Regulation of the Cerebral Circulation*, ed. Ch. Owman & J. E. Hardebo, pp. 407–18. Stockholm: Elsevier.

Edvinsson, L., Uddman, R. & Juul, R. (1990). Peptidergic innervation of the cerebral circulation. Role in subsrachnoid hemorrhage in man. *Neurosurgical Reviews*, **13**, 265–72.

Feindel, W., Penfield, W. & McNaughton, F. (1960). The tentorial nerves and localisation of intracranial pain in man. *Neurology*, **10**, 555–63.

Ferrari, M. D. (1991). Treatment of migraine attacks with sumatriptan. *New England Journal of Medicine,* **325**, 316–21.

Foote, S. L., Bloom, F. E. & Aston-Jones, G. (1983). Nucleus locus coeruleus: new evidence of anatomical and physiological specificity. *Physiological Reviews,* **63**, 844–914.

Friberg, L., Skyhoj-Olsen, T., Roland, P. E. & Lassen, N. A. (1987). Focal ischaemia caused by instability of cerebrovascular tone during attacks of hemiplegic migraine. *Brain,* **110**, 917–34.

Friberg, L., Nicolic, I., Olesen, J., Iversen, H., Sperling, B., Lassen, N. A. & Tfelt-Hansen, P. (1991). Interictal rCBF studies with ^{133}Xe or ^{99m}T$_c$-HMPAO and SPECT in patients suffering from migraine with aura. In *Migraine and other headaches: the vascular mechanisms,* ed. J. Olesen, pp. 53–9. New York: Raven Press.

Goadsby, P. J. (1985). Brainstem activation of the adrenal medulla in the cat. *Brain Research,* **327**, 241–8.

Goadsby, P. J. (1991). Characteristics of facial nerve elicited cerebral vasodilatation determined with laser Doppler flowmetry. *American Journal of Physiology,* **260**, R255–62.

Goadsby, P. J. & Duckworth, J. W. (1987). Effect of stimulation of trigeminal ganglion on regional cerebral blood flow in cats. *American Journal of Physiology,* **253**, R270–4.

Goadsby, P. J. & Duckworth, J. W. (1989). Low frequency stimulation of the locus coeruleus reduces regional cerebral blood flow in the spinalized cat. *Brain Research,* **476**, 71–7.

Goadsby, P. J., Edvinsson, L. & Ekman, R. (1988). Release of vasoactive peptides in the extracerebral circulation of man and the cat during activation of the trigeminovascular system. *Annals of Neurology,* **23**, 193–6.

Goadsby, P. J., Edvinsson, L. & Ekman, R. (1990). Vasoactive peptide release in the extracerebral circulation of humans during migraine headache. *Annals of Neurology,* **28**, 183–7.

Goadsby, P. J. & Gundlach, A. L. (1991). Localization of [3H]-dihydroergotamine binding sites in the cat central nervous system: relevance to migraine. *Annals of Neurology,* **29**, 91–4.

Goadsby, P. J., Lambert, G. A. & Lance, J. W. (1982). Differential effects on the internal and external carotid circulation of the monkey evoked by locus coeruleus stimulation. *Brain Research,* **249**, 247–54.

Goadsby, P. J., Lambert, G. A. & Lance, J. W. (1983). Effects of locus coeruleus stimulation on carotid vascular resistance in the cat. *Brain Research,* **278**, 175–83.

Goadsby, P. J., Lambert, G. A. & Lance, J. W. (1985). The mechanism of cerebrovascular vasoconstriction in response to locus coeruleus stimulation. *Brain Research,* **326**, 213–7.

Goadsby, P. J. & Lance, J. W. (1990). Physiopathologie de la migraine. *Le revue du praticient,* **40**, 389–93.

Goadsby, P. J. & MacDonald, G. J. (1985). Extracranial vasodilatation mediated by VIP. *Brain Research,* **239**, 285–8.

Goadsby, P. J., Seylaz, J. & Mraovitch, S. (1991). Hypercapnic but not neurogenic cortical vasodilatation is blocked by spreading depression in rat. In *Migraine and Other Headaches,* ed. J. Olesen, pp. 181–5. New York: Raven Press.

Goadsby, P. J. & Zagami, A. S. (1990). Thalamic processing of craniovascular pain in the cat: a 2-deoxyglucose study. *Proceedings of the Society for Neuroscience,* **16**, 1144.

Goadsby, P. J. & Zagami, A. S. (1991). Stimulation of the superior sagittal sinus increases metabolic activity and blood flow in certain regions of the brainstem and upper cervical spinal cord of the cat. *Brain*, **114**, 1001–11.

Goadsby, P.J., Zagami, A.S., Donnan, G.A., Symington, G., Anthony, M., Bladin, P.F. & Lance, J.W. (1991). A double blind placebo controlled cross over study of sumatriptan in the treatment of acute migraine attacks. *Lancet*, **ii**, 782–783.

Hachinski, V. C., Olesen, J., Norris, J. W., Larsen, B., Enevoldsen, E. & Lassen, N. A. (1977). Cerebral hemodynamics in migraine. *Canadian Journal of the Neurological Sciences*, **4**, 229–45.

Hamblin, M. W., Ariani, K., Adriaenssens, P. I. & Ciaranello, R. D. (1987). [^{3}H]Dihydroergotamine as a high-affinity, slowly dissociating radioligand for $5HT_{1b}$ binding sites in rat brain membranes: evidence for guanine nucleotide regulation of agonist affinity states. *Journal of Pharmacology and Experimental Therapeutics*, **243**, 989–1001.

Hansen, A. J., Quistorff, B. & Gjedde, A. (1980). Relationship between local changes in cortical blood flow and extracellular K^+ during spreading depression. *Acta Physiologica Scandinavica*, **109**, 1–6.

Harer, C. & von Kummer, R. (1991). Cerebrovascular CO2 reactivity in migraine: assessment by transcranial Doppler ultrasound. *Journal of Neurology*, **238**, 23–6.

Headache Classification Committee of the International Headache Society (1988). Classification and diagnostic criteria for headache disorders, cranial neuralgias and facial pain. *Cephalalgia*, **8** (Suppl. 7), 1–96.

Herold, S., Gibbs, J. M., Jones, A. K. P., Brooks, D. J., Frackowiak, R. S. J. & Legg, N. J. (1985). Oxygen metabolism in migraine. *Journal of Cerebral Blood Flow and Metabolism,* **5** (Suppl.), S445–6.

Jones, T. H., Morawetz, R. B., Crowell, R. M., Marcoux, F. W., Fitzgibbon, S. J., De Girotami, V. & Ojeman, R. G. (1981). Threshold of focal cerebral ischemia in awake monkeys. *Journal of Neurosurgery*, **54**, 773–8.

Juul, R., Edvinsson, L., Gisvold, S. E., Ekman, R., Brubakk, A. O. & Fredriksen, T. A. (1990). Calcitonin gene-related peptide in subarachnoid haemorrhage in man. Signs of activation of the trigemino-cerebrovascular system? *British Journal of Neurosurgery*, **4**, 171–80.

Kalberer, F., Schreier, E. & Zehnder, K. (1971). The pharmacokinetics of 3H-DHE 45–ms in rat, rabbit, cat and dog. In *Dihydroergotamine-mesylate (DHE) 45-ms* (internal document), Basel: Sandoz.

Kimball, R. W., Friedman, A.P. & Vallejo, E. (1960). Effect of serotonin in migraine patients. *Neurology (Minneap)*, **10**, 107–11.

Kraig, R. P. & Nicholson, C. (1978). Extracellular ionic variations during spreading depression. *Neuroscience*, **3**, 1045–59.

Kupfermann, I. (1985). Hypothalamus and limbic system II: motivation. In *Principles of Neural Science*, eds. E.R. Kandel & J.H. Schwartz, pp. 626–35. Amsterdam: Elsevier.

Lambert, G. A. & Duckworth, J. W. (1986). Comparative effects of ergotamine and DHE on craniovascular sensation and reactivity. *Proceedings of the Australasian Society of Clinical and Experimental Pharmacologists*, **20**, 232.

Lambert, G. A., Goadsby, P. J., Zagami, A. S. & Duckworth, J. W. (1988). Comparative effects of stimulation of the trigeminal ganglion and the superior sagittal sinus on cerebral blood flow and evoked potentials in the cat. *Brain Research*, **453**, 143–9.

Lambert, G. A., Zagami, A., Bogduk, N. & Lance, J. W. (1991). Cervical spinal cord neurons receiving sensory input from the cranial vasclature. *Cephalalgia,* **11**, 75–85.

Lance, J.W. (1982). *Mechanism and Management of Headache*, 4th edn. London: Butterworth Scientific.

Lance, J. W., Lambert, G. A., Goadsby, P. J. & Duckworth, J. W. (1983). Brainstem influences on cephalic circulation: experimental data from cat and monkey of relevance to the mechanism of migraine. *Headache*, **23**, 258–65.

Lashley, K. S. (1941). Patterns of cerebral integration indicated by the scotomas of migraine. *Archives of Neurology and Psychiatry*, **46**, 331–9.

Lauritzen, M. (1984). Long-lasting reduction of cortical blood flow of the rat brain after spreading depression with preserved autoregulation and impaired CO_2 response. *Journal of Cerebral Blood Flow and Metabolism*, **4**, 546–54.

Lauritzen, M., Jorgensen, M. B., Diemer, N. H., Gjedde, A. & Hansen, A. J. (1982). Persistent oligaemia of rat cerebral cortex in the wake of spreading depression. *Annals of Neurology*, **12**, 469–74.

Lauritzen, M. & Olesen, J. (1984). Regional cerebral blood flow during migraine attacks by Xenon-133 inhalation and emission tomography. *Brain*, **107**, 447–61.

Lauritzen, M., Skyhoj-Olsen, T., Lassen, N. A. & Paulson, O. B. (1983). The changes of regional cerebral blood flow during the course of classical migraine attacks. *Annals of Neurology*, **13**, 633–41.

Lauritzen, M., Skyhoj-Olsen, T., Lassen, N. A. & Paulson, O. B. (1983). Regulation of regional cerebral blood flow during and between migraine attacks. *Annals of Neurology*, **14**, 569–72.

Leao, A. A. P. (1944*a*). Spreading depression of activity in cerebral cortex. *Journal of Neurophysiology*, **7**, 359–90.

Leao, A. A. P. (1944*b*). Pial circulation and spreading activity in the cerebral cortex. *Journal of Neurophysiology*, **7**, 391–6.

Markowitz, S., Saito, K. & Moskowitz, M. A. (1987). Neurogenically mediated leakage of plasma proteins occurs from blood vessels in dura mater but not brain. *Journal of Neuroscience*, **7**, 4129–36.

Markowitz, S., Saito, K. & Moskowitz, M. A. (1988). Neurogenically mediated plasma extravasation in dura mater: effect of ergot alkaloids. A possible mechanism of action in vascular headache. *Cephalalgia*, **8**, 83–91.

Marshall, W. H. (1959). Spreading cortical depression of Leao. *Physiological Reviews*, **39**, 239–88.

Mathew, N. T., Hrastnik, F. & Meyer, J. S. (1976). Regional cerebral blood flow in the diagnosis of vascular headache. *Headache*, **15**, 252–60.

Matsuyama, T., Shiosaka, S., Matsumoto, M., Yoneda, S., Kimura, K., Abe, H., Hayakawa, T., Inoue, H. & Tohyama, M. (1983). Overall distribution of vasoactive intestinal polypeptide-containing nerves on the wall of the cerebral arteries: an immunohistochemical study using whole-mounts. *Neuroscience*, **10**, 89–96.

Moore-Ede, M. C. (1983). The circadian timing system in mammals: two pacemakers preside over many secondary oscillators. *Federation Proceedings*, **42**, 2802–8.

Moskowitz, M. A., Reinhard, J. F., Romero, J., Melamed, E. & Pettibone, D. J. (1979). Neurotransmitters and the fifth cranial nerve: is there a relation to the headache phase of migraine? *Lancet*, **ii**, 883–4.

Mraovitch, S., Calando, Y. & Seylaz, J. (1989). Long-lasting cerebral blood flow and metabolic changes within the limbic and brainstem regions following cortical spreading depression in rat. *Journal of Cerebral Blood Flow and Metabolism*, **9**, S508.

Mraovitch, S., Lasbennes, F., Calando, Y. & Seylaz, J. (1986). Cerebrovascular changes elicited by electrical stimulation of the centromedian-parafasicular complex in the rat. *Brain Research*, **380**, 42–53.

Norris, J. W., Hachinski, V. C. & Cooper, P. W. (1975). Changes in cerebral blood flow during a migraine attack. *British Medical Journal*, **3**, 676–7.

O'Brien, M. D. (1971). Cerebral blood flow changes in migraine. *Headache*, **10**, 139–43.

Olesen, J., Friberg, L., Skyhoj-Olsen, T., Iversen, H. K., Lassen, N. A., Andersen, A. R. & Karle, A. (1990). Timing and topography of cerebral blood flow, aura and headache during migraine attacks. *Annals of Neurology*, **28**, 791–8.

Olesen, J., Larsen, B. & Lauritzen, M. (1981*a*). Focal hyperemia followed by spreading oligemia and impaired activation of rCBF in classic migraine. *Annals of Neurology*, **9**, 344–52.

Olesen, J., Lauritzen, M., Tfelt-Hansen, P., Henriksen, L. & Larsen, B. (1982). Spreading cerebral oligemia in classical- and normal cerebral blood flow in common migraine. *Headache*, **22**, 242–8.

Olesen, J., Tfelt-Hansen, P., Henriksen, L. & Larsen, B. (1981*b*). The common migraine attack may not be initiated by cerebral ischemia. *Lancet*, **ii**, 438–40.

Peatfield, R. C. & Rose, F. C. (1991). A prospective study of unilateral classical migraine. *New Advances in Headache Research: 2*, pp. 35–8. London: Smith-Gordon.

Penfield, W. & McNaughton, F. L. (1940). Dural headache and the innervation of the dura mater. *Archives of Neurology and Psychiatry*, **44**, 43–75.

Penfield, W. & Perot, P. (1963). The brain's record of auditory and visual experience. *Brain*, **86**, 595–696.

Raskin, N. H., Hosobuchi, Y. & Lamb, S. (1987). Headache may arise from perturbation of brain. *Headache*, **27**, 416–20.

Raskin, N. H. & Knittle, S. C. (1976). Ice cream headache and orthostatic symptoms in patients with migraine headache. *Headache*, **16**, 222–5.

Raskin, N. H. & Schwartz, R. K. (1980). Icepick-like pain. *Neurology*, **30**, 203–5.

Ray, B. S. & Wolff, H. G. (1940). Experimental studies on headache. Pain sensitive structures of the head and their significance in headache. *Archives of Surgery*, **41**, 813–56.

Saito, K. & Moskowitz, M. A. (1989). Contributions from the upper cervical dorsal roots and trigeminal ganglia to the feline circle of Willis. *Stroke*, **20**, 524–6.

Sakai, F. & Meyer, J. S. (1978). Regional cerebral hemodynamics during migraine and cluster headaches measured by the 133-Xe inhalation method. *Headache*, **18**, 122–32.

Sakai, F. & Meyer, J. S. (1979). Abnormal cerebrovascular reactivity in patients with migraine and cluster headache. *Headache*, **19**, 257–66.

Sakas, D. E., Moskowitz, M. A., Wei, E. P., Kontos, H. A., Kano, M. & Ogilvy, C. (1989). Trigeminovascular fibers increase blood flow in cortical grey matter by axon-dependent mechanisms during severe hypertension or siezures. *Proceedings of the National Academy of Sciences, USA*, **86**, 1401–5.

Saxena, P. R. (1991). 5-HT in migraine – an introduction. *Journal of Neurology*, 238, S36–7.

Sicuteri, F., Testi, A. & Anselmi, B. (1961). Biochemical investigations in headache: increases in hydroxyindoleacetic acid excretion during migraine attacks. *International Archives of Allergy*, **19**, 55–8.

Simard, D. & Paulson, O. B. (1973). Cerebral vasomotor paralysis during migraine attack. *Archives of Neurology*, **29**, 207–9.

Skinhøj, E. (1973). Hemodynamic studies within the brain during migraine. *Archives of Neurology*, **29**, 95–8.

Skinhøj, E. & Paulson, O. B. (1969). Regional cerebral blood flow in the internal carotid artery distrbution during migraine. *British Medical Journal*, **3**, 569–70.

Skyhoj-Olsen, T. (1990). Migraine with and without aura: the same disease due to cerebral vasospasm of different intensity. A hypothesis based on CBF studies during migraine. *Headache*, **30**, 269–72.

Skyhoj-Olsen, T., Friberg, L. & Lassen, N. A. (1987). Ischemia may be the primary cause of the neurological deficits in classic migraine. *Archives of Neurology*, **44**, 156–61.

Skyhoj-Olsen, T. & Lassen, N. A. (1989). Blood flow and vascular reactivity during attacks of classic migraine- limitations of the Xe133 intraarterial technique. *Headache*, **29**, 15–20.

Sramka, M., Brozek, G., Bures, J. & Nadvornik, P. (1977). Functional ablation by spreading depression: possible use in human stereotactic neurosurgery. *Applied Neurophysiology*, **40**, 48–61.

Staehelin-Jensen, T., Voldby, B., Olivarius, B. F. & Jensen, F. T. (1981). Cerebral hemodynamics in familial hemiplegic migraine. *Cephalalgia*, **1**, 121–5.

Symonds, C. (1952). Migrainous variants. *Transactions of the Medical Society of London*, **67**, 237–50.

Tomida, S., Wagner, H. G., Klatzo, I. & Nowak, T. S. (1989). Effect of acute electrode placement on regional CBF in the gerbil: A comparison of blood flow measured by hydrogen clearance, [3H]-nicotine, and [14C]-iodo- antipyrine techniques. *Journal of Cerebral Blood Flow and Metabolism*, **9**, 79–86.

Trojaborg, W. & Boysen, G. (1973). Relation between EEG, regional cerebral blood flow and internal carotid artery pressure during carotid endarterectomy. *Electroenephalography and Clinical Neurophysiology*, **34**, 61–9.

White, J. C. & Sweet, W. H. (1955). *Pain: Its Mechanisms and Neurosurgical Control,* pp. 520–2. Springfield: Springfield.

Wolff, H. G. (1963). *Headache and Other Head Pain.* New York: Oxford University Press.

Young, P. A., McCasland, J. S., Woolsey, T.A., Rhoades, R. W. & Jacquin, M. F. (1990). 2-DG labelling patterns in the trigeminal brainstem complex following selective vibrissal stimulation in the adult hamster. *Proceedings of the Society for Neuroscience*, **16**, 223.

Zagami, A. S. & Lambert, G. A. (1990). Stimulation of cranial vessels excites nociceptive neurones in several thalamic nuclei of the cat. *Experimental Brain Research*, **81**, 552–66.

Zagami, A. S., Goadsby, P. J. & Edvinsson, L. (1990). Stimulation of the superior sagittal sinus in the cat causes release of vasoactive peptides. *Neuropeptides*, **16**, 69–75.

20

Serotonin, sumatriptan and migraine

P. P. A. HUMPHREY, W. FENIUK, M. J. PERREN AND A.W. OXFORD

Glaxo Group Research Limited,
Park Road,
Ware, Hertfordshire SG12 0DP,
UK

Is migraine a vascular disease?

For more than half a century migraine had been firmly considered a vascular disease (for review see Wolff, 1963). However, the concept was strongly challenged in the 1980s on the basis of two significant findings. First, the group of Olesen observed that during migraine without aura (or common migraine as it was known), there are no changes in cerebral blood flow, although there was a spreading oligaemia in patients with preceding neurological symptoms (Olesen *et al.*, 1981; Lauritzen & Olesen, 1984; Olesen, 1985). Parenthetically, it would now seem that it is the extracerebral, intracranial vessels which are important (see below) and these are even more difficult to study in man in vivo than the intracranial resistance vessels which control the blood supply to the brain itself.

Secondly, Raskin's group provided evidence for a localized neurological deficit in pain transmission in migraineurs (Raskin, 1981). He pointed out that migraineurs were more susceptible to pain induced by ingestion of cold substances, such as ice cream, and that the pain was frequently felt in a similar unilateral site to the normal migraine headache (Raskin & Knittle, 1976). Furthermore, patients with intractable nonhead pain, who had electrodes implanted into the region of the periaqueductal grey area, experienced severe migraine-like headaches (Raskin, Hosobuchi & Lamb, 1987). These observations led to a view that a pathological disturbance within the brain itself could lead to head pain. However, the periaqueductal grey area is close to the dorsal raphé nucleus and vascular changes are likely to have been involved in the headache produced by the electrode implantation (Goadsby *et al.*, 1985). Nevertheless, it is most likely that migraineurs do have a defective (hypersensitive) pain pathway localised to the anatomical location of their migrainous headaches. This is consistent with the well-known observation that reserpine will induce migraine-like headaches in migraineurs but not in normal individuals (Kimball & Friedman, 1961; Anthony, Hinterberger & Lance, 1969; Lance *et al.*, 1989).

Notwithstanding, the neuroanatomical studies of Moskowitz have clearly implicated the trigeminal (Vth cranial) nerve in the transmission of head pain. Since the afferent terminals of this nerve densely innervate the large intracranial blood vessels, it seems likely that the pain signal originates from a vascular source (Moskowitz et al., 1979; Mayberg et al., 1981; Moskowitz, Henrikson & Markowitz, 1986). Furthermore, the discovery of sumatriptan (see below), a vasoactive drug which penetrates the blood brain barrier poorly, has again focused thinking on the importance of cranial blood vessels in the genesis of head pain (Humphrey & Feniuk, 1991).

The involvement of 5-hydroxytryptamine (5-HT)

The strong belief that 5-HT was somehow involved in the pathogenesis of migraine was a key factor behind the discovery of sumatriptan. It was this conviction that led us to systematically characterize the receptors for 5-HT, particularly those in the vasculature (Apperley, Humphrey & Levy, 1976; Apperley et al., 1980; Humphrey, 1983, 1984; Bradley et al., 1986; Humphrey & Feniuk, 1987).

5-Hydroxytryptamine was first implicated in the pathophysiology of migraine by the work of Sicuteri, who showed that the metabolite of 5-HT, 5-hydroxyindole acetic acid (5-HIAA), was excreted in increased amounts in urine during the period of a migraine headache (Sicuteri, Testi & Anselmi, 1961). Further detailed studies by Lance's group confirmed that disturbances of plasma levels of 5-HT undoubtedly occurred. Furthermore, reserpine induced migraine-like headaches in migraineurs and this correlated with a fall in plasma (platelet) 5-HT content (Curran, Hinterberger & Lance, 1965; Anthony et al., 1969). However, it was not clear why reserpine induces migraine-like headaches in migraineurs, but not in nonmigraineurs. Nevertheless, these studies provided sound evidence for 5-HT's involvement, if only as an epiphenomenon. This, together with the observation that 5-HT was apparently a much more effective vasoconstrictor of cranial rather than peripheral vessels (Toda & Fjita, 1973), led us to focus on vascular 5-HT receptor pharmacology.

Another observation which reinforced the view that 5-HT receptors were involved in migraine was the clinical work of Lance and colleagues on the use of 5-HT receptor antagonists in the prophylaxis of migraine (Lance, Anthony & Somerville, 1970). The key conclusion was that methysergide was much superior to all the others. This led us to concentrate on the pharmacology of methysergide in relation to 5-HT receptors. For a review of the current status of 5-HT receptors see Hen (1992). Our approach led to the identification of a then uncharacterized 5-HT receptor type (Apperley et al., 1977, 1980; Feniuk

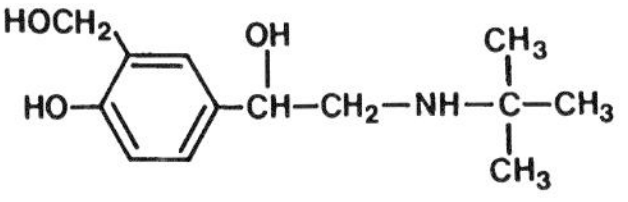

Salbutamol

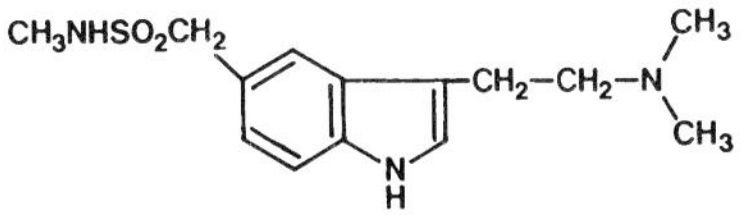

Sumatriptan

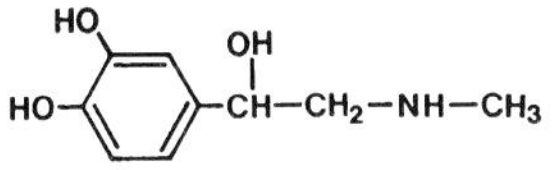

Adrenaline

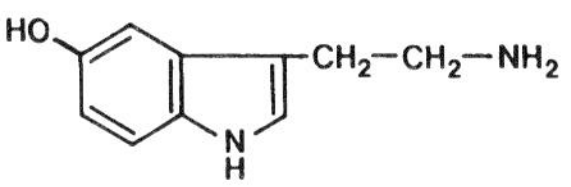

Serotonin (5-HT)

Fig. 20.1. Comparative chemical structures of salbutamol and sumatriptan with their respective 'parent' neurotransmitter.

et al., 1985). Intriguingly, at this receptor methysergide is an *agonist*, albeit a weak partial agonist, in contrast to its potent *antagonist* action at 5-HT_2 receptors (Apperley *et al.*, 1980; Watts, Feniuk & Humphrey, 1981).

Another key observation was that 5-HT given intravenously could abort a migrainous headache, whether it be a spontaneous attack or reserpine-induced (Anthony *et al.*, 1969). This was an important confirmation of an earlier observation that intravenous 5-HT had a beneficial effect in treating spontaneous migrainous headache (Kimball, Friedman & Valeejo, 1960). It seemed likely that 5-HT was acting as a vasoconstrictor to ameliorate the headache, as is the case for ergotamine, noradrenaline and other vasoactive agents (Lance, 1973; Humphrey *et al.*, 1990*a*). We, therefore, argued that an agent which would mimic the desirable effect(s) of 5-HT (possibly carotid vasoconstriction) but not mimic the undesirable effects (including generalized vasoconstriction, bronchoconstriction, platelet aggregation) would be of benefit in the acute treatment of migraine (Humphrey *et al.*, 1989*a*,*b*). It was speculated that a full agonist, at the novel receptor (later called 5-HT_1-like) identified from studies with methysergide, might be such an agent. This idea became particularly attractive as evidence accrued that the receptor mainly occurred in some but not all cranial vessels (see below). The concept also had some appeal from the standpoint that such an agonist could be thought of as a 'replacement' for the abnormally reduced blood levels of the natural chemical mediator, 5-HT, itself. However, at an early stage in the project we rejected the hypothesis that 5-HT was causative in migraine (Humphrey *et al.*, 1990*a*; Humphrey, 1991) and simplified our drug discovery hypothesis. It was thus no longer a *sine qua non* that there was a primary involvement of 5-HT in the disease process. We simply proposed that a selective 5-HT_1-like receptor agonist would produce a

Table 20.1. *Sumatriptan and Dihydroergotamine Interactions with Neurotransmitter Receptor Sites*

Receptor	K_i values (nM)	
Serotoninergic	Sumatriptan	Dihydroergotamine
5-HT_{1D}	17±3	19±3
5-HT_{1A}	100±20	1.2±0.2
5-HT_{2C}	>10 00	39±10
5-HT_{2}	>10 000	78±20
5-HT_{3}	>10 000	>10 000
Adrenergic		
Alpha $_1$	>10 000	6.6±0.9
Alpha $_2$	>10 000	3.4±0.5
Beta	>10 000	960±30
Dopaminergic		
Dopamine$_1$	>10 000	700±100
Dopamine$_2$	>10 000	98±10
Other Sites		
Muscarinic	>10 000	>10 000
Benzodiazepine	>10 000	>10 000

Source: Data from Peroutka and McCarthy, 1989.
Note: K_i values are dissociation constants from various radio-labelled ligand binding experiments at different brain receptors. Values greater than 10 000 indicate that there was less than 50% displacement at the very high concentration of sumatriptan of 10 μM.

selective vasoconstriction of those particular vessels distended and dilatated during the headache period. This was analogous to the conceptual thinking behind the anti-asthma drug, salbutamol, which mimics the desirable effects (bronchodilatation) of the natural hormone, adrenaline, without mimicking its undesirable effects on the heart (see Fig. 20.1). Salbutamol has such a profile because it activates β_2- but not β_1-adrenoceptors (Brittain, Jack & Richie, 1970). However, in the same way that falls in adrenaline levels are not responsible for asthma, so changes in plasma levels of 5-HT are unlikely to be causative in migraine.

Screening of chemical analogues of 5-HT in the search for a selective 5-HT_1-like receptor agonist began as early as 1976. Over the years we identified some key compounds which were unsuitable for development as drugs, but became prototypical drug tools (Humphrey *et al.*, 1989*a*,*b*; 1990*a*,*b*). Eventually, we identified sumatriptan (Fig. 20.1), a remarkably selective 5-HT_1-like receptor agonist which, unlike the parent compound, 5-HT, has no action at 5-HT_2, 5-HT_3 or 5-HT_4 receptors (Feniuk, Humphrey & Connor, 1992), or at receptors for other neurotransmitters (see Table 20.1).

Sumatriptan

Sumatriptan is a highly selective agonist for a 5-HT_1-like receptor subtype localised mainly in the vasculature of certain large cranial blood vessels (Humphrey et al., 1989a,b). Thus, sumatriptan causes contraction of isolated large cerebral arteries from a number of species, including man (Connor, Feniuk & Humphrey, 1989; Parsons et al., 1989) and human isolated dural arteries (Humphrey et al., 1991a) with little effect in most isolated peripheral vascular preparations (Humphrey et al., 1988). Sumatriptan has been shown to selectively constrict the carotid artery bed of anesthetised dogs and cats by an action at 5-HT_1-like receptors without affecting the peripheral circulation (Feniuk, Humphrey & Perren, 1989; Feniuk & Humphrey, 1989). Despite a carotid vasoconstrictor action, sumatriptan does not modify cerebral blood flow in animals or in man (Perren, Feniuk & Humphrey, 1989; Friberg, 1989; Friberg et al., 1991), probably because its action is restricted to large conducting vessels and the meningeal vasculature which do not regulate blood flow to the brain.

Since the pain of migraine is probably of vascular origin, it is likely that the mechanism of action of sumatriptan in the alleviation of migraine headache is attributable to its ability to constrict those cranial blood vessels which are abnormally distended (and inflamed) during a migraine attack (Humphrey & Feniuk, 1991). Whether the primary site of action is at the level of dural or large cerebral blood vessels remains to be determined, but, since sumatriptan does not readily cross the blood–brain barrier (Humphrey et al., 1990b; Sleight et al., 1990; Humphrey et al., 1991b,c), it is theoretically more likely to reach 5-HT_1-like receptors on the vascular smooth muscle of dural vessels than those on cerebral blood vessels where the endothelial cell junctions are presumably tighter. Interestingly, sumatriptan has been shown to constrict potently the isolated perfused meningeal vasculature in the dura taken from human cadavers (Humphrey et al., 1991a). A recent clinical study by Friberg et al. (1991) has shown that vasodilatation of the middle cerebral artery on the headache side, during a migraine attack, is reversed to normal following treatment with sumatriptan. Sumatriptan had little or no effect either on the cerebral artery on the headache free side or on cerebral blood flow. This raises the possibility that some localized disruption of blood–brain barrier function may occur in dilated, inflamed intracranial arteries during a migraine attack, thus allowing selective penetration of sumatriptan through the cerebrovascular intima to the smooth muscle.

It has long been known that mechanical and electrical stimulation of intracranial blood vessels will cause pain (Ray & Wolff, 1940). More recently,

Nichols et al. (1990) have shown that distension of cerebral arteries causes pain in the orbital and temporal regions, areas which are innervated by the trigeminal nerve. Cranial blood vessels, such as large cerebral conducting vessels and meningeal arteries, are innervated by a dense perivascular network of sensory nerves arising from the trigeminal ganglion (Moskowitz et al., 1986; Moskowitz, 1984, 1987). It has been suggested that the pain of migraine results from a sterile neurogenic inflammation of cranial blood vessels innervated by the trigeminal nerve (Markowitz, Saito & Moskowitz, 1987; Moskowitz et al., 1989). Thus antidromic stimulation of trigeminal nerve fibres leads to the release of sensory neuropeptide transmitters, such as substance P and CGRP, at peripheral nerve endings and increased vascular permeability in the blood vessel wall, protein extravasation and edema. Experimental studies in anesthetized rats and guinea-pigs have shown that sumatriptan inhibits plasma protein extravasation in dura mater induced by electrical stimulation of the trigeminal ganglion (Buzzi & Moskowitz, 1990). Such an effect could result from either a localized vasoconstrictor effect of sumatriptan on meningeal vessels or an inhibitory effect on neuropeptide release from perivascular nerve terminals. $5-HT_1$-like receptors similar to those demonstrated on blood vessels have been shown to exist on peripheral nerves where they cause inhibition of neurotransmitter release (Feniuk, Humphrey & Watts, 1979). Interestingly, sumatriptan attenuates the release of CGRP following trigeminal stimulation in the rat (Buzzi et al., 1991). In man, CGRP levels in jugular blood are elevated following both trigeminal stimulation and during a migraine attack (Goadsby, Edvinsson & Ekman, 1988, 1990). Following treatment with sumatriptan, alleviation of the migraine headache is accompanied by a reduction in the circulating levels of CGRP (Goadsby & Edvinsson, 1991). However, these data neither prove nor disprove the theory that sumatriptan is acting predominantly by a neuronal mechanism (see Buzzi et al., 1991). It could be argued that neuropeptides are released as a consequence of trigeminal nerve activation and that abolition of the headache by a vasoconstrictor mechanism is sufficient to prevent activation and consequent neuropeptide release.

Conclusions

Studies with sumatriptan are helping to unravel the precise mechanisms involved in the genesis of a migraine attack. Although the debate about the vascular versus neuronal hypotheses of migraine will no doubt continue, there seems consensus that sumatriptan is acting at the level of the blood vessel wall. As $5-HT_1$-like receptors are largely localised to the cranial vasculature, the

Serotonin, sumatriptan and migraine 329

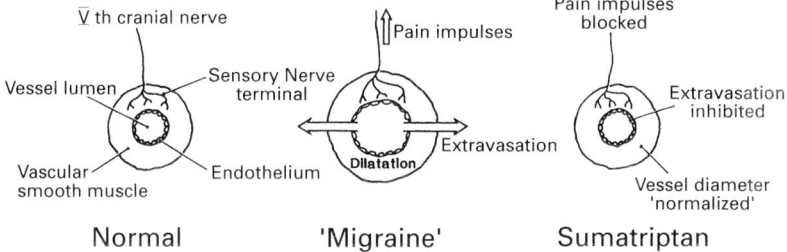

Fig. 20.2. Hypothetical mechanism of action of sumatriptan. During a migrainous headache, intracranial blood vessels may become distended and oedematous due to activation of trigeminal (Vth cranial) nerve terminals and the subsequent release of neuropeptides. Sumatriptan activates vascular 5-HT$_1$ receptors to constrict the affected vessels, thereby opposing the extravasation and consequent pain. An additional inhibitory action on the sensory nerve terminals could concomitantly lead to a reduction in the release of neuropeptides. Humphrey & Feniuk (1991) reproduced with permission.

action of sumatriptan is largely restricted to those vessels believed to be distended and oedematous during a migraine headache (see Fig. 20.2). Whether the primary site of action of sumatriptan involves the blood vessels or perivascular nerves, or both, remains to be determined. Both mechanisms could be important. Further clinical studies involving the use of sumatriptan in headaches induced by various vasodilator agents such as nitroglycerin, alcohol, calcium channel blockers and potassium channel agonists, may help to define further both the mechanism of action of sumatriptan and the pathophysiology of migraine (see Olesen & Iversen, this volume). If sumatriptan acts exclusively at the level of the trigeminal nerve, it may be possible to develop a novel agonist acting at a non-5-HT receptor to mimic this effect. Such agents might include a somatostatin receptor agonist or histamine H$_3$ receptor agonist (Sicuteri *et al.*, 1984; Arrang *et al.*, 1991).

References

Anthony, M., Hinterberger, H. & Lance, J. W. (1969). The possible relationship of serotonin to the migraine syndrome. *Research and Clinical Studies in Headache*, **2**, 29–59.

Apperley, E., Feniuk, W., Humphrey, P. P. A. & Levy, G. P. (1980). Evidence for two types of excitatory receptor for 5-hydroxytryptamine in dog isolated vasculature. *British Journal of Pharmacology*, **68**, 215–24.

Apperley, E., Humphrey, P. P. A. & Levy, G. P. (1976). Receptors for 5-hydroxytryptamine and noradrenaline in rabbit isolated ear artery and aorta. *British Journal of Pharmacology*, **58**, 211–21.

Apperley, E., Humphrey, P. P. A. & Levy, G. P. (1977). Two types of excitatory receptor for 5-hydroxytryptamine in dog vasculature? *British Journal of Pharmacology*, **61**, 465P.

Arrang, J. M., Garbarg, M., Schwartz, J. C., Lipp, R., Stark, H., Schunack, W. & Lecomte, J. M. (1991). The Histamine H_3-receptor : pharmacology, roles and clinical implications studied with agonists. In *New Perspectives in Histamine Research*, pp. 55–67. Basel: Birkhäuser Verlag.

Bradley, P. B., Engel, G., Feniuk, W., Fozard, J. R., Humphrey, P. P. A., Middlemiss, D. N., Mylecharane, E. J., Richardson, B. P. & Saxena, P. R. (1986). Proposals for the classification and nomenclature of functional receptors for 5-hydroxytryptamine. *Neuropharmacology*, **25**, 563–76.

Brittain, R. T., Jack, D. & Richie, A. C. (1970). Recent β-adrenoreceptor stimulants. *Advances in Drug Research*, **5**, 197–253.

Buzzi, M. G., Carter, W. B., Shimizu, T., Heath, H. & Moskowitz, M. A. (1991). Dihydroergotamine and sumatriptan attenuate levels of CGRP in plasma in rat superior sagittal sinus during electrical stimulation of the trigeminal ganglion. *Neuropharmacology*, **30**, 1193–200.

Buzzi, M. G. & Moskowitz, M. A. (1990). The antimigraine drug, sumatriptan (GR43175), selectively blocks neurogenic plasma extravasation from blood vessels in dura mater. *British Journal of Pharmacology*, **99**, 202–6.

Connor, H. E., Feniuk, W. & Humphrey, P. P. A. (1989). Characterisation of 5-HT receptors mediating contractions of canine and primate basilar artery using GR43175, a selective 5-HT_1-like receptor agonist. *British Journal of Pharmacology*, **96**, 379–87.

Curran, D. A., Hinterberger, H. & Lance, J. W. (1965). Total plasma serotonin 5-hydroxyindoleacetic acid and *p*-hydroxy-*m*-methoxymandelic acid excretion in normal and migrainous subjects. *Brain*, **88**, 997–1008.

Feniuk, W. & Humphrey, P. P. A. (1989). Mechanisms of 5-hydroxytryptamine-induced vasoconstriction. In *The Peripheral Actions of 5-Hydroxytryptamine*, ed. J. R. Fozard. pp. 100–122, Oxford: Oxford University Press.

Feniuk, W., Humphrey, P. P. A. & Connor, H. E. (1992). The pharmacology of sumatriptan and its mode of action in migraine. In *Frontiers in Headache Research*, eds. J. Olesen & P. R. Saxena. New York: Raven Press, vol. 2, pp. 213–219.

Feniuk, W., Humphrey, P. P. A. & Perren, M. J. (1989). The selective carotid arterial vasoconstrictor action of GR43175 in anaesthetised dogs. *British Journal of Pharmacology*, **96**, 83–90.

Feniuk, W., Humphrey, P. P. A., Perren, M. J. & Watts, A. D. (1985). A comparison of 5-hydroxytryptamine receptors mediating contraction in rabbit aorta and dog saphenous vein: evidence for different receptor types obtained by use of selective agonists and antagonists. *British Journal of Pharmacology*, **86**, 697–704.

Feniuk, W., Humphrey, P. P. A. & Watts, A. D. (1979). Presynaptic inhibitory action of 5-hydroxytryptamine in dog isolated saphenous vein. *British Journal of Pharmacology*, **67**, 247–54.

Friberg, L. (1989). Effects of a new 5-HT_1-like receptor agonist GR43175 on the regional brain tissue perfusion and intra-cerebral artery tone regulation in healthy volunteers. *Cephalalgia*, **9** (Suppl. 10), 359–60.

Friberg, L., Olesen, J., Iversen, H. K. & Sperling, B. (1991). Migraine pain associated with middle cerebral artery dilatation: reversal by sumatriptan. *Lancet*, **338**, 13–17.

Goadsby, P. J. & Edvinsson, L. (1991). Sumatriptan reverses the changes in calcitonin gene-related peptide seen in the headache phase of migraine. *Cephalalgia*, **11** (Suppl. 11), 3–4.

Goadsby, P. J., Edvinsson, L. & Ekman, R. (1988). Release of vasoactive peptides in the extracerebral circulation of humans and the cat during activation of the trigeminovascular system. *Annals of Neurology*, **23**, 193–6.

Goadsby, P. J., Edvinsson, L. & Ekman, R. (1990). Vasoactive peptide release in the extracerebral circulation of humans during migraine headache. *Annals of Neurology*, **28**, 183–7.

Goadsby, P. J., Piper, R. D., Lambert, G. A. & Lance, J. W. (1985). The effect of activation of the nucleus raphe dorsalis on carotid blood flow. *American Journal of Physiology*, **248**, R257–62.

Hen, R. (1992). Of mice and men: commonalities among 5-HT receptors. *Trends in Pharmacological Sciences*, **13**, 160–5.

Humphrey, P. P. A. (1983). Pharmacological characterisation of cadiovascular 5-hydroxytryptamine receptors. In *Vascular Neuroeffector Mechanisms, IVth International Symposium*, eds. Bevan, J. A., Godfraind, T., Maxwell, R. A. & Vanhoutte, P. M., pp. 237–42. New York: Raven Press.

Humphrey, P. P. A. (1984). Peripheral 5-hydroxytryptamine receptors and their classification. *Neuropharmacology*, **23**, 1503–10.

Humphrey, P. P. A. (1991). 5-Hydroxytryptamine and the pathophysiology of migraine. *Journal of Neurology*, **238**, S38–S44.

Humphrey, P. P. A., Connor, H. E., Stubbs, C. M. & Feniuk, W. (1991*b*). Effect of sumatriptan on pial vessel diameter in vivo. In *Migraine and Other Headaches: The Vascular Mechanisms*, ed. J. Olesen, pp. 335–8. New York: Raven Press.

Humphrey, P. P. A. & Feniuk, W. (1987). Pharmacological characterisation of functional neuronal receptors for 5-hydroxytryptamine. In *Neuronal Messengers in Vascular Function*, ed. A. Nobin, C. Owman & B. Arneklo-Nobin, pp. 3–19. The Netherlands: Elsevier Science Publishers.

Humphrey, P. P. A. & Feniuk, W. (1991). Mode of action of the anti-migraine drug sumatriptan. *Trends in Pharmacological Sciences*, **12**, 444–6.

Humphrey, P. P. A., Feniuk, W., Marriott, A. S., Tanner, R. J. N., Jackson, M. R. & Tucker, M. L. (1991*c*). Preclinical studies on the anti-migraine drug, sumatriptan. *European Neurology*, **31**, 282–90.

Humphrey, P. P. A., Feniuk, W., Motevalian, M., Parsons, A. A. & Whalley, E. T. (1991*a*). The vasoconstrictor action of sumatriptan on human isolated dura mater. In *Serotonin: Molecular Biology, Receptors and Functional Effects*, eds. J. R. Fozard & P. R. Saxena, pp. 421–9. Basel: Birkhäuser Verlag.

Humphrey, P. P. A., Feniuk, W. & Perren, M. J. (1990a). 5-HT in migraine: Evidence from 5-HT$_1$-like receptor agonists for a vascular aetiology. In *Migraine: A Spectrum of Ideas*, eds. M. Sandler & Collins, G. M., pp. 147–72. Oxford: Oxford University Press.

Humphrey, P. P. A., Feniuk, W., Perren, M. J., Beresford, I. J. M., Skingle, M. & Whalley, E. T. (1990*b*). Serotonin and migraine. *Annals of New York Academy of Sciences*, **600**, 587–98.

Humphrey, P. P. A., Feniuk, W., Perren, M. J., Connor, H. E. & Oxford, A. W. (1989a). The pharmacology of the novel 5-HT$_1$-like receptor agonist, GR43175. *Cephalalgia*, **9** (Suppl. 9), 23–33.

Humphrey, P. P. A., Feniuk, W., Perren, M. J., Connor, H. E., Oxford, A. W., Coates, I. H. & Butina, D. (1988). GR43175, a selective agonist for the 5-HT$_1$-like receptor in dog isolated saphenous vein. *British Journal of Pharmacology*, **94**, 1123–32.

Humphrey, P. P. A., Feniuk, W., Perren, M. J., Oxford, A. W. & Brittain, R. T. (1989*b*). Sumatriptan succinate. *Drugs of the Future*, **14**, 35–9.

Kimball, R. W. & Friedman, A. P. (1961). Studies on the pathogenesis of migraine. *Recent Advances in Biology and Psychiatry*, **3**, 200–6.

Kimball, R. W., Friedman, A. P. & Valeejo, E. (1960). Effect of serotonin in migraine patients. *Neurology*, **10**, 107–11.

Lance, J. W. (1973). *The Mechanism and Management of Headache*. 2nd edn. London: Butterworth Publishers.

Lance, J. W., Anthony, M. & Somerville, B. (1970). Comparative trial of serotonin antagonists in the management of migraine. *British Medical Journal*, **2**, 327–30.

Lance, J. W., Lambert, G. A., Goadsby, P. J. & Zagami, A. S. (1989). 5-Hydroxytryptamine and its putative aetiological involvement in migraine. *Cephalalgia*, **9** (Suppl. 9), 7–13.

Lauritzen, M. & Olesen, J. (1984). Regional cerebral blood flow during migraine attacks by Xenon-133 inhalation and emission tomography. *Brain*, **107**, 447–61.

Markowitz, S., Saito, K. & Moskowitz, M. A. (1987). Neurogenically mediated leakage of plasma protein occurs from blood vessels in dura mater but not brain. *Journal of Neuroscience*, **7**, 4129–36.

Mayberg, M., Langer, R. S., Zervas, N. T. & Moskowitz, M. A. (1981). Perivascular meningeal projections format trigeminal ganglia: possible pathways for vascular headaches in man. *Science*, **213**, 228–30.

Moskowitz, M. A. (1984). Neurobiology of vascular head pain. *Annals of Neurology*, **16**, 157–68.

Moskowitz, M. A. (1987). Sensory connections to cephalic blood vessels and their possible importance to vascular headaches. In *Advances in Headache Research*, ed. F. Clifford Rose, pp. 81–6. London: John Libbey.

Moskowitz, M. A., Buzzi, M. G., Sakas, D. E. & Kinnik, M. D. (1989). Pain mechanisms underlying vascular headaches. *Reviews in Neurology (Paris)*, **145**, 181–93.

Moskowitz, M. A., Henrikson, B. M. & Markowitz, S. (1986). Experimental studies on the sensory innervation of the cerebral blood vessels. *Cephalalgia*, **7** (Suppl. 4), 63–6.

Moskowitz, M. A., Reinhard, J. F., Romero, J. & Pettibone, D. J. (1979). Neurotransmitters and the fifth cranial nerve: is there a relation to headache phase of migraine? *Lancet*, **ii**, 883–5.

Nichols, F. T., Mawad, M., Mohr, J. P., Stein, B., Hilal, S. & Michelsen, W. J. (1990). Focal headache during balloon inflation in the internal carotid and middle cerebral arteries. *Stroke*, **21**, 555–9.

Olesen, J. (1985). Migraine and regional cerebral blood flow. *Trends in Neurosciences*, **8**, 318–20.

Olesen, J., Tfelt-Hansen, P., Henriksen, L. & Larsen, B. (1981). The common migraine attack may not be initiated by cerebral ischaemia. *Lancet*, **ii**, 438–40.

Parsons, A. A., Whalley, E. T., Feniuk, W., Connor, H. E. & Humphrey, P. P. A. (1989). 5-HT_1-like receptors mediate 5-hydroxytryptamine-induced contraction of human isolated basilar artery. *British Journal of Pharmacology*, **96**, 434–49.

Peroutka, S. J. & McCarthy, B. G. (1989). Sumatriptan (GR43175) interacts selectively with 5-HT_{1B} and 5-HT_{1D} binding sites. *European Journal of Pharmacology*, **163**, 133–6.

Perren, M. J., Feniuk, W. & Humphrey, P. P. A. (1989). The selective closure of feline carotid arteriovenous anastomoses by GR43175. *Cephalalgia*, **9** (Suppl. 9), 41–6.

Raskin, N. H. (1981). Pharmacology of migraine. *Annual Review of Pharmacology and Toxicology*, **21**, 463–78.

Raskin, N. H., Hosobuchi, Y. & Lamb, S. (1987). Headache may arise from perturbation of brain. *Headache* **27**, 416–20.

Raskin, N. H. & Knittle, S. C. (1976). Ice cream headache and orthostatic symptoms in patients with migraine. *Headache*, **16**, 222–5.

Ray, B. S. & Wolff, H. G. (1940). Experimental studies on headache: Pain-sensitive structures of the head and their significance in headache. *Archives of Surgery*, **41**, 813–56.

Sicuteri, F., Geppetti, P., Marabini, S. & Lembeck, F. (1984). Pain relief by somatostatin in attacks of cluster headache. *Pain*, **18**, 359–65.

Sicuteri, F., Testi, A. & Anselmi, B. (1961). Biochemical investigations in headache: Increase in the hydroxyindoleacetic acid excretion during migraine attacks. *International Archives of Allergy*, **19**, 55–8.

Sleight, A. J., Cervenka, A. & Peroutka, S. J. (1990). In vivo effects of sumatriptan (GR43175) on extracellular levels of 5-HT in the guinea pig. *Neuropharmacology*, **29**, 511–13.

Toda, N. & Fjita, Y. (1973). Responsiveness of isolated cerebral and peripheral arteries to serotonin, norepinephrine and transmural electrical stimulation. *Circulation Research*, **33**, 89–104.

Watts, A. D., Feniuk, W. & Humphrey, P. P. A. (1981). A pre-junctional action of 5-hydroxytryptamine and methysergide on noradrenergic nerves in dog isolated saphenous vein. *Journal of Pharmacy and Pharmacology*, **33**, 515–20.

Wolff, H. S. (1963). *Headache and Other Head Pain*. Oxford: Oxford University Press.

21
The treatment of primary headache

F. CLIFFORD ROSE AND
*London Neurological Centre,
London, England*

M. ANTHONY
*Institute of Neurological Sciences,
The Prince Henry and Prince of Wales Hospitals, Sydney, Australia*

The International Headache Society has identified 12 main categories of headache, and defined the criteria for the diagnosis of each (Headache Classification Committee of the International Headache Society, 1988), as a basis for studies on pathophysiology, organization of clinical trials and management of patients.

The first three categories of headache comprise: migraine, tension-type headache and cluster headache (including its variant, chronic paroxysmal hemicrania). These three categories are the commonest forms of headache encountered in clinical practice, and are often referred to as 'idiopathic headache', as no underlying pathology or causative mechanism has been identified so far. Since such headaches are amenable to therapy with drugs, the treatment of acute attacks and their prevention form the theme of this chapter. The definition of migraine and cluster headache is given in the chapter on Unilateral Headache (Anthony, this volume).

Tension-type headache is defined as recurrent headache lasting minutes to days. The pain is typically pressing/tightening in quality, of mild or moderate intensity, bilateral in location and does not worsen with routine physical activity. Nausea is absent but photophobia or phonophobia may be present. Occasionally, patients with such headaches also suffer from episodes of migraine and this combination had been called 'tension-vascular headache' in the past.

Nonpharmacological treatment of headache

Several measures have been used in the past for the treatment of migraine. These have included psychological counselling, behavioural management, relaxation therapy, biofeedback of various types, hypnotherapy, transcutaneous nerve stimulation, acupuncture, chiropractic treatment and physiotherapy to the neck. None of these has produced any lasting benefit, most probably because remissions and exacerbations in migraine are often, though not always, linked to the emotional state of patients and their ability to cope with decisions and commitments.

The same comments apply to the management of tension-type headache, where the above measures may modify the frequency and severity of the attacks for a period but alone are rarely sufficient to render the patient headache-free (Anthony & Lance, 1992).

In the case of cluster headache none of the above measures has been found useful in suppressing the pain of the attack which has a rapid onset, and is of short duration and intense severity. The same comments apply to the prevention of attacks.

Pharmacological treatment of headache: drugs acting on 5HT receptors

Drugs that affect endogenous 5-HT fall into two main categories.

1. Antagonists of 5-HT$_2$ receptors. They are most useful in migraine prophylaxis (e.g. methysergide).
2. Agonists of 5-HT$_1$-like receptors. They are effective in the treatment of the acute attack of migraine and cluster headache (e.g. sumatriptan).

Ergotamine

Chemically, ergotamine is a lysergic acid molecule with an amino-alochol, amino-propanol radical on the carboxyl group. In general, in small doses the drug acts as an alpha-adrenergic and 5-HT$_2$ agonist, whilst in large doses it behaves as an antagonist (Berde & Sturmer, 1978).

Dihydroergotamine (DHE) is the hydrogenated product of ergotamine, and has properties which are only quantitatively different from the parent compound. Thus DHE has less intrinsic vasoconstrictor activity than ergotamine, yet its inhibitory effects on the CNS, facilitatory effect on the release of sympathetic transmitters and alpha-adrenoceptor antagonism are all greater (Berde, 1972).

The constrictor actions of ergotamine and DHE are more marked in the external than in the internal carotid territory, and this is true both in the monkey and in the human (Spira, Mylecharane & Lance, 1976; Edmeads, Hachinski & Norris, 1976).

Clinical use

The use of ergotamine is restricted to the treatment of the acute attack of migraine and is said to be effective in about 75% of cases. In the acute attack of cluster headache its effectiveness is much less. Studies in migraine patients suggest that a good therapeutic response can be obtained with peak plasma levels of 0.2 ng/ml within 1 hour of its administration and this is achieved if ergotamine is administered parenterally or inhaled rather than when taken by the oral or rectal route (Tfelt-Hansen, Ibraheem & Paalzow, 1982). The drug is usually combined with caffeine which increases its absorption and enhances its vasoconstrictor effects.

Side-effects

The most common are generalized muscle aches, nausea, vomiting, drowsiness and confusion. Peripheral vasoconstriction and intermittent claudication occur occasionally after prolonged and frequent use. Daily use of ergotamine leads to rebound headaches, as the effect of the drug ceases a few hours after it is taken.

Methysergide

This compound is a semisynthetic derivative of the naturally occurring ergometrine and consists of a lysergic acid nucleus and an amide side chain. In addition to its powerful 5-HT$_2$ receptor antagonism, the drug has certain additional pharmacological properties such a weak oxytocic effect, an anticonvulsant action, varying from species to species, and vasoconstrictor action on the carotid vascular bed of the monkey, most marked in the external carotid territory (Saxena, 1972; Fozard, 1975).

Clinical use

Methysergide usually reduces the frequency of migraine headache by more than 50% in about 65% of patients, one-third of whom become almost headache free (Lance, Anthony & Somerville, 1970). The dose required for this degree of improvement varies between 4 and 6 mg methysergide daily, in divided doses.

Side-effects

About 10% of patients are unable to tolerate the drug because of side effects, mainly gastrointestinal and cardiovascular. An additional 30% have unpleasant symptoms in the form of epigastric discomfort or muscle cramps, which are mild and transient, disappearing in a few days or weeks (Curran, Hinterberger & Lance, 1967). The major side-effects are:

1. Vascular constriction in the limbs (Curran *et al.*, 1967), resulting in muscle cramps on exercise, at rest or even as pallor and coldness of the extremities with reduced or absent pulses. Although usually dose related, vasoconstrictive phenomena may also appear in patients taking as little as 1 mg of the drug, suggesting individual idiosyncrasy (Curran *et al.*, 1967).
2. Epigastric discomfort and nausea, particularly in patients with a previous history of peptic ulceration.
3. Other less common symptoms include vomiting, diarrhea, constipation, drowsiness, leg edema, mental confusion, arthralgias, falling hair and weight gain.
4. Fibrotic reactions, mostly in the retroperitoneal space but fibrosis of the pleura and cardiac valves has also been reported (Graham, 1967). Since the introduction of one month's rest from treatment in every five to permit resolution of any impending fibrosis, few such cases have been reported.

Pizotifen

This drug has a cycloheptathiophene nucleus with a side chain resembling that of cyproheptadine. Like methysergide, the drug causes vasoconstriction and potentiation of 5-HT, noradrenaline or histamine on the isolated human temporal or rabbit ear artery (Aellig, 1983). Pizotifen also acts as a potent H_1 receptor antagonist (Leysen *et al.*, 1981).

Clinical use

Whilst not as effective as methysergide for migraine prevention, pizotifen is widely used due to its lack of significant side effects. Rates of improvement in different trials vary from 40% at doses of 1.5 mg daily to 70% with higher doses (3 mg daily) (Anthony & Lance, 1972).

Side-effects

The most frequent are drowsiness and increased appetite resulting in weight gain. Other less common side effects include mental depression, vertigo and aching muscles.

Cyproheptadine

This drug is chemically related to pizotifen which it resembles in nearly all of its pharmacological effects. It possesses similar antiserotonin and antihistaminic properties.

Clinical use

The drug is used as an antiserotonin agent for migraine prevention. It produces improvement rates of 43%, but has been displaced by pizotifen, which is somewhat more effective and better tolerated (Lance *et al.*, 1970). It is also an appetite stimulant via a hypoglycemic effect (Cerdan, Acosta & Jolin, 1976).

Side-effects

These are identical to those of pizotifen.

Amitriptyline

This tricyclic antidepressant, the prototype of similar drugs used in the treatment of depression, has the ability to block monoamine uptake within the CNS, with specific selectivity towards 5-HT receptors (Fozard, 1982).

Amitriptyline is effective in the treatment of migraine, either alone (Couch & Hassanein, 1979) or in combination with propranolol (Mathew, 1981). It is also useful in the management of tension headaches. Side-effects are mostly drowsiness, dry mouth and blurred vision, affecting a significant proportion of patients. Other tricyclic drugs which can be used instead of amitriptyline include dothieprin, imipramine, clomipramine and nortriptyline.

Sumatriptan

Sumatriptan (Imigran) is a specific agonist of the $5-HT_{1D}$ receptor and selectively constricts the carotid arterial bed of cats and dogs, including arteriovenous anastomoses. It produces the same increase in carotid vascular resistance as ergotamine but does not increase total peripheral resistance, nor does it decrease heart rate and cardiac output like ergotamine (Humphrey *et al.*, 1991). In the anesthetized rat and guinea pig, i.v. sumatriptan inhibits plasma extravasation from vessels of the dura mater produced by electrical stimulation of the trigeminal nerve (Moskowitz, 1987), and this suggests that 5-HT-like receptors are present in such vessels. Of the 5-HT-like receptor subtypes, the $5-HT_{1D}$

receptor is the one most commonly found in the human brain and cerebral circulation, and recent studies have demonstrated that sumatriptan is a potent and selective 5-HT_{1D} receptor agonist (Peroutka, 1990). For further details, see Humphrey *et al.* (this volume).

Clinical use

Sumatriptan has been administered by now to over 4000 patients in placebo-controlled clinical trials either orally or subcutaneously, in single or repeated doses for the relief of acute attacks of migraine headache. The response end-point in all trials was the reduction of a severe or moderate headache to a mild or no headache state. Its effectiveness has been compared to oral Cafergot (ergotamine 2 mg plus caffeine 200 mg) and aspirin with metoclopramide. The optimal subcutaneous dose of sumatriptan was found to be 6 mg and the oral dose 100 mg. Response rates for oral sumatriptan were 67%, compared with 27% in the placebo group, at the end of 2 hours (The Oral Sumatriptan Dose-Defining Study Group, 1991), whereas subcutaneous sumatriptan gave a response rate of 77% compared to 26% of the placebo group (The Sumatriptan Auto-injector Study Group, 1991). Comparison of sumatriptan with cafergot produced response rates of 66% and 48% respectively (The Multinational Oral Sumatriptan and Cafergot Comparative Study Group, 1991), whilst the comparison with the combination of aspirin and metoclopramide gave response rates of 65% and 34% respectively (Legg, 1990). The majority of patients who responded to sumatriptan became free of headache.

Side-effects

A salient feature of the side-effects of sumatriptan is their short duration (usually minutes). Their incidence is dose related and include flushing of the face, tingling of the body, pressure in the head and various parts of the body, dizziness, malaise and fatigue (Brown *et al.*, 1991). Increases in both systolic and diastolic blood pressure have been observed occasionally, and six patients reported transient symptoms of tightness in the chest without evidence of myocardial ischemia. A unique feature of the drug is its efficacy even in cases of established severe headache and the paucity and mildness of its side-effects.

Beta adrenoceptor antagonists

The relief of migraine by beta-adrenoceptor antagonists (BAA) was a chance finding when propranolol given to patients who happened to be suffering from angina and migraine was found to relieve both conditions (Wykes, 1968). BAA possess several properties apart from beta blockade. However, one property that is essential for migraine prevention is the absence of intrinsic sympathomimetic activity, as it has been found that drugs with such activity are ineffective in migraine (Shanks, 1984).

In view of these findings, it seems that the antimigraine effect of BAA may include the inhibition of beta-mediated modulation of noradrenergic transmission from neurone to neurone and/or neurone to blood vessel.

Propranolol

Clinical use

The drug has been assessed in many double-blind trials for migraine prevention. The doses used varied from 60 mg to 160 mg daily, with response rates varying from 55 to 81%. Its use in patients with classical migraine may lead to permanent neurological deficit during an attack (Prendes, 1980; Gilbert, 1982).

Side-effects

These include fatigue, insomnia, vivid dreams, postural hypotension, dizziness, diarrhea, nausea, abdominal cramps, weight gain, visual disturbances, numbness of limbs and depression, roughly in that order. Their occurrence ranges from 4% to 29%. Generally, these side-effects are mild and well tolerated (Diamond *et al.*, 1982).

Metoprolol

Clinical use

The drug reduced significantly the frequency, duration and severity of migraine attacks when used in a double-blind trial (Anderson *et al.*, 1983). Its relative cardioselectivity implies that it would be a suitable drug for migraine prevention in patients suffering from asthma, where nonselective beta-blockers are contraindicated but, even so, it should be prescribed with caution.

Timolol and atenolol

Both drugs are also effective in migraine prevention with responses similar to propranolol in controlled trials (Hakkarainen & Kangasniemi, 1981; Forssman, Lindblad & Zbornicova, 1983).

Calcium channel antagonists

This is a heterogeneous group of drugs and their mode of action is to prevent the entry of calcium into the cell (vascular smooth muscle, cardiac muscle or neurone) and its intracellular release from storage sites, such as the mitochondria or sarcoplasm.

How such drugs prevent migraine is a matter for conjecture. They are known to prevent constriction of cranial vessels, and Meyer and colleagues (1985) suggested that depletion of intracellular calcium stores modulates extreme vasodilatation. In the present state of knowledge, it would be reasonable to suggest that the antimigrainous action of these drugs is due not only to inhibition of cerebral vasoconstriction but also to prevention of neuronal hypoxia and the platelet release reaction (Peatfield, Fozard & Rose, 1986).

Flunarizine

This drug has the ability to prevent vascular constriction, and has the additional property of preventing cerebral hypoxia, as well as being a histamine receptor (H_1) and dopamine antagonist (Amery *et al.*, 1981).

Clinical Use

Flunarizine has been studied in several clinical trials and in comparison with pizotifen, propranolol or other calcium channel blockers (Louis, 1981; Amery Cears & Aerts, 1985; Cerbo *et al.*, 1986; Sorge *et al.*, 1988; Ludin, 1989). It is generally agreed that flunarizine is at least as effective as pizotifen or propranolol in migraine prevention, in a dose of 10 mg at bedtime.

Side-effects

The most prominent side-effects include weight gain and somnolence, similar in type and incidence to pizotifen. Dryness of mouth, dizziness and hypotension were other less frequent side-effects. Flunarizine also produced side-effects similar in type and incidence to propranolol (Lucking *et al.*, 1988).

Nimodipine

Of all the calcium channel blockers, nimodipine has the most pronounced effects on cerebral vessels with least systemic effects and absence of hypotension. The drug is lipid soluble and therefore crosses the blood–brain barrier with ease

(Kerckhoff & Drewes, 1985), thus exerting a significant effect both on the cerebral arteries and neurones. Its spasmolytic and neuroprotective effects have been demonstrated in its effectiveness in preventing cerebral vasospasm in patients with subarachnoid hemorrhage (Ohman & Keiskanen, 1988) and reducing morbidity and mortality in patients with ischemic stroke (Gelmers *et al.*, 1988).

Clinical use

Initial uncontrolled studies had shown the drug to be useful in migraine prevention (Meyer & Hardenberg, 1983; Meyer *et al.*, 1985; Solomon, 1985), as did subsequent placebo-controlled, double-blind studies (Gelmers, 1983, 1985; Havanka-Koanniaine, Hokkanen & Myllyla, 1987; Bussone *et al.*, 1987; Battistella *et al.*, 1990). However, in two recent double-blind placebo-controlled trials in classical and common migraine (migraine with and without aura, respectively), the effectiveness of the drug was no greater than that of placebo (Migraine Nimodipine European Study Group, MINES 1989*a*, *b*).

Side-effects

These are few and infrequent. Abdominal discomfort and a feeling of flushing have been reported occasionally.

Verapamil

Cerebral blood flow is not significantly decreased in migrainous patients receiving verapamil and yet there are significant decreases in cerebrovascular resistance, indicating relaxation of small arteries and arterioles (Meyer *et al.*, 1985).

Clinical use

In several clinical trials, the effectiveness of verapamil in migraine and cluster headache, has been shown to range from 56 to 83% (Meyer *et al.*, 1985; Jonsdottir, Meyer & Rodgers, 1987; Prusinski & Kozubski, 1987; Solomon & Diamond, 1987; Solomon, 1989).

Side-effects

The incidence of side-effects is high, ranging from 33 to 50%. The commonest is constipation, whilst the remainder are made up of other GIT disturbances, flushing, postural hypotension and fatigue.

Nifedipine

This drug is chemically related to nimodipine, but has more potent relaxant effects on smooth muscle af the peripheral vessels. As a result, it lowers blood pressure quite effectively. Its main advantage in the treatment of migraine is the rapid abolition of neurological prodromal symptoms of an attack by 1 or 2 drops of the drug under the tongue. Systemic administration of the drug has been found to be effective in the treatment of attacks of classical migraine (migraine with aura) and cluster headache, producing response rates of 92 and 84% respectively at a dose of 40 mg thrice daily, though this at the expense of a high incidence of side effects of 53 to 64.5% (Meyer *et al.*, 1985; Jonsdottir *et al.*, 1987). The commonest side effects are postural hypotension, precordial sensations, flushing and constipation. Because of their high incidence, the drug has not proven popular in migraine prophylaxis.

Nonsteroidal anti-inflammatory drugs

This is a group of heterogeneous compounds, mostly chemically unrelated, but sharing certain therapeutic actions and side-effects. Their therapeutic activities depend upon the inhibition of biosynthesis of prostaglandins and related substances, by inhibiting the activity of the cyclo-oxygenase (Peatfield *et al.*, 1986). Apart from their anti-inflammatory properties, these drugs have significant analgesic effects.

Clinical use

Whilst the three main properties of this group of drugs are analgesic, antipyretic and anti-inflammatory, it is probably the analgesic property that is useful in the control of headaches.

Aspirin

The effectiveness of the drug in the control of acute attacks of headaches is well established, although this was proven by a clinical trial only comparatively recently (Graffenried & Hill, 1978). In addition, it is likely that regular aspirin intake is capable of preventing headache attacks (O'Neil & Mann, 1978; Baldrati *et al.*, 1983).

Paracetamol

The substance is the active metabolite of phenacetin. However, its antiinflammatory activity is weak and seldom clinically useful. Either alone or in combination with codeine in various amounts, the drug is widely used in the treatment of migraine and other forms of headaches and has gained wide acceptance. The drug is well tolerated and has none of the side-effects of aspirin, although over-dosage may cause potentially fatal acute hepatic necrosis.

Other nonsteroidal anti-inflammatory drugs (NSAIDs)

Those used so far have included indomethacin, sulindac, diclofenac, naproxen, brufen, ketoprofen, tenoprofen, mefanamic acid, flufenamic acid, phenylbutazone, fenoxicam and pyroxicam. Like aspirin, they inhibit cyclo-oxygenase and prevent formation of various prostaglandins, including thromboxane A2, which is a potent platelet aggregating agent. Their side-effects are similar, and include:

1. Gastrointestinal effects: epigastric discomfort, nausea, vomiting, exacerbation of peptic ulcers, erosive gastritis and gastrointestinal hemorrhage.
2. Hematological effects: marrow depression and agranulocytosis occur occasionally but far less frequently than gastrointestinal disturbances.
3. Hypersensitivity reactions: mostly skin rashes, itching and urticaria, and sometimes acute attacks of asthma.
4. Effects on the central nervous system: drowsiness headache, dizziness, fatigue, depression, insomnia and psychosis, have all been reported with these agents.

Clinical use

The usual dose, frequency, method of administration, indications and contraindications to the use of these drugs are similar to those used in the treatment of arthritis.

Monoamine oxidase inhibitors

These drugs inhibit the monoamine oxidase (MAO) which catalyses the degradation of monoamines, particularly 5-HT, circulating levels of which have been shown to fall during the migraine attack (Anthony, Hinterberger &

Lance, 1967). Such changes almost certainly reflect similar changes in brainstem neurones of the endogenous pain control system, which is known to modulate pain appreciation, at least from the head and neck (Lance *et al.*, 1983).

The commonly used MAOIs are either:

1. Hydrazine derivatives – phenelzine ('Nardil') and nialamide ('Niamid') or
2. Phenylcyclopropylamines – tranylcypromine (Parnate). A combination of tranylcypromine of 10 mg and trifluoperazine 1 mg ('Parstelin') is used occasionally in psychoneurotic disorders.

Clinical use

In spite of the theorectical advantages of this group of drugs, their use continues to be limited, because of fear of potentially serious side-effects. So far, four trials have been reported, three using nialamide (Niamid), 25 mg thrice daily and one using phenelzine (Nardil) 15 mg thrice daily. In only the first and last studies were figures of improvement provided (53% and 80% respectively). In the last study, serial estimations of plasma serotonin showed a 50% increase in levels during the period of treatment, though no strict correlation was found between plasma levels and clinical response (Anthony & Lance, 1992).

Side-effects

MAOIs produce side-effects which are mostly related to the nervous system. They include agitation, insomnia, increased anxiety, tremor and increased frequency of convulsions in patients with epilepsy.

Orthostatic hypotension, difficulty with micturition and ejaculation, blurred vision and impotence have been reported. Potentially serious side-effects are paroxysmal hypertension following ingestion of foodstuffs containing tyramine and dopamine, e.g. cheese, vegetable or meat extracts, red wines, broad beans, chicken livers, etc, and rapid irreversible hypotension and coma following the administration of pethidine and other alkaloid analgesics in susceptible patients.

Moclobemide

Moclobemide ('Aurorix') is a recently introduced reversible inhibitor of monoamine oxidase A (MAO-A). The drug has a short half-life (1–4 hours) and, whilst metabolized in the liver, it is excreted almost completely by the kidneys. Unlike the previously described irreversible, 'suicide', MAO inhibitors (e.g. phenelzine, tranylcypromine, etc) it produces either a weak or no reaction with tyramine (e.g. after eating cheese), due to its short half-life and its enzyme selectivity, which allows MAO-B to deaminate the ingested tyramine. So far, moclobemide has been used as an antidepressant in lieu of MAO-A inhibitors or tricyclic compounds (Priest, 1990). No information exists as to its use in treating recalcitrant migraine.

Corticosteroids

It is generally agreed that steroids have no place in the treatment of migraine, other than the exceptional instance of prolonged unresponsive attacks (status migrainosus), for which they can be given either orally or parenterally for 3 to 5 days, allegedly with beneficial effects (Lance, 1982; Blau, 1987). However, steroids are effective in the prevention of recurrent attacks of cluster headache (Kudrow, 1978). Attacks of cluster headache are associated with a rapid rise of whole-blood histamine (Anthony & Lance, 1969). Anthony and Daher (1990) have demonstrated that this rise can be suppressed by prednisone 60–75 mg daily or dexamethasone 6–8 mg daily. Lower doses (<30 mg prednisone or <3 mg dexamethasone) failed to suppress recurrent attacks and the rise in blood histamine that accompanies them.

Clinical use

In an open trial of 77 patients with cluster headache (Kudrow, 1978), short-term prednisone therapy of 40 mg daily produced a marked reduction in headache frequency, and a similar result was reported by another group (Couch & Ziegler, 1978). Local injection of steroids into the region of the greater occipital nerve ipsilateral to the headache may arrest attacks of cluster headache (Anthony, this volume).

Side effects

The side-effects of corticosteroids are so numerous and well known that they need not be listed here.

Lithium

This monovalent cation was introduced into the treatment of headache, and in particular chronic cluster headache by Ekbom (1974). It is thought to act by impairing monoaminergic transmission (Eadie & Tyrer, 1985). Lithium is most commonly given in the form of its carbonate salt, the usual dose being 0.75–1.5 g daily, aiming to maintain serum levels of the drug at 0.7–1.2 mmol/l,

The treatment of primary headache 347

although lower levels of 0.3–0.8 mmol/l have been reported to be effective in cluster headache (Manzoni *et al.*, 1983).

Clinical use

Several trials were summarized by Ekbom (1987) on the use of lithium in the treatment of cluster headache and overall the drug produced a remission rate in 85% in chronic cases and 68% in patients with the episodic variety.

Side effects

Fine tremor of the hands, polyuria, thirst and ankle oedema are common when blood levels are within the therapeutic range. High serum concentrations may produce muscle weakness and gastrointestinal disturbances such as anorexia, nausea, vomiting and diarrhoea. Long-term complications include hypothyroidism and nodular goitre as well as oliguric renal failure and diabetes insipidus (Williams & Györy, 1976).

Miscellaneous drugs

Phenothiazines

The plethora of clinically useful properties include antinausea and antiemetic effects, through blockade of dopamine D_1 receptors at central nervous system synapses (Baldessarini, 1985). These properties have proven very useful in the treatment of the acute attack of migraine.

Thiethylperazine

The dose is 10 mg intramuscularly, orally or as a suppository.

Prochlorperazine

The intramuscular injection contains 12.5 mg of the drug, the suppository 25 mg, and the tablet form 5 mg.

Metoclopramide

It has been suggested that the slow intravenous injection of 10 mg of the drug may arrest an acute attack of migraine (Hughes, 1977), but other studies have not con-

Table 21.1. *Treatment of the acute attack of migraine*

General measures
1. Sit in a comfortable chair in a quiet, darkened, cool room.
2. Do not drink coffee or orange juice.
3. Do not move excessively and do not read or watch television.

Drug treatment
1. Administer a simple analgesic and wait at least 1 hour.
2. If no better after 1 hour, administer an oral ergotamine preparation and repeat in 1 hour, if required.
3. If vomiting is frequent or pain severe or both, administer 'Cafergot' suppository. Repeat in 2 hours if necessary.
4. Sumatriptan 100 mg orally or 6 mg subcutaneously. Either can be repeated in 4 hours or 30 min, respectively, if required.
5. If attack non–responsive to above, administer 1 mg dihydroergotamine and 10 mg metoclopramide both intramuscularly.
6. Reserve intramuscular pethidine 100 mg, pentazocine 45–60 mg or codeine phosphate 60 mg for severe and prolonged headaches.

firmed this observation (Fozard, 1982). A dose of 10 mg given orally will relieve the nausea of an attack but not the pain (Slettnes & Sjaastad, 1977), and will enhance the absorption of aspirin given at the same time (Ross-Lee *et al.*, 1982).

Side-effects

Acute extrapyramidal reactions and Parkinsonian signs occur occasionally with all of the above drugs.

Domperidone

The drug has similar actions to metoclopramide but does not cross the blood–brain barrier, and therefore is less likely to cause extrapyramidal side-effects, such as acute dystonic reactions (Wilkinson, 1987). In a controlled clinical trial domperidone 30 mg was found to arrest migraine, with or without aura, if given 6 to 24 hours before the headache, during the period of premonitory symptoms (Amery & Waelkens, 1983).

Anxiolytic and hypnotic drugs

The most effective drugs for such purposes are the benzodiazepines. They possess both of the above properties and, because of their safety, are widely used in clinical practice.

Table 21.2. *Trigger factors in migraine*

1. Stress
Emotional
   Excitement (short, unexpected)
   Expectation
   Sudden news (pleasant or unpleasant)
   Relaxation after stress (weekends)
Physical
   Physical exercise (sport)
   Sudden exertion (sex headaches, vigorous sport)
   Head trauma
   Heat exposure
   Bright lights
   Dry winds ('Sharav', 'Foehn')
   Weather change
   Menstruation
   Sleeping in
2. Food
   Alcohol
   Chocolate
   Cheese
   Citrus fruit
   Fasting
   Food preservatives
3. Drugs
   Estrogens
   Estrogen-containing contraceptives
   Nitrates
   Monosodium glutamate
4. Underlying diseases
   Cervical spondylosis
   Uncontrolled hypertension
   Mental depression

Sleep frequently relieves migraine and emotional stress and anxiety do induce attacks in many migrainous patients. A short-acting benzodiazepine (oxazepam, alprazolam, diazepam), may bring a severe attack to an end by inducing sleep.

Details of headache treatment

Migraine

The acute attack
This treatment is outlined in Table 21.1.

Table 21.3. *Drugs for migraine prevention*

Drugs	Dose	Response rate	Common side effects	Common contraindications
Beta adrenoceptor antagonists				
Propranolol	40–80 mg b.d.–t.d.s.	65%	Insomnia	Asthma
Metoprolol	50–100 mg t.d.s.		Tiredness	Heart failure
			Hypotension	Diabetes
Atenolol	50–100 mg daily		Vivid dreams	
5-HT Antagonists				
Pizotifen	1.5–3.0 mg nightly	50%	Drowsiness	Nil
			Weight gain	
Methysergide	1.0–2.0 mg t.d.s.	65%	Gastric irritation	Peptic ulceration
			Peripheral vascular or	Peripheral vascular or
			coronary ischemia	coronary disease
				Severe hypertension
Calcium channel antagonists				
Flunarizine	10 mg nightly	60%	Abdominal discomfort	Beta adrenoceptor
				blockers
Nimodipine	40 mg t.d.s.		Weight gain	Diabetes
			Hypotension	Cardiac failure
Verapamil	40–80 mg t.d.s.			
Nifedipine	10–20 mg t.d.s.			
Tricyclic drugs				
Amitriptyline	25–75 mg nightly	60%	Drowsiness	Heart failure
Dothiepin	25–75 mg nightly		Agitation	Old age
Doxepin	25–75 mg nightly			

Monoamine oxidase inhibitors			
Phenelzine	15 mg t.d.s.	80%	Hypertensive crises
Tranylcypromine	10 mg t.d.s.		Severe hypotension (due to dietary or drug interactions)
			Mental disorder
NSAIDs			
Naproxen	500 mg t.d.s.	50%	Gastric irritation
Diclofenac	50 mg t.d.s.		Marrow depression
Ketoprofen SR	100–200 mg daily		Hypersensitivity reactions
			Peptic ulceration
			Hematological disease (e.g. thrombocytopenia)
Piroxicam	10–20 mg daily		
Corticosteroids			
Prednisone	50–75 mg daily	Variable	Gastric irritation
Dexamethasone	4–8 mg daily		Mental confusion
			Peptic ulceration
			Psychic disorders
Lithium (for Cluster only)			
Lithium carbonate (Lithicarb, Priadel)	0.25–0.5 G t.i.d.	68-85%	Tremor
			Ankle oedema
			Gastric symptoms
			Psychic disorder (other than bipolar illnesses)

Table 21.4. *Treatment of acute attacks of cluster headache*

1. Effervescent aspirin or paracetamol
2. Oral ergotamine tartrate (Cafergot, Ergodryl)
3. Subcutaneous sumatriptan (Imigran) – 6 mg. This can be repeated in 30 min if required
4. Intranasal instillation of lignocaine (4%)
5. Perineural injection of lignocaine into region of ipsilateral greater occipital nerve
6. Oxygen inhalation – 8–10 l/min for 10–30 min
   (Speediest, most effective and safest method of treating an attack, see Anthony & Drummond, 1985.)

* Most patients can get relief from simple analgesics.
* Oral ergotamine should be used at hourly intervals up to three doses. Otherwise 'Cafergot' suppository or intramuscular DHE should be employed.
* Whenever available, sumatriptan should be used in preference to ergotamine, 100 mg orally or 6 mg subcutaneously. Alkaloid analgesics should be reserved for those cases who fail to respond to the above regimes.

Prevention of attacks

Trigger factors are given in Table 21.2 and treatment in Table 21.3.

* It is important to eliminate trigger factors as far as possible, treat underlying diseases, and particularly eliminate oral contraceptives and exogenous hormones, if they can be identified as responsible for aggravating the patient's headaches.
* Drugs for prevention of attacks should be used in the order shown in Table 21.3. Both beta-blockers and 5-HT antagonists should be given with a night dose of a tricyclic drug, if they do not produce a satisfactory response when given alone.
* Neurectomy of the ipsilateral greater occipital nerve can be employed in patients with frequent and severe unilateral headaches that fail to respond to preventive medication.

Cluster headache

The acute attack.

Treatment is outlined in Table 21.4.

Table 21.5. *Prevention of attacks of cluster headache*

1. Oral ergotamine tartrate: 1 mg twice daily
   Response rate: 30–40%
2. Parenteral dihydroergotamine (DHE) - 1 mg b.d. intramuscularly
   Response rate: 50%
3. Methysergide: 1–2 mg, 4th hourly
   Response rate: 26–50% (Krabbe, 1989)
4. Lithium carbonate: 0.25–0.5 g t.d.s.
   Response rate: 65–87% (Kudrow, 1978)
5. Corticosteroids
   Prednisone 60–75 mg daily, or dexamethasone 6-8 mg daily
   Response rate: 65–80% (Kudrow, 1980)
6. Calcium channel antagonists
   Verapamil or nimodipine – as in migraine
7. Perineural injection of methylprednisolone (Depomedrol) into region of ipsilateral greater occipital nerve
   In 32 patients with cluster headache, such injection produced relief in 28 (87%), ranging from 5 to 56 days, mean period of relief being 18 days (Anthony, 1987)
8. Surgical Occipital Neurectomy of the ipsilateral GON – the procedure produced relief in all of the 12 patients in whom it was performed, with relief ranging from 3 months to 4 years (Anthony, 1987)
9. Thermocoagulation of the Gasserian Ganglion
   Response rate: 74% (Mathews & Hurt, 1988)

* Analgesic drugs and ergotamine preparations are generally ineffective, although they may be useful in the minority of patients with mild or prolonged attacks.
* Whenever available, subcutaneous sumatriptan should be used, as in migraine.
* Anesthesia of the ipsilateral sphenopalatine ganglion with 4% lignocaine drops on a cottonwool bud inserted into the posterior part of the middle concha may prove effective. Otherwise, injection of 4–5 ml lignocaine 1% into the region of the ipsilateral greater occipital nerve will arrest an attack within 5–10 min in most patients (Anthony, 1987).
* However, the simplest, safest and least expensive method of arresting an attack is the inhalation of 100% oxygen for 10–30 min.

Prevention of attacks.

Preventive therapy is outlined in Table 21.5.

* Oral ergotamine should be tried first, because of its simplicity of administration. Methysergide should be given if ergotamine fails. DHE intramuscularly is used after oral medications prove to be ineffective.

Table 21.6. *Treatment of tension-type headache*

1. Psychological counselling, behaviour modification, relaxation therapy biofeedback - helpful but not usually sufficient to control headaches
2. Tricyclic antidepressants: very helpful amitriptyline, diothepin, imipramine, clomipramine, and nortriptyline
3. NSAIDs – in patients whose headaches are associated with disease of the cervical spine
4. MAO Inhibitors – phenelzine or tranylcypromine – as in migraine

Once a satisfactory medication has been found, it should be continued for at least 6 months, before the dose is reduced and the drug eventually stopped.

* Calcium channel antagonists have not proven as effective as originally thought, and theoretically nimodipine is the preferred one.
* Steroids are certainly the most effective of the oral drugs, but the high doses required to suppress headaches cannot be continued for more than 10–15 days. However, steroid injections into the ipsilateral greater occipital nerve are simple, safe and equally effective, and obviate the need for daily oral drugs.
* Surgical procedures are to be reserved for those with chronic cluster headache.

Tension-type headaches

* Tricyclic drugs are the treatment of choice. They should be changed until a suitable one is found for the patient, producing the least side-effects and maximal benefit.
* In anxious patients, anxiolytic drugs may be more effective. Drug therapy should be accompanied, at least in the initial stages, with non-pharmacological measures, as outlined in Table 21.6.
* MAO inhibitors should be used as a last resort.

References

Aellig, W. A. (1983). Influence of pizotifen and ergotamine on the venoconstrictor effect of 5-hydroxytryptamine and noradrenative in man. *European Journal of Pharmacology,* **25**, 759–62.

Amery, W. K., Cears, L. I. & Aerts, T. J. L. (1985). Flunarizine, a calcium entry blocker in migraine prophylaxis. *Headache,* **25**, 70–4.

Amery, W. K. & Waelkens, J. (1983). Prevention of the last chance: an alternative pharmacologic treatment of migraine. *Headache,* **23**, 27–8.

Amery, W. K., Wanquier, A., Van Neuten, J. M., De Clerk, F., Van Reepripts, J. V. & Janssen, P. A. J. (1981). The antimigrainous pharmacology of flunarizine (R 14 950), a calcium antagonist. *Drugs under Experimental and Clinical Research* **7**, 1–10.

Anderson, P. G., Dahl, S., Hansen, J. H., Hedman, C., Kristensen, T.N. & Olivarius de F. (1983). Prophylactic treatment of classical and non-classical migraine with metoprolol - a comparison with placebo. *Cephalalgia*, **3**, 207–12.

Anthony,. M. (1987). The role of the occipital nerve in unilateral headache. In *Advances in Headache Research*, ed. F. C. Rose, pp. 257–62. London: John Libbey.

Anthony, M. & Daher, N. B. (1990). Mechanism of action of steroids in cluster headache. *8th International Migraine Symposium London*, September 25–28.

Anthony, M. & Drummond, P. D. (1985). Extracranial vascular responses of sublingual nitroglycerine and oxygen inhalation in cluster headache patients. *Headache*, **25**, 70–4.

Anthony, M. & Lance J. W. (1969). Whole blood histamine and plasma serotonin in cluster headache. *Archives of Neurology*, **25**, 225–31.

Anthony, M. & Lance, J. W. (1972). Current concepts in the pathogenesis and interval treatment of migraine. *Drugs*, **3**, 153–8.

Anthony, M. & Lance, J. W. (1992). Headache. In *Drug Therapy in Neurology*, ed. M. J. Eadie, pp. 335–74. London: Churchill Livingstone.

Anthony, M., Hinterberger, H. & Lance, J. W. (1967). Plasma serotonin in migraine and stress. *Archives of Neurology*, **16**, 544–52.

Baldessarini, R. (1985). Drugs in the treatment of psychiatric disorders. In *Goodwin and Gilman's The Pharmacological Basis of Therapeutics,* 7th edn., ed. Gilamn, A. G., Goodman, L. S., Rall, T. W. & Murad, F. pp. 387–412. New York: Macmillan.

Baldrati, A., Cortelli, P., Procaciantti, A., Gamerini, G., D'Alessandro, R., Baruzzi, A. M. & Sacquegna, T. (1983). Propranolol and acetylsalicylic acid in migraine. *Acta Neurologica Scandinavica*, **67**, 181–6.

Battistella, P. A., Ruffini, R., Moro, R., Fabiani, M., Bertoli, S., Antolini, A. & Zacchello, F. (1990). A placebo-controlled crossover trial of Nimodipine in pediatric migraine. *Headache*, **30**, 264–8.

Berde, B. (1972). Recent progress in the elucidation of the mechanism of action of ergot compounds used in migraine. *Medical Journal of Australia*, **2**, 15–16.

Berde, B. & Sturmer, E. (1978). Introduction to the pharmacology of ergot alkaloids and related compounds as a basis of their therapeutic application. In *Ergot alkaloids and related compounds. Handbuch der Experimentellen Pharmacologie*, vol. 49, eds. B. Berde & H.O. Schild, pp. 1–28. Berlin: Springer Verlag.

Blau, J. N. (1987). A Clinicotherapeutic approach to migraine. In *Migraine: Clinical, Therapeutic, Conceptual and Research Aspects*, ed. J.N. Blau, pp. 185–204. London: Chapman & Hall.

Brown, E. G., Endersby, C. A., Smith, R. N. & Talbot, J. C. C. (1991). The safety and tolerability of sumatriptan: an overview. *European Neurology*, **31**, 339–44.

Bussone, G., Baldini, G., D'Andrea, G., Cananzi, F., Frediani, L., Caresia, L. & Boiardi, A. (1987). Nimodipine versus flunarizine in common migraine: a controlled pilot trial. *Headache*, **27**, 76–9.

Cerbo, R., Casacchia, M., Formisano, R., Feliciani, M., Cusimano, G., Buzzi, M. G. & Agnoli, A. (1986). Flunarizine-Pizotifen single dose double-blind crossover trial in migraine prophylaxis. *Cephalalgia*, **6**, 15–8.

Cerdan, A., Acosta, M. & Jolin, T. (1976). Long-term administration of pizotifen to migraine patients: effects on oral glucose tolerance test and insulin levels. *Headache*, **15**, 126–8.

Couch, J. R. & Hassanein, R. S. (1979). Amitriptyline in migraine prophylaxis. *Archives of Neurology*, **36**, 695–9.

Couch, J. R. & Ziegler, D. K. (1978). Prednisone therapy for cluster headache. *Headache*, **18**, 219–21.

Curran, D. A., Hinterberger, H. & Lance, J. W. (1967). Methysergide. *Research and Clinical Studies in Headaches*, **1**, 74–122.

Diamond, S., Kudrow, L., Stevens, J. & Shapiro, B. D. (1982). Long-term study of propranolol in the treatment of migraine. *Headache*, **22**, 268–71.

Eadie, M. J. & Tyrer, J. H. (1985). *The Biochemistry of Migraine*, pp. 135–66. Lancaster: MTP Press.

Edmeads, J., Hachinski, V. C. & Norris, J. W. (1976). Ergotamine and the cerebral circulation. *Hemicrania*, **7**, 6–10.

Ekbom, K. (1974). Lithium via kroniska symptom on cluster headache. *Opresc Medicine*, **19**, 148–56.

Ekbom, K. (1987). Treatment of cluster headache. In *Migraine: Clinical, Therapeutic, Conceptual and Research Aspects*, ed. J.N. Blau, pp. 223–37. London: Chapman & Hall.

Forssman, B., Lindblad, C. J. & Zbornicova, V. (1983). Atenolol for migraine prophylaxis. *Headache*, **23**, 188–90.

Fozard, J. R. (1975). The animal pharmacology of drugs used in the treatment of migraine. *Journal of Pharmacy and Pharmacology*, **27**, 297–321.

Fozard, J. R. (1982). Basic mechanism of antimigraine drugs. In *Advances in Neurology*, eds. M. Critchley, A. P. Friedamn, S. Gorini & F. Sicuteri, pp.295–307. New York: Raven Press.

Gelmers, H. J. (1983). Nimodipine: a new calcium antagonist in the prophylactic treatment of migraine. *Headache*, **23**, 106–9.

Gelmers, H. J. (1985). Calcium channel blockers in the treatment of migraine. *American Journal of Cardiology*, **55**, 139B–43B.

Gelmers, J. A., Gorter, K., De Weerdt, C. J. & Wiezer, H. J. A. (1988). A controlled trial of nimodipine in acute ischaemic stroke. *New England Journal of Medicine*, **318**, 203–7.

Gilbert, G.J. (1982). An occurrence of complicated migraine during propranolol therapy. *Headache*, **22**, 81–3.

Graffenried, B. V. & Hill, R. C. (1978). Headache as a model for testing mild analgesics. *Research and Clinical Studies in Headache*, **6**, 103–9.

Graham, J. R. (1967). Cardiac and pulmonary fibrosis during methysergide therapy for headache. *American Journal of Medical Science*, **254**, 1–12.

Hakkarainen, H. & Kangasniemi, P. (1981). Timolol maleate in the prophylactic treatment of classic and common migraine. In *Proceedings of Timolol International Symposium*, Stockholm, Sweden, pp. 443–4.

Havanka-Koanniaine, H., Hokkanen, E. & Myllyla, V. V. (1987). Efficacy of nimodipine in comparison with pizotifen in the prophylaxis of migraine. *Cephalalgia*, **7**, 7–13.

Headache Classification Committee of the International Headache Society (1988). Classification and diagnostic criteria for headache disorders, cranial neuralgias and facial pain. *Cephalalgia*, **8** (Suppl.), 9–96.

Hughes, J. B. (1977). Metoclopramide in migraine *Medical Journal of Australia*, **2**, 580.

Humphrey, P. P. A., Feniuk, W., Marriott, A. S., Tanner, R. J. N., Jackson, M. R. & Tucker, M. L. (1991). Preclinical studies of the antimigraine drug, sumatriptan. *European Neurology*, **31**, 282–90.

Jonsdottir, M., Meyer, J. & Rodgers, R. (1987). Efficacy side effects and tolerances compared during headache – treatment with three different calcium blockers. *Headache*, **27**, 364–9.

Kerckhoff van den W, & Drewes, L. R. (1985). The transfer of Ca-antagonists nifedipine and nimodipine across the blood brain barrier and their regional distribution in vivo. *Journal of Cerebral Blood Flow and Metabolism*, **5** (Suppl. 1), 459–60.

Krabbe, A. (1989). Limited efficacy of methysergide in cluster headache: a clinical experience *Cephalalgia*, 9 (Suppl.) **10**, 404–5.

Kudrow, L. (1978). Comparative results of prednisone, methysergide and lithium therapy in cluster headache. In *Current Concepts in Migraine Research*, ed. R. Greene, pp. 159–63. New York: Raven Press.

Kudrow, L. (1980). *Cluster Headaches. Mechanisms and Management*, pp.113–18. Oxford: Oxford University Press.

Lance, J. W. (1982). *Mechanism and Management of Headache*, p. 198. London: Butterworths.

Lance, J. W., Anthony, M. & Somerville, B. (1970). Comparative trial of serotonin antagonists in the treatment of migraine. *British Medical Journal*, **2**, 327–32.

Lance, J. W., Lambert, G. A., Goadsby, P. J. & Duckworth, J. W. (1983). Brainstem influences on the cephalic circulation: experimental data from cat and monkey of relevance to the mechanism of migraine. *Headache*, **23**, 258–65.

Legg, N. J. (1990). Oral sumatriptan compared to aspirin with metoclopramide. *Proceedings 8th Migraine Trust International Symposium London*, September 25th-28th, 22–3.

Leysen, J. E., Awouters, F., Kennis, L., Ladunon, P. M., Vandenberk, J., & Jansen, P. A. J. (1981). Receptor binding of R 41468, a novel antagonist of 5-HT receptors. *Life Science*, **28**, 1015–22.

Louis, P. (1981). A double-blind, placebo-controlled prophylactic study of flunarizine (Sibelium) in migraine. *Headache*, **21**, 235–9.

Lucking, C. H., Oestreich, N., Schmidt, R. & Soyka, D. (1988). Flunarizine vs propranolol in the prophylaxis of migraine: two double-blind comparative studies in more than 400 patients. *Cephalalgia*, **8** (Suppl. 8), 21–6.

Ludin, H.-P. (1989). Flunarizine and propranolol in the treatment of migraine. *Headache*, **29**, 219–22.

Manzoni, G. C., Bono, G., Lafranci, M., Micicli, G., Terzano, M. C. & Nappi, G. (1983). Lithium carbonate in cluster headache: assessment of its short and long-term therapeutic efficacy. *Cephalalgia*, **3**, 109–14.

Mathew, N. T. (1981). Prophylaxis of migraine and mixed headache. A randomized control study. *Headache*, **21**, 105–9.

Mathew, N. T. & Hurt, W. (1988). Percuteneous radiofrequency trigeminal gangliolysis in intractable cluster headache. *Headache*, **28**, 328–31.

Meyer, J. S. & Hardenberg, J. (1983). Clinical effectiveness of calcium entry blockers in prophylactic treatment of migraine and cluster headaches. *Headache*, **23**, 313–32.

Meyer, J. S., Nance, M., Walker, M., Zetusky, W. J. & Dowell, R. E. (1985). Migraine and cluster headache treatment with calcium antagonists supports a vascular pathogenesis. *Headache*, **25**, 358–67.

MINES (1989*a*). Migraine Nimodipine European Study Group (1989). European multicenter trial of nimodipine in the prophylaxis of classic migraine (Migraine with aura). *Headache*, **29**, 639–42.

MINES (1989b). Migraine Nimodipine European Study Group (1989). European multicenter trial of nimodipine in the prophylaxis of common migraine (Migraine without aura). *Headache*, **29**, 633–8.

Moskowitz, M. A. (1987). Sensory connections to cephalic blood vessels and their possible importance to vascular headache. In *Advances in Headache Research*, ed. F.C. Rose, pp. 81–6. London: John Libbey.

Ohman, J. & Keiskanen, O. (1988). Effect of nimodipine on the outcome of patients after aneurysmal subarachnoid haemorrhage and surgery. *Journal of Neurosurgery*, **69**, 683–6.

O'Neil, B. P. & Mann, J. D. (1978). Aspirin prophylaxis in migraine. *Lancet*, **ii**, 1179–81.

Peatfield, R. C., Fozard, J. R. & Rose, F. C. (1986). Drug Treatment of Migraine. In *Handbook of Clinical Neurology*, vol. 4, eds. P.J. Vinken, G.W. Bruyn & H. L. Klawans, pp. 173–216. Amsterdam: Elsevier.

Peroutka, S. J. (1990). The pharmacology of current antimigraine drugs. *Headache*, **30**, 5–11.

Prendes, J. L. (1980). Considerations on the use of propranolol in complicated migraine. *Headache*, **20**, 93–5.

Priest, R.G. (1990). Moclobemide and the reversible inhibitors of monoamine oxidase antidepressants. *Acta Psychiatrica Scandinavica*, **82** (Suppl. 360), 39–41.

Prusinski, A. & Kozubski, W. (1987). Use of verapamil in the treatent of migraine. *Wiad Lek*, **1**, 40, 734–8.

Ross-Lee, L., Hazelwood, V., Tyrer, J. H. & Eadie, J. S. (1982). Aspirin treatment of migraine attacks: plasma drug level data. *Cephalalgia*, **2**, 9–14.

Saxena, P. R. (1972). The effects of antimigraine drugs on the vascular responses evoked by 5-hydroxytryptamine and related biogenic substances on the external carotid bed of dogs. Possible pharmacological implications to their antimigraine action. *Headache*, **12**, 44–54.

Shanks, R. G. (1984). Mechanism of action of beta-adrenoceptor antagonists in migraine. In *Migraine and Beta-blockade*, eds. J. D. Carroll, V. Pfaffenrath & O. Sjaastad, pp. 45–53. Sweden: A.B. Hassle.

Slettness, O. & Sjaastad, O. (1977). Metoclopramide during attacks of migraine. In *Headache: New Vistas*, ed. F. Sicuteri, pp. 201–4. Biomedical Press: Florence.

Solomon, G. (1985). Comparative efficacy of calcium antagonist drugs in the prophylaxis of migraine. *Headache*, **25**, 368–71.

Solomon, G. (1989). Verapamil in migraine prophylaxis – a five year review. *Headache*, **29**, 425–7.

Solomon, G. D. & Diamond, S. (1987). Veramapil in migraine prophylaxis – comparison of dosages. *Clinical Pharmacology & Therapeutics*, **1**, 202–4.

Sorge, F., De Simone, R., Marano, E., Nolano, M., Orefice, G. & Carricri, P. (1988). Flunarizine in prophylaxis of childhood migraine: a double-blind, placebo-controlled crossover study. *Cephalalgia*, **8**, 1–6.

Spira, P. J., Mylecharane, E. J. & Lance, J. W. (1976). The effect of humoral agents and antimigraine drugs on the cranial circulation of the monkey. *Research and Clinical Studies in Headache*, **4**, 37–75.

Tfelt-Hansen, P., Ibraheem, J. J. & Paalzow, L. (1982). Clinical Pharmacology of ergotamine studies with an HPLC method. In *Advances in Migraine Research and Therapy*, ed. F.C. Rose, pp. 173–9. New York: Raven Press.

The Multinational Oral Sumatriptan and Cafergot comparative Study Group (1991). A randomized, double-blind comparison of sumatriptan and cafergot in the acute treatment of migraine. *European Neurology*, **31**, 314–22.

The Oral Sumatriptan Dose-Defining Study Group (1991). Sumatriptan - an oral dose-defining study. *European Neurology*, **31**, 300–5.

The Sumatriptan Auto-injector Study Group 1991. Self-treatment of acute migraine with subcutaneous sumatriptan using an auto-injector device. *European Neurology*, **31**, 323–31.

Williams, W. O. & Györy, A. Z. (1976). Aspects of the use of lithium for the non-psychiatrist. *Australian and New Zealand Journal of Medicine*, **6**, 233–42.

Wilkinson, M. (1987). Drug therapy during migraine attacks. In *Migraine: Clinical, Therapeutic, conceptual and Research Aspects*, ed. N.J. Blau, pp. 205–13. London: Chapman & Hall.

Wykes, P. (1968). The treatment of angina pectoris with coexistent migraine. *Practitioner*, **200**, 702–4.

Part V

Viral and immunological disorders

22
Neurovirology: the evolution of new challenges
R. T. JOHNSON
The Johns Hopkins University School of Medicine,
Baltimore,
Maryland, USA

Neurovirology presents an ever-changing face. The high burst rates of virus replicative cycles and the high mutational rate, particularly of the ribonucleic acid (RNA) viruses, assure that 'new' or altered viruses will continue to emerge. The evolution of new virus strains and their ability to be sustained are largely dependent upon the size of the host population and the environmental and social interactions of this population. The continual population increase of this global village and modern transportation that presses us functionally closer together will accelerate the evolution of new viruses and aid their perpetuation within both animal and human populations.

DNA viruses

Deoxyribonucleic acid (DNA) viruses, such as the herpes viruses and the papovaviruses, have probably been human parasites for millennia. This is assumed because these viruses have acquired novel adaptations to survive in small and isolated population clusters. They have the ability to remain latent following initial infection and to reactivate years later when new nonimmune members have been born into the society. The ultimate adaptation, however, is the ability to establish latency in neurones, where the host cell survives for the lifespan of the host (Hope-Simpson, 1965). Herpes simplex viruses, type 1 and 2, and varicella-zoster virus are examples of neurotropic viruses that have acquired this remarkable strategy. Childhood infections, such as herpetic gingivostomatitis or chickenpox, are followed by retrograde transport of virus to the sensory ganglia where the viral DNA persists in neurones (Steiner *et al.*, 1988; Gilden *et al.*, 1987). Years later reactivation leads to anterograde spread of virus back to the mucosa or skin, where active reinfection causes cold sores or shingles and allows spread of virus to the next generation. Spread centripetally into the central nervous system can cause life-threatening disease, and this

Table 22.1. *Societal changes that enhance the evolution and spread of neurotropic viruses*

Providing an adequate pool of susceptibles	Increasing global population; Increasing human contacts (travel)	Measles, picornaviruses
Altering forms of human or animal contact		
Societal mores	Increased sexual contacts	HSV2, HIV
	Day care with early exposure	Measles, picornaviruses
	Altering woods for suburbs and recreation	La Crosse encephalitis virus
	Agricultural clearing or irrigation	Western encephalitis virus, arenaviruses
	Global movement of animals and animal products	Rabies, bovine spongiform encephalopathy
Medical practices	Blood transfusions	Cytomegalovirus, HIV, HTLV
	Immunosuppressive therapy	HSV, varicella-zoster, JC virus, cytomegalovirus
	Organ transplants (infected donor)	JC virus, rabies
	Antiviral drugs	Resistant HSV, HIV

Notes: Abbreviations
HSV = herpes simplex virus.
HIV = human immunodeficiency virus.
HTLV = human T cell leukemia virus.

occurs on rare occasions during primary infection or reactivation. In nonimmune infants, type 1 or type 2 herpes simplex virus can cause a diffuse usually fatal encephalitis, in immune adults type 1 herpes simplex virus can cause a localized temporal lobe encephalitis, and in the immunodeficient patients herpes simplex viruses or varicella-zoster virus can cause severe neurological infections. However, natural selection would favour non-neurovirulent strains of herpes viruses, since central nervous system infections are 'deadend' infections.

With social changes in the 1960s and 70s, herpes simplex virus, type 2, an unusual virus that causes genital herpes and maintains latency in the sacral ganglia, became a far more common disease with more frequent spread to cause fatal diffuse encephalitis in infants and with reactivations associated with recurrent meningitis or radiculopathies in adults (Bergstrom et al., 1990) (Table 22.1). Genetic alteration of these large complex viruses occurs, but is not important for their worldwide persistence as pathogens. Indeed, studies of these viruses may lead to better diagnosis and treatment or even the production of vaccines, but their total elimination from the population will probably never be possible.

Human papovaviruses, JC and BK, are also well adapted to persistence in human populations. Asymptomatic infection usually occurs in childhood. The viruses remain latent being reactivated during events such as a pregnancy and are excreted in urine (Arthur & Shah, 1989). BK virus may cause no disease thus representing optimal host adaption. JC virus in immunosuppressed persons can be neurotropic with selective infection and lysis of oligodendrocytes causing progressive multifocal leukoencephalopathy. This is an emerging disease due to an old but silent virus; the disease first became more frequent with iatrogenic immunosuppression for autoimmune diseases and transplantation, and in the past decade it has become quite common as a complication of the new human immunodeficiency virus (HIV) infections.

RNA viruses

RNA viruses have higher mutational rates because the lack of fidelity in the replication of RNA, in contrast to DNA in which enzymatic proofreading occurs. The rate of mutation of DNA is estimated at about one error for the replication of 10^9 base pairs. The error rate from RNA to RNA is approximately 10^4, about 100 000 times greater than the mutation rate for DNA viruses (Steinhauer & Holland, 1987). Some RNA viruses have specific areas of their genome where rates of mutation are particularly frequent. Furthermore, some viruses (e.g. influenza viruses, arenaviruses and bun-

Table 22.2. *Mechanisms for viral survival in host populations*

Virus	Mechanism
DNA viruses	Latency and activation after new nonimmune hosts born
	Seldom cause fatal disease (host survival)
RNA viruses	
Arboviruses	Mutation/spread in animal populations
Zoonotic viruses	Mutation/spread in animal populations
Human viruses	
Picornaviruses	Mutations which supply new viruses
Influenza	Mutations and reassortment allow reinfection
Measles	'New' virus that requires large human populations
Retroviruses	Latency and high mutation rate
HIV 1 and 2	Additional low transmissibility, long incubation period, and long infectivity period
HTLV-I	All of above plus low disease penetrance

yaviruses) have multipartite genomes so that reassortment can readily occur if two viruses infect the same cell. Many of the RNA viruses are animal viruses that only incidentally spread to man as a deadend host. Some animal viruses also cycle through an obligatory replicative cycle in arthropods. Mutations of viruses are continually appearing and disappearing within the cycle between animals or between arthropods and animals. At times, variants arise that involve man. For example, an intramolecular recombinant apparently occurred between Eastern equine encephalitis and a Sindbis-like virus to produce Western equine encephalitis virus, which causes epidemic disease in the Western United States and Canada (Hahn *et al.*, 1988). In the 1960s, an apparent new virulent La Crosse virus of the California group of arboviruses appeared in the Midwestern US. Furthermore, the encroachment of suburbs into forests and recreational activities increased the exposure to the virus in its woodland animal-treehole mosquito cycle (Johnson, Lepow & Johnson, 1968). Furthermore, vectors may change. A new threat of arthropod-borne disease has arisen in the United States with the introduction of the *Aedes albopictus* mosquito, imported into the United States from Southeast Asia in used automobile tires (Francy *et al.*, 1990). A variant of an old virus may appear *de novo* in a new area such as the Kyasanur Forest virus, a variant of the Siberian tick-borne encephalitis virus which appeared abruptly in west-central India in 1957 (Webb & Rao, 1961). In retrospect, the virus may have flown in on ticks aboard birds migrating across the Himalayas.

The zoonotic neurotropic viruses pose different problems (Table 22.2). In island populations such as England, Guam or Hawaii, where rabies had been limited to dogs and cats, quarantine methods have excluded the virus, but in large continental land masses, sylvatic cycles of rabies exist, as in the United States where the virus is common among raccoons, bats, and foxes, so that rabies remains a public health threat (Steele, 1988). Australia is unique in being the only continent that has not harboured rabies, yet because of the large population of susceptible animals, an ever present threat remains. Ironically, a person entering the country during the incubation period of rabies poses no significant threat since secondary human cases rarely, if ever, occur. On the other hand, an infected animal entering the country could permanently seed the virus into sylvatic animal populations.

Arenaviruses cause natural life-long infections of rodents, and some strains can cause neurological infections and hemorrhagic fevers in humans. Lassa fever virus, for example causes a fatal hemorrhagic disease in Africa, and in hospital settings spread from patients to health care workers occurs from infected blood (Holmes *et al.*, 1990). However, the natural rodent host of this virus does not exist in the United States, so aside from imported human cases infecting a few health care workers, epidemic disease would be unlikely to occur unless there was a major shift in virus ecology.

RNA viruses that spread from man to man require different strategies to be sustained. The picornaviruses, the orthomyxoviruses (the influenza viruses), and the paramyxoviruses require an initial population size to be maintained. The picornaviruses, however, are constantly changing, and new serotypes continue to arise that may be more or less pathogenic or neuroinvasive in man. For example, in 1969 acute hemorrhagic conjunctivitis due to echovirus 70 appeared in Africa and spread through India and Southeast Asia, only to largely disappear. This infection was complicated with a number of cases of clinical poliomyelitis (Hung & Kono, 1979).

Influenza viruses are usually not regarded as neurotropic, but they have been associated with Reye's syndrome, with postinfectious encephalomyelitis, and possibly with direct encephalitis reported during the Asian influenza epidemic of 1957. Influenza viruses have a unique mechanism of reinfecting humans: they undergo antigen drift with point mutations and major antigenic shifts with reassortments. The major antigen shifts of Asian influenza of 1957 and in Hong Kong influenza of 1968 were due to reassortment with the human influenza virus acquiring segments of avian influenza viruses coding for new surface polypeptides (Scholtissek *et al.*, 1978). The source of these new major shifts has been Asia where large populations of people, ducks, and pigs live in close proximity, thus permitting the dual infection of a single cell of one of the three hosts with two different influenza viruses.

Speculation has linked vonEconomo's disease to influenza virus. The epidemic of sleeping sickness swept across Europe between 1915 and 1920, and appeared first in the United States in 1918 (coincident with the pandemic of swine influenza). Certainly its slow international spread does not resemble influenza, but at that period of time, influenza viruses spread more slowly because of the pace of international human movement. Whether or not von Economo's disease will ever be tied to influenza virus is unknown (Johnson, 1982). However, new pandemic of influenza viruses will appear; most will cause little neurological disease but we must be alert since some new influenza viruses may produce new diseases and others may give rise to increased rates of Reye's syndrome or postinfectious encephalomyelitis.

Measles virus has remained relatively antigenically stable over a number of years, but none the less it provides an instructive lesson on new and emerging viruses. Measles (the only human morbillivirus) appears, from RNA homology studies to have evolved from a virus similar to African rhinderpest. Calculations estimate that a population of over 250 000 persons is needed to maintain measles virus, so the virus could not have been sustained in early or isolated human populations. Indeed, measles was not described by the early Greek or Roman physicians, but was first recorded by Rhazes of Baghdad, who dated its appearance in the Arab world to the 6th century. The virus moved into Europe with the Saracen invasion in the 8th century and caused widespread disease with high death rates. Even in this century, virgin population epidemics have had high death rates in infants and adults over age 40. The virus is highly infectious in nature, has a relatively short incubation period, and confers permanent immunity. Therefore in the Middle Ages an army had to be mounted to cycle the virus during the long trek across the Pyrenees into Europe (Johnson, Griffin & Moench, 1988*b*). In today's world of rapid transport, it might spread in one day by car, or in an hour by plane.

Measles still remains the commonest cause of human demyelinating disease. Measles encephalomyelitis complicates about 1:1000 cases in persons over 2 years of age. With increasing vaccine failures, infection will increase in older persons, and highest rates of complications can be anticipated. The history of measles, which evolved from an animal virus in Africa and spread in humans to cause immunodeficiency, neurological complications and death, shows an ancient example of a pattern seen in the last decade with HIV.

Retroviruses

Retroviruses represent both new and newly recognized human neuropathogens. Retroviruses pose the most interesting strategies of viral survival in human populations; they combine the capacity of latency of DNA viruses with the high mutational rate of the RNA viruses. Retroviruses contain a reverse transcriptase and after entering the cell, a DNA copy is made of the genomic RNA. This proviral DNA integrates into the host cell and can remain latent for life without symptoms, can cause cell transformation (oncornaviruses) or can continue to replicate genomic RNA and virus at low levels with frequent mutations (lentiviruses). With lentiviruses, latency and continued replication allows the generation of new mutants in a single host animal. For example, equine infectious anemia virus causes an acute febrile illness with viremia and hemolysis. Disease abates as neutralizing antibody is formed against the virus, but weeks or months later the horse suffers a new bout of fever and viraemia. The virus isolated at that time is not neutralized by the antibody formed against the original virus. This sequence can recur repeatedly over years or until the death of the horse (Payne *et al.*, 1987).

All known lentiviruses also cause subacute encephalitis. Visna virus of sheep, the prototype virus of this family, has been studied in the greatest detail. The promonocytes of the bone marrow are latently infected, and although they contain the virus genome little, if any, viral protein or infectious virus are produced. The same is true of the monocytes circulating in the blood. However, as the monocytes move into specific tissues, lungs, joints, breast or brain, they mature into macrophages and biochemical changes in the cells favour the active production of infectious virus. Antibodies are formed and, over periods of time, sequential viruses are isolated which are not neutralized by earlier antibody (Narayan *et al.*, 1987).

How long HIV has existed in human populations is a matter of speculation. HIV did not spread worldwide until the late 1970s and clinical disease was first recognized in 1981 with the clusters of Karposi's sarcoma and pneumocystis pneumonia in gay men. In 1985 the virus was recovered from brain, peripheral nerve and spinal fluid (Levy *et al.*, 1985; Ho et al., 1985), viral DNA and RNA were shown in inflammatory nodules in the brain (Shaw et al., 1985) and intrathecal antibody synthesis was demonstrated (Resnick *et al.*, 1985). These findings established that HIV infected the nervous system. A wide spectrum of disease has now been associated with HIV including occasional cases of acute meningitis or Guillain–Barré syndrome occurring at the time of initial seroconversion, chronic meningitis during the seropositive state, mononeuritis with vasculitis during the AIDS-related complex phase, and high rates of dementia, myelopathies and sensory neuropathies occurring at the time of the acquired immunodeficiency syndrome (AIDS) with opportunistic infection. AIDS is clearly a new disease and the causative virus is capable of producing multiple different neurological syndromes (Johnson, McArthur & Narayan, 1988c).

Studies of cerebrospinal fluid in healthy gay men whose time of seroconversion is known show evidence that two-thirds have a pleocytosis, increased protein and/or oligoclonal bands suggesting infection (McArthur et al., 1988). Thus of the 10 million people infected worldwide, we can assume that over 5 million have central nervous system infection making HIV in only a decade the commonest viral infection of the central nervous system. The pathogenesis of these many complications are unknown, but because of their different times of occurence during infection and their different associated pathologies, it is assumed that the pathogenesis of varied complications may be different. The acute demyelinating neuropathy and vasculitis with mononeuritis multiplex resemble immune-mediated diseases, and their occurrence before severe immunosuppression suggests a mechanism involving deregulation of normal immune responses. On the other hand, the vacuolar myelopathy and the dementia occur late and their development correlates with the presence of virus and virus load. This suggests a direct effect of the virus found in macrophages and microglial cells of the brain. There has been considerable speculation that the neurological damage may be due either to a viral protein or a monokine incited by the virus infection of these macrophage-derived cell populations (Johnson et al., 1988c).

The example of new recognition of a retrovirus causing neurological disease is the association between human T-cell leukemia virus, type I (HTLV-I), and tropical spastic paraparesis. This clinical disease has been recognized for many years, and this virus probably has been causing infection for many years in populations in the south of Japan and in numerous tropical and semitropical latitudes. The persistence of this virus in human populations is dependent upon the chronicity of the infection and the fact that only about 1% of those infected develop either the T cell leukemia or the neurological disease. Many are infected by breast milk, but others are infected by sexual contact, or by blood transfusion. Since the onset of the tropical spastic paraparesis does not occur until the middle life, this is indeed a long incubation period (Roman, Roman & Osame, 1990). Although the pathology found in the biopsy of a single case during the acute phase showed acute vasculitis (Johnson et al., 1988a), later hyalinization of the vessels with necrosis and demyelination are found with minimal inflammation. Whether the virus replicates in any cells other than T cells is unknown. These observations certainly lead to the question of why less than 1:100 who are infected develop the disease, why incubation is as long as 40 years, why the disease localizes in the thoracic spinal cord, and why over the years the disease becomes relatively quiescent despite the fact that there is ongoing high levels of intrathecal antibody synthesis suggesting that there is antigenic stimulation by virus within the nervous system.

Table 22.3. *Predictions*

HIV 1	will infect 20 million worldwide by AD 2000
HIV 2	will spread into other continents
Influenza	new recombinants will cause pandemics
Measles	increasing cases will occur in developed countries with early childhood exposure and vaccine failures
Japanese encephalitis	will continue to spread into new regions of Asia
Other arboviruses	new virulent strains will appear and strains will appear in new locations
Arenaviruses	new pathogens will be found with agricultural encroachments on rodent habitats
Rabies	will involve new sylvatic hosts and spread into new regions

The unexpected will happen

Summary

Of the many viruses that cause neurological diseases in humans, in the past many may have entered human populations and either exhausted susceptible populations or were inefficient in spread and disappeared. Such occurrences might explain the great plague of ancient Athens, the English sweating disease or von Economo's disease in the early part of this century. Our enlarging population and rapid transportation favour the perpetuation of similar new agents.

Some viruses have adapted to persist in small populations, such as some DNA viruses by the mechanism of latency or as HTLV-I by its low level of virulence and long infectious, incubation period. With other viruses, such as the zoonotic agents and arboviruses, man is a deadend host and not involved with the normal cycle of the virus, but man will have increasing involvement as he changes the water table, builds houses in forested areas, imports new vectors and otherwise tinkers with the environment, however well meaningly (Table 22.3). Viruses that are spread from man to man will continue to evolve and change. Over the next decades we will see new picornaviruses, with possibly interesting new clinical syndromes, HIV-1 and probably HIV 2 as well, will continue to spread, a new pandemic of influenza is overdue, and a resurgence of measles in developed countries is under way. Unless we discover novel methods of control, these changes can be anticipated; but possibly the most interesting prediction is that the unexpected will happen. Who would have predicted a decade ago that an unknown human lentivirus would be the commonest nervous system infection in 1992?

Acknowledgements

The author wishes to acknowledge the support of the Hamilton Rhoddis Foundation and the National Institutes of Health National Institutes of Neurologic Disease PO1 NS26643.

References

Arthur, R. R. & Shah, K. V. (1989). Occurrence and significance of papovaviruses BK and JC in the urine. In *Progress in Medical Virology*, vol. 36, ed. J. L. Melnick, pp. 42–61. Basel: Karger.

Bergstrom, T., Vahlne, A., Alestig, K., Jeansson, S., Forsgren, M. & Lycke, E. (1990). Primary and recurrent herpes simplex virus type 2-induced meningitis. *Journal of Infectious Diseases*, **162**, 322–30.

Francy, D. B., Karabatsos, N., Wesson, D. M., Moore, Jr., C. G., Lazuick, J. S., Niebylski, M. L., Tsai, T. F. & Craig, Jr., G. B. (1990). A new arbovirus from Aedes albopictus, an Asian mosquito established in the United States. *Science*, **250**, 1738–40.

Gilden, D. H., Rozenman, Y., Murray, R., Devlin, M. & Vafai, A. (1987). Detection of varicella-zoster virus nucleic acid in neurones of normal human thoracic ganglia. *Annals of Neurology*, **22**, 377–80.

Hahn, C. S., Lustig, S., Strauss, E. G. & Strauss, J. H. (1988). Western equine encephalitis virus is a recombinant virus. *Proceedings of the National Academy of Sciences, USA*, **85**, 5997–6001.

Ho, D. D., Rota, T. R., Schooley, R. T., Kaplan, J. C., Allan, J. D., Groopman, J. E., Resnick, L., Felsenstein, D., Andrews, C. A. & Hirsch, M. S. (1985). Isolation of HTLV-III from cerebrospinal fluid and neural tissues of patients with neurologic syndromes related to the acquired immunodeficiency syndrome. *New England Journal of Medicine*, **313**, 1493–7.

Holmes, G. P., McCormick, J. B., Trock, S. C., Chase, R. A., Lewis, S. M., Mason, C. A., Hall, P. A., Brammer, L. S., Perez-Oronoz, G. I., McDonnell, M. K., Paulissen, J. P., Schonberger, L. B. & Fisher-Hoch, S. P. (1990). Lassa fever in the United States: investigation of a case and new guidelines for management. *New England Journal of Medicine*, **323**, 1120–3.

Hope-Simpson, R. E. (1965). The nature of herpes zoster. A long-term study and a new hypothesis. *Proceedings of the Royal Society of Medicine*, **58**, 9–20.

Hung, T. P. & Kono, R. (1979). Neurological complications of acute hemorrhagic conjunctivitis. In *Handbook of Clinical Neurology*, ed. P.J. Vinken & G.W. Bruyn, pp. 595–623. Amsterdam: North Holland.

Johnson, K. P., Lepow, M. L. & Johnson, R. T. (1968). California encephalitis. I. Clinical and epidemiological studies. *Neurology*, **18**, 250–4.

Johnson, R. T. (1982). *Viral Infections of the Nervous System*. New York: Raven Press.

Johnson, R. T., Griffin, D. E., Arregui, A., Mora, C., Gibbs, Jr., C. J., Cuba, J. M., Trelles, L., & Vaisberg, A. (1988*a*). Spastic paraparesis and HTLV-I in Peru. *Annals of Neurology*, **23**, S151–5.

Johnson, R. T., Griffin, D. E., Hirsch, R. L., Wolinsky, J. S., Rodenbeck, S., Lindo de Soriano, I. & Vaisberg, A. (1984). Measles encephalomyelitis: clinical and immunological studies. *New England Journal of Medicine*, **310**, 137–41.

Johnson, R. T., Griffin, D. E. & Moench, T. R. (1988*b*). Pathogenesis of measles immunodeficiency and encephalomyelitis: parallels to AIDS. *Microbial Pathogenesis*, **4**, 169–74.

Johnson, R. T., McArthur, J. C. & Narayan, O. (1988c). The neurobiology of human immunodeficiency virus infections. *Federation of American Societies for Experimental Biology Journal*, **2**, 2970–81.

Levy, J. A., Shimabukuro, J., Hollander, H., Mills, J. & Kaminsky, L. (1985). Isolation of AIDS-associated retroviruses from cerebrospinal fluid and brains of patients with neurological symptoms. *Lancet*, **ii**, 586–8.

McArthur, J. C., Cohen, B. A., Farzedegan, H., Cornblath, D. R., Selnes, O. A., Ostrow, D., Johnson, R. T., Phair, J. & Polk, B. F. (1988). Cerebrospinal fluid abnormalities in homosexual men with and without neuropsychiatric findings. *Annals of Neurology*, **23**, S34–7.

Narayan, O., Clements, J., Kennedy-Stoskopf, S., Sheffer, D. & Royal, W. (1987). Mechanisms of escape of visna lentiviruses from immunological control. In *Contributions to Microbiology and Immunology*, vol. 8, ed. J. M. Cruse & R. E. Lewis, Jr., pp. 60–76. Basel: Karger.

Payne, S. L., Salinovich, O., Nauman, S. M., Issel, C. J. & Montelaro, R. C. (1987). Course and extent of variation of equine infectious anemia virus during parallel persistent infections. *Journal of Virology*, **61**, 1266–70.

Resnick, L, diMarzo-Veronese, F., Schupbach, J., Tourtelotte, W. W., Ho, D. D., Muller, F., Shapshak, P., Vogt, M., Groopman, J. E., Markham, P. D. & Gallo, R. C. (1985). Intra- blood–brain-barrier synthesis of HTLV-III-specific IgG in patients with neurologic symptoms associated with AIDS or AIDS-related complex. *New England Journal of Medicine*, **313**, 1498–504.

Roman, G. C., Roman, L. N. & Osame, M. (1990). Human T lymphotropic virus type 1 neurotropism. In *Progress in Medical Virology*, vol. 37, ed. J. L. Melnick, pp. 190–210. Basel: Karger.

Scholtissek, C., Rohde, W., von Hoyningen, V. & Rott, R. (1978). On the origin of the human influenza virus subtypes H2N2 and H3N2. *Virology*, **87**, 13–20.

Shaw, G. M., Harper, M. E., Hahn, B. H., Epstein, L. G., Gajdusek, D. C., Price, R. W., Navia, B. A., Petito, C. K., O'Hara, C. J., Cho, E.-S., Oleske, J. M., Wong-Staal, F. & Gallo, R. C. (1985). HTLV-III infections in brains of children and adults with AIDS encephalopathy. *Science*, **227**, 177–82.

Steele, J. H. (1988). Rabies in the Americas and remarks on global aspects. *Reviews of Infectious Diseases*, **10**, S585–97.

Steiner, I., Spivack, J. G., O'Boyle, H., Lavi, E. & Fraser, N. W. (1988). Latent herpes simplex virus type 1 transcription in human trigeminal ganglia. *Journal of Virology*, **62**, 3493–6.

Steinhauer, D. A. & Holland, J. J. (1987). Rapid evolution of RNA viruses. *Annual Review of Microbiology*, **41**, 409–33.

Webb, H. E. & Rao, R. L. (1961). Kyasanur Forest disease: a general clinical study in which some case with neurological complications were observed. *Transactions of the Royal Society of Tropical Medicine and Hygiene*, **55**, 284–98.

23
Multiple sclerosis: current concepts and Australian studies

J. G. McLEOD
Department of Medicine,
University of Sydney,
Sydney, NSW Australia

The historical aspects of multiple sclerosis have been well described by Compston (1988, 1990a). The earliest pathological descriptions were those of Cruveilhier in 1835 and Carswell in 1838. Credit is usually given to Charcot (1868) for the clear recognition of the clinicopathological entity. The history of multiple sclerosis in Australia has been reviewed by Frith (1989) who found that the first published description of the disease in the Australian literature was in 1875 and that the first clinical cases were reported in 1886. The first Australian clinicopathological report was that of Flashman and Latham in 1915.

Pathology

The earliest lesion in multiple sclerosis is a breakdown in the blood–brain barrier around small blood vessels in the central nervous system white matter particularly in periventricular regions, brainstem, spinal cord and optic nerves. Lymphocytes and macrophages invade the white matter and cause myelin destruction and death of oligodendrocytes (Prineas, 1975; Adams, Poston & Buk, 1989). Blood–brain barrier breakdown causes leakage of plasma proteins into surrounding brain tissue (Gay & Esiri, 1991), and these pathological changes are paralleled by the appearance of early lesions seen on magnetic resonance imaging (MRI) using gadolinium enhancement before other MRI abnormalities are present (Kermode *et al.*, 1990). This process is followed by extensive gliosis by astrocytes and limited remyelination at the edges of the plaques where some inflammatory activity may continue in chronic lesions (Prineas, 1985). Immunoglobulins are present within the plaques which are surrounded by both CD8+ and CD4+ T lymphocytes (Rodriguez, 1989; Compston, 1990*a*).

Etiology

Both environmental and genetic factors clearly play a role in the etiology of multiple sclerosis but their relative contributions remain uncertain. This topic has recently been reviewed by Hughes (1992).

Environmental factors

It is well recognized that, in general, the frequency of multiple sclerosis increases with distance from the equator (Kurtzke, 1985; Acheson, 1985). However, genetic factors may influence the distribution of multiple sclerosis in white populations; for example, it has been proposed that people of Nordic descent are more susceptible to multiple sclerosis than some other racial groups and that their migrations across Europe and to other countries are at least in part responsible for the distribution of multiple sclerosis. (Compston, 1990a,b). Epidemiological studies in Australia and New Zealand are of considerable importance since they survey a range of latitudes in a predominantly white population derived mainly from the United Kingdom. Variable genetic factors are therefore largely eliminated (Miller *et al.*, 1990).

Early studies in Australia demonstrated a relationship between multiple sclerosis frequency and latitude (McCall *et al.*, 1968). A more recent regional point prevalence survey based on the national census day in 1981 confirmed the presence of a relationship between increasing prevalence of multiple sclerosis and increasing southerly latitude, the frequency increasing sevenfold from tropical Queensland (11.8:100 000 population) to Hobart in the south of Australia (prevalence 75.6:100 000) (Fig. 23.1). Other regions of Australia were intermediate in frequency between these extremes, appropriate to their latitude (Hammond *et al.*, 1989*a*). Independent verification of the gradient of multiple sclerosis frequency with latitude was obtained from mortality statistics for the decade 1971 to 1980 which showed an increase in mortality from 0.36:100 000 in Queensland to 0.81:100 000 in Tasmania (Hammond *et al.*, 1989*b*) (Table 23.1). Several epidemiological studies have also been conducted in New Zealand and, when the Australian and New Zealand prevalence data are combined, an impressive correlation between Southern latitude and prevalence of multiple sclerosis is evident (Miller *et al.*, 1990) (Fig. 23.2).

Could the relationship between frequency of multiple sclerosis and latitude be explained by genetic factors? Compston (1990*a,b*) has carefully considered the arguments for both genetic and environmental factors determining the distribution of multiple sclerosis and there is no question that genetic factors play an important role. Studies in the United Kingdom (Swingler & Compston,

Table 23.1. *Multiple sclerosis mortality at different latitudes in Australia 1971--1980*

Area	Latitude (°S)*	Patients	Rate+
Queensland	25.1	71	0.36
Western Australia	31.4	40	0.39
New South Wales	33.6	222	0.46
South Australia	34.8	72	0.75
Tasmania	42.0	32	0.81

Notes: * Mean latitude calculated from population distribution.
+ Adjusted to age distribution of 1981 population.

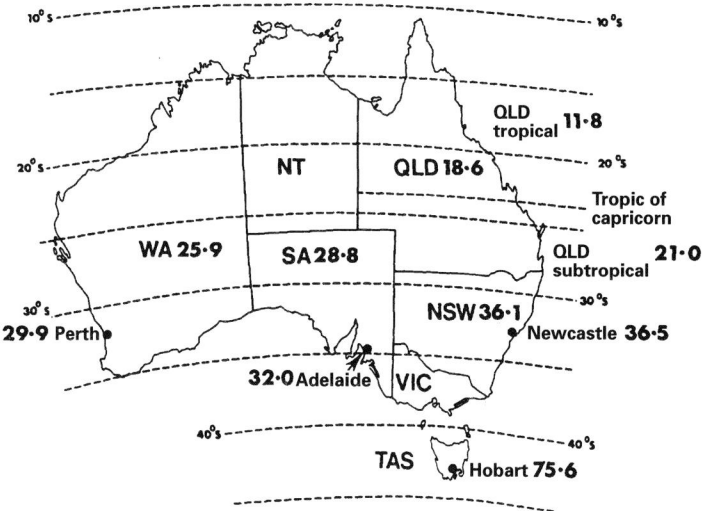

Fig. 23.1. Standardized prevalence (cases/100 000 population) of multiple sclerosis in Australia on June 30, 1981. Hammond S. (1989a) reproduced with permission.

1986) and New Zealand (Skegg *et al.*, 1987) have suggested that the frequency gradient of multiple sclerosis with latitude found in both countries may be determined, at least in part, by a greater susceptibility to multiple sclerosis among people of Scottish origin. However, this does not appear to apply in Australia since the proportion of Scottish migrants among migrants from the United Kingdom and Ireland was similar in the different survey areas (Table 23.2). Moreover, the proportion of the population whose names began with 'Mc' or 'Mac' in the telephone directories of the survey areas was examined as

Table 23.2. *Proportion of United Kingdom and Ireland migrants by individual country of origin in each survey area of Australia in 1981*

Survey area	Latitude (°S)	Scotland	England	Wales	Ireland	Total
Queensland						
Above Tropic of Capricorn SD	18.9	13.9%	76.7%	2.4%	7.0%	18 584
Below Tropic of Capricorn SD	26.9	13.2%	79.6%	1.5%	5.7%	121 496
Western Australia	31.3	11.0%	82.0%	2.2%	4.8%	186 143
Perth SD	32.0	10.8%	82.3%	2.1%	4.8%	152 595
Newcastle SDi	32.9	19.6%	71.9%	4.4%	4.1%	19 151
South Australia	34.9	11.5%	82.0%	2.1%	4.4%	152 091
Adelaide SD	34.9	10.9%	82.7%	2.0%	4.4%	128 405
Hobart SD	42.8	11.6%	81.6%	2.4%	4.4%	10 792

Notes: Abbreviations: SD = statistical division; SDi = statistical district.
Source: From Hammond *et al.*, 1989a.

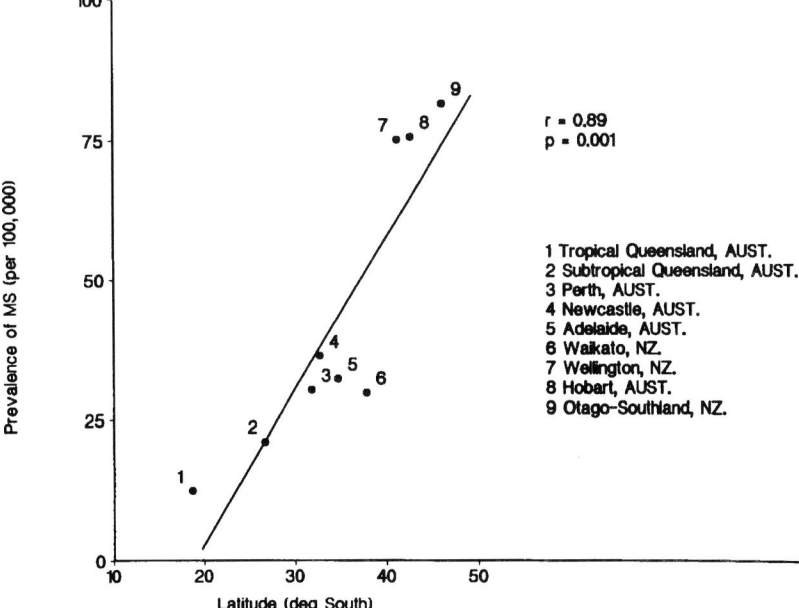

Fig. 23.2. The relationship of prevalence of multiple sclerosis with latitude in Australia and New Zealand. Miller *et al.* (1990) reproduced with permission.

a crude index of Scottish ancestry in the overall population, as suggested by Skegg and colleagues (1987), and there is no evidence for a greater concentration of these people in southern latitudes in Australia (Hammond *et al.*, 1989*a*). In addition, the frequency of HLA-DR2, the genetic marker which is most frequently associated with multiple sclerosis among Caucasians, and common in the Scottish population of the United Kingdom (Swingler & Compston, 1986), was not significantly different in the different parts of Australia that were surveyed (Hammond *et al.*, 1989*a*). A final point in favour of environmental factors is that the prevalence rate in Australian people originating from the United Kingdom is considerably less than that observed in most parts of the British Isles (Hammond *et al.*, 1988*a, b*; Compston, 1990*b*).

Migration studies, which have shown that a subject moving from a high to a low risk zone in early life has a smaller risk of developing multiple sclerosis than a person moving in later adult life, provide further support for the influence of environmental factors in the disease (Dean & Kurtzke, 1971; Alter, Kahana & Loewenson, 1978). Studies of the migrant population in Australia support these findings although they suggest that the risk of acquisition of multiple sclerosis may extend over a wider age range than was previously thought. (Hammond *et al.*, 1988*a*). Further strong evidence for the presence of an environmental factor in the etiology of multiple sclerosis was the outbreak in the Faroe Islands following World War II, suggesting a point source epidemic (Kurtzke & Hyllested, 1986).

Is the frequency of multiple sclerosis increasing? This issue has been addressed in a number of studies and whether or not the increase is apparent or real remains undecided. Three major cities in Australia, Perth, Newcastle and Hobart, were resurveyed in 1981 after a 20-year interval. The increased prevalence in all three cities was attributed to better case ascertainment, increased recognition of the less severely disabled patient, increased survival time and differential immigration of a population at higher risk of developing multiple sclerosis than the indigenous population (Hammond *et al.*, 1988*a*). The finding that mortality rates from multiple sclerosis have declined would contribute to the rise in prevalence (Hammond *et al.*, 1989*b*; Williams, Jones & McKeran, 1991). A recent study in Norway showed that the prevalence of MS had increased threefold and the average annual incidence rate had almost doubled over a twenty year period (Midgard, Riise & Nyland, 1991) and an increase has also been reported in one area surveyed in the United States (Wynn *et al.*, 1990).

Viral infection is one of the environmental factors implicated by epidemiological studies. However, all attempts to isolate a virus in culture, or by inoculation of brain material into primates, have been unsuccessful. Antibody titres to measles virus are raised in the serum of multiple sclerosis patients compared

Multiple sclerosis 379

with controls, but the IgG oligoclonal bands in the cerebrospinal fluid (CSF) are not directed against the measles virus. A number of viruses are known to cause demyelination (canine distemper, visna, JHM, Semliki Forest and Theiler's virus), but these have not been implicated in multiple sclerosis (Rodriguez, 1989). HTLV-1 is associated with tropical spastic paraparesis but, in spite of early expectations, a number of studies have shown that it is not associated with multiple sclerosis (Ehrlich *et al.*, 1991).

Genetic factors

There is no doubt that genetic factors play an important part in determining the risk of developing multiple sclerosis (Compston, 1990*a,b*). Evidence that supports this contention includes the following:

1. About 15%–20% of patients have an affected relative (Myrianthopoulos, 1985).
2. Children and siblings of multiple sclerosis patients have a risk of developing the disease that is 30 to 50 times greater than that of the general population (Sadovnick & Baird, 1988).
3. The concordance rate of multiple sclerosis among monozygotic twins is 25.9% but only 2.3% in dizygotic twins (Ebers *et al.*, 1986).
4. A number of racial groups have a low risk of developing multiple sclerosis including the Australian aboriginal, Eskimos, Africans, gypsies, Orientals and Arabs (Acheson, 1985; Hammond *et al.*, 1988*a*; Yu *et al.*, 1989).
5. There is a strong association of multiple sclerosis with the HLA system, particularly in Caucasian populations (Jersild *et al.*, 1975; Stewart & Kirk, 1983). Studies in Australia, Europe and North America confirm that there is an association with the HLA-A3, HLA-B7 and HLA-DR2 antigens. In Australia, 62.5% of patients are HLA-DR2 positive compared with 17.6% of controls. (Stewart *et al.*, 1977) (Fig. 23.3). In the Orkney Islands and Aberdeen, there was no association with DR2 due to the unusually high frequency of this antigen in the normal population, although there was an association with another class II antigen, DQwl. (Francis *et al.*, 1987*a*). There was no DR2 association in Japanese, Hungarian gypsies, Israelis or Chinese (Stewart & Kirk, 1983). An association of DR4 in Jordanian arabs has been reported (Kurdi *et al.*, 1977).
6. An association with certain alpha-1 antitrypsin phenotypes encoded on chromosome 14 has been reported (McCombe *et al.*, 1985). Gm allotypes are also located on chromosome 14 but these are associated weakly or not at all with multiple sclerosis.

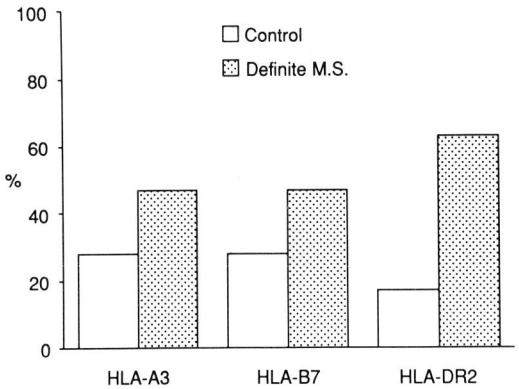

Fig. 23.3. Frequency of HLA antigens (-A3, -B7 and -DR2) in control subjects and patients with multiple sclerosis in Sydney, Australia.

These observations leave little doubt that the susceptibility to multiple sclerosis is genetically determined. No single susceptibility gene for multiple sclerosis has so far been localized but recent studies suggest that a susceptibility locus may be linked to the T cell receptor–chain complex. (Seboun *et al.*, 1989; Lee *et al.*, 1991).

Immunological factors

There is strong circumstantial evidence that there is an underlying disturbance of immune function in multiple sclerosis that is genetically determined. Evidence for abnormal immune function includes:
1. The pathology of the lesion with the presence of T lymphocytes, plasma cells and macrophages in a perivascular distribution (Prineas, 1985). Both CD4 and CD8 T lymphocytes are present at the edge of lesions and in the adjacent normal appearing white matter (Traugott, Reinherz & Raine, 1983). In chronic disease, Class II antigen is increased on endothelium and on astrocytes associated with active lesions (Traugott, 1987).
2. The presence of oligoclonal IgG in the CSF.
3. The presence of activated T lymphocytes in the blood and CSF.
4. The demonstration by some workers of reduction of the number of suppressor/cytotoxic (CD8) T lymphocytes during relapses.
5. Elevation of soluble interleukin-2 receptors (IL-2R) in active disease (Adachi, Kumamoto & Araki, 1990).
6. The presence of T cells reactive to myelin basic protein in the peripheral blood of MS patients (Allegretta *et al.*, 1990).

Multiple sclerosis

7. The presence of antibodies to myelin-oligodendrocyte glycoprotein in CSF patients with multiple sclerosis (Xiao, Linington & Link, 1991).
8. The presence of antioligodendrocyte and antigangliocyte antibodies in the CSF (Traugott & Raine, 1981).
9. The pathological and immunological findings in relapsing experimental allergic encephalomyelitis that closely resemble those of multiple sclerosis.

In spite of the strong evidence for disturbance of immune function in multiple sclerosis, no specific antigen has yet been identified.

Diagnosis and classification

There is no specific test for multiple sclerosis, the diagnosis being based primarily on the clinical features, with evidence of two or more distinct lesions in the central nervous system white matter, when there is no satisfactory alternative explanation (McDonald & Halliday, 1977). Electrophysiological studies and MRI may be used to localize subclinical lesions, and detection of oligoclonal bands in the CSF also strengthens the diagnosis. The disease may be classified in different ways (McDonald & Halliday, 1977), but the one most commonly used clinically is that of Rose and colleagues (1976), based on the Schumacher criteria, in which individual cases are categorized as clinically definite, probable or possible. Now that the reliability of laboratory investigations is well established, these may be taken into account also in the classification of the disease, particularly in research protocols (Poser *et al.*, 1983).

Clinical features

The clinical profile of multiple sclerosis differs in certain racial groups, but in Australia it is remarkably similar to that in other predominantly Caucasian populations in the Northern hemisphere (Confavreux, Aimard & Devie *et al.*, 1980; Hammond *et al.*, 1988*a*; Phadke, 1990).

Females are affected about twice as commonly as males and the age of onset is most frequently between the ages of 20 and 50 years. The mean age of onset of 31.9 years in the Australian study is almost identical to that of 31.3 years in France (Confavreux *et al.*, 1980). Nevertheless, about 5% of cases present under the age of 20 and about 5% over the age of 50 (Hammond *et al.*, 1988*b*). The most common presenting symptoms are those involving multiple systems (24%), sensory (24%), pyramidal (16.2%) and visual (14.8%). Less common presenting symptoms are vertigo, ataxia, bowel and bladder disturbances, trigeminal neuralgia, facial palsy and mental changes.

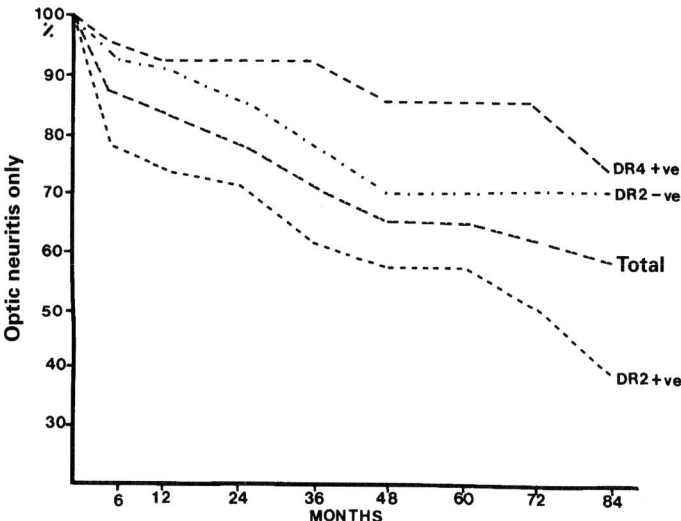

Fig. 23.4. Actuarial curves showing the percentage of patients who have not developed multiple sclerosis at intervals after initial attack of optic neuritis. Hely *et al.* (1986) reproduced with permission.

About 80% of patients have an exacerbating-remitting clinical course in the early stages of their disease, the remaining 20% having a progressive course from outset. Patients with an exacerbating-remitting course may later become progressive. Patients with a relapsing and remitting course have an earlier age of onset (29 years) than those with a progressive onset (37 years). The mean duration of disease from onset to death is greater than 25 years, about 45% of patients having a benign course (i.e. having only mild to moderate disability after a period of 15 years). Over 50% of patients in the community with multiple sclerosis are fully ambulatory (Hammond *et al.*, 1988*b*). Cognitive and memory disturbances are relatively common in multiple sclerosis, particularly in the later stages of the disease (Beatty *et al.*, 1989; Petersen & Kokmen, 1989). The clinical profile has been compared in medium and high frequency prevalent zones in Australia. The clinical profiles in both frequency zones were very similar, although male patients in the hotter climate of Queensland showed a greater tendency to develop progressive disease and hence more disability than those in the more southerly placed cities of Perth, Newcastle and Hobart (Hammond *et al.*, 1988*b*).

Risk of developing multiple sclerosis after a single demyelinating episode

The early risk of multiple sclerosis following an attack of optic neuritis has been extensively studied. Most studies in Europe and the United Kingdom have found that the incidence of multiple sclerosis after optic neuritis is 50% or more whereas the frequency is lower in American studies. A long term follow-up study from London demonstrated that 57% of patients developed multiple sclerosis after a mean follow-up time of 11.6 years and the probability of developing multiple sclerosis 15 years after the initial episode of optic neuritis was 75% (Francis *et al.*, 1987b). In an Australian follow-up study of 82 patients, all of whom had abnormal visual evoked potentials in the symptomatic eye, 32% developed clinically definite or probable multiple sclerosis after a mean follow-up time of 5 years; actuarial analysis predicted that 42% would develop multiple sclerosis by 7 years. There was an increased risk of multiple sclerosis in patients with HLA-DR2 and HLA-B7 tissue types (Hely *et al.*, 1986*a*) (Fig. 23.4). In some patients, visual evoked potentials returned to normal (Hely *et al.*, 1986*b*). MRI reveals multifocal white matter lesions indistinguishable from those in multiple sclerosis in 50%–70% of adults with clinically isolated optic neuritis and those with abnormal MRI are more likely to develop multiple sclerosis (Miller *et al.*, 1988). MRI may therefore have some predictive value (Ormerod *et al.*, 1986, 1987).

A similar follow-up study to the studies of patients with optic neuritis has been undertaken to determine the risk of developing multiple sclerosis after clinically isolated lesions of the brainstem or spinal cord. Progression to multiple sclerosis was seen in 57% of patients who had a brainstem syndrome and 42% of those who had a spinal cord syndrome after mean intervals of 15 and 16 months respectively (Miller *et al.*, 1989). The presence of oligoclonal bands in the CSF and multifocal white matter abnormalities on MRI predict an increased risk of developing multiple sclerosis (Paty *et al.*, 1988; Miller *et al.*, 1989).

Precipitating and aggravating factors

Pregnancy

There is an increased risk of relapse during the later stages of pregnancy and early postpartum period but this increased risk is offset by a reduced risk during the first two trimesters so that overall, pregnancy does not appear to influence the course of the disease (Fig. 23.5) (Frith & McLeod, 1988; Birk *et al.*, 1990).

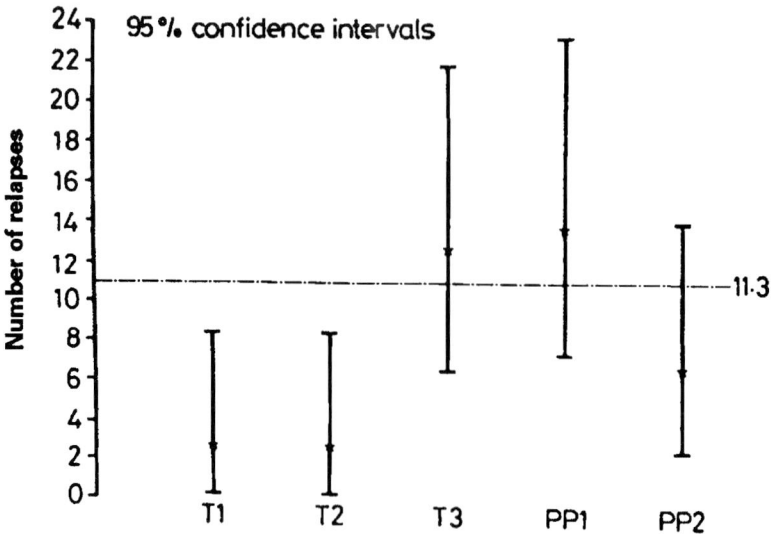

Fig. 23.5. Relapses during pregnancy in 52 women with clinically definite multiple sclerosis. The number of relapses (*) in each trimester or postpartum period with 95% confidence units (vertical bars) is compared with the reported number of relapses (horizontal dotted line) for each trimester and postpartum period. The expected value (11.3) was derived from the relapse rate for the same women when not pregnant. With a nonpregnancy relapse rate of 0.53/year, in the 106.25 years of pregnancy 56.3 relapses were reported (i.e. 11.3 relapses in each 3 months of pregnancy and postpartum period). T1 = 1st trimester, T2 = 2nd trimester, T3 = 3rd trimester, PP1 = 1st 3 months postpartum, PP2 = 2nd 3 months postpartum. Frith and McLeod (1988) reproduced with permission.

Trauma

The influence of trauma on the onset, relapse and course of multiple sclerosis remains controversial. Some studies suggested an increased likelihood of having a relapse following trauma (McAlpine, Lumsden & Acheson, 1972). However, other studies have shown no association between onset or deterioration of multiple sclerosis and trauma, except in the specific case of electrical injury (Bamford *et al.*, 1981; Sibley *et al.*, 1991).

Stressful life events

Sibley (1988) found no relationship between stressful life events and relapses or progression of disability. However, severe life-threatening events may predispose to the onset or relapses of multiple sclerosis (Grant *et al.*, 1989).

Viral infections
Sibley (1988) found an association between viral infections and relapses of multiple sclerosis.

Diagnostic tests
Although the diagnosis of multiple sclerosis depends on clinical evaluation, a number of investigations are helpful in providing confirmation of the diagnosis or in establishing the presence of clinically silent lesions.

Cerebrospinal fluid (CSF)
Approximately 90% of patients with clinically definite multiple sclerosis and 30%–40% of those with possible multiple sclerosis have oligoclonal bands in their CSF (Ebers, 1984).

Evoked potentials
In clinically definite multiple sclerosis, visual evoked potentials are abnormal in about 90% of patients, somatosensory evoked potentials are abnormal in about 70% of patients and brainstem evoked potentials in about 50% of patients. Lower frequencies of abnormalities are found when the diagnosis is less certain (Walsh *et al.*, 1982; Matthews, 1985). Follow-up studies indicate that the abnormalities progress is parallel with the clinical disability (Walsh *et al.*, 1982).

Computerized tomography (CT)
CT scanning is of most value in excluding other diagnoses. Contrast enhanced lesions may be seen in about 10%–25% of patients with clinically active multiple sclerosis. However, these can be demonstrated better with magnetic resonance imaging (see below).

Magnetic resonance imaging (MRI)
MRI is the preferred imaging technique for diagnosis of multiple sclerosis demonstrating abnormalities in about 95% of patients with clinically definite disease. Lesions are best detected on T2-weighted images and are most commonly seen in periventricular regions but are also seen elsewhere in the white

matter including the cerebellum, brainstem and cervical cord. However, white matter lesions are common in healthy subjects over the age of 50 and caution is required in interpreting MRI findings in patients in the older age groups (Paty, McFarlind & McDonald, 1991).

As well as having an important role in the diagnosis of multiple sclerosis, MRI has played a significant part in understanding the dynamics and pathogenesis of the disease. Using gadolinium enhancement, there is breakdown in the blood–brain barrier before new lesions appear on MRI or clinically (Kermode *et al.*, 1990). These observations provide strong evidence that vascular changes are a very early event in the development of new lesions and are consistent with the pathological findings. New lesions, and expanding and contracting asymptomatic lesions, have been observed on serial MRI scans and demonstrate that multiple sclerosis is an active process even in the absence of clinical expression of the disease (Willoughby *et al.*, 1989; Harris *et al.*, 1991). Patients with primary progressive MS have fewer lesions than patients with a relapsing and remitting course or secondary progressive disease (Thompson *et al.*, 1991), so that there is a difference in the dynamics of disease activity in these two forms of multiple sclerosis.

Transcranial electromagnetic stimulation

Transcranial electromagnetic stimulation may provide useful evidence of lesions in the descending motor pathways in a high percentage of patients (Eisen & Shytbel, 1990).

Management

General

With the use of diagnostic tests, it is now possible to diagnose MS earlier in the course of the disease. As a rule, the diagnosis, nature and course of the disease should be discussed with the patient and emphasis laid on the long-term course and generally favourable outcome as most patients fear that they will become rapidly disabled.

Treatment of acute attacks

The effectiveness of a course of corticotropin (ACTH) was established by a placebo-controlled trial (Rose *et al.*, 1970). The synthetic hormone tetracosactrin (Synacthen) and oral corticosteroids seem to be equally effective but have not been subjected to rigorous clinical trial. More recently, intravenous methylprednisolone, 500 mg daily for five days, has been shown to be effective in inducing rapid remission (Milligan, Newcombe & Compston, 1987). Although these forms of treatment shorten the duration of attacks, they have no long-term effect on the course or prognosis of the disease.

Immunotherapy for long-term management

A number of immunosuppressive and immunopotentiating agents have been used to influence the course of the disease. The results of controlled clinical trials have been disappointing since they have demonstrated that most forms of treatment are of marginal benefit, at best. Immunotherapy has included corticosteroids, antilymphocyte globulin, azathioprine, cyclophosphamide, cyclosporin, interferons, copolymer 1 and transfer factor (Compston, 1989; Goodin, 1991). Hyperbaric oxygen has proved to be of no benefit (Kindwall *et al.*, 1991). A recent trial of plasma exchange in chronic progressive multiple sclerosis demonstrated some benefit (Khatri *et al.*, 1991), although this was not the case in another study (Canadian Co-operative Multiple Sclerosis Study Group, 1991). A double-blind, three-year placebo-controlled trial of alpha-interferon and transfer factor demonstrated no benefit (Austims Research Group, 1989). Cyclophosphamide and plasma exchange failed to show any benefit in progressive multiple sclerosis (Canadian Co-operative Multiple Sclerosis Study Group, 1991). A randomized double-blind trial of 3 mg/kg of azathioprine in relapsing and remitting multiple sclerosis has demonstrated modest benefit (Goodkin *et al.*, 1991). Cop 1 (a series of random basic copolymers of L-alanine, L-glutamic acid, L-lysine and L-tyrosine) is immunologically cross-reactive with myelin basic protein and is able to suppress experimental allergic encephalomyelitis. A placebo-controlled double-blind randomized trial of Cop 1 in chronic progressive multiple sclerosis demonstrated no significant benefit (Bornstein *et al.*, 1991). An open trial of OKT3 (a monoclonal antibody that is pan-T-cell reactive) may have resulted in disease suppression but the toxic effects of the agent were unacceptable for routine clinical use (Weinshenker *et al.*, 1991). A recent study of cyclosporin in the treatment of chronic progressive multiple sclerosis demonstrated some delay in the progression of disability but the benefits have to be weighed against the toxic effects of the drug (Rudge *et al.*, 1989). At present, interest in immunotherapy is being focused on the use of antibodies to T cell receptors, e.g. anti-CD4 monoclonal antibodies (Editorial, 1991).

Treatment of complications of multiple sclerosis

Although treatment designed to alter the course of multiple sclerosis has so far proved disappointing, a great deal can be done to alleviate the symptoms and complications of multiple sclerosis (Poser, 1985). Spasticity may be improved with the use of drugs such as baclofen, diazepam and dantrolene. Bladder dysfunction may be helped by drug treatment. Precipitancy, urgency and frequency of micturition can be improved by the use of anticholinergic drugs and cholinergic drugs may be of value for poorly contracting bladders. Psychiatric complications, of which depression is the most common, may require treatment with tricyclic antidepressants or other drugs. Paroxysmal disturbances such as paroxysmal ataxia and dysarthria, trigeminal neuralgia, sensory disturbances and tonic seizures may respond to treatment with carbamazepine. Rehabilitation, psychological therapy and social support are essential components of management.

References

Acheson, E. D. (1985). The epidemiology of multiple sclerosis. In *McAlpine's Multiple Sclerosis*, ed. W. B. Matthews, E. D. Acheson, J. R. Batchelor & R. O. Weller, pp. 3–46. Edinburgh: Churchill Livingstone.

Adachi, K., Kumamoto, T. & Araki, S. (1990). Elevated soluble interleukin-2 receptor levels in patients with active multiple sclerosis. *Annals of Neurology*, **28**, 687–91.

Adams, C. W. M., Poston, R. N. & Buk, S. J. (1989). Pathology, histochemistry and immunocytochemistry of lesions in acute multiple sclerosis. *Journal of the Neurological Sciences*, **92**, 291–306.

Allegretta, M., Nicklas, J., Sriram, S.& Albertini, R. J. (1990). T cells responsive to myelin basic protein in patients with multiple sclerosis. *Science*, **247**, 718–21.

Alter, M., Kahana, E. & Loewenson, R. (1978). Migration and risk of multiple sclerosis. *Neurology*, **28**, 1089–93.

Austims Research Group (1989). Interferon-alpha and transfer factor in the treatment of multiple sclerosis: a double-blind, placebo-controlled trial. *Journal of Neurology, Neurosurgery, and Psychiatry*, **52**, 566–74.

Bamford, C. R., Sibley, W. A., Thies, C., Laguna, J. F., Smith, M. S. & Clark, K. (1981). Trauma as an etiologic and aggravating factor in multiple sclerosis. Neurology, **31**, 1229–34.

Beatty, W. W., Goodkin, D. E., Monson, N. & Beatty, P. A. (1989). Cognitive disturbances in patients with relapsing-remitting multiple sclerosis. *Archives of Neurology*, **46**, 1113–19.

Birk, K., Ford, C., Smeltzer, S., Ryan, D., Miller, R. & Rudick, R. A. (1990). The clinical course of multiple sclerosis during pregnancy and the puerperium. *Archives of Neurology*, **47**, 738–42.

Bornstein, M. B., Miller, A., Slagle, S., Weitzman, M., Drexler, E., Keilson, M., Spada, V., Weiss, W., Appel, S., Rolak, L., Harati, Y., Brown, S., Arnon, R., Jacobsohn, I., Teitelbaum, D. & Sela, M. (1991). A placebo-controlled, double blind, randomised, two-center, pilot trial of Cop 1 in chronic progressive multiple sclerosis. *Neurology*, **41**, 533–9.

Canadian Co-operative Multiple Sclerosis Study Group. (1991). The Canadian co-

operative trial of cyclophosphamide and plasma exchange in progressive multiple sclerosis. *Lancet*, **337**, 441–6.
Charcot, J. M. (1868). Histologie de la sclérose en plaques. *Gazette des hôpitaux civils et militaires (Paris)*, **41**, 554–5; 557–8; 566.
Compston, A. (1988). The 150th anniversary of the first depiction of the lesions of multiple sclerosis. *Journal of Neurology, Neurosurgery, and Psychiatry*, **51**, 1249–52.
Compston, D. A. S. (1989). The management of multiple sclerosis. *Quarterly Journal of Medicine*, **70**, 93–101.
Compston, D. A. S. (1990*a*). The dissemination of multiple sclerosis. The Langdon-Brown Lecture 1989. *Journal of the Royal College of Physicians of London*, **24**, 207–18.
Compston, A. (1990*b*). Risk factors for multiple sclerosis: race or place? *Journal of Neurology, Neurosurgery, and Psychiatry*, **53**, 821–3.
Confavreux, C., Aimard, G. & Devie, M. (1980). Course and prognosis of multiple sclerosis assessed by the computerized data processing of 349 patients. *Brain*, **103**, 281–300.
Dean, G. & Kurtzke, J. F. (1971). On the risk of multiple sclerosis according to age at immigration to South Africa. *British Medical Journal*, **3**, 725–9.
Ebers, G. C. (1984). Oligoclonal banding in MS. *Annals of the New York Academy of Science*, **436**, 206–12.
Ebers, G. C., Bulman, D. E., Sadnovick, A. D., Paty, D. W., Warren, S., Hader, W., Murray, T. J., Seland, T. P., Duquette, P., Grey, T., Nelson, R., Nicolle, M. & Brunet, D. (1986). A population based study of multiple sclerosis in twins. *New England Journal of Medicine*, **315**, 1638–42.
Editorial (1991). Where to hit MS. *Lancet*, **337**, 765–7.
Ehrlich, G. D., Glaser, J. B., Bryz-Gornia, V., Maes, J., Waldmann, T. A., Poiesz, B. J., Greenberg, S. J. and the HTLV-MS Working Group. (1991). Multiple sclerosis, retroviruses, and PCR. *Neurology*, **41**, 335–43.
Eisen, A. & Shytbel, W. (1990). Clinical experience with transcranial magnetic stimulation. *Muscle and Nerve*, **13**, 995–1011.
Flashman, J. F. & Latham, O. (1915). A contribution to the study of the aetiology of disseminated sclerosis. *Medical Journal of Australia*, **2**, 265–9.
Francis, D. A., Batchelor, J. R., McDonald, W. I., Hing, S. N., Dodi, I. A., Fielder, A. H. L., Hearn, J. E. C. & Downie, A. W. (1987*a*). Multiple sclerosis in North East Scotland. An association with HLA-DQw1. *Brain*, **110**, 181–96.
Francis, D. A., Compston, D. A. S., Batchelor, J. R. & McDonald, W. I. (1987*b*). A reassessment of the risk of multiple sclerosis developing in patients with optic neuritis after extended follow-up. *Journal of Neurology, Neurosurgery, and Psychiatry*, **50**, 758–65.
Frith, J. A. (1989). History of multiple sclerosis. An Australian perspective. *Clinical and Experimental Neurology*, **25**, 7–16.
Frith, J. A. & McLeod, J. G. (1988). Pregnancy in multiple sclerosis. *Journal of Neurology, Neurosurgery, and Psychiatry*, **51**, 495–8.
Gay, D. & Esiri, I. M. (1991). Blood–brain barrier damage in acute multiple sclerosis plaques. An immunocytological study. *Brain*, **114**, 557–72.
Goodin, D. S. (1991). The use of immunosuppressive agents in the treatment of multiple sclerosis: a critical review. *Neurology*, **41**, 980–5.
Goodkin, D. E., Bailly, R. C., Teetzen, M. L., Herstgaard, D. & Beatty, W. F. (1991). The efficacy of azathioprine in relapsing-remitting multiple sclerosis. *Neurology*, **41**, 20–5.
Grant, I., Brown, G. W., Harris, T., McDonald, W. I., Patterson, T. & Trimball, M. R.

(1989). Severely threatening events and marked life difficulties preceding onset or exacerbation of multiple sclerosis. *Journal of Neurology, Neurosurgery, and Psychiatry*, **52**, 8–13.

Hammond, S. R., English, D. R., de Wytt, C., Maxwell, I. C., Millingen, K. S., Stewart-Wynne, E. G., McLeod, J. G. & McCall, M. G. (1988*b*). The clinical profile of multiple sclerosis in Australia: A comparison between medium and high frequency prevalence zones. *Neurology*, **38**, 980–6.

Hammond, S. R., English, D. R., de Wytt, C., Hallpike, J. F., Millingen, K. S., Stewart-Wynne, E. G., McLeod, J. G. & McCall, M. G. (1989a). The relationship between multiple sclerosis frequency and latitude within Australia. In *Multiple Sclerosis Research*, ed. M.A. Battaglia, pp. 171–7. Amsterdam: Elsevier Science Publishers.

Hammond, S. R., English, D. R., de Wytt, C., Hallpike, J. F., Millingen, K. S., Stewart-Wynne, E. G., McLeod, J. G. & McCall, M. G. (1989b). The contribution of mortality statistics to the study of multiple sclerosis in Australia. *Journal of Neurology, Neurosurgery, and Psychiatry*, **52**, 1–7.

Hammond, S. R., McLeod, J. G., Millingen, K. S., Stewart-Wynne, E. G., English, D., Holland, J. T. & McCall, M. G. (1988a). The epidemiology of multiple sclerosis in three Australian cities. Perth, Newcastle and Hobart. *Brain*, **111**, 1–25.

Harris, J. O., Frank, J. A., Patronas, N., McFarlin, D. E. & McFarland, H. F. (1991). Serial gadolinium-enhanced magnetic resonance imaging scans in patients with early, relapsing-remitting multiple sclerosis: implications for clinical trials and natural history. *Annals of Neurology*, **29**, 548–55.

Hely, M. A., McManis, P. G., Doran, T. J., Walsh, J. C. & McLeod, J. G. (1986*a*). Acute optic neuritis: A prospective study of risk factors for multiple sclerosis. *Journal of Neurology, Neurosurgery, and Psychiatry*, **49**, 1125–30.

Hely, M. A., McManis, P. G., Walsh, J. C. & McLeod, J. G. (1986*b*). Visual evoked responses and ophthalmological examination in optic neuritis. A follow-up study. *Journal of Neurological Sciences*, **75**, 275–83.

Hughes, R. A. C. (1992). Pathogenesis of multiple sclerosis. *Journal of the Royal Society of Medicine*, **88**, 373–6.

Jersild, C., du Pont, B., Fog, T., Platz, P. & Svejgaard, A. (1975). Histocompatibility determinants in multiple sclerosis. *Transplantation Reviews*, **22**, 148–63.

Kermode, A. G., Thompson, A. J., Tofts, P., MacManus, D. G., Kendall, B. E., Kingsley, D. P. E., Moseley, I. F., Rudge, P. & McDonald, W. I. (1990). Breakdown of the blood–brain barrier precedes symptoms and other MRI signs of new lesion in multiple sclerosis. Pathogenetic and clinical implications. *Brain*, **113**, 1477–89.

Khatri, B. O., McQuillen, M. P., Hoffmann, R .G., Harrington, G. J. & Schmoll, D. (1991). Plasma exchange in chronic progressive multiple sclerosis: A long-term study. *Neurology*, **41**, 409–14.

Kindwall, E. P., McQuillen, M. P., Khatri, B. O., Gruchow, H. W. & Kindwall, M. L. (1991). Treatment of multiple sclerosis with hyperbaric oxygen. Results of a national registry. *Archives of Neurology*, **48**, 195–9.

Kurdi, A., Ayesh, I., Abdallat, A., Maayta, U., McDonald, W. I., Compston, D. A. S. & Batchelor, J. R. (1977). Different B lymphocyte alloantigens associated with multiple sclerosis in Arabs and North Europeans. *Lancet*, **i**, 1123–5.

Kurtzke, J. F. (1985). Epidemiology of multiple sclerosis. In *Handbook of Clinical Neurology, vol. 3 (47) Demyelinating Diseases*, ed. J.C. Koetrier, pp. 259–87. Amsterdam: Elsevier Science Publishers.

Kurtzke, J. F. & Hyllested, K. (1986). Multiple sclerosis in the Faroe Islands II. Clinical update, transmission, and the nature of MS. *Neurology*, **36**, 307–28.

Lee, S. J., Wucherpfenni, J. K. W., Brod, S. A., Benjamin, D., Weiner, H. L. & Hafler, D. A. (1991). Common T-cell receptor V usage in oligoclonal T lymphocytes derived from cerebrospinal fluid and blood of patients with multiple sclerosis. *Annals of Neurology*, **29**, 33–40.

Matthews, W. B. (1985). Laboratory diagnosis. In *McAlpine's Multiple Sclerosis*, ed. W. B. Matthews, E. D. Acheson, J. R. Batchelor & R. O. Weller, pp. 167–209. Edinburgh: Churchill–Livingstone.

McAlpine, D., Lumsden, C. E. & Acheson, E. D. (1972). *Multiple Sclerosis. A Reappraisal*. London: Churchill–Livingstone.

McCall, M. G., Brereton, T. le G., Dawson, A., Millingen, K., Sutherland, J. M. & Acheson, E. D. (1968). Frequency of multiple sclerosis in three Australian cities – Perth, Newcastle, and Hobart. *Journal of Neurology, Neurosurgery, and Psychiatry*, **31**, 1–9.

McCombe, P. A., Clark, P., Frith, J., Hammond, S., Feeney, D., Pollard, J. D. & McLeod, J. G. (1985). Alpha-1 antitrypsin phenotypes in demyelinating diseases. *Annals of Neurology*, **18**, 514–16.

McDonald, W. I. & Halliday, A. M. (1977). Diagnosis and classification of multiple sclerosis. *British Medical Bulletin*, **33**, 4–8.

Midgard, R., Riise, T. & Nyland, H. (1991). Epidemiological trends in multiple sclerosis in More and Romsdal, Norway: a prevalence/incidence study in a stable population. *Neurology*, **41**, 887–92.

Miller, D. H., Hammond, S. R., McLeod, J. G., Purdie, G. & Skegg, D. C. G. (1990). Multiple sclerosis in Australia and New Zealand: are the determinants genetic or environmental? *Journal of Neurology, Neurosurgery, and Psychiatry*, **53**, 903–5.

Miller, D. H., Ormerod, I. E. C., McDonald, W. I., MacManus, D. G., Kendall, B. E., Kingsley, D. P. E. & Moseley, I. F. (1988). The early risk of multiple sclerosis after optic neuritis. *Journal of Neurology, Neurosurgery, and Psychiatry*, **51**, 1569–71.

Miller, D. H., Ormerod, I. E. C., Rudge, P., Kendall, B. E., Moseley, I. F. & McDonald, W. I. (1989). The early risk of multiple sclerosis following isolated acute syndromes of the brainstem and spinal cord. *Annals of Neurology*, **26**, 635–9.

Milligan, N. M., Newcombe, R. & Compston, D. A. S. (1987). A double blind controlled trial of high dose methylprednisolone in patients with multiple sclerosis: 1: Clinical effects. *Journal of Neurology, Neurosurgery, and Psychiatry*, **50**, 511–6.

Myrianthopoulos, N. C. (1985). Genetic aspects of multiple sclerosis, In *Handbook of Clinical Neurology, vol. 3, (47): Demyelinating Diseases*, ed. J.C. Koetsier, pp. 289–317. Amsterdam: Elsevier Science Publishers.

Ormerod, I. E. C., McDonald, W. I., du Boulay, G. H., Kendall, B. E., Moseley, I. F., Halliday, A. M., Kakigi, R., Kriss, A. & Peringer, E. (1986). Disseminated lesions at presentation in patients with optic neuritis. *Journal of Neurology, Neurosurgery, and Psychiatry*, **49**, 124–7.

Ormerod, I. E. C., Miller, D. H., McDonald, W. I., du Boulay, E. P. G. H., Rudge, P., Kendall, B. E., Moseley, I. F., Johnson, G., Tofts, P. S., Halliday, A. M., Bronstein, A. M., Scaravilli, F., Harding, A. E., Barnes, D. & Zilkha, K. J. (1987). The role of NMR imaging in the assessment of multiple sclerosis and isolated neurological lesions. A quantitative study. *Brain*, **110**, 1579–616.

Paty, D. W., Oger, J. J. F., Kastrukoff, L. F., Hashimoto, S. A., Hooge, J. P., Eisen, A. A., Eisen, K. A., Purves, S J., Low, M. D., Brandejs, V., Robertson, W. D. & Li, D. K. B. (1988). MRI in the diagnosis of MS: a prospective study with comparison of clinical evaluation evoked potentials, oligoclonal banding, and CT. *Neurology*, **38**, 180–5.

Paty, D. W., McFarlind, E. & McDonald, W. I. (1991). Magnetic resonance imaging and laboratory aids in the diagnosis of multiple sclerosis. *Annals of Neurology*, **29**, 3–5.

Petersen, R. C. & Kokmen, E. (1989). Cognitive and psychiatric abnormalities in multiple sclerosis. *Mayo Clinic Proceedings*, **64**, 657–63.

Phadke, J. G. (1990). Clinical aspects of multiple sclerosis in North East Scotland with particular reference to its course and prognosis. *Brain*, **113**, 1597–628.

Poser, C. M., Paty, D. W., Scheinberg, L., McDonald, W. I., Davis, F. A., Ebers, G. C., Johnson, K. P., Sibley, W. A., Silberberg, D. H. & Tourtelotte, W. W. (1983). New diagnostic criteria for multiple sclerosis: guidelines for research protocols. *Annals of Neurology*, **13**, 227–31.

Poser, S. (1985). Management of patients with multiple sclerosis. In *Handbook of Clinical Neurology, vol. 3, (47): Demyelinating Diseases.*, ed. J. C. Koetsier, pp. 147–86. Amsterdam: Elsevier Science Publishers.

Prineas, J. W. (1975). Pathology of the early lesion in multiple sclerosis. *Human Pathology*, **6**, 531–54.

Prineas, J. W. (1985). The neuropathology of multiple sclerosis. In *Handbook of Clinical Neurology, vol. 3, (47): Demyelinating Diseases*, ed. J. C. Koetsier, pp. 213–57. Amsterdam: Elsevier Science Publishers.

Rodriguez, M. (1989). Multiple sclerosis: basic concepts and hypothesis. *Mayo Clinic Proceedings*, **64**, 570–6.

Rose, A. S., Ellison, G. W., Myers, L. W. & Tourtellotte, W. W. (1976). Criteria for the clinical diagnosis of multiple sclerosis. *Neurology*, **26**, 20–2.

Rose, A. S., Kuzma, J. W., Kurtzke, J. F., Namerow, N. S., Sibley, W. A. & Tourtellotte, W. W. (1970). Co-operative study in the evaluation of therapy in multiple sclerosis – ACTH vs placebo: final report. *Neurology*, **20**, 1–59.

Rudge, P., Koetsier, J. C., Mertin, J., Mispelblom Beyer, J. O., van Walbeek, H. K., Clifford-Jones, R., Harrison, J., Robinson, K., Mellein, B., Poole, T., Stokvis, J. C. J. M. & Timonen, P. (1989). Randomised double blind controlled trial of cyclosporin in multiple sclerosis. *Journal of Neurology, Neurosurgery, and Psychiatry*, **52**, 559–65.

Sadovnick, A. D. & Baird, P. A. (1988). The familial nature of multiple sclerosis: Age-corrected empiric recurrence risks for children and siblings of patients. *Neurology*, **38**, 990–1.

Seboun, E., Robinson, M. A., Doolittle, T. H., Ciulla, T.A., Kindt, T. J. & Hauser, S.L. (1989). A susceptibility locus for multiple sclerosis is linked to the T cell receptor chain complex. *Cell,* **57**, 1095–1100.

Sibley, W. A., Bamford, C. R., Clark, K., Smith, M. S. & Laguna, J. F. (1991). A prospective study of physical trauma and multiple sclerosis. *Journal of Neurology, Neurosurgery, and Psychiatry*, **54**, 584–9.

Sibley, W. A. (1988). Risk factors in multiple sclerosis – implications for pathogenesis. In *A Multidisciplinary Approach to Myelin Disease*, ed. G. Serlupi Crescenzi, pp. 227–32. New York: Plenum.

Skegg, D. C. G., Corwin, P. A., Craven, R. S., Malloch, J. A., Pollock, M. (1987). Occurrence of multiple sclerosis in the North and South of New Zealand. *Journal of Neurology, Neurosurgery, and Psychiatry*, **50**, 134–9.

Stewart, G. J., Basten, A. Guinan, J., Bashir, H. V., Cameron, J. & McLeod, J. G. (1977). HLA-Dw2, viral immunity and family studies in multiple sclerosis. *Journal of the Neurological Sciences*, **32**, 153–67.

Stewart, G. J. & Kirk, R. L. (1983). The genetics of multiple sclerosis: the HLA system and other genetic markers. In *Multiple Sclerosis*, ed. J. F. Hallpike, C. W. M. Adams & W. W. Tourtellotte, pp. 97–128. Baltimore: Williams & Wilkins.

Swingler, R. J. & Compston, D. A. S. (1986). The distribution of multiple sclerosis in the United Kingdom. *Journal of Neurology, Neurosurgery, and Psychiatry*, **49**, 1115–24.

Thompson, A. J., Kermode, A. G., Wicks, D., MacManus, D. G., Kendall, B. E., Kingsley, D. P. E. & McDonald, W. I. (1991). Major differences in the dynamics of primary and secondary progressive multiple sclerosis. *Annals of Neurology*, **29**, 53–62.

Traugott, U. & Raine, C. S. (1981). Antioligodendrocyte antibodies in cerebrospinal fluid of multiple sclerosis and other neurologic diseases. *Neurology*, **31**, 695–700.

Traugott, U., Reinherz, E. L. & Raine, C. S. (1983). Multiple sclerosis: distribution of T cells, T-cell subsets, and Ia-positive macrophages in lesions of different ages. *Journal of Neuroimmunology*, **4**, 201–21.

Traugott, U. (1987). Multiple sclerosis: relevance of Class I and Class II MHC-expressing cells in lesion development. *Journal of Neuroimmunology*, **16**, 285–302.

Walsh, J. C., Garrick, R., Cameron, J. & McLeod, J. G. (1982). Evoked potential changes in clinically definite multiple sclerosis. A two year follow-up study. *Journal of Neurology, Neurosurgery, and Psychiatry*, **45**, 494–500.

Weinshenker, B. G., Bass, B., Karlik, S., Ebers, G. C. & Rice, G. P. A. (1991). An open trial of OKT3 in patients with multiple sclerosis. *Neurology*, **41**, 1047–52.

Williams, E. S., Jones, D. R. & McKeran, R. O. (1991). Mortality rates from multiple sclerosis: geographical and temporal variations revisited. *Journal of Neurology, Neurosurgery, and Psychiatry*, **54**, 104–9.

Willoughby, E. W., Grochowski, E., Li D. K. B., Ojer, J., Kastrukoff, L. F. & Paty, D. W. (1989). Serial magnetic resonance scanning in multiple sclerosis: A second prospective study in relapsing patients. *Annals of Neurology*, **25**, 43–9.

Wynn, D. R., Rodriguez, M., O'Fallon, W. M. & Kurland, L. T. (1990). A reappraisal of the epidemiology of multiple sclerosis in Olmsted County, Minnesota. *Neurology*, **40**, 780–6.

Xiao, B.-G., Linington, C. & Link, H. (1991). Antibodies to myelin-oligodendrocyte glycoprotein in cerebrospinal fluid from patients with multiple sclerosis and controls. *Journal of Neuroimmunology*, **31**, 91–6.

Yu, Y. L., Woo, E., Hawkins, B. R., Ho, H. C. & Huang, C.-Y. (1989). Multiple sclerosis among Chinese in Hong Kong. *Brain*, **112**, 1445–67.

Part VI
Epilepsy

24

Partial epilepsies

R. A. MACKENZIE

Comprehensive Epilepsy Programme,
The Prince Henry Hospital,
Little Bay, Sydney NSW, Australia

The word 'epilepsy' derives from a Greek word meaning 'to seize' and refers to a patient being seized by an epileptic attack. In the medical context, 'epilepsy' is used to refer only to those conditions of chronic recurrent epileptic seizures that can be considered epileptic disorders. Because there are many types of epileptic disorders, it is more correct to refer to them as 'epilepsies'. Seizures are termed partial when behavioural or EEG evidence indicates they begin in a part of the brain limited to one hemisphere; they are called generalized when they appear to begin bilaterally. Partial seizures are further subdivided into simple partial seizures, which do not have impairment of consciousness and complex partial seizures (CPS) which do have impairment of consciousness. Simple partial seizures can progress to CPS and may have motor, sensory, autonomic and psychic features that are remembered as an 'aura' prior to loss of consciousness.

The treatment of simple and complex partial epilepsy is far from successful in controlling most patients' seizures (Rodin, 1968). Optimal control of epilepsy is when the patient experiences no seizures (including auras) and has no adverse side-effects from the treatment. New antiepileptic drugs have not appreciably altered the success rate for controlling seizures (Elwes *et al.*, 1984). However, there have been significant recent advances in our understanding of the nature of epilepsy, the mechanism of action of anticonvulsants and the role of surgery. It is intended in this chapter to briefly review these developments, especially as they apply to the partial epilepsies.

Mechanisms of action of anticonvulsants

In the last five years, significant advances have occurred in understanding the mechanisms of action of anticonvulsant drugs. In most cases, these mechanisms appear to involve ion channels that control the balance between physiological and pathological neuronal firing. A consistent feature of neuronal activity during a seizure is sustained repetitive firing (SRF) of action potentials (Macdonald, McLean & Skerritt, 1985). Phenytoin limits SRF by acting on the voltage-dependent sodium channel but does not inhibit the ability of a neurone to fire single action potentials (McLean & Macdonald, 1983). This selective effect on repetitive but not single firing of a neurone might explain why phenytoin limits the pathologic firing of neurones during a seizure with minimal effects on the physiological function of neurones. Carbamazepine and valproic acid also limit SRF at therapeutically relevant concentrations; phenobarbital and benzodiazapines only reduce SRF at supratherapeutic concentrations.

The anticonvulsant mechanism of phenobarbital and the benzodiazepines probably acts through the augmentation of inhibitory synaptic function. The major inhibitory neurotransmitter in the brain is gamma aminobutyric acid (GABA), which acts on postsynaptic receptors to decrease neuronal excitability (Cooper, Bloom & Roth, 1986). These receptors respond to GABA by opening neuronal ion channels permeable to chloride, thereby stabilizing the cell membrane near the resting potential. Phenobarbital and the benzodiazepines augment the inhibitory response of neurones to GABA through distinct mechanisms that increase the overall length of time for which the chloride ion channel can operate (Twymann, Rogers & Macdonald, 1989). These effects of phenobarbital and benzodiazepines occur at therapeutically relevant concentrations and probably contribute to their anticonvulsant action.

There are new anticonvulsant drugs under investigation which may have their primary action on calcium channels. For example, flunarizine has been demonstrated to be effective against CPS with or without secondary generalized seizures in patients (Overweg *et al.*, 1984). This drug has been shown to be an effective calcium channel antagonist, blocking a transient low threshold calcium current in rat hypothalamic neurones (Akaike, Kostyuk & Osipchuk, 1989) but it is not clear whether this is the reason for its anticonvulsant action.

While all of these anticonvulsant drugs have been shown to be effective, there are clearly a number of patients especially those with CPS, who are refractory to drug therapy. Hopefully, new anticonvulsant drugs that are under development will have actions on new neurotransmitter receptors or channels. For example, considerable effort has been directed towards developing compounds which are antagonists of excitatory amino acid transmission. Hopefully, with new anticonvulsant drugs patients currently intractable will be controlled medically.

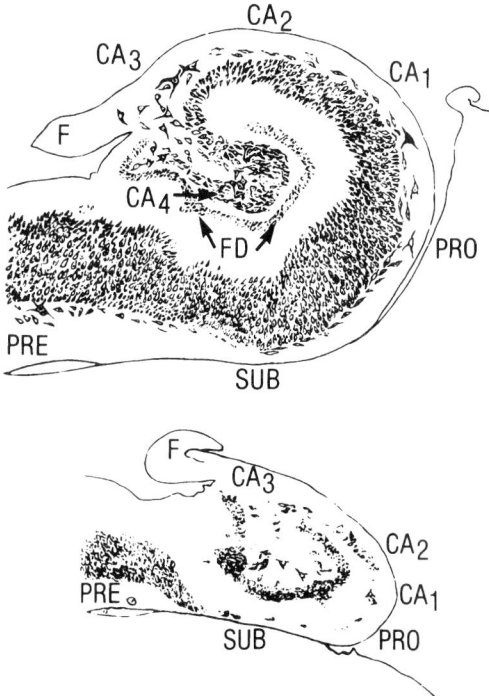

Fig. 24.1. Wood-carved plate by Bratz (1899) of cells in normal (top) and atrophied (bottom) hippocampus. In the normal specimen, note the large CA3 pyramids, properly oriented Ammon's horn (CA) to presubiculum (PRE) and smaller granule cells of fascia dentata (FD). In the epileptic hippocampus note maximum cell loss in CA3, CA1 and prosubiculum (PRO) but relative preservation of CA2 and presubiculum (PRE). Subiculum (SUB) also shows damage. Babb and Brown (1987) reproduced with permission.

Mesial temporal sclerosis: cause or consequence of epilepsy?

'Mesial temporal sclerosis' (MTS) refers to a loss of neurones and a glial proliferation involving the hippocampus, amygdala and mesial temporal cortex. It is the most frequent pathological abnormality found in anterior temporal lobectomy specimens removed for intractable CPS (Babb & Brown, 1987). Most commonly it is the sole lesion but it can be found in association with other focal pathology such as glioma or hamartoma. When originally described in 1880 by Sommer in the brains of patients who died in institutions after having epilepsy for many years, it was assumed that the lesion was the cause of the patient's seizures. However, Pfleger in the same year suggested that severe seizures could result in hypoxia and cause these lesions to develop. Bratz (1899) found unilateral hippocampal atrophy in 50% of autopsied epileptic brains. Babb & Brown (1987) have reproduced some of his original wood-carved plates (Fig. 24.1) which demonstrate beautifully the selective damage to pyramidal and granule cells of CA3, CA1 and prosubiculum with relative sparing of cells in CA4, CA2 and subiculum. This pattern is currently found in temporal lobe resections and autopsies of patients having suffered from temporal lobe epilepsy.

The use of animal models has firmly established that sustained seizure activity can directly damage neurones (Meldrum & Brierley, 1973; Meldrum, Horton & Brierley, 1974). Thus sustained generalized seizures in primates will damage the hippocampus, cerebellum and neocortex. The brains of children dying a few hours or days after prolonged seizures show a pattern of damage similar to that observed in experimental animals after prolonged generalized seizures; i.e. there is necrosis of nerve cells in the hippocampus (CA1 predominantly, but also CA3 and endfolium, commonly asymmetrically), in the amygdala, in the neocortex and cerebellum (Zimmerman, 1938). Similar patterns of damage are seen in adult patients dying after status epilepticus (Meldrum & Corsellis 1984). Lesions of the hippocampus indistinguishable from those found in chronic epilepsy can also be found following other stresses including cerebral hypoxia (e.g. strangulation) or viral infections involving the limbic system (Meldrum, 1991).

The reasons for regarding MTS as a cause of temporal lobe epilepsy were stated by Meldrum (1991) as follows. Clinico-pathological studies of epilepsy of all types show that MTS is found in patients with CPS and not in patients without CPS. In temporal lobectomy specimens, MTS is the commonest and usually the sole lesion found. When this lesion is found the prognosis for postoperative seizure relief is extremely good, but if no lesion is found in the removed specimen, seizures usually persist postoperatively. There is also a correlation between removal of the lesion and suppression of seizures in terms of the antero-posterior extent of the lesion: the prognosis is better when the sclerosis is confined to the anterior part of the lobe (and therefore completely removed) than when it is more posterior (and incompletely removed).

The success of selective amygdalohippocampectomy in certain cases could indicate that the seizure activity originates in these regions, but it is also possible that they are necessary for development and clinical expression of the seizure. Another study correlating hippocampal pathology with clinical history (Sagar & Oxbury, 1987) found that, of 32 temporal lobectomy specimens, 13 showed severe cell loss in CA1 and in the endfolium and in the dentate gyrus; all 13 had experienced their first seizure in the first two years of life. Population studies showed that the risk of recurrent seizures by age 20 years is 1% for the general population, 3% in those having had a brief febrile convulsion and 8% after febrile status lasting more than 30 minutes (Annegers *et al.*,

Partial epilepsies 401

1979). The risk of developing CPS is also enhanced when more than one febrile convulsion occurs in a 24-hour period.

The simplest and most coherent interpretation of the clinical and experimental data is that MTS is most commonly the result of a prolonged seizure (which may be febrile and may be unilateral) early in life (i.e. before 4 years of age). In some cases a developmental lesion (hamartoma or other alien tissue lesion) is the cause of the early convulsion. The MTS through an as yet unidentified process is the focal cause of the subsequent CPS.

Kindling: relevance to human epilepsy

This subject was recently reviewed by McNamara (1991). The term 'kindling' refers to a phenomenon in which periodic, focal application of initially subconvulsive electric stimulations to a brain structure eventually results in intense limbic and clonic motor seizures (Goddard, McIntyre & Leech, 1969). Once established, this enhanced sensitivity may persist in the absence of additional stimulations for the life of the animal. The initial stimulus elicits minimal or no change in behaviour or brain electrical activity measured with an electroencephalograph (EEG). Subsequent stimulations induce electrographic seizures or 'after discharges' principally localised initially to the stimulated structure. Repeated stimulations produce progressive lengthening and propagation of after discharge coinciding with expression of behavioural seizures as classified by Racine (1972). Repeated induction of after discharge appears a necessary and sufficient condition for induction of kindling. Either chemical or electrical stimuli can be used to induce kindling. Electrical stimuli are simply the most convenient means of triggering an after discharge and thus are the most commonly used method for induction of kindling. Thus, the essence of kindling development is that seizures (after discharges) beget a lasting propensity for longer and more widespread electrographic seizures accompanied by more intense behavioural seizures and a lowered after discharge threshold (McNamara, 1991). Kindling can be induced by stimulation of many, but not all, sites in the brain. Stimulation of multiple limbic, neocortical and basal ganglia structures results in kindling. The amygdala is a structure commonly used for induction of kindling, in part because of the relatively few (approximately 13) stimulations required to produce kindling. Actual epilepsy or spontaneous (not simply stimulation induced) seizures can be reliably induced if kindled animals are subjected to hundreds of stimulation induced kindled seizures (Pinel & Rovner, 1978). Moreover, spontaneous seizures persist for as long as seven months following termination of the stimulations, suggesting that epilepsy itself is longlasting and perhaps permanent in this condition.

Similarities of kindled and human CPS have led to speculation that kindling or kindling-like processes may actually contribute to epileptogenesis in humans. These include: similar EEG activity recorded from intracerebral electrodes during kindled 'seizures' and human CPS; strikingly similar behaviours in kindled seizures and in human CPS with secondary generalization; and similar sensitivities to conventional anticonvulsants. This last point is striking in that carbamazepine, phenobarbital, phenytoin and valproic acid are effective whereas ethosuximide and trimethadione are ineffective against both human complex partial and kindled seizures (McNamara et al., 1989). Recently, animals stimulated to produce spontaneous seizures were shown to have small but significant loss of neurones in the hippocampus; increasing the number of stimulus-evoked seizures resulted in striking loss of neurones in the hilar (CA4) region of hippocampus (Cavazos & Sutula, 1989). This result is important because it shows that isolated, periodic seizures (in contrast to status epilepticus) are sufficient to cause neurone death and provide an explanation for the neurone loss seen in the human condition. Also, it shows a clear parallel between kindling and at least one form of severe CPS in humans.

Another argument for the presence of kindling-like processes in humans is the occasional finding of secondary foci in a subset of human epileptic patients. Some patients with epilepsy due to a localized lesion such as a tumour or hamartoma develop evidence of two sites of seizure initiation (or 'foci'), one in the vicinity of the lesion and another in the contralateral hemisphere. The evidence for a secondary focus in some instances consists of epileptiform abnormalities recorded in the EEG between seizures and in other instances consists of behavioural and EEG seizure patterns distinct from those of the primary focus and indicative of a focus in the opposite hemisphere (Morrell, 1979).

Engel and Cahan (1986) summarize the evidence for the clinical relevance of kindling as follows: 'Although there is no evidence from clinical data to invoke kindling mechanisms as a means of creating an epileptogenic lesion in epileptic patients, progressive changes can be demonstrated that could be explained by kindling in the human brain. Trans-synaptic electrophysiological changes are similar in amygdaloid kindled animals and in human limbic epilepsy and initial behavioural manifestations suggest the same structures are involved (amygdala and hippocampus); however, subsequent spread is stereotyped with amygdaloid kindling and variable in humans. 'All or none' seizure generalization in amygdaloid kindled animals does not occur in human epilepsy, but this could reflect the efficacy of anticonvulsant drugs. Spontaneous seizures arise from secondary sites in patients with limbic seizures for many years. Consequently, kindling mechanisms may enhance epileptogencity in distant brain structures in humans once an epileptic condition exists. Kindling of distant sites could underlie progressive worsening of a partial epileptic condition, and perhaps contribute to the appearance of interictal behavioural disturbances.'

Relationship of mossy fibre sprouting and complex partial epilepsy

It has been known for some years that, in experimental animals, intravenous injection of kainic acid produces hippocampal damage with neuronal loss typical of Ammon's horn sclerosis. In addition to the loss of pyramidal neurones and hippocampal degeneration, Nadler (1981) noticed a morphologic rearrangement of mossy fibres, the axons of dentate granule cells. Whereas mossy fibres in normal animals innervate hippocampal pyramidal cells, the reorganization that occurred in the kainic acid model was an aberrant growth (or sprouting) of these fibres into the supragranular dendritic zone of the dentate gyrus. These mossy fibres were therefore situated in such a manner that they could potentially reinnervate the dentate gyrus cells (Fig. 24.2). Since the mossy fibres themselves release an excitatory transmitter, these aberrant fibres, if functionally active, might result in recurrent excitation of dentate cells. In this way an input that would normally elicit a single action potential in the granule cells might produce sustained repetitive firing. This type of recurrent excitation could play a key role in amplifying excitatory neurotransmission, and thereby initiate a seizure and/or amplify seizure propagation. Interestingly, activation of the mossy fibres of dentate granule cells caused repetitive firing of dentate granule cells in kainic acid treated animals (Tauck & Nadler, 1985), but only single firing of granule cells of normal animals. These findings suggest that the sprouted mossy fibres in the kainic acid models are functional.

The same morphologic changes were recently identified in a kindling model: just as was apparent in the kainic acid model, kindled animals also exhibited a sprouting of mossy fibres into the supragranular zone of the dentate gyrus (Sutula *et al.*, 1988). The functional significance of this morphologic change in the kindled animals is unknown. However, the existence of a potential recurrent excitatory circuit within the dentate gyrus of kindled animals suggests that it could play a role in the propagation of generalized tonic-clonic seizures in the fully kindled state.

These findings from animal models were extended to humans by the identification of mossy fibre sprouting within the supragranular dentate gyrus of hippocampal specimens surgically removed from patients with CPS (Sutula *et al.*, 1989). Although it has not yet been proved that this morphologic change in human patients is functional, the data from the kainic acid model provide cir-

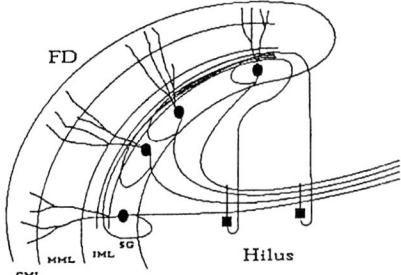

Fig. 24.2. Schematic diagram showing reciprocal innervation between principal neurones of the hilus (squares) and dentate granule cells of the hippocampus formation (circles). In the normal situation (top), granule cell axons (mossy fibres) synapse on principal neurones of the hilus and CA3 while hilar neurones make synapses on proximal dendrites of the granule cells. Hippocampal sclerosis (bottom) is associated with cell loss and decreased hilar input to granule cells, causing sprouting of mossy fibres of surviving granule cells which then make contact with the vacated post-synaptic membrane, creating monosynaptic recurrent excitatory circuits. FD: fascia dentata; OM: outer molecular layer; MML: middle molecular layer; IML: inner molecular layer; SG: stratum granulosum. Engel (1990) reproduced with permission.

cumstantial evidence that this mossy fibre sprouting in humans could play a role in CPS initiation and propagation. The outcome of temporal lobectomy in one group of patients with CPS strengthens the potential clinical evidence of mossy fibre sprouting: this study demonstrated that the presence of mossy fibre sprouting in patients with CPS correlated with the decrease in the severity of the seizures following temporal lobectomy (de Lanerolle et al., 1989). It is possible, though not yet proven, that removal of the morphologically rearranged mossy fibres was one factor that produced a beneficial outcome of surgery in these patients. In other words, mossy fibre sprouting may not be just a patho-

Partial epilepsies 405

logical hallmark of CPS: this morphologic rearrangement may have functional significance as the pathophysiological focus underlying CPS in some patients.

Mechanism of epileptogenesis

The electrical events underlying epilepsy have recently been reviewed by Engel (1990). He points out that, whether the pathological substrate of partial epilepsy is hippocampal sclerosis, cicatrix formation, tumour or other structural defects, neuronal cell loss appears to be a consistent feature. The cause of cell death depends upon the aetiology of the disease process, but creates a situation where synaptic reorganization occurs. Axon sprouting also occurs, and this would increase the efficacy of recurrent circuits. Sprouting in epileptic brain has so far been described only for excitatory mossy fibres because these axons contain zinc, which is easily stained by histochemical techniques. Engel suggests that degeneration of normal presynaptic terminals elsewhere induces similar sprouting in other systems, perhaps even including inhibitory ones. New recurrent inhibitory circuits might then account for the enhanced inhibition revealed by electrophysiological studies of human epileptic hippocampus (Babb & Crandall, 1976; Babb, Wilson & Isokawa-Akesson, 1987). Such synaptic reorganization would also be expected to exert trophic affects which might alter channel distribution over neuronal membranes, regulate receptor number and function, and induce other molecular processess that influence synaptic efficacy. These changes might presumably predispose to hypersynchronization of surviving neurones in the region of cell loss. In addition, the gliosis that invariably accompanies neuronal death would contribute to these functional disturbances by producing changes in the ionic microenvironment. Intermittent hypersynchronous discharges from the damaged tissue might then cause changes in adjacent and distant structures via a mechanism similar to those produced by kindling stimulation in experimental animals (Goddard *et al*., 1969). Once an epileptogenic region has matured, the status is one of partial neuronal deafferentation, but with increased recurrent excitatory and inhibitory circuits. The surviving neurones therefore are relatively hypoactive but hypersynchronous, accounting for the appearance of intermittent high amplitude EEG spike-and-wave transients (Fig. 24.3).

Strong inhibitory influences exist to maintain the interictal state, such as the after-hyperpolarization which contributes to the wave of the EEG spike and wave, and an inhibitory surround created by activation of inhibitory neurones within the epileptic focus that send axons outside the focus. Engel believes that 'interictal spikes' can be viewed as fragments of spike-and-wave ictal events with surrounding inhibitory cells preventing more than brief sustained repeti-

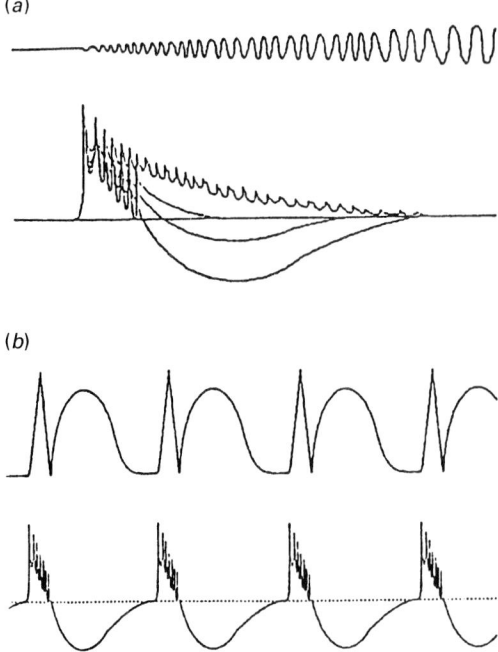

Fig. 24.3. (a) The lower trace represents an intracellular recording of a paroxysmal depolarization shift (PDS), demonstrating how the after-hyperpolarization gradually disappears to become an after-depolarization, giving rise to continuous high frequency action potential discharge. The upper trace shows the simultaneous EEG recording increasing amplitude and slowing frequency as more and more neurones are recruited into this process and develop increasing synchrony ("Recruiting rhythm").
(b) The lower trace shows an intracellular recording of recurrent PDSs interspersed with episodes of hyperpolarization. In the upper trace the EEG records the PDS as a spike followed by a depolarization wave due to the activity of surrounding inhibitory cells. Both partial and generalized ictal events can be associated with repetitive hypersynchronous discharge such as the classical 3/s spike-and-wave pattern of petit mal absences. Engel (1990) reproduced with permission.

with surrounding inhibitory cells preventing more than brief sustained repetitive firing. Ictal onset occurs when there is a breakdown of the protective hyperpolarization which follows the paroxysmal depolarization shift, eventually creating a prolonged depolarization and rapidly repetitive unit firing, associated with low-voltage fast activity on the EEG As more neurones become recruited in this process, the EEG becomes of higher amplitude, while increasing synchronization results in slower EEG frequencies, giving rise to the classical recruiting rhythm. A second type of ictal onset consists of high amplitude wave discharges that occur during the interictal state. The prototype of this form of ictal event is the typical petit mal absence, but similar electrographic seizures also occur in localised areas of the brain in association with partial seizures. In both instances, the synaptic inhibitory events that give rise to the slow wave of the spike and wave complex are the key to synchronization. Inhibitory feedback provides the most powerful mechanism for synchronization in the normal brain (Andersen & Sears, 1964) and in partial epilepsy contributes to recruitment of adjacent neurones into the hypersynchronous ictal discharge.

Surgery for epilepsy

Anterior temporal lobectomy became an established procedure after the pioneering work of Penfield and Flanigin (1950). Later, portions of the frontal lobe, and to a lesser extent other cerebral areas, were resected (Rasmussen, 1969). Epilepsy surgery has been carried out in several centres in Australia for more than 20 years (Davis, 1974). The Neuropsychiatric Institute (NPI) was established at Prince Henry Hospital in 1977 as a tertiary referral centre for the investigation and treatment of difficult neurological and psychiatric disorders. From 1983 the unit has evolved a comprehensive assessment and management programme for intractable epilepsy. To date, the unit has assessed over 500 patients, including depth electrode recordings in 150 leading to surgery in some 100 patients. The outcome in patients operated in the first six years was reported recently (Mackenzie *et al.*, 1990). A brief outline of the procedures followed is given here (see Fig. 24.4).

Before entering the inpatient programme, all patients undergo a thorough outpatient evaluation of their seizure disorder. This consists of a detailed history and examination, including the recall of old hospital records, witness accounts of seizures and a review of past and present EEGs, computed tomography (CT) scans and more recently magnetic resonance imaging (MRI) scans. All patients undergo an adequate trial of appropriate anticonvulsants with serum levels maintained in the therapeutic range. Patients are considered for surgical evaluation when the seizures seem likely to have partial onset, are refractory to anticonvulsant medication and seriously interfere with social, educational and employment opportunities. The patients need to be sufficiently motivated and mature to complete the entire evaluation and to co-operate with the test procedures.

In the first phase of in-hospital evaluation, patients are monitored continuously with a video camera and simultaneous 16-channel scalp and sphenoidal electrode recordings are taken. Patients or relatives can sound an event alarm,

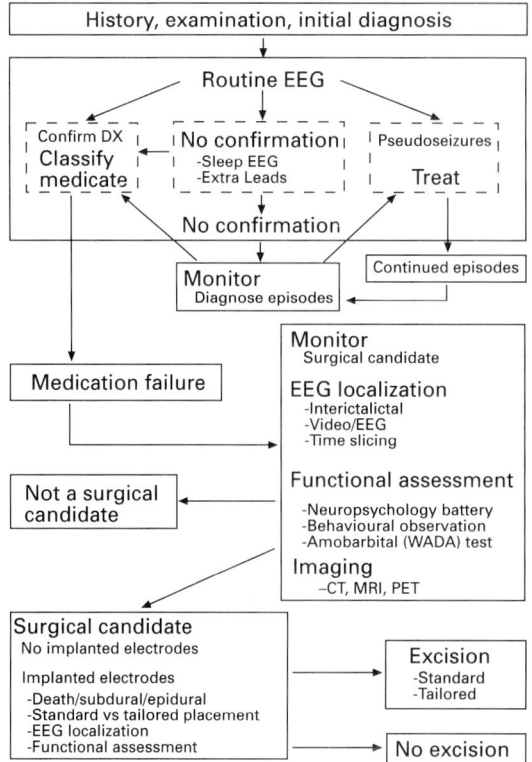

Fig. 24.4. Flow diagram of the decision processes involved in evaluating patients with a history of intractable complex partial seizures. Lesser, Fisher and Kaplan (1989) reproduced with permission.

and staff observe all activity either directly or at desk television monitors. The tapes are played back to patients and relatives to ensure that these are typical seizures. This is especially important when anticonvulsant doses are reduced, since atypical seizures can sometimes occur under these circumstances.

During this phase, further neuropsychological tests are carried out, including bilateral internal carotid artery injections of sodium amytal to establish language laterality and to assess the ability of each hemisphere to support memory. Recently we have injected a radio-isotope at the time of at least one typical seizure, after which a single photon emission computed tomogram (SPECT) scan is performed. This technique is being increasingly refined and is now claimed to detect an increase in cerebral blood flow at the site of seizure onset in up to 97% of cases studied (Rowe *et al.*, 1989, 1991).

Partial epilepsies 409

Phase I assessment may last two or three weeks. During this time the medical, technical and nursing staff are able to observe ictal and interictal behaviour, as well as note the patient's interaction with other patients, relatives and friends. Knowledge is gained of patient awareness, understanding, sophistication, motivation and insight, and this is considered when making a final decision about the patient's suitability for surgery.

Patients may be excluded from further consideration of surgery because of seizure type or the presence of severe neuropsychological deficits. Currently, some 50% proceed directly to surgical treatment when all tests localize the epileptic focus to a resectable area or if corpus callosum section is indicated. In the other 50%, further EEG recordings are necessary to be certain of the site of seizure onset. When only confirmation of temporal lobe onset is required, foramen ovale electrodes may be inserted transcutaneously with the guidance of an image intensifier (Weiser, Elger & Stodieck, 1985). If an extratemporal onset is suspected, subdural strip electrodes may be inserted via bilateral frontal and temporal burrholes. If precise localization of onset within a temporal lobe is required (e.g. for selective surgery), then depth electrodes are inserted bilaterally via frontal burrholes with additional posterior temporal depth electrodes being inserted through laterally placed drill holes. Ictal recordings are then obtained again, these sensitive techniques showing onset of localised epileptiform disturbance often 20 to 60 seconds prior to onset of the clinical seizure.

Surgical procedures

Anterior temporal lobectomy is removal of the anterior portion of the temporal lobe 5 cm back from the temporal pole in the dominant hemisphere or 6 cm in the nondominant hemisphere: the amygdala and the anterior portion of the hippocampus are also removed. This operation is performed when surface EEG recordings show temporal lobe ictal onset or depth electrode studies suggest regional rather than mesial temporal onset. This is the time-honoured and still most commonly performed surgical procedure for CPS.

Selective amygdalohippocampectomy is the removal of amygdala, uncus and hippocampus by an approach through the Sylvian fissure. This is performed when depth electrode studies show stereotyped focal ictal activity in amygdala and/or hippocampus precede the clinical ictus. It has also been performed when the electrical onset is not identified but mesial temporal electrodes on one side show isolated pronounced ictal activity during the seizures. This procedure has been popularised by Yasargil (Weiser & Yasargil, 1984) and is being increasingly performed by specialized epilepsy centres. It is claimed that selective amygdalohippocampectomy has a lower incidence of neuropsychological deficits compared to anterior temporal lobectomy, especially for dominant hemisphere operations (Nadig, Weiser & Perrett, 1985; Weiser, 1986).

In the first six years of our programme, 130 patients with medically intractable epilepsy underwent the inpatient comprehensive assessment. All had scalp and sphenoidal EEG telemetry and 41 patients (32%) underwent further video and EEG recordings of their seizures with depth (intracerebral) or strip (subdural) electrodes. After these assessments, 46 patients (35%) underwent surgery. There were 20 anterior temporal lobectomies, 10 selective amygdalohippocampectomies, 5 corpus callosotomies and 11 extratemporal cortical excisions (two patients had two procedures).

Surgical outcome

Best results were obtained with temporal lobe surgery. The follow-up period for the 30 patients in our series ranged from six months to six years, with a mean of 2.4 years. Twenty-one patients (70%) are classified as seizure-free (although seven of these are experiencing occasional auras). One patient is having rare seizures and three others have a 90% reduction in seizure frequency. Five have had no improvement. Overall, 25 of the 30 patients (83%) have benefited from temporal lobe surgery. Eleven patients had excision of an extratemporal lesion; two were rendered seizure free while another six had a worthwhile reduction in seizure frequency. Five had corpus callosotomy procedures; two had 90% reduction in frequency while three had 50% reduction.

These results are similar to those obtained by others using similar techniques. Bladin (1987) reported 40 of 62 patients (64.5%) had remained free of seizures two years after anterior temporal lobectomy and overall 80% of patients were improved. Engel (1987) used more ancillary tests including Positron Emission Tomography (PET) scanning and a more extensive depth electrode array and reported 70% of patients free of seizures after two years or more.

Recently it has been claimed that MRI can detect MTS in a high proprtion of patients with CPS (Berkovic *et al.*, 1991). In ten patients who had preoperative MRI evidence of MTS, there was an 80% seizure free rate at a mean post-surgery follow-up time of 33 ± 4 months.

Psychosocial outcome

When considering the results of epilepsy surgery, it is important to focus not simply on the issue of seizure frequency; one must also address the question of whether the patient's life has been changed for the better. This involves a careful psychosocial assessment both preoperatively and postoperatively.

Dodrill (1978) has developed an inventory called the Washington Psychosocial Seizure Inventory (WPSI) which objectively evaluates each of the following areas: family background, emotional adjustment, financial status, adjustment to seizures, medicine and medical management, and overall psychosocial adjustment. The inventory consists of 132 yes/no answers, and cross cultural comparisons have been done. Using WPSI, patients can be evaluated before and after surgery, and objective indices obtained of the nature and extent of psychosocial change.

Dodrill (1987) reported the preoperative and 1 year postoperative average profiles for 60 adults having cortical resection surgery for epilepsy. Taken as a whole, about two-thirds of these patients had experienced a significant improvement (at least a 75% reduction) in seizure frequency. The WPSI showed improvement on every scale of the inventory, and often the improvement reached high levels of statistical confidence. These changes were considerably greater than those shown on the Minnesota Multiphasic Personality Inventory (MMPI) and without an inventory such as the WPSI, it is likely that important and favourable results of the surgery would have been missed.

Dodrill made two other important observations. Firstly, even persons who reported no seizure relief improved psychosocially after surgery. In many, a feeling of satisfaction that 'everything possible had been done' had led to a more ready acceptance of continuing seizures. Second, simply because the patient reports improved psychosocial adjustment does not mean that major changes in life functioning have been made. For example, vocational adjustment scales show that seizures were no longer seen as a barrier to employment, but employability did not improve. In general, people who were employed before surgery were also employed after surgery, and people who were not employed before surgery were not employed after surgery.

Rausch & Crandall (1982) reported that psychosocial improvement was more likely with good family support, no or mild psychopathology, a nonaddictive personality, and a young age. Interestingly, although all of these variables played a role in determining whether or not a patient achieved the highest level of psychosocial functioning postsurgically, it was 'good family support' that was a necessary requirement.

Taylor (1987) described the predicament to be considered before making a decision about epilepsy surgery as follows:

'The results of epilepsy surgery make this an attractive treatment option for patients with epilepsy. However, it is very evident that the patients, at the time of seeking this crucial treatment, may be handicapped by many things which are only remotely, if at all, causally connected to their epilepsy, and which are not usually evaluated in the technical routines of medicine. When patients present themselves they are looking for a way out of a fix, out of their total predicament. Their illness is epilepsy; there may be a disease underlying this illness; but their predicament depends on how these processes impinge on their life. This depends on preexisting intelligence, personality, relationships, family and work environment. Reduction or even relief of seizures may not necessarily relieve their predicament if this is due to factors other than the physical occurrence of epilepsy. It behoves the clinician to make a careful assessment of all of these factors before recommending the patient submit himself to investigation for possible surgical treatment'.

Conclusions

The last ten years have seen a tremendous explosion in our knowledge and understanding of the partial epilepsies. Although the early promise of the control with drugs has not been fulfilled, basic research and the growing sophistication of investigative tools have led to much improved results of surgery for the partial epilepsies, particularly where MTS is found. It is likely that the next ten years will bring even better understanding and more effective treatments for this disorder.

References

Akaike, N., Kostyuk, P. G. & Osipchuk, Y. V. (1989). Dihydropyridine-sensitive low theshold calcium channels in isolated rat hypothalamic neurones. *Journal of Physiology (London)*, **412**, 181–95.

Andersen, P. & Sears, T. A. (1964). The role of inhibition in the phasing of spontaneous thalamocortical discharge. *Journal of Physiology (London)*, **173**, 459–80.

Annegers, J. F., Hauser, W. A., Elveback, L. R. & Kurland, L. T. (1979). The risk of epilepsy following febrile convulsions. *Neurology*, **29**, 297–303.

Babb, T. L. & Brown, W. J. (1987). Pathological findings in epilepsy. In *Surgical Treatment of the Epilepsies*, ed. J. Engel Jr, pp. 511–40. New York: Raven Press.

Babb, T. L. & Crandall, P. H. (1976). Epileptogenesis of human limbic neurones in psychomotor epileptics. *Electroencephalography and Clinical Neurophysiology*, **40**, 225–43.

Babb, T. L., Wilson, C. L. & Isokawa-Akesson, M. (1987). Firing patterns of human limbic neurones during stereoencephalography (SEEG) and clinical temporal lobe seizures. *Electroencephalography and Clinical Neurophysiology*, **66**, 467–82.

Berkovic, S. F., Andermann, F., Olivier, A., Ethier, R., Melanson, D., Robitaille, Y., Kurniecky, R., Peters, T. & Feindel, F. (1991). Hippocampal sclerosis in temporal lobe epilepsy demonstrated by magnetic resonance imaging. *Annals of Neurology*, **29**, 175–82.

Bladin, P. F. (1987) Post-temporal lobectomy seizures. *Clinical and Experimental Neurology*, **24**, 77–83.

Bratz, E. (1899). Ammonshornbefunde der epileptischen. *Archiv für Psychiatrie und Nervenkrankheiten*, **31**, 820–36.

Cavazos, J. E., & Sutula, T. P. (1989). Progressive neuronal loss induced by kindling: A possible mechanism for mossy fibre synaptic reorganization and hippocampal sclerosis. *Epilepsia*, **30**, 702 (Abstract).

Cooper, J. R., Bloom, F. E. & Roth R. H. (1986). Central nervous system drug effects. In *The Biochemical Basis of Neuropharmacology*, 5th edn. New York: Oxford University Press.

Dasheiff, R. M. (1989). Epilepsy Surgery: Is it an effective treatment? *Annals of Neurology*, **25**, 506–9.

Davis, E. (1974). Temporal lobectomy for epilepsy – a follow-up. *Proceeding of the Aust.ralian Association of Neurologists*, **11**, 245–6.

de Lanerolle, N. C., Kim, J. H., Robbins, R. J. & Spencer, D. D. (1989). Hippocampal interneuron loss and plasticity in human temporal lobe epilepsy. *Brain Research*, **495**, 387–95.

Dodrill, C. B., (1978). A neuropsychological battery for epilepsy. *Epilepsia*, **19**, 611–23.

Dodrill, C. B. (1987). Commentary: Psychological evaluation. In *Surgical Treatment of the Epilepsies*, ed. J. Engel Jr., pp. 197–201. New York: Raven Press.

Elwes R. D. S., Johnson, A. L., Shorvon, S. D. & Reynolds, E. H. (1984). The prognosis for seizure control in newly diagnosed epilepsy. *New England Journal of Medicine*, **311**, 944–7.

Engel, J. Jr. (1987). Outcome with respect to seizures. In *Surgical Treatment of the Epilepsies*, ed. J. Engel Jr., pp. 553–71. New York: Raven Press.

Engel, J. Jr. (1990). Functional explorations of the human epileptic brain and their therapeutic implications. *Electroencephalography and Clinical Neurophysiology*, **76**, 296–316.

Engel, J. Jr. & Cahan, L. (1986). Potential relevance of kindling to human partial epilepsy. In *Kindling 3*, ed. J. A. Wada, pp. 37–51. New York: Raven Press.

Goddard, G. V., McIntyre, D. C. & Leech, C. K. (1969). A permanent change in brain function resulting from daily electrical stimulation. *Experimental Neurology*, **25**, 295–330.

Hauser, A. (1987). Postscript: How should outcome be determined and reported? In *Surgical Treatment of the Epilepsies*, ed. J. Engel Jr., pp. 573–9. New York: Raven Press.

Lesser, R. P., Fisher, R. S. & Kaplan, P. (1989). The evaluation of patients with intractable complex partial seizures. *Electroencephalography and Clinical Neurophysiology*, **73**, 381–8.

Macdonald, R. L., McLean, M. J. & Skerritt, J. H. (1985). Anticonvulsant drug mechanisms of action. *Federation Proceedings*, **44**, 2634–9.

Mackenzie, R. A., Matheson, J. M., Smith, J. S. & Dwyer, M. (1990). Surgery for refractory epilepsy. *Medical Journal of Australia*, **153**, 69–76.

McLean, M. J. & Macdonald R. L. (1983). Multiple actions of phenytoin on mouse spinal cord neurones in cell culture. *Journal of Pharmacoogy and Expeimental Theraputics*, **227**, 779–89.

McNamara, J. O. (1991). Kindling: relevance to human epilepsy. In *Clinical Epilepsy, Course No. 344*, ed. T. Pedley, American Academy of Neurology.

McNamara, J. O., Rigsbee, L. C., Butler, L. S. & Shin, C. (1989). Intravenous phenytoin is an effective anticonvulsant in the kindling model. *Annals of Neurology*, **26**, 675–8.

Meldrum, B. S. (1991). Mesial temporal sclerosis: Cause or consequence of epilepsy? In *Clinical Epilepsy, Course No. 344*, ed. T. Pedley, American Academy of Neurology.

Meldrum, B. S. & Brierley, J. B. (1973). Prolonged epileptic seizures in primates: ischaemic cell change and its relation to ictal physiological events. *Archives of Neurology*, **28**, 10–7.

Meldrum, B. S. & Corsellis, J. A. N. (1984). Epilepsy. In *Greenfield's Neuropathology*, eds. J. H. Adams, J. A. N. Corsellis & L. W. Luchen, pp. 921–50. London: Edward Arnold.

Meldrum, B. S., Horton, R. W., & Brierley, J. B. (1974). Epileptic brain damage in adolescent baboons following seizures induced by allylglycine. *Brain*, **97**, 417–28.

Morrell, F. (1979). Human secondary epileptogenic lesions. *Neurology*, **29**, 558 (Abstract).

Nadig, T., Weiser, H. G. & Perrett, E. (1985). Learning and memory performance before and after unilateral selective amygdalohippocampectomy. In *Brain Plasticity, Learning & Memory*, eds. B. E. Schmitt, P. Dalrymple & J. C. Alford, pp. 397–403. New York: Plenum Press.

Nadler, J. V. (1981). Minireview. Kainic acid as a tool for the study of temporal lobe epilepsy. *Life Sciences*, **29**, 2031–42.

Overweg, J., Binne, C. D., Meijer, J. W. A., Meinardi, H., Nuijten, S. T. M., Schmaltz, S. & Wauquier, A. (1984). Double-blind placebo-controlled trial of flunarizine as add-on therapy in epilepsy. *Epilepsia*, **25**, 217–22.

Penfield, W. & Flanigin, H. (1950). Surgical therapy of temporal lobe seizures. *A.M.A. Archives of Neurology and Psychiatry*, **64**, 491–500.

Pinel, J. P. J. & Rovner, L. I. (1978). Experimental epileptogenesis: kindling-induced epilepsy in rats. *Experimental Neurology*, **58**, 190–202.

Racine, R. J. (1972). Modification of seizure activity by electrical stimulation: I. After-discharge threshold. *Electroencephalography and Clinical Neurophysiology*, **32**, 269–79.

Rasmussen, T. (1969). The role of surgery in the treatment of focal epilepsy. *Clinics in Neurosurgery*, **16**, 288–314.

Rausch, R. & Crandall, P. H. (1982). Psychological status related to control of temporal lobe seizures. *Epilepsia*, **23**, 191–202.

Rodin, E. A. (1968). *The Prognosis of Patients with Epilepsy*. Springfield, Ill.: Thomas.

Rowe, C. C., Berkovic, S. F., Austin, M. C., McKay, W. J. & Bladin, P. F. (1991). Patterns of post-ictal cerebral blood flow in temporal lobe epilepsy: qualitative and quantitative analysis. *Neurology*, **41**, 1096–103.

Rowe, C. C., Berkovic, S. F., Sia, S. T. B., Austin, M., McKay, W. J., Kalnins, R. M. & Bladin, P. F. (1989). Localisation of epileptic foci with post-ictal single photon emission computed tomography. *Annals of Neurology*, **26**, 660–8.

Sagar, H. J. & Oxbury, J. M. (1987). Hippocampal neuron loss in temporal lobe epilepsy: correlation with early childhood convulsions. *Annals of Neurology*, **22**, 334–40.

Sutula, T., Cascino, G., Cavazos, J., Parada, I. & Ramirez, L. (1989). Mossy fibre synaptic reorganization in the epileptic human temporal lobe. *Annals of Neurology*, **26**, 321–30.

Sutula, T., Xiao-Xian, H., Cavazos, J. & Scott, G. (1988). Synaptic reorganization in the hippocampus induced by abnormal functional activity. *Science*, **239**, 1147–50.

Tauck, D. & Nadler, J. V. (1985). Evidence of functional mossy fibre sprouting in hippocampus formation of kainic-acid treated rats. *Journal of Neuroscience*, **5**, 1016–22.

Taylor, D. C. (1987). Psychiatric and social issues in measuring the input and outcome of epilepsy surgery. In *Surgical Treatment of the Epilepsies*, ed. J. Engel Jr., pp. 485–503. New York: Raven Press.

Twymann, R. E., Rogers, C. J. & Macdonald R. I. (1989). Differential regulation of gamma-amino-butyric acid receptor channels by diazapam and phenobarbital. *Annals of Neurology*, **25**, 213–20.

Weiser, H. G. (1986). Selective amygdalohippocampectomy: indications, investigative technique and results. In *Advances & Technical Standards in Neurosurgery*, Vol. 13, ed. L. Symon, J. Brihaye, B. Guidetti, S. Loew, J.D. Miller, H. Nornes, E. Pasztor, B. Pertuiset & M.G. Yasargil, pp. 39–133. Vienna: Springer-Verlag.

Weiser, H. G. & Yasargil, M. G. (1984). Selective amygdalohippocampectomy as a surgical treatment of mesiobasal limbic epilepsy. *Surgical Neurology*, **17**, 445–57.

Weiser, H. G., Elger, C. E. & Stodieck, S. R. G. (1985). The 'foramen ovale electrode': a new recording method for the pre-operative evaluation of patients suffering from mesiobasil limbic temporal lobe epilepsy. *Electroencephalography and Clinical Neurophysiology*, **61**, 314–22.

Zimmerman, H.M. (1938). The histopathology of convulsive disorders in children. *Journal of Paediatrics*, **13**, 859–90.

25
The changing face of antiepileptic drug therapy

M. J. EADIE

Department of Medicine,
The University of Queensland,
St Lucia, Queensland, Australia

This chapter traces the more important changes that have occurred in the drug therapy of epilepsy over the last 30 years. Subsequently, the present situation regarding anticonvulsant therapy is reviewed.

Antiepileptic drugs in use since 1960

If one reopens the neurological and therapeutic textbooks of one's professional youth, and sees what drugs were then discussed in relation to the treatment of epilepsy, one finds in the 7th (1952) edition of F. M. R. Walshe's *Diseases of the Nervous System* (1952) mention of phenobarbitone, the hydantoinates, and tridione, and an indication that the use of bromides had virtually been abandoned. F. M. Forster (1955), writing on *The Epilepsies and Convulsive Disorders* in the second of the three volumes of the first edition of Baker's *Clinical Neurology*, mentioned the same drugs but added paramethadione, methoin, phenylethylurea, primidone and phensuximide and mentioned some minor agents which were then still in use (bromides, methylphenobarbitone, glutamic acid, benzedrine and the ketogenic diet).

Over the intervening years, some of these drugs have continued in widespread use: others have failed to stand the test of time and have disappeared from the marketplace; new agents have been introduced – some have already been found wanting and have departed the scene, or are in the process of departing; others have taken on increasingly important therapeutic roles, often supplanting their predecessors in so doing.

Hydantoins

The first hydantoin to be used in humans, phenytoin (30 years ago usually called diphenylhydantoin, or referred to by its brand name of 'Dilantin') was introduced into therapeutics half a century ago (Merritt & Putnam, 1938). Despite a formidable array of recorded adverse effects and an unusual pattern of pharmacokinetics which makes its dosage difficult to handle in the individual patient, phenytoin still remains one of the major antiepileptic agents in current use.

One of its congeners, methoin ('Mesantoin') had considerable efficacy as an anticonvulsant, and more recently proved an interesting tool for investigating stereo-specific drug metabolism and the inheritance of a relative inability to metabolise one of the drug's enantiomers. However, the propensity of the N-desalkylated metabolite of the drug for causing aplastic anemia has seen methoin disappear from therapeutic use. Structurally related molecules have either been insufficiently effective (e.g. deltoin) or have caused bone marrow toxicity.

Another hydantoin, ethotoin, was regarded as a safe but relatively ineffective agent, and has faded from the Australian marketplace, almost unnoticed. The little pharmacokinetic data available for this substance suggest that it has probably never been used optimally. Ethotoin could well be resurrected and have a limited pattern of use in those intolerant to phenytoin.

Barbiturate anticonvulsants

Phenobarbitone, N-methylphenobarbitone and primidone are very old drugs which continue in reasonably widespread use despite the general disfavour in which the whole family of barbiturates is regarded. These drugs are effective anticonvulsants, though they need to be used with care to minimize their sedative-type adverse effects.

Oxazolidinediones

The various 'diones' were effective only in treating absence seizures. They have now disappeared from use, having been supplanted by the succinimides, which are both safer and more effective. Evidence had also appeared that troxidone was a teratogen, which hastened the demise of this agent. Paramethadione and other 'diones' were never widely used.

Succinimides

The succinimide family of drugs was beginning to come into use 30 years ago. The first ones to be introduced, phensuximide ('Milontin') and methsuximide ('Celontin'), were probably little more effective than troxidone, though they were less toxic. However, ethosuximide ('Zarontin'), introduced a little later than its congeners, was distinctly more efficacious. It was also safer than troxidone, and for some years came to be the treatment of choice for absence seizures though now, arguably, it has been supplanted by valproate. Phensuximide and methsuximide continue to be marketed. They seem to have no advantage over ethosuximide and appear to be rarely used in clinical practice.

Sulthiame

The sulphonamide derivative sulthiame was introduced in the 1960s for the treatment of partial seizures and possibly for myoclonic ones. The drug achieved some popularity for a time, but its use has been declining for several years since it was shown to be less effective than phenytoin in a therapeutic trial which had a reasonably satisfactory cross-over design (Green et al., 1974). It was also realised that the apparent additive antiseizure effect sulthiame showed when combined with phenytoin might depend on a pharmacokinetic interaction which increased plasma and tissue concentrations of the latter substance.

Very little clinical pharmacokinetic study of sulthiame has been carried out, and it is possible that the drug has not been used optimally. However, there seems little interest in perpetuating its presence in the marketplace.

Carbamazepine

Carbamazepine ('Tegretol') began its commercial career as an anticonvulsant in 1962. It took about 10 years to find its definitive place in the treatment of all varieties of partial seizures and in the convulsive seizures of generalized epilepsy. The use of the drug has expanded in more recent years so that, in many places, it now rivals phenytoin as the most widely used antiepileptic agent in contemporary medical practice.

Valproate

Valproate, supplied as the free acid or the sodium salt, began to be marketed as an anticonvulsant in the 1960s. It has proved a very effective agent for all varieties of generalised epilepsy, though its role in treating partial seizures is less certain. It appears to prevent the generalization of partial seizure discharges, but the literature is less clear as to its capacity for suppressing the local manifestations of partial seizure activity. There have been some concerns over the use of the drug, particularly because of the very occasional but potentially very serious hepatotoxicity associated with its intake (some 112 deaths reported worldwide to 1988 – Scheffner *et al.*, 1988), and the small risk (roughly 1%) of spina bifida in the offspring if a mother takes the drug in pregnancy (Nau & Hendrickx, 1987). None the less, valproate has found a place as one of the major anticonvulsants in contemporary medical practice.

Changes in therapeutic practice since 1960

As well as the changes in the availability of antiepileptic drugs since 1960, there have been changes in the way these drugs have been used. These changes have resulted from developments in our understanding of epilepsy and of clinical pharmacology. The main alterations have depended on the following:

A better understanding of epilepsy and its natural history

The increasingly widespread use of the International League Against Epilepsy's Classification of Epileptic Seizures (Commission on Classification and Terminology, 1981) has made it easier to compare the findings of different studies and has also provided a framework for the better categorization of the individual patient's seizures on the basis of clinical and electroencephalographic data. This has proved important, because there is evidence that there is a relative specificity of efficacy of particular antiepileptic drugs for particular types of seizure disorder (see below).

In recent years, several studies of the natural history of various types of seizure disorder have appeared. For the patient with a single unprovoked convulsive seizure, there is a 30% to 60% risk of a further attack within 12 months if anticonvulsant therapy is not offered (e.g. Elwes, Chesterman & Reynolds, 1985). It is known that, the more seizures a patient has before anticonvulsants are taken, the less is the chance of ultimate cure of the seizure disorder. Therefore there is an increasingly strong argument for treatment to be commenced after the first seizure, without waiting for a second episode to occur.

A clearer realisation that epilepsy may be cured by prolonged seizure control

Thirty years ago anticonvulsant therapy was usually prescribed with the aim of keeping the patient free from major seizures, rather than from all seizures. If the seizures did not recur after the therapy was withdrawn, this was regarded as an additional dividend, perhaps an almost undeserved one. Gradually the reali-

Table 25.1. *Correlations between the effectiveness of commonly used antiepileptic drugs and types of epileptic seizure*

Drug	Seizure type			Partial seizures (Simple/complex)
	Generalized epilepsy			
	Absences	Myoclonic	Convulsive (Tonic-clonic, etc)	
Carbamazepine	0	0	+	+
Clonazepam	+	+	+	±
Corticotrophin	0	+#	0	0
Ethosuximide	+	0	0	0
Methylpheno barbitone	0	+‡	+	+
Phenobarbitone	0	+‡	+	+
Phenytoin	0	+	+	+
Primidone	0	+‡	+	+
Valproate	+	+	+	±

Notes: # - in infantile spasms, only.
‡ - in juvenile myoclonic seizures, only.
+ = effective: ± = probably effective: 0 = ineffective.

sation has developed that prolonged full control of all clinical manifestations of epilepsy may lead to the 'dying out' of the underlying epileptic process itself, providing this process is not based on active progressive pathology. Thus seizure disorders may be cured, providing an additional argument for prompt efficient treatment aimed not so much at keeping the epilepsy quiet, but at suppressing it fully.

Results, such as those of Oller-Daurella and Oller (1987), indicate an approximately 75% chance of cure of epilepsy by the time anticonvulsants are withdrawn after 5 years of full seizure control.

A recognition that certain anticonvulsants are relatively specific for certain types of epileptic seizure

Accumulating clinical experience, and the results of studies in experimental animal models of particular types of epileptic seizure, have led to correlations between seizure type and particular antiepileptic drugs which are likely to be effective in these varieties of seizure (Table 25.1). These correlations are not perfect, but they are certainly good enough to prove very useful in therapeutic practice. Their use avoids the old treatment practice of blindly trying out one antiepileptic drug after another, without any guidance as to the chance of success from the use of any particular agent.

The application of clinical pharmacological concepts to the treatment of epilepsy

Methods for measuring antiepileptic drugs at the concentrations at which they occur in biological fluids during the drug treatment of epilepsy have become increasingly available. This has opened up a new area of applied pharmacological science, and the results have had a major influence on the efficiency of the drug treatment of epilepsy. Leaving aside the numerous scientific questions which have been explored and answered by the use of these methods, there have been several important practical dividends.

Quite early, 'therapeutic' ranges of plasma concentrations for the various anticonvulsants were defined, not utilizing a rigid statistical approach but rather a commonsense and semi-informal one. Particular ranges of plasma concentrations of the various anticonvulsant drugs were found to be associated with a reasonable chance of controlling epileptic seizures in patients, without producing an undue incidence of those adverse effects which were customarily regarded as due to overdosage from the drug in question. When knowledge of these 'therapeutic' ranges was applied to populations of patients treated with the then customary doses of anticonvulsant drugs, it rapidly became obvious that many patients were being undertreated, either by virtue of prescribed underdosage or because of their noncompliance. Where these problems could be remedied, many patients' seizure disorders came under better, or complete control, for the first time ever. It also became clear that some patients were being overtreated by conventional anticonvulsant doses, and that these patients achieved a better state of health if they received lower doses of the drug in question.

However, problems began to arise when clinicians came to regard these 'therapeutic' ranges in the same light as the normal physiological ranges for endogenous substances such as urea or glucose in plasma, ignoring the fact that it is not physiological to have drugs present in plasma in the first place. Anticonvulsant doses were often altered to achieve plasma drug levels in the 'therapeutic' range, regardless of whether the patients' seizures were already controlled, or whether adverse effects were present. Clinically absurd courses of action were sometimes taken in the name of scientific medicine. Patients who were clinically well, and whose seizures were obviously fully controlled but who had 'sub-therapeutic' or 'toxic' drug levels, had their anticonvulsant doses respectively increased or reduced to improve already perfect results, or to correct non-existent toxicity. Depending on the direction of the dosage change, they then became toxic or their seizures relapsed. At the time of writing, many prescribers have not come to terms with the realisation that the 'therapeutic' range is a population parameter which provides little more than a rough general preliminary guide to anticonvulsant dosage. Achieving 'therapeutic' range plasma levels of a given drug should never be allowed to override clinical commonsense. Data have now appeared, showing that the 'therapeutic' range is not a general property of the drug in question. The ranges for phenytoin, carbamazepine and phenobarbitone differ, depending on whether the drugs are used to treat primarily generalized or partial seizures (Schmidt, Einicke & Haenez, 1986).

The clinical pharmacological approach has helped rationalize the drug therapy of epilepsy in other ways. Thus it has become clear that most of the anticonvulsants are sufficiently slowly eliminated for single doses of the drug to maintain plasma and tissue concentrations within acceptable limits for between 12 and 24 hours (and sometimes longer). This realization has allowed most of the anticonvulsants to be taken once or twice daily, rather than three or four times daily, as was the customary previous practice. This less frequent intake enhances compliance, with consequent improvement in seizure control, since several studies have shown that deliberate or accidental noncompliance is a major, or perhaps the major factor causing failure of drug therapy to prevent further seizures in many patients (Eisler & Mattson, 1975; Desai *et al.*, 1978).

Recognition that some anticonvulsants are metabolized to molecules which also possess antiepileptic activity has also led to a more sensible approach to therapy. N-methylphenobarbitone and primidone are both metabolized to phenobarbitone, itself an effective anticonvulsant. There is clearly little sense in prescribing more than one of these agents in a given patient, so long as an adequate dose of the prescribed drug has been given. The widely used anticonvulsant carbamazepine forms an epoxide metabolite which is effective as an active anticonvulsant and in relieving tic douloureux (Bertilsson & Tomson, 1986). This metabolite achieves relatively higher plasma (and tissue) concentrations when carbamazepine is taken with phenytoin or phenobarbitone than when carbamazepine is used as monotherapy. Because of this effect, the therapeutic range of plasma levels of carbamazepine needs to be lower in those taking the drug with phenytoin or phenobarbitone than in those taking it as monotherapy. Overdosage manifestations occur at lower plasma carbamazepine concentrations in patients taking the drug with phenytoin or barbiturate anticonvulsants than in those taking it alone.

Measurement of plasma anticonvulsant concentrations at different stages of pregnancy and the postpartum period has shown that, relative to drug dose, plasma drug levels fall as pregnancy advances. The levels rise again after childbirth (Lander & Eadie, 1991). At least in the case of phenytoin, this phenomenon appears to be due to increased elimination of the drug during pregnancy (Dickinson *et al.*, 1989) and is the result of increased metabolism of the drug to only one of its several known metabolites, viz. 5-*p*-hydroxyphenyl-5-phenylhydantoin (personal data). Adjustment of anticonvulsant dosage to maintain plasma drug concentrations at (satisfactory) prepregnancy values throughout pregnancy largely obviates the previously documented problem of seizures becoming more frequent in these circumstances.

The facility for carrying out routine measurements of plasma anticonvulsant concentrations has been widely available for at least a decade, and has had a very significant impact in bettering the drug treatment of epilepsy, though it could be argued that the approach has not always been applied as wisely as it might have been.

An increased consciousness of the adverse effects of anticonvulsant therapy

With the worldwide attempt to put the drug therapy of epilepsy on an increasingly scientific basis, a heightened awareness has developed of the considerable variety of adverse effects, some obvious, others subtle, that may be produced by the antiepileptic drugs. The attempt to achieve full control, and ultimately cure, of seizure disorders in humans has often involved the treating practitioner in cautiously pushing anticonvulsant doses towards the individual patient's clinical threshold of toxicity, and sometimes to beyond this point. There has been a growing realization that antiepileptic drug doses should not be kept at a level where adverse effects of the therapy, whatever their clinical expression may be in the individual, come to be a greater actual or potential burden for him or her than further seizures would be. Acceptance of this principle sets the limit to anticonvulsant dosage in clinical practice.

The issue of anticonvulsant-associated teratogenesis remains a special and contentious case of a possible adverse effect, or class of adverse effect, of this group of agents. After two decades of investigations there is still no unambiguous evidence that the anticonvulsants which are commonly used at the present time are teratogens, with the possible exception of valproate; for this drug, the risk of fetal malformation (specifically spina bifida) is very small, though probably genuine (Nau & Hendrickx, 1987). The problem has been to distinguish between malformations due to the drugs used to treat epilepsy and malformations related to the various pathological processes which may result in epilepsy which requires drug therapy.

Table 25.2. *Commencing doses of anticonvulsants, and therapeutic range of plasma anticonvulsant concentrations. The doses are intended as rough guides to the dose likely to be needed when the patient has become tolerant to any initial sedative effects of therapy*

Drug	Approximate starting dose (mg/kg/day)	'Therapeutic' range plasma level mg/l	μ mole/l
Carbamazepine	5–10	6-12 (4-8[#1])	25–50 (16–32[#1])
Clonazepam	0.015	0.025–0.075	0.08–0.24
Corticotrophin	c. 2–4 (i.u.)	-	–
Ethosuximide	30	50–100	300–700
Methylphenobarbitone	5[#2]	10–30[#4,5]	45–130[#4,5]
Phenobarbitone	2.5[#2]	10–30[#5]	45–130[#5]
Phenytoin	6[#2] 11[#3]	10–20[#5]	45–80[#5]
Primidone	10	10–30[#4,5]	45–130[#4,5]
Valproate	c. 20–30	50–100	300–600

Notes: #1: if taken with phenytoin or phenobarbitone.
#2: adults.
#3: children.
#4: as phenobarbitone.
#5: range to control convulsive seizures: range higher for partial seizures.

The contemporary practice of anticonvulsant therapy

Once a patient has had his or her first epileptic seizure or seizures, it is now fairly widely accepted among neurologists that an attempt should be made to (i) classify the seizure type, (ii) determine the cause (which may prove to be an entity which needs treatment in its own right e.g. a glioma) and (iii) assess the risk of further seizures if the disorder is left untreated, and the disadvantages such seizures are likely to hold for the patient, and for society more generally. Unless there appears to be very little risk of further seizures, or little chance that such seizures would be an appreciable handicap to the patient, anticonvulsant therapy will usually be commenced after the initial assessment. The choice of an appropriate anticonvulsant drug is determined from the correlation between seizure type and therapeutic agent shown in Table 25.1. The drug is prescribed in a dose likely to achieve an initial steady-state plasma level towards the lower part of the therapeutic range for the type of seizure disorder present, the decision as to dosage perhaps being guided by data such as that set down in Table 25.2. However, all the anticonvulsants other than phenytoin, valproate and probably ethosuximide tend to produce sedation when they are first used. It therefore often proves better to commence therapy with some 50% of the anticipated definitive drug dose, and to then build up the dose to its expected value over the next 4 to 8 weeks, so long as adverse effects have not occurred in the meantime, necessitating an alteration in the dosage or in the drug prescribed.

If seizures continue, even if the plasma drug levels are in the therapeutic range, the drug dose may be increased cautiously until adverse effects preclude any further dose increase or the desired therapeutic effect is obtained. If maximum tolerated doses of the first anticonvulsant agent fail to control a patient's seizures, a second potentially appropriate anticonvulsant may be substituted and its dose adjusted to try to achieve seizure control without producing unacceptable adverse effects. The process of balancing drug and dosage against adverse effects and continuing seizures continues, as necessary, until the desired benefits are attained or all appropriate anticonvulsants have been tried as monotherapy. If this latter state has been reached, combinations of potentially appropriate anticonvulsants may have to be tried. Unfortunately, these often fail (Schmidt, 1982). It then becomes increasingly unlikely that full seizure control will ever be possible with currently available therapy. In these circumstances, the best compromise between seizure control and adverse effects must be sought, and maintained.

Should full seizure control occur, the effective therapy is continued in full dosage for 3 to 5 years. After that period of freedom from seizures, an attempt may be made to assess the risk of relapse in the individual if treatment were withdrawn, and the possible disadvantage further seizures might hold for the sufferer and his or her family. In the light of such considerations, a decision may then be taken to attempt a gradual withdrawal of the therapy, or to continue it in full dosage until a more propitious time.

During the long course of anticonvulsant therapy, steady-state plasma antiepileptic drug levels may be measured at intervals (i) as a preliminary guide to the potential adequacy of therapy till enough time has elapsed for this to be established from the clinical response, (ii) as a guide to anticonvulsant dose adjustment, (iii) as an aid in interpreting possible adverse effects of therapy, (iv) as a way of encouraging compliance with therapy, and (v) as a help in interpreting pharmacokinetic interactions between coadministered drugs, and thus allowing appropriate decisions to be taken to remedy clinical problems which may arise in these circumstances.

As already stressed, the value of the plasma anticonvulsant concentration should never be regarded as the final arbiter of the adequacy of the therapy of epilepsy – the clinical state of the patient should always remain the ultimate criterion for optimal management.

The future

Despite the continuing attempts of the past 30 years to put the drug treatment of epilepsy on an increasingly sound scientific basis, and to thereby improve the control of seizure disorders, apparently optimal therapy still fails to keep a significant number of patients free from seizures. Some of this failure appears due to the belated instigation of therapy, and some to non-compliance on the part of patients. In theory, these problems could be remedied by better education and better organization of medical services. However, in other instances therapeutic failure seems a consequence of the intrinsic characteristics of the patient's seizure disorder or the limitations of the currently available antiepileptic drugs. Greater knowledge of epileptogenesis and of the natural history of different types of seizure disorders might improve this situation, and surgical resection of epileptogenic areas of cerebrum may come to play an increasing role in improving the problem (see Mackenzie, this volume). The availability of new anticonvulsants may also help. Some drugs which modify brain gamma-aminobutyrate activity (e.g. vigabatrin) are already beginning to be marketed and the availability of others is imminent. Agents which modify neuro-transmission at the N-methyl-D-asparate and other types of excitatory aminoacid receptors are being studied experimentally and it seems not unreasonable to hope that advances in neurochemistry will ultimately lead to the availability of new and more efficient anticonvulsants.

The endeavours of the past three decades have undoubtedly improved the medical treatment of epilepsy and the quality of life of many of those who suffer from seizures. Paradoxically, this improvement in outcome has also produced a keener awareness of how much remains to be achieved.

References

Bertilsson, L. & Tomson, T. (1986). Clinical pharmacokinetics and pharmacological effects of carbamazepine and carbamazepine-10,11-epoxide. An update. *Clinical Pharmacokinetics,* **11**, 177–98.

Commission on Classification and Terminology of the International League Against Epilepsy (1981). Proposal for revised clinical and encephalographic classification of epileptic seizures. *Epilepsia,* **22**, 489–501.

Desai, B. T., Riley, T. L., Porter, R. J. & Penry, J. K. (1978). Active non-compliance as a cause of uncontrolled seizures. *Epilepsia,* **19**, 447–52.

Dickinson, R. G., Hooper, W. D., Wood, B., Lander, C. M. & Eadie, M. J. (1989). The effect of pregnancy in humans on the pharmacokinetics of stable isotope labelled phenytoin. *British Journal of Clinical Pharmacology,* **28**, 17–27.

Eisler, J. & Mattson, R. H. (1975). Compliance in anticonvulsant drug therapy. *Epilepsia,* **16**, 203.

Elwes, R. D. C., Chesterman, P. & Reynolds, E. H. (1985). Prognosis after a first untreated tonic-clonic seizure. *Lancet,* **ii**, 752–3.

Forster, F. M. (1955). The epilepsies and convulsive disorders. In *Clinical neurology*, ed. A. B. Baker, pp. 1036–74. London: Cassell & Co.

Green, J. R., Troupin, A. S., Halpern, L. M., Friel, P. & Kanarell, P. (1974). Sulthiame: evaluation as an anticonvulsant. *Epilepsia,* **15**, 329–49.

Lander, C.M. & Eadie, M.J. (1991). Plasma antiepileptic drug concentrations during pregnancy. *Epilepsia*, **32**, 257–66.

Merritt, H. H. & Putnam, T. J. (1938). Sodium diphenylhydantoinate in the treatment of convulsive disorders. *Journal of the American Medical Association*, **111**, 1068–73.

Nau, H. & Hendrickx, A. G. (1987). Valproic acid teratogenesis. ISI Atlas of Science: *Pharmacology*, **50**, 52–6.

Oller-Daurella, L. & Oller, L. (1987). Suppression of antiepileptic treatment. *European Neurology*, **27**, 106–13.

Scheffner, D., Konig, St., Rauterberg-Rutland, I., Kochen, W., Hoffman, W. J. & Unkelbach, St. (1988). Fatal liver failure in 16 children with valproate therapy. *Epilepsia*, **29**, 530–42.

Schmidt, D. (1982). Two antiepileptic drugs for intractable epilepsy with complex-partial seizures. *Journal of Neurology, Neurosurgery and Psychiatry*, **45**, 1119–24.

Schmidt, D., Einicke, I. & Haenez, F. (1986). The influence of seizure type on the efficacy of plasma concentrations of phenytoin, phenobarbital and carbamazepine. *Archives of Neurology*, **43**, 263–5.

Walshe, F. M. R. (1952). *Diseases of the nervous system*, 7th edn, pp. 119–35. Edinburgh: Churchill Livingstone.

Part VII

Clinical myology and conclusion

26

The changing face of neuroscience

LORD WALTON OF DETCHANT

13 Norham Gardens,
Oxford OX2 6PS
UK

Neuroscience: its changing face and its impact on clinical neurology

In a lecture on 'The changing face of neurology' delivered 20 years ago I endeavoured to show then how numerous developments in the treatment of infection, neuropharmacology, clinical neurophysiology, neuroradiology, biochemistry and histopathology had transformed the practice of neurological medicine. To show how the face of neurology had changed, I also projected photographs of the great Hughlings Jackson, one of the principal founders of the science of cerebral localization, and of my own teacher and mentor, the late Dr Henry Miller. But the discoveries and developments which I highlighted then have paled into insignificance when one takes note of the almost incredible developments in neuroscience of the last 20 years, of which I can mention very few. The conquest of many once devastating infective disorders such as smallpox is now largely a matter of history, but as new vaccines and antibiotics have been discovered, so nature has had a habit of posing new challenges. AIDS represents one such infective disorder which presents the medical profession with yet another difficult mountain to climb. Burgeoning developments in neuropharmacology have transformed our management of diseases like Parkinsonism and epilepsy. Perhaps, above all, new techniques of imaging, including CT and NMR scanning, the use of sophisticated ultrasonic methods including the Doppler method, and many more have revealed images of the nervous system and of the effects of pathological processes upon it undreamt of a few short years ago.

While many of these new techniques are expensive, both in capital costs and revenue consequences, the benefit to patients and the avoidance of the suffering associated with older methods such as pneumoencephalography and ventriculography have outweighed the cost to society if assessed in purely financial terms; one only wishes that governments and others responsible for financing health care could see it thus. Steroids have also proved strikingly beneficial in patients with many autoimmune disorders of the nervous and neuromuscular systems, while new and improved methods of intensive care have saved many individuals who would formerly have died from head injury or severe paralysing neurological illness. One must not forget, either, that scientific method has helped neurologists to develop new and improved methods of rehabilitating patients recovering from acute neurological disorders and of improving the lot, lifestyle and longevity of many of those suffering from progressive and as yet incurable conditions. The vast range of new appliances and of physical and communication aids developed by bioengineering expertise has also been striking. As I have always told medical students, in neurology there are many incurable diseases but none are untreatable (Walton, 1986).

There is yet another truism, in my view totally valid but not always capable of convincing the providers of research moneys. It is that today's discovery in basic laboratory science is tomorrow's practical development in patient care. In no field is this more evident than in molecular biology and molecular genetics. It would clearly be impossible to try to highlight even the major developments in these sciences which have influenced clinical practice. Therefore, I propose to discuss first some of the neurological diseases in which the causal gene has recently been located, secondly, to comment on new knowledge relating to Alzheimer's disease, and thirdly to refer briefly to new developments in multiple sclerosis and other neurological disorders, before finally turning to exciting developments in my own field of research interest, neuromuscular disease, with particular reference to the Duchenne and Becker varieties of muscular dystrophy.

Molecular genetics and neurological disease

Gene mapping

As mapping of the human genome (the HUGO project) proceeds apace through international collaborative effort, we learn almost every month that yet another gene has been located, and many have been fully characterized and sequenced. Table 26.1 presents a selective list of some important neurological disorders in which gene mapping has been achieved within the last few years. In relatively few of them do we yet know the nature of the product which is missing because of the genetic defect. Of course, it is possible that in some instances the abnormal gene causes the accumulation of a specific biochemical agent which has a positive, rather than negative, harmful effect. Nevertheless, these discoveries have inevitably moved towards identifying the ultimate cause of these disorders so that we may anticipate that in some, effective methods will soon be devised of circumventing the genetic defect, either through gene therapy or through

The changing face of neuroscience 433

Table 26.1. *Some gene locations in neurological disease*

Autosomal dominant	Chromosome
Von Hippel–Lindau disease	3p
Huntington's chorea	4pter–p16.2
Spinocerebellar atrophy	6p24–p3
Tuberous sclerosis	9q
Generalized familial dystonia	9q32–q34
Neurofibromatosis (peripheral)	17q11.2
Gerstmann–Straussler syndrome	20pter–p12
Familial Alzheimer's disease	21q21–q22.1
Neurofibromatosis (central)	22q11–q13.1
Autosomal recessive	
Gaucher's disease	1q21
Gangliosidosis GM1	3p21–cen
Gangliosidosis GM2	
Tay Sachs	15q23–q24
Sandhoff	5q13
Ataxia telangiectasia	11q22–q23
Wilson's disease	13q14.11
Metachromatic leucodystrophy	22q13.31–qter
X-linked recessive	
Pelizaeus–Merzbacher disease	Xq21.33–q22
Lesch–Nyhan disease	Xq26
Adrenoleucodystrophy	Xq28

some other means. The subject has been reviewed recently by Rosenberg and Pettegrew (1991) and by Brice and Mallet (1991).

Alzheimer's Disease

As knowledge accumulates, it becomes increasingly clear that Alzheimer's disease, whether occurring in the presenium or in the senium, is not a single entity. Furthermore, despite the increasing sophistication of diagnostic techniques, including CT and NMR imaging and PET scanning, along with refined psychometric tests, clinical diagnosis is still imprecise and, according to various estimates, is probably correct in no more than 80% of cases of dementia. This problem may well have prejudiced the conclusions drawn from some epidemiological studies based upon clinical diagnosis. Nevertheless, knowledge of pathogenesis is accumulating apace. It is agreed that the histological hallmarks of the disease are neurofibrillary tangles and senile plaques in the cerebral cortex. Tomlinson, Blessed and Roth (1968, 1970) showed that similar changes were found in the brains of nondemented elderly people and also demonstrated a quantitative relationship between the severity of dementia and the incidence and ubiquity of these changes. Later, Bowen, Smith, White and Davison (1976) and Perry, Perry, Blessed and Tomlinson (1978) showed that acetylcholine, as well as choline acetyltransferase activity, were markedly reduced in the cerebral cortex. More recently it has been shown that paired helical filaments in the plaques are associated with hyperphosphorylated tau protein (Deary & Whalley, 1988) and that amyloid precursor protein (APP) is also present; a mutation in this may be a major factor in precipitating cytoskeletal breakdown. The fact that the β-amyloid gene is situated on chromosome 21 is of considerable interest since in patients with Down's syndrome the cerebral changes of Alzheimer's disease commonly develop in the late twenties and thirties. More recently, much interest was aroused by the finding of a specific point mutation within the β-amyloid gene on chromosome 21 in some patients with dominantly inherited Alzheimer's disease (Goate *et al.*, 1991; Hardy, 1991) but point mutations in both exons 7 and 17 within that gene have now been identified in different families (Hardy, 1991; Higgins, 1991). The finding of aluminium silicate within senile plaques (Candy *et al.*, 1986) raised the question as to whether ingested aluminium might be a pathogenetic factor.

A recent conference concluded that aluminium cannot be regarded as the cause, but it was accepted that it might play a part in the pathological cascade of changes following the initial process leading to amyloid deposition (Walton, 1991*a*). Nevertheless, similar deposits may be found in degenerating cells in the brains of patients with the amyotrophic lateral sclerosis/Parkinsonism-dementia complex seen on Guam and also in some patients with the punch-drunk syndrome (traumatic encephalopathy) (Perl & Good, 1991). Just as calcium is often laid down in degenerating or necrotic tissue, it is possible that the deposition of aluminium silicate in degenerating neurones is an epiphenomenon (Yates & Mann, 1986). While Wisniewski (1991) suggested that Alzheimer's disease may be a specific form of cerebral amyloidosis, which may or may not be genetically determined, other participants adduced indirect and circumstantial evidence supporting the view that aluminium ingestion might contribute to the pathological process. In experimental studies Petersen (1991) had shown that aluminium injected by micropipette in minute quantities into pancreatic cells reduced their calcium signal, but this reduction could be inhibited by the addition of silicic acid. Birchall (1991) pointed out that silicon in drinking water or in foods could potentially inhibit the toxic effect of aluminium (Birchall & Chappell, 1988). Epidemiological studies have suggested that there was a higher incidence of Alzheimer's disease in populations exposed to higher concentrations of aluminium in drinking water (Martyn *et* *al.*, 1989). Costa (1991) referred to the work of Rifat in Ontario, Canada who had studied goldminers exposed to aluminium-containing powder in the atmosphere; those at risk of inhalation showed a higher incidence of cognitive impairment than did a control group. McLachlan (1991) reported that in a trial of treatment in Alzheimer's disease using desferrioxamine, a chelating agent presumed to be capable of extracting aluminium from the brain, video-recorded evidence relating to the tasks of daily living and studies of cognitive performance suggested that the treatment had delayed disease progression when compared with a control group (McLachlan *et al.*, 1991). Farrar *et al.* (1991) raised the interesting possibility that Alzheimer's disease might be due to a deficiency of transferrin binding; as transferrin binds not only iron but also aluminium, if such binding were defective, this might release greater quantities of aluminium for deposition in the tissues, including the brain (Farrar *et al.*, 1990). The mechanism might be comparable to that in Wilson's disease in which a deficiency of caeruloplasmin (the copper-binding protein) leads to the abnormal deposition of copper.

While circumstantial evidence has therefore suggested that aluminium might be a pathogenetic factor, others have pointed out that there is no evidence of a higher incidence of Alzheimer's disease in regular users of antacid preparations containing the metal (Colin-Jones *et al.*, 1989), that tea also contains high concentrations, some of which (contrary to views previously expressed) is bioavailable and is absorbed (Koch *et al.*, 1988) and that there is no evidence, either, to indicate any relationship between Alzheimer's disease and exposure to high concentrations of aluminium in cosmetics and aerosols. Admittedly, the appropriate epidemiological studies have not been done. It is also clear that aluminium encephalopathy (Hughes, 1989) and dialysis dementia (Ward, 1991) are different clinically and pathologically from Alzheimer's Disease. Nevertheless, the conference agreed that it is a wise precaution to restrict the amount of aluminium in drinking water to no more than the current accepted EEC/WHO maximum concentration in soya-based infant feeds (Bishop, McGraw & Ward, 1989), to restrict aluminium intake in patients with Alzheimer's disease and probably also in the elderly, but not as yet to make any firm recommendations about the use of aluminium cooking utensils. Some participants suggested that it may be unwise to cook rhubarb in aluminium pans as this may extract some of the aluminium from the utensil. Others had already given up drinking tap water (in which the aluminium is often derived from aluminium sulphate added to the water for leaching purposes) in favour of drinking bottled water, which contains little or none, and had also discontinued the use of aluminium cooking utensils. That, however, was not the majority view.

Finally, what of treatment? Numerous trials of treatment of Alzheimer's disease have been mounted using anticholinesterase preparations such as physostigmine and acetylcholine precursors including choline and lecithin. Most recently tacrine (tetrahydroaminoacridine) has been used by Eagger, Levy and Sahakian (1991) and seemed to produce an improvement in key outcome measures roughly equivalent to the deterioration which might have occurred over 6–12 months. However, pramiracetam, shown to have a cognitive-enhancing effect in animal models of learning and memory, has proved ineffective (Claus *et al.*, 1991). Loss of high affinity agonist binding to M1 muscarinic receptors may account for the failure of cholinergic replacement therapy (Flynn, Weinstein & Mash, 1991). Further rigorous controlled trials of other similar treatments and of chelating agents such as desferrioxamine are needed. Perhaps at last in this tragic disease we are beginning to see some faint light at the end of the tunnel.

Some other neurological diseases

How do we stand in relation to some of the commoner crippling neurological disorders and to what extent have developments in molecular genetics and in neuroimmunology extended our knowledge? There is now general agreement (Matthews *et al.*, 1991) that, in multiple sclerosis, there is an underlying genetic susceptibility (which may be related to the individual's HLA antigen status). Against such a background an environmental factor, which may be one of several different viruses, triggers an autoimmune process. Myelin-reactive T cells then act upon the myelin in the central nervous system but also increase the permeability of the blood–brain barrier, leading to perivascular inflammatory cell infiltration which in turn increases the extent of the demyelination. Nevertheless, treatment of the disease with steroids, however administered, and with immunosuppressive agents such as azathioprine, has usually given disappointing results. However, at a recent symposium a number of new approaches were discussed (*Lancet*, 1991), including techniques of blocking T cell function as, for example, by the use of chimeric anti-CD4, which inhibits T cell activation and induces tolerance to antigens. The possibility of vaccination with autologous attenuated T cell clones was also raised; clearly, in multiple sclerosis, hopes of more effective treatment soon being introduced are being justifiably raised. Other aspects of the epidemiology and pathogenesis of multiple sclerosis are discussed by McLeod (this volume).

In Parkinson's disease treatment has been transformed within the last 20 years by levodopa and its analogues and subsequently by various dopamine agonists, but as yet we know little about the primary pathological process which causes degeneration of cells in the substantia nigra. Nevertheless, the introduction of ubiquitin staining, indicating abnormal protein degradation in nerve cells, has helped us to understand better the pathological substrate of Lewy body dementia which is commonly associated with that disease but which may also cause dementia in some individuals who do not manifest the clinical features of Parkinsonism. Ubiquitin staining has also demonstrated immunoreactive inclusions in anterior horn cells in patients with motor neurone disease (amyotrophic lateral sclerosis) (Leigh *et al.*, 1991), a finding which may shed light upon the pathogenesis of this mysterious disorder. Many years ago we showed that the number of anterior horn cells in the spinal cord diminishes progressively with increasing age and this process is associated with evidence of progressive denervation, especially in some lower limb muscles (Tomlinson, Walton & Rebeiz, 1969; see also Jennekens, 1982). This led us to suggest that motor neurone disease might represent an acceleration of normal ageing, thus raising a possible analogy with Alzheimer's disease. But clearly we are far from being able to identify the primary cause of this tragic condition. As yet neither neuroimmunology nor molecular genetics has shed light upon its etiology.

On the other hand, recent work has shown that a transmissible disorder due to what was once called a 'slow virus infection' may nevertheless be genetically determined. In all the spongiform encephalopathies such as scrapie, bovine spongiform encephalopathy, Kuru and Creutzfeldt–Jakob disease, infectivity is associated with an isoform of the host-encoded prion protein (PrP) (Kretzschmar *et al.*, 1991). Such a mutation has now been identified in the Gerstmann–Strässler–Scheinker syndrome and has been located at codon 102 (pProline-leucine) (Kretzschmar *et al.*, 1991). Until recently it had not been thought that a genetic factor was operative in Creutzfeldt–Jakob disease. The high incidence of this disorder in Libyan Jews had been tentatively attributed to their fondness for eating sheep's brains and eyeballs. However, a codon 200 lysine mutation of the prion protein gene has been described in Libyan Jews with this disease, confirming the importance of a genetic factor in pathogenesis (Hsiao *et al.*, 1991). In these cases there is accumulation in the brain of an amyloid protein (not always in plaques), unlike the non-infective variety seen in Alzheimer's disease, Down's syndrome, the Guamanian ALS-Parkinsonism–dementia complex and congophilic angiopathy (Brown, Goldfarb & Gajdusek, 1991).

Neuromuscular disorders

The remarkable increase in knowledge of the genetic basis of many neuromuscular disorders is exemplified by the number of these conditions in which the

Table 26.2. *Neuromuscular disorders: gene location**

Disease	Mode of inheritance	Gene location
Muscular dystrophies		
Duchenne/Becker (DMD/BMD)	XR	Xp21
Emery–Dreifuss	XR	Xq28
Facioscapulohumeral (FSH)	AD	4q35–qter
Limb girdle (LG)	AR	15
Myotonic syndromes		
Myotonic dystrophy (Steinert)	AD	19q13.2–q13.3
Hereditary myasthenia	AR;AD	
Acetylcholine receptors		
x subunit		2q24–q32
b subunit		17p11–p12
g subunit		2q33–qter
Congenital myopathies		
Myotubular (centronuclear)	XR	Xq28
Central core disease	AD	19q12–q13.2
Metabolic myopathies		
Glycogenoses		
Type II (Pompe's disease; acid maltase deficiency)	AR	17q
Type V (McArdle's disease; phosphorylase deficiency)	AR	11q
Type VII (Tarui's disease; phosphofructokinase deficiency)	AR	1q
Hyperkalemic periodic paralysis	AD	17q
Malignant hyperthermia	AD	19q13.1–q13.3
Neurogenic syndromes		
Spinal muscular atrophy (SMA) Werdnig–Hoffmann Kugelberg–Welander	AR	5q11.2–q13.3
Spinal muscular atrophy (Kennedy)	XR	Xq21.3–q22
Hereditary motor and sensory neuropathy (HMSN) (Charcot–Marie–Tooth; peroneal muscular atrophy)		
HMSN Ib	AD	1q21.2–q23
HMSN Ia	AD	17p11.2–q23
HMSN	XD	Xq13
Familial amyloid neuropathy	AD	18q11.2–q12.1
Friedreich's ataxia	AR	9cen–q21

Source: *Reproduced with permission from *Neuromuscular Disorders*, Vol. 1, No. 1, edited by Prof. V. Dubowitz, Permagon Press, UK.
Notes: XR = X-linked recessive; AD = autosomal dominant; AR = autosomal recessive; XD = X-linked dominant.

location of the causal gene has been identified within the last few years. Table 26.2 lists those disorders in which such knowledge was available in early 1991; it is more than likely that since the Table was prepared other genes have been located.

Duchenne and Becker muscular dystrophy

Much of my early work into the clinical and genetic aspects of the muscular dystrophies was concerned with delineating accurately the clinical and genetic varieties of muscular dystrophy and in distinguishing these by electrophysiological, biochemical and pathological techniques from the many other forms of myopathy which were progressively identified, including not least inflammatory myopathies like polymyositis and dermatomyositis. But we then had no clue as to the nature and location of the gene or genes responsible for Duchenne and Becker dystrophy, even though we knew that both were X-linked. At that time, too, we had no accurate method of detecting female carriers who were likely to pass the disease on to their sons, though we were using an indirect technique based upon estimation of serum creatine kinase activity which gave us, at best, an approximation of the risk.

A major breakthrough came in 1975 when Cullen and Fulthorpe, using electron microscopy, studied the earliest structural changes occurring in the muscle fibres of patients with preclinical Duchenne dystrophy. With standard histological methods and light microscopy we had previously demonstrated the waxy hyaline fibres seen prominently in transverse sections which many previous workers, including my teacher, Raymond Adams, had concluded were due to fixation artefact. However, Cullen and Fulthorpe examined first, under phase-contrast illumination, unfixed muscle fibres and found multiple contraction bands which were clearly responsible for those hyaline changes. In plastic-embedded fibres stained with toluidine blue, such areas of hypercontraction might be extremely focal, involving only a small segment of a transverse section of a fibre, usually just beneath the sarcolemma; they also showed later with specific stains a marked excess of calcium within these areas. They concluded that in Duchenne dystrophy there might be a defect in the plasma membrane allowing the ingress of calcium from the extracellular space and that this in turn might activate calcium-activated neutral proteases which were then, at least in part, responsible for the breakdown and digestion of the muscle fibres. Pennington (1988) confirmed that such calcium-activated proteases were indeed present in excess. Almost simultaneously, Mokri and Engel (1975) also using electron microscopy, demonstrated focal defects in the plasma membrane which they called 'the delta lesion'; hence it seemed that the mechanism of fibre breakdown in muscular dystrophy was becoming more clearly understood.

But the greatest impetus to research has resulted from work in molecular biology. In 1987, following upon work in many laboratories throughout the world, including some in Holland, Toronto and Oxford but ultimately through the energy, industry and expertise of Drs Kunkel, Hoffman and their colleagues at the Boston Children's Hospital, the gene responsible for Duchenne muscular dystrophy was finally characterised in the Xp 21 region of the female X chromosome. It has proved to be one of the largest genes known in human genetics with a length of about 2 megabases (Monaco et al., 1986; Hoffman, Brown & Kunkel, 1987). Furthermore, the genes for Duchenne and Becker dystrophy have been shown to be allelic and their clinical expression may depend, at least in part, upon the number and extent of the deletions present within the gene. Nevertheless, this work has provided us with a more accurate means of identifying the female carriers of these X-linked genes in most affected families.

Identification of the gene led to isolation of the missing gene product, a protein now called dystrophin (Hoffman et al., 1987) which is totally absent in almost all cases of Duchenne dystrophy and much reduced in those of the Becker variety. Further work has demonstrated that dystrophin forms a vital structural component of the plasma membrane of the skeletal muscle fibre (Arahata et al., 1989); hence it appears that the earlier observations of Cullen, Fulthorpe, Mokri and Engel upon the pathogenesis of muscle fibre breakdown have been largely vindicated. Presumably, it is the absence of dystrophin which renders the plasma membrane incompetent and which leads on to the breakdown process which I have outlined (Nicholson et al., 1990).

Future research will clearly be directed towards methods of attempting to repair the defect in the muscle fibre membrane or of identifying affected children early in life in the hope of offering them a form of treatment (perhaps, ultimately, gene replacement) which will prevent the membrane from becoming defective. Of great importance in relation to treatment prospects is the fact that a naturally-occurring muscular dystrophy of X-linked inheritance has been clearly identified in the mdx mouse (Bulfield et al., 1984) and yet another has been detected in a strain of golden retrievers and subsequently in rottweilers (Valentine et al.,1990). Current evidence strongly suggests comparability of these afflictions to the human disease. In both species the Duchenne gene transcript is lacking (Cooper et al., 1988) and dystrophin is absent from the muscle cells of affected animals. Cardiomyopathy like that of Duchenne dystrophy occurs in the dogs (Valentine, Cummings & Cooper, 1989), and in carriers of the X-linked canine disorder mosaic expression of dystrophin comparable to that found in the muscle of human female carriers has been noted (Cooper *et al.*, 1990). Plainly experiments in gene replacement will not only be feasible eventually in such animals but will surely become a reality once a method of reintroducing the defective gene or the missing gene product into the skeletal muscle has been devised.

In the meantime, through work being done in Great Britain by Dr Partridge and his colleagues at Charing Cross Medical School, Dr Law in the United States, Dr Karpati in Canada and others, methods have been devised of harvesting large numbers of myoblasts grown in culture from muscle satellite cells derived from normal animals and of introducing these by injection into the affected muscles of dystrophic animals (Morgan, Hoffman & Partridge, 1990; Partridge, 1991). Current evidence suggests that problems of rejection are being overcome and that such myoblasts introduced into diseased muscle can produce dystrophin demonstrable by immunohistochemical techniques. Monoclonal antibodies are available, for example, in the United States from Dr Hoffman's laboratory and in the UK from Professor Harris's muscular dystrophy laboratories in Newcastle. Experiments have been carried out using myoblast transfer in human subjects by Drs Law and Karpati, and it has been shown in some boys with Duchenne muscular dystrophy that such myoblasts can restore dystrophin to some muscle fibres of the extensor digitorum brevis (Law *et al.*, 1990).

Whether myoblast transfer therapy will ever become of practical benefit in human patients is still very open (Walton, 1991*b*). The problems to be overcome are first those of producing a sufficient number of myoblasts of human origin for use in this work; secondly, the ever-present problem of rejection; and thirdly, above all, we must recognise the enormous difficulty inevitably encountered in trying to introduce a sufficient number of myoblasts into many muscles to make the effects of the treatment clinically useful.

A new dimension has been added through work recently reported by Acsadi and colleagues (1991). They quoted previous studies which had shown that, after pure plasma DNA was injected into rodent skeletal and cardiac muscle, the cells expressed reporter genes. They have now been able to demonstrate that a 12-kb full-length human dystrophin cDNA gene and a 6.3-kb Becker-like gene could be expressed in cultured cells and *in vivo*. When the human dystrophin expression plasmids were injected intramuscularly into dystrophin-deficient mdx mice, the human dystrophin proteins were present in the cytoplasm and sarcolemma of myofibres. Those myofibres expressing human dystorphin contained an increased percentage of peripheral nuclei. Whether this treatment will become effective in human subjects is difficult to answer, but nevertheless, these findings give considerable hope for the future.

The final point I wish to raise in relation to Duchenne dystrophy relates to embryo research. As mentioned above, identification of the gene and of its missing protein product has now led to 95–98% accuracy, through the application of DNA studies, in carrier detection in informative families. Such female carriers in the past could only be advised that if they fell pregnant, they should undergo amniocentesis at the fourteenth week and should then have an abortion if the fetus was shown to be male. Now, with the aid of chorionic cell biopsy at eight or nine weeks, it is possible not only to sex the unborn fetus but, using current molecular biological techniques, to demonstrate whether or not, if the fetus is male, the abnormal gene is present. Hence selective abortion of only affected males is now feasible. Even though chorionic cell biopsy does carry some limited clinical risks (MRC Working Party on the Evaluation of Chorion Villus Sampling, 1991), most carriers regard these as wholly acceptable. Of even greater importance is that *in vitro* fertilization and so-called preembryo biopsy is now making preimplantation diagnosis possible. Were it not for the extensive research on animal and human embryos carried out during the last 20 and more years, *in vitro* fertilization in human subjects would never have become possible. Indeed, if the Human Fertilisation and Embryology Act, which received the royal assent after its final passage through both Houses of Parliament in the UK in 1990, had not become law, allowing research on the human embryo up to 14 days after fertilization, all research of vital importance to the infertile and to carriers of the gene responsible for Duchenne dystrophy and those causing many other crippling inherited disorders would have been prevented by law.

When the female ovum, released into the uterus at the time of ovulation, is fertilised by a sperm, the process of cell division begins and within the first few days floating free in the uterus are groups of undifferentiated but pluripotential cells, each forming what I call a conceptus or a pre-embryo rather than an embryo. The term 'pluripotential' means that it is impossible at first to identify which cells will form the membranes within which the fetus will eventually lie and which will later form an identifiable embryo from which a fetus will form. By about the fourth or fifth day, the conceptus becomes a blastocyst (McLaren, 1987) in which there is a nodule or cluster of cells called the inner cell mass from which the embryo later derives, and also an outer ring of cells capable of forming the membranes and the placenta. But no such blastocyst is yet attached to, or embedded in the wall of the uterus and about 80% of those formed are spontaneously aborted. About one in five begins to attach to the uterine wall at about the seventh day, subsequently receiving a blood supply and nourishment from the maternal circulation, and later, at about the fourteenth day, that specific linear arrangement of cells within the basal cell mass which constitutes the primitive streak appears.

Work done in the last few years by Professor Winston at Hammersmith and by others has clearly demonstrated that it is now feasible, without damage to the subsequent development of the embryo, to carry out biopsy by removal of a single cell from the blastocyst at about the fourth or fifth day; that single cell can now be removed from the part of the blastocyst which will ultimately form the membranes and the placenta. Sexing of the conceptus is now commonly performed and I understand that Professor Winston and his colleagues have not only found it possible to identify in such single cells the sex of the conceptus but are also close to being able to determine whether or not the dystrophic gene is present (Winston, 1990). This will make it feasible for such carrier women to have normal sons and noncarrier daughters, a prospect undreamed of a few short years ago.

I fully understand the sincerity of those who believe that human life begins at conception and that any experiment on what they regard as a human life, or what can become one, is to them abhorrent. I am, however, personally satisfied, as are many eminent theologians including the Archbishop of York, Lord Soper, the Rev. Prof. Gordon Dunstan and the Rev. Dr Norman Ford, a distinguished Australian Roman Catholic theologian (Walton, 1990) that individuation of the human embryo does not begin until the primitive streak appears at the fourteenth day. Therefore Parliament was undoubtedly right to accept the Human Fertilization and Embryology Bill (now an Act). The work it is making possible will bring inestimable benefits to human health, not least to the families of patients with Duchenne dystrophy and to those in which many other serious crippling inherited diseases are present.

Conclusions

In a recent notable paper entitled 'The physician scientist: an endangered but far from extinct species' Sir David Weatherall (1991) said that he doubted whether future physician scientists could, or should, be full-time laboratory workers. Rather they should have been exposed to periods of rigorous training in a basic science or clinical research laboratory in order to provide in future the environment in which others, led and advised by them, could pursue good science. He concluded that the physician scientist is not an extinct species but one which will have to continue to diversify in order to survive. The genus must encompass full-time bench workers, hybrids of clinician and scientist, clinicians who critically evaluate the delivery of health care, and those who can evaluate undergraduate and postgraduate education to ensure that it is not neglected in the effort to achieve excellence in clinical practice and research.

In conclusion, hopefully at least some of the information above will have convinced you that burgeoning developments in neuroscience, perhaps above all in molecular genetics, have already had a substantial impact upon practice in clinical neurology. Without doubt, within the next 10 years further developments in this rapidly evolving field will bring new hope to many of our patients. At last they will begin to see a prospect of introducing effective treatment for some of those progressive and crippling neurological disorders which are at present incurable.

References

Acsadi, G., Dickson, G., Love, D. R., Jani, A., Walsh, F. S., Gurusinghe, A., Wolff, J. A. & Davies, K. E. (1991). Human dystrophin expression in mdx mice after intramuscular injection of DNA constructs. *Nature,* **352**, 815–8.

Arahata, K., Hoffman, E. P., Kunkel, L. M., Ishiura, S., Tsukahara, T., Ishihara, T., Sunohara, N., Nonaka, I., Ozawa, E. & Sugita, H. (1989). Dystrophin diagnosis: comparison of dystrophin abnormalities by immunofluorescence and immunoblot analysis. *Proceedings of the National Academy of Sciences, USA,* **86**, 7154–8.

Birchall, J. D. (1991). The role of silicon in aluminium toxicity. In *Alzheimer's Disease and the Environment,* ed. Lord Walton, London: Royal Society of Medicine Services, in press.

Birchall, J. D. & Chappell, J. S. (1988). Aluminium, chemical physiology, and Alzheimer's disease. *Lancet,* **ii**, 1008–110.

Bishop, N., McGraw, M. & Ward, N. (1989). Aluminium in infant formulas. *Lancet,* **i**, 490.

Bowen, D. M., Smith, C. B., White, P. & Davison, A. N. (1976). Neurotransmitter-related enzymes and indices of hypoxia in senile dementia and other abiotrophies. *Brain,* **99**, 459–96.

Brice, A. & Mallet, J. (1991). La génétique moléculaire: une nouvelle approche des neurosciences cliniques. *Revue Neurologique,* **147**, 1–16.

Brown, P., Goldfarb, L. G. & Gajdusek, D. C. (1991). The new biology of spongiform encephalopathy; infectious amyloidoses with a genetic twist. *Lancet,* **337**, 1019–22.

Bulfield, G., Siller, W. G., Wight, P. A. L. & Moore, K. J. (1984). X-chromosome-linked muscular dystrophy (mdx) in the mouse. *Proceedings of the National Academy of Sciences, USA,* **81**, 1189–92.

Candy, J. M., Oakley, A. E., Klinowski, J., Carpenter, T. A., Perry, R. H., Atack, J. R., Perry, E. K., Blessed, G., Fairbairn, A. & Edwardson, J. A. (1986). Alumino-silicates and senile plaque formation in Alzheimer's disease. *Lancet,* **i**, 354–7.

Claus, J. J., Ludwig, C., Mohr, E., Giuffra, M., Blin, J. & Chase, T. N. (1991). Nootropic drugs in Alzheimer's disease: symptomatic treatment with pramiracetam. *Neurology,* **41**, 570–4.

Colin-Jones, D., Langman, M. J. S., Lawson, D. H. & Vessey, M. P. (1989). Alzheimer's disease in antacid users. *Lancet,* **i**, 1453.

Cooper, B. J., Gallagher, E. A., Smith, C. A., Valentine, B. A. & Winand, N. J. (1990). Mosaic expression of dystrophin in carriers of canine X-linked muscular dystrophy. *Laboratory Investigations,* **62**, 171–8.

Cooper, B. J., Winand, N. J., Stedman, H., Valentine, B. A., Hoffman, E. P., Kunkel, L. M., Scott, M. O., Fischbeck, K. H., Kornegay, J. N., Avery, R. J., Williams, J. R., Schmicketl, R. D. & Sylvester, J. E. (1988). The homologue of the Duchenne locus is defective in X-linked muscular dystrophy of dogs. *Nature,* **334**, 154–6.

Costa, P. (1991). The role of aluminium in cognitive functioning. In *Alzheimer's Disease and the Environment*, ed. Lord Walton, pp. 116–24. London: Royal Society of Medicine Services.

Cullen, M. J. & Fulthorpe, J. J. (1975). Stages in fibre breakdown in Duchenne muscular dystrophy. *Journal of the Neurological Sciences*, **24**, 179–200.

Deary, I. J. & Whalley, L. J. (1988). Recent research on the causes of Alzheimer's disease: what causes neuronal death, and why the specific patterns? *British Medical Journal*, **297**, 807–10.

Eagger, S. A., Levy, R. & Sahakian, B. J. (1991). Tacrine in Alzheimer's disease. *Lancet*, **337**, 989–92.

Farrar, G., Altmann, P., Welch, S., Wychrij, O., Ghose, B., Lejeune, J., Corbett, J., Prasher, V. & Blair, J. A. (1990). Defective gallium-transferrin binding in Alzheimer disease and Down syndrome: possible mechanism for accumulation of aluminium in brain. *Lancet*, **335**, 747–50.

Farrar, G., Hodgkins, P., Altmann, P. & Blair, J. A. (1991). A biochemical mechanism for Alzheimer's disease. In *Alzheimer's Disease and the Environment*, ed. Lord Walton, pp. 53–7. London: Royal Society of Medicine Services.

Flynn, D. D., Weinstein, D. A. & Mash, D. C. (1991). Loss of high-affinity agonist binding to M1 muscarinic receptors in Alzheimer's disease: implications for the failure of cholinergic replacement therapies. *Annals of Neurology*, **29**, 256–62.

Goate, A., Chartier-Harlin, M.-C., Mullan, M., Brown, J., Crawford, F., Fidani L., Giuffra, L., Haynes, A., Irving, N., James, L., Mant, R., Newton, P., Rooke, K., Roques, P., Talbot, C., Pericak-Vance, M., Roses, A., Williamson, R., Rossor, M., Owen, M. & Hardy, J. (1991). Segregation of a missense mutation in the amyloid precursor protein gene with familial Alzheimer's disease. *Nature*, **349**, 704–6.

Hardy, J. (1991). The genetics of Alzheimer's disease. In *Alzheimer's Disease and the Environment*, ed. Lord Walton, pp. 9–11. London: Royal Society of Medicine Services.

Higgins, G. A. (1991). The regulation of the amyloid gene in Alzheimer's disease: possible environmental influences. In *Alzheimer's Disease and the Environment*, ed. Lord Walton, pp. 17–20 London: Royal Society of Medicine Services.

Hoffman, E. P., Brown, R. M. & Kunkel, L. M. (1987). Dystrophin: the protein product of the Duchenne muscular dystrophy gene. *Cell*, **51**, 919–28.

Hsiao, K., Meiner, Z., Kahana, E., Cass, C., Kahana, I., Avrahami, D., Scarlato, G., Abramsky, O., Prusiner, S. B. & Gabizon, R. (1991). Mutation of the prion protein in Libyan Jews with Creutzfeldt–Jakob disease. *New England Journal of Medicine*, **324**, 1091–7.

Hughes, J. T. (1989). Aluminium and the human brain. *The Practitioner*, 233, 920–23.

Jennekens, F. G. I. (1982). Disuse, cachexia and ageing. *Skeletal Muscle Pathology*, ed. F. L. Mastaglia & Sir John Walton), Chapter 22, Edinburgh: Churchill Livingstone.

Koch, K. R., Pougnet, M. A. B., de Villiers, S. & Monteagudo, F. (1988). Increased urinary excretion of Al after drinking tea. *Nature*, **333**, 122.

Kretzschmar, H. A., Honold, G., Seitelberger, F., Feucht, M., Wessely, P., Mehraein, P. & Budka, H. (1991). Prion protein mutation in family first reported by Gerstmann, Sträussler, and Scheinker. *Lancet*, **337**, 1160.

Lancet (1991). Where to hit MS. *Lancet*, **337**, 765–67.

Law, P. K., Bertorini, T. E., Goodwin, T. G., Chen, M., Fang, Q., Li, H.-J., Kirby, D. S., Florendo, J. A., Herrod, H. G. & Golden, G. S. (1990). Dystrophin production induced by myoblast transfer therapy in Duchenne muscular dystrophy. *Lancet*, **336**, 114–15.

Leigh, P. N., Whitwell, H., Garofalo, O., Buller, J., Swash, M., Martin, J. E., Gallo, J.-M., Weller, R. O. & Anderton, B. H. (1991). Ubiquitin-immunoreactive intraneuronal inclusions in amyotrophic lateral sclerosis. *Brain*, **114**, 775–88.

McLachlan, D. R. C. (1991). The possible relationship between aluminium and Alzheimer's disease and mechanisms of cellular pathology. In *Alzheimer's Disease and the Environment*, ed. Lord Walton, pp. 42–50. London: Royal Society of Medicine Services.

McLachlan, D. R. C., Dalton, A. J., Kruck, T. P. A., Bell, M. Y., Smith, W. L., Kalow, W. & Andrews, D. F. (1991). Intramuscular desferrioxamine in patients with Alzheimer's disease. *Lancet*, **337**, 1304–8.

McLaren, A. (1987). Can we diagnose genetic disease in pre-embryos? *New Scientist*, 10 December.

Martyn, C. N. (1991). Aluminium in water and its possible relationship to Alzheimer's disease. In *Alzheimer's Disease and the Environment*, ed. Lord Walton, pp. 87–91. London: Royal Society of Medicine Services.

Martyn, C. N., Barker, D. J. P., Osmond, C., Harris, E. C., Edwardson, J. A. & Lacey, R. F. (1989). Geographical relation between Alzheimer's disease and aluminium in drinking water. *Lancet*, **i**, 59–62.

Matthews, W. B., Compston, A., Allen, I. V. & Martyn, C. N. eds. (1991). *McAlpine's Multiple Sclerosis*, 2nd edn, Edinburgh: Churchill-Livingstone.

Mokri, B. & Engel, A. G. (1975). Duchenne dystrophy: electron microscopic findings pointing to a basic or early abnormality in the plasma membrane of the muscle fibre. *Neurology*, **25**, 1111–20.

Monaco, A. P., Neve, R. L., Colletti-Feener, C. Bertelson, C. J., Kurnit, D. M. & Kunkel, L. M. (1986). Isolation of candidate cDNAs for portion of the Duchenne muscular dystrophy gene. *Nature*, **323**, 646–50.

Morgan, J. E., Hoffman, E. P. & Partridge, T. A. (1990). Normal myogenic cells from newborn mice restore normal histology to degenerating muscles of the mdx mouse. *Journal of Cell Biology*, **111**, 2437–49.

MRC Working Party on the Evaluation of Chorion Villus Sampling (1991). Medical Research Council European trial of chorion villus sampling. *Lancet*, **337**, 1491–9.

Nicholson, L. V. B., Johnson, M. A., Gardner-Medwin, D., Bhattacharya, S. & Harris, J. B. (1990). Heterogeneity of dystrophin expression in patients with Duchenne and Becker muscular dystrophies. *Acta Neuropathologica*, **80**, 239–50.

Partridge, T. A. (1991). Myoblast transfer: a possible therapy for inherited myopathies? *Muscle and Nerve*, **14**, 197–212.

Pennington, R. J. T. (1988). Biochemical aspects of muscle disease. *Disorders of Voluntary Muscle*, 5th edn, ed. Sir John Walton, Chapter 13, Edinburgh: Churchill Livingstone.

Perl, D. P. & Good, P.F. (1991). The relationship between aluminium and neurofibrillary tangle formation. In *Alzheimer's Disease and the Environment*, ed. Lord Walton, pp. 60–6 London: Royal Society of Medicine Services.

Perry, E. K., Perry, R. H., Blessed, G. & Tomlinson, B. E. (1978). Changes in brain cholinesterases in senile dementia of Alzheimer's type. *Neuropathology and Applied Neurobiology*, **4**, 273–7.

Petersen, O. H. (1991). The effects of aluminium on cellular calcium homeostasis in pancreatic acinar cells. In *Alzheimer's Disease and the Environment*, ed. Lord Walton, pp. 95–102. London: Royal Society of Medicine Services.

Rosenberg, R. N. & Pettegrew, J. W. (1991). Genetic neurological diseases. *Comprehensive Neurology*, ed. R. N. Rosenberg, Chapter 2, New York: Raven Press.

Tomlinson, B. E., Blessed, G. & Roth, M. (1968). Observations on the brains of non-demented old people. *Journal of the Neurological Sciences*, **7**, 331–56.

Tomlinson, B. E., Blessed, G. & Roth, M. (1970). Observations on the brains of demented old people. *Journal of the Neurological Sciences*, **11**, 205–42.

Tomlinson, B. E., Walton, J. N. & Rebeiz, J. J. (1969). The effects of ageing and of cachexia upon skeletal muscle: a histopathological study. *Journal of the Neurological Sciences*, **9**, 321–46.

Valentine, B. A., Cooper, B. J., Cummings, J. F. & de Lahunta, A. (1990). Canine X-linked muscular dystrophy: morphologic lesions. *Journal of the Neurological Sciences*, **97**, 1–23.

Valentine, B. A., Cummings, J. F. & Cooper, B. J. (1989). development of Duchenne-type cardiomyopathy: morphologic studies in a canine model. *American Journal of Pathology*, **135**, 671–8.

Walton, J. (1986). The science of clinical neurology. *Journal of the Royal Society of Medicine*, **79**, 5–14.

Walton, Lord (1990). Embryo research - why the Cardinal is wrong. *Journal of Medical Ethics*, **16**, 185–86.

Walton, Lord (ed.) (1991a). *Alzheimer's Disease and the Environment*, ed. Lord Walton, London: Royal Society of Medicine Services.

Walton, Lord (1991b). Science and clinical myology. *Conquest*, **180**, 1–10.

Ward, M. K. (1991). Aluminium toxicity in people with impaired renal function. In *Alzheimer's Disease and the Environment*, ed. Lord Walton, pp. 106–116. London: Royal Society of Medicine Services.

Weatherall, D. J. (1991). The physician scientist: an endangered but far from extinct species. *British Medical Journal*, 302, 1002–5.

Winston, R. (1990). Personal communication.

Wisniewski, H.M. (1991). Neuropathology and biochemistry of Alzheimer's disease and aluminium encephalopathy. In *Alzheimer's Disease and the Environment*, ed. Lord Walton, pp. 35–9. London: Royal Society of Medicine Services.

Yates, P. O. & Mann, D. M. A. (1986). Aluminosilicates and Alzheimer's disease. *Lancet*, **i**, 681.

Index

4-aminopyridine, 24–5
acrylamide toxicity, 212
Adie's syndrome, 208
AF-MSA (Shy–Drager syndrome), 160, 207, 238
AIDS, 213, 369–70, 431
alcohol
  alcoholic peripheral neuropathy, 211–12
  in essential tremor, 199
  in physiological tremor, 193
aluminium toxicity, 434–5
Alzheimer's disease, 433–6
amitryptyline, 338, 350
amputation
  congenital, 83
  traumatic, 83
amyloidosis, 209–10
anticonvulsant therapy, 416–27
  mode of action, 397–8
  in paresthesiae, 35
antiphospholipid antibodies, 243–50
arsenic toxicity, 212
aspirin, 343
atenolol, 340, 350
athetosis management, 184, 185
autonomic disorders, 205–22
  classification, 206
  following facial nerve injury, 230–8
  investigations, 213–17
  treatment, 217–18
autonomic innervation of face, 223–42
  nerve injury effects, 230–8

β-adrenergic blockers, in migraine, 313, 339–40, 350
Babinski response, 75
back pain, 39–57
barbiturate anticonvulsants, 398, 417
Bastian, H. C., 3

BDNF (brain derived neurotrophic factor), 130–1
Becker muscular dystrophy, 439–43
Bell, Sir Charles, 3
benzodiazepines
  in epilepsy, 398
  in migraine, 348–9
biofeedback training, in cerebral palsy, 185–7
blood pressure
  disorders, see hypotension
  investigations, 215–16
botulinum toxin
  botulism, 209
  in tremor control, 134
brain derived neurotrophic factor (BDNF), 130–1
brainstem tumours, 208

calcium channel antagonists
  in epilepsy, 398
  in migraine, 341–3, 350
carbamazepine
  in epilepsy, 398, 418, 422
  in paresthesiae, 35
cardiovascular tests, 213–16
cerebellar hypotonia, 89, 99–100
cerebral palsy, 169–70
  spasticity reduction, 183–7
cerebral shock, hypotonia in, 99
cerebrovascular disease, 208
Chagas' disease, 213
Charcot Marie Tooth disease, 200, 211
cholinergic dysautonomia, 208–9
chorea, 161
CIDP (chronic inflammatory demyelinating polyradiculoneuropathy), 211
clasp-knife phenomenon, 106, 120–1
clonus, 106, 121

Index

clozapine, 133
cluster headache, 274–9
  cervical sympathetic lesions in, 234–5
  pupillary disturbance, 225, 234
  treatment of, 335, 346–7, 352–4
computed tomography, 154
  in multiple sclerosis, 385
  see also SPECT
connective tissue diseases, 212–13
corticobasal degeneration, 161
corticospinal tract, 61–74
  movement control and, 69–72
  upper motoneurone lesions, 75–88
corticosteroid therapy
  in migraine, 346, 351
  in multiple sclerosis, 386–7
Creutzfeldt–Jakob disease, 437
crocodile tears, 233
CT, *see* computed tomography
cutaneous receptors
  propioceptive role, 9–11
  reflex pathways from, 119–20
cyproheptadine, 338

3,4-diaminopyridine, 24–5
denervation supersensitivity, 231–2
deprenyl, 132
diabetes
  autonomic neuropathy, 209
  nerve ischaemia in, 31
dihydroergotamine, 315
discography
  cervical, 51
  vertebral, 48–9
discs (intervertebral)
  nerve supply to, 46
  painful, 48–9, 51–2
domperidone, 348
dopamine agonists, 132–3
dopamine therapy
  in dystonia, 148–9, 162–3
  in Parkinson's disease, 132–3
drug therapy
  in Alzheimer's disease, 436
  in dystonia, 148–9, 162–3
  in epilepsy, 397–8, 416–27
  in essential tremor, 199
  in migraine, 335–54
  in multiple sclerosis, 386–8
  in orthostatic hypotension, 217–18
  in Parkinson's disease, 130–4, 436
Duchenne muscular dystrophy, 439–43
dysautonomia, 208–9
dyskinesias
  chorea, 161
  essential tremor, 161–2
  functional imaging, 161–2
  Huntington's disease, 161
  neuroacanthocytosis, 161
dystonia, 139–40
  dopa-responsive, 148–9, 162–3
  dystonia–Parkinsonism, 149–50
  functional imaging, 157–8
  genetics, 140–53

Eaton–Lambert myasthenic syndrome, 24–5
echovirus-79, 367
EGF (epidermal growth factor), 130–1
electromyography
  paraspinal, 42
  tremor studies, 196–7
encephalitis (viral)
  herpes simplex-induced, 365
  Japanese, 370
  lentivirus-induced, 369
  measles-induced, 368
  post-influenza, 367
  Western equine, 366
epidermal growth factor (EGF), 130–1
epilepsy, 80, 397
  animal studies, 401–5
  anticonvulsant mechanisms, 397–8
  brain pathology, 399–400
  drug therapy, 397–8, 416–27
  partial epilepsies, 397–415
  surgery for, 407–12
ergotamine, 313, 315, 335–6
ethnic considerations
  in dystonia inheritance, 140–53
  in multiple sclerosis etiology, 379
ethosuximide, 418
ethotoin, 417

face, autonomic innervation, 223–42
facet syndrome, 51
fibroblast growth factors, 131
Fisher's syndrome, 237
flunarizine, 341, 350, 398
Friedreich's ataxia, 107, 211
fusimotor system, 89–101

gene mapping, 432–3, 438
genetics, *see* inheritance
Gerstmann–Sträussler–Scheinker syndrome, 437
glutamate antagonists, 134
Guillain–Barré syndrome, 29, 210, 237
  HIV and, 369
gustatory sweating, 232–3

5-hydroxytryptamine receptors, role in headache, 323–33, 335–9
harlequin syndrome, 223, 237

head, autonomic innervation, 223–42
headache/migraine
    acupuncture for, 294, 335
    antiphospholipid antibodies and, 243, 247
    cervicogenic, 274
    cluster, *see* cluster headache
    definitions, 268, 303, 334
    due to referred pain, 47–8, 270
    experimental, 253–65, 310–12
    functional imaging, 261–3, 305, 309–10
    hemicrania continua, 279
    neural pathways, 284–302
    neuropharmacology, 312–15
    occipital neuralgia, 47–8, 52, 269–74
    pathophysiology, 266–7, 303–22
    post-traumatic, 279–80
    treatment, 312–13, 334–59; 5-HT agonists/antagonists, 323–33, 335–9
    trigger factors, 349
    unilateral, 266–83
    vascular mechanisms, 254–6
heart rate investigations, 213–15
hemiparesis/hemiplegia, 75–85
    spinal cord lesions, 106–24
hemodialysis (chronic), 211
herpes simplex viruses, 363–5
histamine-induced headache, 256–8, 278–9
HIV (human immunodeficiency virus), 213, 244, 365, 369–71
Holmes–Adie syndrome, 107, 236–7, 238
Horner's syndrome, 224, 225, 231, 232, 237
human immunodeficiency virus (HIV), 213
Huntington's disease, 161
hyperreflexia, following spinal cord transection, 106, 109–18
hyperventilation, paresthesiae due to, 20, 33–5
hypotension (postural)
    in amyloidosis, 210
    in autonomic neuropathy, 209
    in CNS disorders, 207, 208
    in diabetes, 209
    in Guillain-Barré syndrome, 210
    in infections, 213
    in metabolic disorders, 211
    in Parkinson's disease, 207, 208
    in porphyria, 210
    in pure autonomic failure, 207
    in Shy–Dräger syndrome, 207
    in spinal cord injuries, 208
    in toxic neuropathies, 212
    treatment, 217–18
hypotonia, 99–100

idiopathic polyneuritis (Guillain–Barré syndrome), 29, 210, 237, 369
imaging techniques, 154–68, 431

computed tomography, 154
    functional techniques, 154–68
    nuclear magnetic resonance, 154
Imigran (sumatriptan), 313, 315, 323–33, 338–9
immunotherapy, in multiple sclerosis, 387
implants, fetal mesencephalic, 133, 135
indomethacin, 276, 279
infections
    antiphospholipid antibodies in, 244
    autonomic dysfunction in, 213
    see also viral infections
influenza viruses, 367–8
inheritance
    Creutzfeldt–Jakob disease, 437
    dystonia, 143–50
    gene mapping, 432–3, 438
    hereditary neuropathies, 210–11
    multiple sclerosis, 379
investigations
    antiphospholipid antibody detection, 243–4
    of autonomic function, 213–17
    imaging techniques, 154–68, 431
    in multiple sclerosis, 385–6
ischemia (nerve), 31–2

Jackson, John Hughlings, 431
joint receptors, 8–9

kinesthesia, 3–19

lacrimation disorders, 233–4
Lassa fever, 367
lazabemide, 132
lazeroids, 132
lentiviruses, 369–71
    HIV, 213, 244, 365, 369–71
leprosy, 213
leukaemia virus, 370
    multiple sclerosis and, 379
levodopa therapy
    in dystonia, 148–9, 162–3
    in Parkinson's disease, 132–3, 436
Lhermitte's symptom, 20
lithium, 28, 33
    in headache treatment, 346–7, 351
'lubag' (dystonia–Parkinsonism), 149–50
Lyme disease, 244

Madopar HBS, 133
magnetic resonance imaging, 154
    in epilepsy, 410
    in multiple sclerosis, 383, 385–6
malignancy
    brainstem tumours, 208
    neuropathy and, 212

Index

MAO-inhibitors, 130, 132, 344–6, 351
measles virus, 368
mercury toxicity, 212
metabolic disorders, autonomic dysfunction in, 211–12
metal poisoning, 212
methoin, 417
methsuximide, 417–18
methysergide, 313, 336–7, 350
metoclopramide, 347–8
metoprolol, 340, 350
migraine, *see* headache/migraine
Miller, Henry, 431
moclobemide, 345–6
monoamine oxidase inhibitors, 130, 132, 344–6, 351
motoneurone lesions, 75–85
motor neurone disease, 79, 437
in spinal lesions, 122–3
movement control
corticospinal tract and, 69–72
fusimotor system and, 89–101
tonic stretch reflexes and, 170–83
movement disorders
functional imaging, 157–64
see also specific disorder
MRI, *see* magnetic resonance imaging
multiple sclerosis, 374–93, 436
autonomic dysfunction, 208
classification, 381
clinical features, 381–3
diagnosis, 381
diagnostic tests, 385–6
etiology, 375–81, 436
management, 29–30, 386–8
pathology, 374
multiple system atrophy (MSA), 160
muscle atrophy, in upper motoneurone lesions, 123
muscle fatigue, in motoneurone lesions, 83–4
muscle receptors, 4–8
muscle spasms, spinal spasticity, 106, 120–2
muscle spindles, 89–101
muscle stiffness
central control, 170–2
stretch reflex and, 180–3
muscle strain, back strain, 49–50
muscle tone, 97–9
abnormalities, 99–101
muscular dystrophies, 439–43
gene location, 438

neck pain, 47, 48, 51, 52
nerve compression
paresthesiae due to, 20, 31
root compression in spinal pain, 39–40

nerve conduction, 21–8
nerve injury, autonomic facial nerves, 230–8
nerve ischemia, 31–2
neuroacanthocytosis, 161
neuroma, paresthesiae due to, 20
neuromuscular disorders, 83, 437–43
neuromyotonia, 20
neuropathies
autonomic, 208–9
CIDP, 211
hereditary, 107, 200, 210–11
Riley–Day syndrome, 210
Swanson type, 210
toxic, 212
neurovirology, 363–73
nialamide, 345
nifedipine, 343, 350
nimodipine, 341–2, 350
nitroglycerine-induced headache, 258–63
NSAIDs in migraine control, 343–4, 351
nuclear magnetic resonance, *see* magnetic resonance imaging

occipital neuralgia, 47–8, 52, 269–74
ocular manifestations
denervation supersensitivity, 231–2
Holmes–Adie syndrome, 107, 236–7, 238
Horner's syndrome, 224, 225, 231, 232, 237
in multiple sclerosis, 383
tonic pupils, 236–8
olivopontocerebellar degeneration (OPCA), 160
optic neuritis, multiple sclerosis and, 383
orthostatic hypotension, *see* hypotension
ouabain, 28, 32

PAF (pure autonomic failure), 207
pain
neck, 47, 48, 51, 52
radicular, 39–40
referred, 40–1, 47
somatic, 40–1
spinal, 39–57
pan-dysautonomia, 208–9
papovaviruses, 365
paracetamol, 344
paramethadione, 417
paresthesiae, pathophysiology, 20–38, 40
Parkinson's disease, 100, 130–8, 436–7
autonomic failure with, 207
functional imaging, 158–61
parotidectomy, 225
perhexiline maleate toxicity, 212
peripheral neuropathy, tremor in, 200
pernicious anemia, 211

PET (positron emission tomography), 154–64
  in migraine studies, 305
phenelzine, 345, 351
phenobarbital, 398, 417, 422
phenothiazines, 347
phenoxybenzamine, 184
phensuximide, 417–18
phenytoin
  in epilepsy, 398, 416–17, 422, 423
  in paresthesiae, 35
'pins-and-needles', 31
pizotifen, 313, 337, 350
poliomyelitis, 367
porphyria, 210
postural hypotension, see hypotension
Pourfour Du Petit syndrome, 225
prenatal detection
  in Duchenne dystrophy, 442
  in dystonia, 148
primidone
  in epilepsy, 417, 422
  in essential tremor, 199
prochlorperazine, 347
progressive supranuclear palsy (PSP), 159–60
propioception, 3–19
propranolol
  in essential tremor, 199
  in migraine, 340, 350
puffer-fish poisoning, 22
pure autonomic failure (PAF), 207

rabies, 367
radiculopathy, 39–40
referred pain, 40–1, 47–8
Renshaw cells, 119
retroviruses, 368–71
Reye's syndrome, 367
rigidity, 89, 100–1
Riley–Day syndrome, 210
RNA viruses, 365–8
ropinerol, 133
Ross' syndrome, 235

salivary gland
  innervation, 226–8
  parotidectomy, 225
scorpion venom, 30
serotonin, role in headache, 323–33, 335–9
Sherrington, C. S., 3
shock syndromes, 99–100
Shy–Dräger syndrome, 160, 207, 238
Sinemet CR, 133
skin, cutaneous proprioception, 9–11
SLE, 244
somatic pain, 40–1

spasticity, 89, 100–1
  in cerebral palsy, 170, 183–7
  spinal spasticity, 106–29
  tendon jerk reflexes in, 97
SPECT (single photon emission computed tomography), 154–64
  in epilepsy studies, 408
  in headache studies, 261–3
spinal cord
  injuries, 106–24, 208, 225
  normal motor circuitry, 107–9
spinal pain, 39–57
spinal shock, hypotonia in, 99, 123
spinal transplants, 124
Steele–Richardson–Olszewski syndrome, 159–60
stiff-man syndrome, 89
strains, back strain, 49–50
stretch reflexes
  hyperreflexia, 106, 109–18
  role in cerebral palsy, 183–7
  tonic stretch reflex, 169–90
striatonigral degeneration, 160
stroke
  antiphospholipid antibodies in, 243–50
  recovery from, 75–85
sulthiame, 418
sumatriptan, 313, 315, 323–33, 338–9
surgical approaches
  in epilepsy, 407–12
  in essential tremor, 162
  in Parkinson's disease, 134–5, 162
Swanson type neuropathy, 210
sweat tests, 214, 216
sweating abnormalities
  in amyloidosis, 210
  Charcot–Marie–Tooth disease, 211
  in facial nerve injuries, 223–38
  in Guillain–Barré syndrome, 210
  in infections, 213
  in multiple sclerosis, 208
  in pan-dysautonomia, 208, 212
  in pure autonomic failure, 207
  in Shy–Dräger syndrome, 207
syphilis, 244
syringobulbia/syringomyelia, 225
systemic lupus erythematosis, 244

tabes dorsalis, 107, 200, 208
tendon jerks, 89, 91–7, 174–5
  hyperreflexia, 106, 109–18
tetanic stimulation, 32–3
tetraethyl ammonium, 30
tetrodotoxin, 22
thallium toxicity, 212
thiethylperazine, 347

Index

thrombosis, *see* stroke
timolol, 340
tonic vibration reflex, 97–9
transplants
   dopamine-producing tissue, 133, 135
   spinal, 124
tranylcypromine, 345, 351
tremor, 191–201
   botulinum toxin treatment, 134
   essential, 161–2, 199–200
   in Parkinson's disease, 134, 161
   physiological, 191–8
tricyclic drugs, 338, 350
trifluoperazine, 345
troxidone, 417
tumours, brainstem, 208

upper motoneurone lesions, 75–85
   spinal cord lesions, 106–24

valproate
   in epilepsy, 398, 418–19, 423
   in paresthesiae, 35
Valsalva's manoeuvre, 215, 216
varicella-zoster virus, 363
verapamil, 342, 350
vibration, reflex effects of, 97–9, 169–70
vigabatrin, 426
vincristine toxicity, 212
viral infections, 363–73
   demyelinating, 378–9
   HIV, 213, 244, 365, 369–71
   multiple sclerosis and, 385, 436
visna virus, 369
vitamin B12 deficiency, 211
von Economo's disease, 368

Wernicke's encephalopathy, 205, 208, 211
Wilson's disease, 163